AF410831

Climate Change, Carbon Capture, Storage and CO$_2$ Mineralisation Technologies

Climate Change, Carbon Capture, Storage and CO$_2$ Mineralisation Technologies

Editors

Nikolaos Koukouzas
Pavlos Tyrologou
Petros Koutsovitis

MDPI • Basel • Beijing • Wuhan • Barcelona • Belgrade • Manchester • Tokyo • Cluj • Tianjin

Editors
Nikolaos Koukouzas
Chemical Process and Energy
Resources Institute
(CERTH/CPERI)
Greece

Pavlos Tyrologou
Chemical Process and Energy
Resources Institute
(CERTH/CPERI)
Greece

Petros Koutsovitis
Chemical Process and Energy
Resources Institute
(CERTH/CPERI)
Greece

Editorial Office
MDPI
St. Alban-Anlage 66
4052 Basel, Switzerland

This is a reprint of articles from the Special Issue published online in the open access journal *Applied Sciences* (ISSN 2076-3417) (available at: https://www.mdpi.com/journal/applsci/special_issues/CCS_CO2_Mineralisation).

For citation purposes, cite each article independently as indicated on the article page online and as indicated below:

LastName, A.A.; LastName, B.B.; LastName, C.C. Article Title. *Journal Name* **Year**, *Article Number*, Page Range.

ISBN 978-3-03943-641-5 (Hbk)
ISBN 978-3-03943-642-2 (PDF)

Contents

About the Editors

Nikolaos Koukouzas holds a BSc degree in Geology with an MSc and PhD degrees in Industrial Mineralogy from Leicester University, UK. He received his bachelor's degree in Geology from the National University of Athens in 1987. He continued his postgraduate studies in the UK, receiving a 4-year grant from the State Scholarship Foundation of Greece, and completed his PhD in 1995. Dr. Koukouzas is currently Director of Research in the Centre for Research and Technology Hellas (CERTH) and Director of the Laboratory of the Solid Fuels in CERTH, supervising a team of scientists with offices in Athens, Thessaloniki, and Ptolemais, Greece. He is Member of the Executive Committee of the Monitoring and Assessment Exercise of the Research Fund for Coal and Steel, National Delegate in the Government Group of Zero Emission Power Plant Technology Platform (ETP-ZEP), and National Delegate in the Policy and Technical Group of the Carbon Sequestration Leadership Forum – CSL coordinated by US DoE. His research expertise is focused on coal geology, industrial mineralogy, coal geochemistry, fossil fuel characterization, coal combustion byproducts, coal slope stability, waste and mine water treatment, and CO_2 storage and monitoring. Dr. Koukouzas has served as Project Manager with scientific responsibility of numerous research projects, with over 25 years' experience on energy and environment issues. He has 250 publications corresponding to 1650 citations and a h-index of 21.

Pavlos Tyrologou is Research Fellow at CERTH, holds a BSc (4 years with Hons) degree in Geology and Applied (Engineering) Geology from Glasgow University, an MSc in Applied Environmental Geology from Cardiff University, and a PhD in Environmental Geotechnical Engineering from Imperial College London which was fully supported by a scholarship award from The Institute of Materials, Minerals and Mining, UK. He is professionally accredited in the UK both as a Chartered Geologist and Eurogeologist in the field of Engineering Geology. Pavlos has 15 years of consultancy experience in various public infrastructure design/works in Greece, United Kingdom, Madagascar, and Sint Maarten. He is an expert witness for both the Athens Public Prosecutor's Office and Athens Court of Justice. He has evaluated more than 65 research and business projects as independent evaluator for both the European Commission and Greek Government. Pavlos, due to his wide expertise, serves as Coordinator in the European Federation of Geologists (EFG) on the Natural Hazard and Climate Change Panel, and is a member of the panel of experts on soil protection. On the same note, he is the Greek representative within the EFG council through the Association of Greek Geologists. He is a member of a) the Greek Society for Soil Mechanics and Geotechnical Engineering, b) Geological Society of London, c) International Society for Rock Mechanics, and d) Member of International Society for Soil Mechanics and Geotechnical Engineering, in which he serves on the technical committee of the Engineering Practice of Risk Assessment and Management. He also has extensive research experience in several European H2020 projects, including proposal writing, inception and execution (CHPM30, UNEXMIN, SLOPES, KINDRA, INTRAW, COALTECH2051, StrategyCCUS, LEILAC2, RECPP) and in various tasks including project management and research activities. His current research interests are focused on climate change and related activities and how these are related to security issues such as state fragility and vulnerability.

Petros Koutsovitis is Assistant Professor at the University of Patras, Department of Geology. He holds a PhD in the fields of Mineralogy, Petrology and Geochemistry from the University of Athens, Greece. Dr. Koutsovitis has been a Post-doctoral researcher at the University of Vienna (under an FWF-funded project) and was later employed at the Greek Institute of Geology and Mineral Exploration, within the frame of ESTMAP (Horizon 2020) and YPOTHER (NSRF-funded) projects. He continued his research activities at the Centre for Research & Technology (CERTH), participating in EU-funded Horizon2020 and RFCS project's (STRATEGY-CCUS, COAL2GAS & COALBYPRO), as well as at the University of Athens as an Academic Educational Fellow. Dr. Koutsovitis has also participated in the INTRAW Horizon 2020 programme, in cooperation with the European Federation of Geologists (EFG). He has published several papers in peer reviewed Journals and has been awarded by the Academy of Athens for his research.

Editorial

Special Issue "Climate Change, Carbon Capture, Storage and CO_2 Mineralisation Technologies"

Nikolaos Koukouzas [1], Pavlos Tyrologou [1] and Petros Koutsovitis [2,*]

[1] Chemical Process & Energy Resources Institute, Centre for Research & Technology Hellas (CERTH), 15125 Maroussi, Greece; koukouzas@certh.gr (N.K.); tyrologou@certh.gr (P.T.)

[2] Section of Earth Materials, Department of Geology, University of Patras, 26504 Patras, Greece

* Correspondence: pkoutsovitis@upatras.gr; Tel.: +30-2610-997598

Received: 12 October 2020; Accepted: 14 October 2020; Published: 23 October 2020

This Special Issue presents sixteen scientific papers that explore the application of carbon capture and storage technologies, mitigating the effects of climate change. Emphasis has been given on mineral carbonation techniques that combine innovative applications to emerging problems and needs. The aim of this issue is to contribute to the knowledge of the ongoing research regarding climate change and CCS technological applications, focusing on carbon capture and storage practices. Climate change is a global issue that is interrelated with the energy and petroleum industry. In this scope, there is an increasing demand for new low cost and energy efficient techniques that reduce the CO_2 emissions. The use of fossil fuels is the primary source of CO_2 emissions, which is one of the main greenhouse gases. Carbon capture and storage (CCS) is regarded as one of the most efficient technologies that allows carbon intensive industries to continue to operate with lower CO_2 emissions. CCS offers double benefits combining the reduction in greenhouse gas emissions with the direct use of captured carbon for enhanced oil recovery (EOR). Mineral carbonation is a permanent and secure CCS and sequestration technology that gives the solution in cases of smaller to medium emitters. It is based on the in situ (injecting CO_2 into the earth's surface) or ex situ (chemical reactor systems) production of carbonate minerals through the chemical reaction of CO_2 with Ca, Mg, and Fe-silicate minerals.

In this Special Issue, Drexel et al. [1] investigated the wettability alteration by carbonated water injection (CWI) on a coquina carbonate rock analogue of a pre-salt reservoir, and its consequences in the flow of oil. Pore-scale simulations showed that CWI altered the wettability of the carbonate rock from neutral to water-wet. Zarogiannis et al. [2] employed a systematic approach to identify the impact that operating parameter variations and different solvents exert on multiple CO_2 capture performance indicators. Moita et al. [3] examined experimentally the mineral carbonation of CO_2 in plutonic ophiolitic mafic rocks through a set of laboratory experiments on cumulate gabbro and gabbro-diorite specimens from the Sines Massif (Portugal). Tectonically related deformation processes upon coal affecting their nanopore structure and their relation to CO_2 adsorption are presented by Wang and Long [4]. The subject of ophiolitic-related rocks from southern Portugal and their potential for mineral carbonation of CO_2 was brought forward by Pedro et al. [5] confirming that they exhibit the appropriate mineralogical and geochemical features for this purpose. An interesting study is presented in this Special Issue to investigate the suitability of metallurgical slugs of different chemical and mineralogical composition as clinker substitute in cement manufacture [6]. This study revealed that the most critical parameter in the compressive strength development of the slag cements is the mineralogical composition of the slag.

Zhou et al. [7] present a research study on the role of atomic stress in calcite nucleation, based on molecular dynamics simulations on amorphous Ca-carbonate gels. The research results indicate that the gelation reaction strongly depends on the development of local molecular stresses within the Ca and C precursors, which progressively get released upon gelation. Numerical simulations on CO_2 bubbles dissolving into the seawater, as well as on the diffusion of dissolved CO_2 by ocean flows, indicate that all leaked and rising CO_2 bubbles are dissolved into the seawater before reaching the free surface [8]. The results of this research study provide important outcomes on the behavior and environmental risk of CO_2 leakage. Kim et al. [9] present an experimental study for the development of a carbon-capturing concrete. The research was based on the use of blast-furnace slag instead of cement. CO_2-adsorption and diffusion properties of metal-organic frameworks (CoRE-MOFs) was investigated based on machine learning methods and Monte Carlo/molecular dynamics simulations [10]. The simulation results provide valuable guidelines for the synthesis of new MOFs in experiments that capture low-concentration CO_2 directly from the air. Fedunik-Hofman et al. [11] investigated the kinetic parameters of $CaCO_3$/CaO reaction systems in energy storage and carbon capture. The research outcomes indicate a strong association between the experimental conditions, material properties, and the kinetic method with the kinetic parameters. Chemical reactions between synthetic sandstones, formation water and CO_2 were investigated based on experimental studies and numerical modelling [12]. The research results provide new geochemical insights on the dissolution mechanisms of CO_2 under high-pressure/temperature conditions.

Microporous carbonspheres for CO_2 adsorption were designed and prepared by deploying potassium oxalate monohydrate and ethylene diamine (EDA) [13]. The study revealed that carbonization temperature increase results in an increase in the specific surface area of the subsequent CO_2 adsorption. Wang et al. [14] provide a case study in Qinghai (China) that aims to investigate: (a) the change in carbon footprint (CF) caused by agrochemical and agricultural energy inputs, (b) the contributions of various inputs to the total carbon footprint (TCF), and (c) the different changing trends between carbon intensity in output value (CV) and carbon intensity in area (CA) for the period 1995–2016. Air purification tests on dusty and clean samples by deploying different light sources and setups, were conducted on photocatalytic pavement blocks from a 7 year service bicycle lane in Poland [15]. The research outcomes show that samples maintained their nitric oxide removal capability with 4–45% reduction rate based on the light source and their surface cleanliness. Gal et al. [16] investigated the soil-gas concentrations and flux at the TOTAL Lacq-Rousse carbon capture and storage (CCS) pilot site, in southern France. The research reveals that near surface gases are naturally produced and they are not associated with the ascending CO_2 from the storage reservoir.

The guest editors remain positive that the readers will enjoy the articles presented in this beneficiary Special Issue of "Climate Change, Carbon Capture, Storage and CO_2 Mineralisation Technologies".

Funding: This research received no external funding.

Acknowledgments: We are grateful to all authors for their contributions and making this Special Issue a success. Many thanks to all reviewers that provided responsible suggestions and comments. We would also like to express our gratitude to all members of the editorial team of Applied Sciences and especially to Wing Wang for their professional support.

Conflicts of Interest: The authors declare no conflict of interest.

References

1. Drexler, S.; Hoerlle, F.; Godoy, W.; Boyd, A.; Couto, P. Wettability Alteration by Carbonated Brine Injection and Its Impact on Pore-Scale Multiphase Flow for Carbon Capture and Storage and Enhanced Oil Recovery in a Carbonate Reservoir. *Appl. Sci.* **2020**, *10*, 6496. [CrossRef]
2. Zarogiannis, T.; Papadopoulos, A.I.; Seferlis, P. Off-Design Operation of Conventional and Phase-Change CO_2 Capture Solvents and Mixtures: A Systematic Assessment Approach. *Appl. Sci.* **2020**, *10*, 5316. [CrossRef]
3. Moita, P.; Berrezueta, E.; Abdoulghafour, H.; Beltrame, M.; Pedro, J.; Mirão, J.; Miguel, C.; Galacho, C.; Sitzia, F.; Barrulas, P.; et al. Mineral Carbonation of CO_2 in Mafic Plutonic Rocks, II—Laboratory Experiments on Early-Phase Supercritical CO_2-Brine-Rock Interactions. *Appl. Sci.* **2020**, *10*, 5083. [CrossRef]
4. Wang, L.; Long, Z. Evolutions of CO_2 Adsorption and Nanopore Development Characteristics during Coal Structure Deformation. *Appl. Sci.* **2020**, *10*, 4997. [CrossRef]

5. Pedro, J.; Araújo, A.A.; Moita, P.; Beltrame, M.; Lopes, L.; Chambel, A.; Berrezueta, E.; Carneiro, J. Mineral Carbonation of CO_2 in Mafic Plutonic Rocks, I—Screening Criteria and Application to a Case Study in Southwest Portugal. *Appl. Sci.* **2020**, *10*, 4879. [CrossRef]

6. Tzevelekou, T.; Lampropoulou, P.; Giannakopoulou, P.P.; Rogkala, A.; Koutsovitis, P.; Koukouzas, N.; Petrounias, P. Valorization of Slags Produced by Smelting of Metallurgical Dusts and Lateritic Ore Fines in Manufacturing of Slag Cements. *Appl. Sci.* **2020**, *10*, 4670. [CrossRef]

7. Zhou, Q.; Du, T.; Guo, L.; Sant, G.; Bauchy, M. Role of Internal Stress in the Early-Stage Nucleation of Amorphous Calcium Carbonate Gels. *Appl. Sci.* **2020**, *10*, 4359. [CrossRef]

8. Jeong, S.-M.; Ko, S.; Sean, W.-Y. Numerical Prediction of the Behavior of CO_2 Bubbles Leaked from Seafloor and Their Convection and Diffusion near Southeastern Coast of Korea. *Appl. Sci.* **2020**, *10*, 4237. [CrossRef]

9. Kim, S.; Park, C. Durability and Mechanical Characteristics of Blast-Furnace Slag Based Activated Carbon-Capturing Concrete with Respect to Cement Content. *Appl. Sci.* **2020**, *10*, 2083. [CrossRef]

10. Deng, X.; Yang, W.; Li, S.; Liang, H.; Shi, Z.; Qiao, Z. Large-Scale Screening and Machine Learning to Predict the Computation-Ready, Experimental Metal-Organic Frameworks for CO_2 Capture from Air. *Appl. Sci.* **2020**, *10*, 569. [CrossRef]

11. Fedunik-Hofman, L.; Bayon, A.; Donne, S.W. Comparative Kinetic Analysis of $CaCO_3$/CaO Reaction System for Energy Storage and Carbon Capture. *Appl. Sci.* **2019**, *9*, 4601. [CrossRef]

12. Yu, Z.; Yang, S.; Liu, K.; Zhuo, Q.; Yang, L. An Experimental and Numerical Study of CO_2–Brine-Synthetic Sandstone Interactions under High-Pressure (P)–Temperature (T) Reservoir Conditions. *Appl. Sci.* **2019**, *9*, 3354. [CrossRef]

13. Staciwa, P.; Narkiewicz, U.; Sibera, D.; Moszyński, D.; Wróbel, R.J.; Cormia, R.D. Carbon Spheres as CO_2 Sorbents. *Appl. Sci.* **2019**, *9*, 3349. [CrossRef]

14. Wang, X.; Zhang, Y. Carbon Footprint of the Agricultural Sector in Qinghai Province, China. *Appl. Sci.* **2019**, *9*, 2047. [CrossRef]

15. Witkowski, H.; Jackiewicz-Rek, W.; Chilmon, K.; Jarosławski, J.; Tryfon-Bojarska, A.; Gąsiński, A. Air Purification Performance of Photocatalytic Concrete Paving Blocks after Seven Years of Service. *Appl. Sci.* **2019**, *9*, 1735. [CrossRef]

16. Gal, F.; Pokryszka, Z.; Labat, N.; Michel, K.; Lafortune, S.; Marblé, A. Soil-Gas Concentrations and Flux Monitoring at the Lacq-Rousse CO_2-Geological Storage Pilot Site (French Pyrenean Foreland): From Pre-Injection to Post-Injection. *Appl. Sci.* **2019**, *9*, 645. [CrossRef]

Publisher's Note: MDPI stays neutral with regard to jurisdictional claims in published maps and institutional affiliations.

Article

Wettability Alteration by Carbonated Brine Injection and Its Impact on Pore-Scale Multiphase Flow for Carbon Capture and Storage and Enhanced Oil Recovery in a Carbonate Reservoir

Santiago Drexler *, Fernanda Hoerlle, William Godoy, Austin Boyd and Paulo Couto

LRAP/COPPE, Universidade Federal do Rio de Janeiro, Rio de Janeiro 21941-594, Brazil; fernandahoerlle@petroleo.ufrj.br (F.H.); wmgodoy@petroleo.ufrj.br (W.G.); austin@petroleo.ufrj.br (A.B.); pcouto@petroleo.ufrj.br (P.C.)
* Correspondence: santiago.drexler@petroleo.ufrj.br

Received: 14 August 2020; Accepted: 9 September 2020; Published: 17 September 2020

Abstract: Carbon capture and storage is key for sustainable economic growth. CO_2-enhanced oil recovery (EOR) methods are efficient practices to reduce emissions while increasing oil production. Although it has been successfully implemented in carbonate reservoirs, its effect on wettability and multiphase flow is still a matter of research. This work investigates the wettability alteration by carbonated water injection (CWI) on a coquina carbonate rock analogue of a Pre-salt reservoir, and its consequences in the flow of oil. The rock was characterized by routine petrophysical analysis and nuclear magnetic resonance. Moreover, micro-computed tomography was used to reconstruct the pore volume, capturing the dominant flow structure. Furthermore, wettability was assessed by contact angle measurement (before and after CWI) at reservoir conditions. Finally, pore-scale simulations were performed using the pore network modelling technique. The results showed that CWI altered the wettability of the carbonate rock from neutral to water-wet. In addition, the simulated relative permeability curves presented a shift in the crossover and imbibition endpoint values, indicating an increased flow capacity of oil after CWI. These results suggest that the wettability alteration mechanism contributes to enhancing the production of oil by CWI in this system.

Keywords: CCS; carbonated water injection; CO_2-EOR; pore network modelling; relative permeability

---

1. Introduction

CO_2 emissions from the combustion of fuels and their impact on the environment are among the greatest global concerns of our time. Maintaining continuous economic growth while reducing emissions has been a challenging goal for decades [1]. Carbon capture and storage (CCS) is critical to achieve the target of 26 Gton of CO_2 per year by 2030 [2]. Within this context, CCS has been raising well-deserved attention in recent years. According to a recent publication [3], CCS may account for up to 33% of the total carbon abatement by 2050. With the current global emissions of the order of 42 Gton of CO_2 per year, it is evident that research efforts to develop and deploy CCS technologies are a priority nowadays [4].

Different separation and storage processes have been used to reduce CO_2 emissions. A recently published review analyzed the state-of-the-art of the different CO_2 capture, storage, and utilization methods [5]. Capture can take place before (pre-combustion) or post-combustion. Pre-combustion methods consist of removing CO_2 from a stream rich in this gas prior to the combustion reaction. For example, low-temperature phase separation has been used to energy- and cost-efficiently remove CO_2 from pressurized syngas [6].

On the other hand, post-combustion methods are based on absorption, adsorption, or membranes. Chemical absorption with different solvents (mainly amine solutions) is a well-known process that is widely used in natural gas processing [7]. A recent study reviewed the status of different solvents for chemical absorption CCS at the industrial scale [8]. Moreover, adsorption of CO_2 in solid materials has been used for around 30 years [9]. The use of traditional adsorbents (such as alumina and zeolites) and newer structures (such as metal–organic frameworks and polymers) was reviewed elsewhere [10–15]. Finally, the use of organic and inorganic membranes has been implemented for gas separation in both pre- and post-combustion processes [9,16–19].

After capture, CO_2 is injected in geological formations where it should be safely stored for at least 10,000 years [5]. According to Bui et al. [9], enhanced oil recovery (EOR) [2] and injection in saline formations [20] are the only storage methods with ongoing onshore and offshore projects at the commercial scale. By contrast, other storage methods, such as injection in depleted locations [21], ocean storage [22], and mineralization [23], are promising alternatives still in the development stage. Furthermore, Zhang et al. [5] discussed other alternatives including enhanced gas recovery [24], enhanced coalbed methane recovery [25], and applications in shale formations [26].

CO_2 injection, a mature EOR method, can incrementally improve oil recovery from 5 to 20% of the original oil in place mainly through mechanisms such as viscosity reduction, oil swelling, and miscible displacement [27]. CO_2-EOR accounts for an additional oil production of around 300,000 bbl per day [9]. Furthermore, the over 13 large-scale CO_2-EOR projects active in 2018 represented a capture capacity of 26 Mtpa, and the number of active projects is expected to increase during the next several years [28].

Carbonated water injection (CWI) is an alternative CO_2-EOR method that consists of injecting CO_2 dissolved in the injection water [29]. As carbonated brine has a higher viscosity than CO_2, CWI presents a more favorable mobility ratio that results in a greater sweeping efficiency compared to CO_2 injection [30]. In addition, the dissolution of CO_2 reduces the pH of the aqueous phase due to the formation of carbonic acid, and this reduction plays an important role in oil–brine–rock interactions. CO_2 dissolution was reported to shift the wettability of carbonate rocks toward more water-wet [31–35]. This is caused by several complex phenomena that depend on the composition of oil, brine, and rock. These effects include element changes on the rock surface by chemical reaction [36], surface charge modification [37], desorption of adsorbed oil on the surface of the rock [38], and destabilization of polar compounds in the oil [39], and they will be further discussed throughout this work.

CO_2-EOR projects have been applied on the offshore Pre-salt carbonate fields of Brazil for over a decade [40] to improve oil recovery while avoiding the emissions from the high CO_2 volume in the produced gas reported in some of the fields [41,42]. However, given the recent development of these fields and their complex rock–fluid characteristics, the impact of CO_2 dissolution on the flow properties and ultimate recovery is still a subject of active research.

In this context, this work presents the experimental measurement of the effect of CO_2 dissolution on the wettability of a carbonate rock analogous to some Pre-salt reservoir fields through contact angle measurement at reservoir conditions. Wettability was altered from intermediate to water-wet by the injection of CO_2. In addition, the rock was characterized through routine core analysis and its pore structure was evaluated and represented using nuclear magnetic resonance and micro-computed tomography (microCT). This porous medium, which captures the dominating flow system, was used to perform pore network modelling of the immiscible flow under the capillary limit. The simulations allowed the construction of the pore-scale imbibition relative permeability curves for both systems, before and after CO_2 dissolution, showing evidence of the effect of the wettability shift in the flow capacity of oil and brine. The obtained plots show an increase in the crossover saturation, indicating an incremental increase in the relative permeability of oil for a greater range of saturations.

To the best the author's knowledge, this is the first work applying experimental measurements of wettability in pore-scale flow simulations of flow for carbonated brine EOR in complex carbonate systems analogous to the Pre-salt reservoirs. The results suggest an incremental improvement in the flow of oil, which contributes to increased production. However, the goal of this study is to analyze the separate effect of wettability alteration on the immiscible flow in the porous medium. Other effects such as viscosity reduction and mineral dissolution should be considered when performing oil recovery numerical calculations.

2. Materials and Methods

The composition of the formation brine characteristic from a Pre-salt field is displayed in Table 1. Brine samples were prepared by dissolution of salts (Sigma Aldrich, St. Louis, MO, USA, purity over 99.9%) in deionized water. In addition, CO_2-saturated fluids were produced by injecting the required mass of CO_2 (Praxair, São Paulo, Brazil, 99.9% purity) in oil or brine in a high-pressure Hastelloy cell.

Table 1. Formation brine composition and total dissolved solids (TDS).

Ions	Concentration [mg/L]
Na^+	57,580
Ca^{2+}	24,250
Mg^{2+}	2120
K^+	1200
Ba^{2+}	24
Sr^{2+}	1260
SO_4^{2-}	54
Cl^-	139,900
TDS	226,388

Table 2 shows the main properties of the dead oil sample for crude oil used in this work. SARA (saturates, aromatics, resins, and asphaltenes) analysis was performed by the IP 143 n-heptane precipitation method (asphaltenes) and medium-pressure liquid chromatography (saturates, resins, and aromatics). Besides, the total acid number (TAN) and total base number (TBN) were calculated using the American Society for Testing and Materials (ASTM) D664 procedures. A full characterization of this Pre-salt oil is presented elsewhere [43].

Table 2. Crude oil characterization.

Crude Oil Properties	
Saturates (wt.%)	64.06
Aromatics (wt.%)	25.98
Resins (wt.%)	8.46
Asphaltenes (wt.%)	1.50
Total Acid Number (mg KOH/g)	0.37
Total Base Number (mg KOH/g)	4.00

Moreover, the rock material used in this work was a carbonate rock from Morro do Chaves Formation, Sergipe-Alagoas Basin. The Morro do Chaves Formation is located in the Northeast region of Brazil. This Formation is composed of coquinas and shales [44]. Coquina is a carbonate rock composed of cemented shell debris and sediments. The shells and shell debris were mechanically sorted and were deposited by a transport agent [45,46]. The studied rock was collected from an outcrop at the Atol quarry, located in Alagoas state in Brazil. This outcrop has been geologically well-studied and -classified in the literature [47–51]. The coquina from this region of Morro do Chaves Formation is composed mainly of whole and fragmented bivalve shells, with a low occurrence of ostracods, and gastropods. The rock matrix is composed of siliciclastic components [48,50].

The pore system of the coquinas from Morro do Chaves Formation was strongly influenced by the diagenetic process, which affected the connectivity of the pores, having a significant influence on permeability but low impact on porosity [48,49]. Previous works [51,52] studied the pore geometry and highlighted the irregularity of pore shapes and sizes. Coquinas can have unimodal or bimodal pore size distribution systems. Pores can vary from micro to meso-pores within the same sample. In addition, there is no linearity in the relation between porosity and permeability of the coquinas from this formation [52]. This heterogeneity makes coquinas challenging rocks for both experimental analysis and computational simulation.

The coquina sample was 3.6 cm in length and 3.5 cm in diameter. The sample was cleaned with methanol and toluene to remove any residual oil or salt that could have been in its porous system. After the sample was cleaned and dried, it was characterized by geological petrographic description, routine petrophysics, microCT, and nuclear magnetic resonance (NMR).

The geological characterization was done with a rock thin section with dimensions of 7.0 cm × 4.5 cm and a thickness of 30 microns. This analysis was done with a Zeiss Axioskop 40 microscope.

The routine petrophysics was done using the Advanced Automated Porosimeter-Permeameter DV-4000, from CSL Capital Management, L.P (Houston, TX, USA). Porosity and pore volume were measured using helium and the permeability was measured using nitrogen. The confining pressure for the measurements was 500 psi.

The microCT images were acquired in the SkyScan 1173 equipment (Bruker, Kontich, Belgium). Image reconstruction was done using the InstaRecon® software (InstaRecon, Champaign, IL, USA), and image processing was done using Avizo® software (Thermo Fisher Scientific, Waltham, MA, USA).

For the NMR experiments, the Oxford MQC—5MHz (Oxford Instruments, Abingdon, UK) equipment was used to acquire the transverse relaxation time (T2) using a CPMG sequence from Coates et al. [53]. For that, the sample was saturated with NaCl synthetic brine with a 50,000 ppm concentration.

Equilibrium contact angle (CA) and interfacial tension (IFT) were measured in a Drop Shape Analysis (DSA) Hastelloy high-pressure system (Kruss, Hamburg, Germany and Eurotechnica, Bargteheide, Germany) equipped with high-precision pumps, pressure indicators, and a temperature-controlling device. All experiments were carried out at 60 °C and 6.895 MPa (1000 psi). Rock samples were cleaned using toluene and methanol, and CA was measured in both clean and aged rocks. The latter were aged for one day in formation brine and 15 days in crude oil at 60 °C. For the tests considering carbonated brine, both fluids (oil and brine) were saturated with CO_2 at test conditions prior to the experiment to reduce mass transfer effects. A complete description of CA and IFT measurement experimental procedures both in the presence and absence of CO_2 can be found in prior works [38,43].

Primary drainage capillary pressure curves were obtained using a Core Lab Instruments A-200 centrifuge. The coquina plugs were saturated in brine and immersed in oil. The speed of rotation was increased in steps (from 500 to 9000 rpm), recording the volume of brine produced from the rock on each step. The speed of rotation was converted to capillary pressure using the equations presented by Hassler and Brunner [54], and the inverse method described in Albuquerque et al. [55] was applied to analyze the data.

The simulations were performed using PORE software, developed by Engineering Simulation and Scientific Software (ESSS) Corporation, Florianopolis, Brazil. PORE constructs a pore network model from 3D microCT images, which are binarized based on a user-selected threshold. The binary image is then developed into a pore network model of pores and throats based on the Maximum Spheres [56,57]. The shapes of the pores and throats are then established based on their Form Factor "G" which is defined as the cross-sectional area divided by the cross-sectional perimeter squared [58]. Euclidean geometrical shapes such as circles, squares, equilateral triangles, and prisms each have a unique form factor, and the pores and throats are individually transformed into the closest matching geometrical shape [59]. Multi-phase flow is then simulated based on the conservation of momentum with increasing pressure steps sequentially from pore to pore through the connecting throats where the radii of curvature in the Young–Laplace equation is controlled by the geometrical shapes of the pores and throats [60,61]. The contact angle and the interfacial tension are user-input parameters and were selected based on lab experiments, and are presented in Section 3.2.

3. Results

This section presents the results for rock characterization, wettability alteration by CO_2 injection, and the effects of CO_2 injection on relative permeability curves.

3.1. Rock Characterization

The coquina was classified as grainstone (Figure 1). It was composed of robust and medium shell fragments, with a size varying from 0.3 to 9.0 mm. The rock was well-selected, with most bioclasts ranging up to 4 mm. It was observed that rare siliciclastic grains were present (3%). There were also rare fragments of disjointed ostracodes (2%) and traces of microcrystalline pyrite. The porosity was classified as intraparticle, interparticle, vug, moldic, intercrystalline, and breccia, in which the predominance was of intraparticle, interparticle, and intercrystalline porosities.

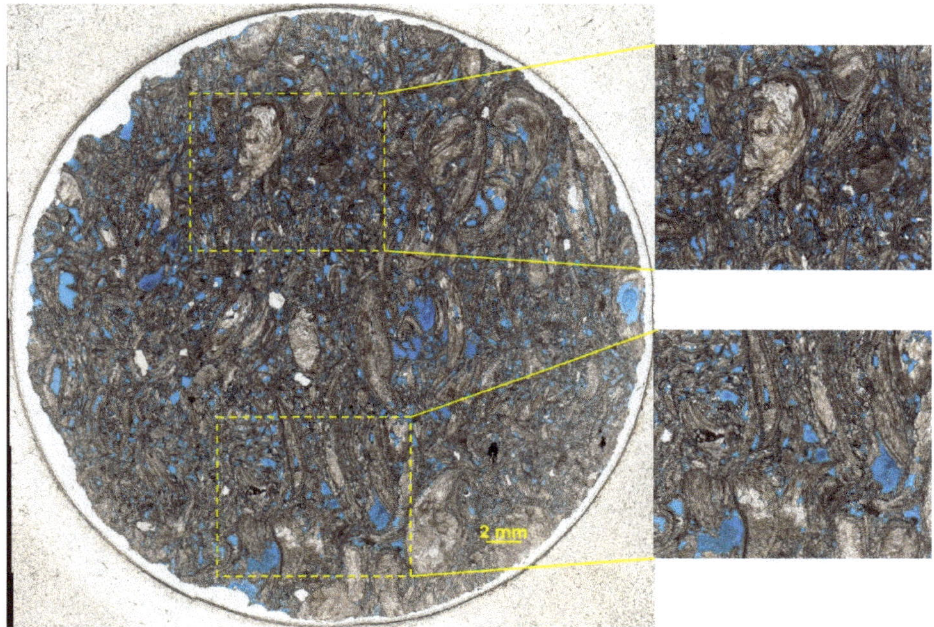

Figure 1. Thin section of coquina sample classified as a grainstone. With the zoomed images on the right, it is possible to observe the different pore types and the heterogeneity of the rock constituents. Magnification of 2.5 times.

The coquina sample was characterized by its petrophysics characteristics. The porosity of the sample was 19.8% and the permeability was 765.1 mD. Other properties can be found in Table 3.

Table 3. Routine petrophysics of the studied coquina sample.

Sample	Pore Volume (cc)	Porosity (%)	Permeability (mD)	Grain Density (g/cc)	Diameter (cm)	Length (cm)	Weight (g)
Coquina	6.8	19.8	765.1	2.69	3.5	3.6	74.3

The sample was digitalized and 4304 microCT images were generated with 9.97 μm/voxel. During the processing, a non-local means filter [62] was used to smooth the image and remove artifacts from the acquisition process. After the image processing, it was possible to identify the different components of the rock. The shells and minerals described in the thin section appeared as different shades of grey. The pore space was identified as black and dark grey (Figure 2).

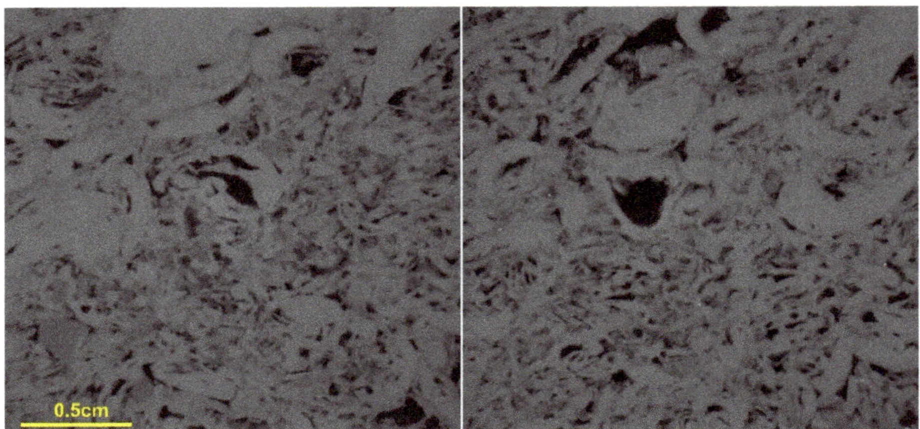

Figure 2. Slices from the micro-computed tomography (microCT) coquina sample with a pixel size of 9.98 μm.

To produce the digital capillary pressure and relative permeability curves, it was necessary to segment the microCT images into pore and rock. For that, the Kittler and Illingworth method [63] was used. With the segmented images, the digital total porosity was calculated, and the value obtained was 15.3% for the whole digital sample (Figure 3). After that, a region of interest (ROI) was defined to be used in the simulation process.

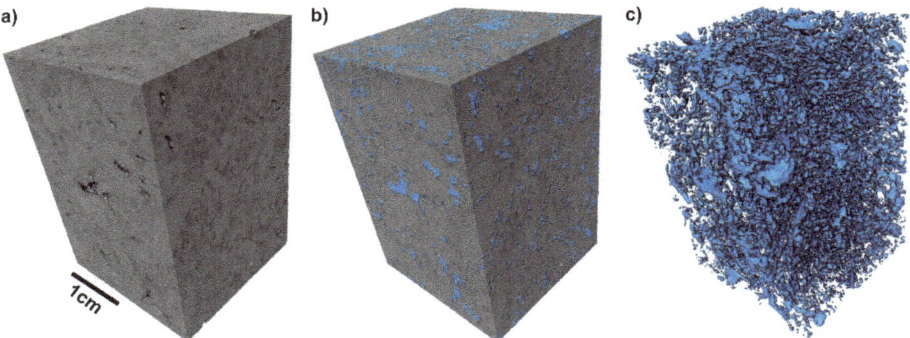

Figure 3. The volume of interest used in the digital sample. (**a**) microCT images of the coquina, and (**b**) rock matrix in grey and segmented pores in blue. (**c**) Segmented pore system.

The NMR data acquisition, interpretation, and transformation of T2 to pore size distribution are discussed elsewhere [51]. It can be observed in Figure 4 that 82% of the pore size distribution had a radius larger than 10 μm, which means that the microCT images at a resolution of 9.97 μm captured the dominant porous space of this sample.

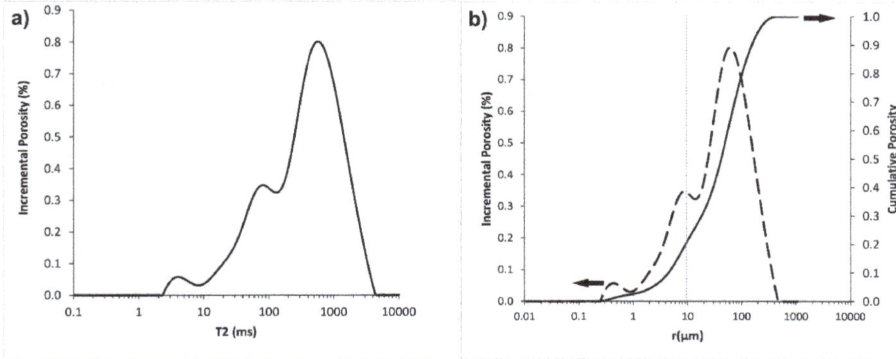

Figure 4. Nuclear Magnetic Resonance (NMR) data from the studied coquina sample. (**a**) shows the T2 distribution and (**b**) illustrates the incremental (dashed curve) and cumulative (full curve) pore size distributions. The vertical dashed line shows the microCT pixel size (9.97 μm). Pores larger than this value are captured in the digital image. By contrast, pores with a radius smaller than this value are under the image resolution.

3.2. Contact Angle Measurements

Equilibrium CA measurement tests were carried out on 15 day aged coquinas both in the absence of CO_2 and using CO_2-saturated fluids. As CA was measured toward the brine phase, lower CA values indicate a greater water-wet behavior. Figure 5 shows screenshots of the CA measurements, giving evidence of the change in wettability caused by CO_2 dissolution in oil and brine.

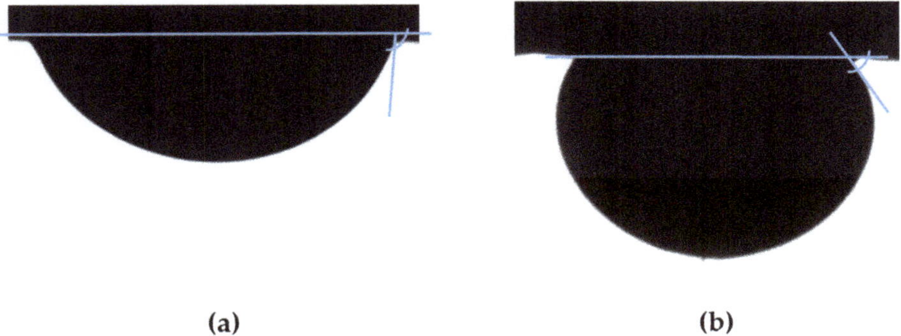

(**a**)   (**b**)

Figure 5. Images illustrating contact angle measurements. (**a**) Measurement on an aged rock with no CO_2. (**b**) Measurement on aged coquina using CO_2-saturated fluids. Images captured after 24 h of experiment.

In addition, CA was measured in unaged coquina samples without CO_2 dissolution. Figure 6 illustrates the results of tests including aged and unaged coquinas in the presence or absence of CO_2. It can be observed that, as expected, the unaged rock was water-wet, and aging shifted the wettability toward neutral-wet. In addition, CO_2 injection reduced the CA, increasing the water-wet behavior of the coquina rock sample. These results will be discussed in Section 4.

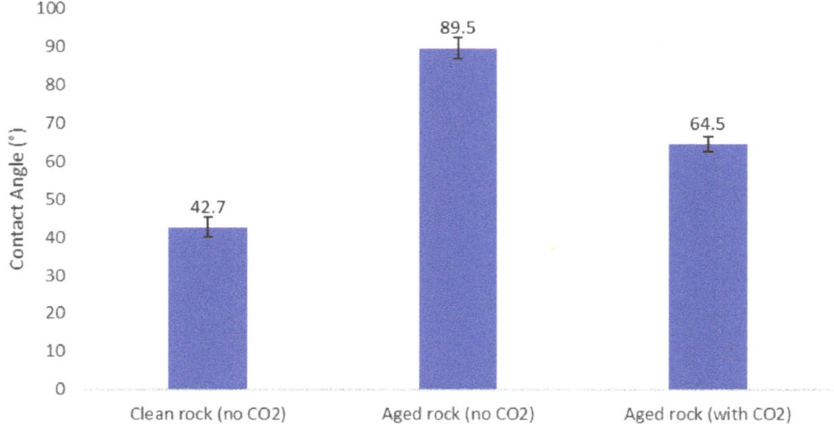

Figure 6. Equilibrium contact angles between Pre-salt crude oil and formation brine on clean and 15 day aged coquinas. The test with CO_2 considers oil and brine saturated with CO_2. Experiment conditions: 60 °C and 6.895 MPa. The reported value is the mean of triplicate measurements, and the error bars represent the standard error of the mean.

3.3. Capillary Pressure and Relative Permeability Curves

The PORE simulations were performed on a sub-cube centered on the lower right quadrant comprising 1000 slices with a width of 1000 pixels and a height of 1000 pixels. A threshold of 56 was used, which yielded a porosity of 15.4% and permeability of 813 mD. At 9.97 microns per pixel resolution, the porosity compares favorably with the NMR porosity at a diameter greater than 10 microns (16.2%), and the computed permeability is close to the measured permeability of 765 mD.

Primary drainage consists of the displacement of water by oil until reaching the connate water saturation. As the rock is fully saturated with brine and has not been in contact with oil prior to the primary drainage, the water-wet conditions of an unaged rock represent this stage [64]. Primary drainage capillary pressure curves were obtained experimentally and by numerical simulations using the procedures described in Section 2. Connate water saturation ($S_{wi} = 0.089$) was obtained experimentally and used as input in the simulations for quality control. The contact angle value obtained in Section 3.3 for the unaged rock (CA = 42.7°) was considered. In addition, the experimentally measured interfacial tension for this oil-brine system with no CO_2 content (IFT = 10.5 mN/m) previously reported by our group [43] was applied. Figure 7 shows good agreement between the experimental and simulated curves.

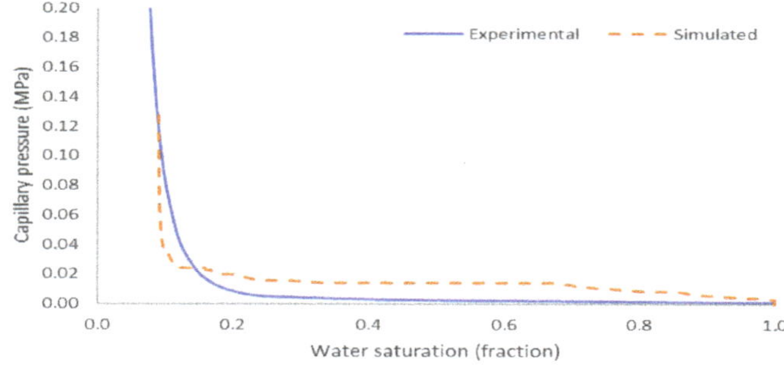

Figure 7. Experimental and numerical simulated primary drainage capillary pressure curves for the coquina sample considered in this work.

Imbibition, consisting of the displacement of oil by water, represents the oil recovery process. It starts with the rock saturated with oil (at connate water saturation) at oil-wet conditions [64]. To compare the effects of water injection and carbonated brine injection, the imbibition relative permeability curves were simulated using the experimentally measured CA and IFT values for each case. CA values are reported in Figure 6 in this work, and IFT values were presented and discussed in previous work by our group [43]. It should be noted that the objective of these curves is to qualitatively analyze the effects of wettability and IFT alteration due to CO_2 injection on multiphase flow in the porous medium. They aim to support discussions regarding the effects of CO_2 on fluid displacement. As the quantitative effects of mineral dissolution and reactive transport were not included in these simulations, they are not intended to provide quantitative input data for reservoir modelling.

Considering the results in Figure 6, the water injection case (with no CO_2) indicated a neutral-wet (CA = 89.5°) rock and an IFT of 10.5 mN/m previously measured by our group [43]. By contrast, the carbonated brine injection case consisted of a water-wet rock (CA = 64.5°, Figure 6) and an IFT of 16.4 mN/m, as reported in the aforementioned study. The results are shown in Figure 8.

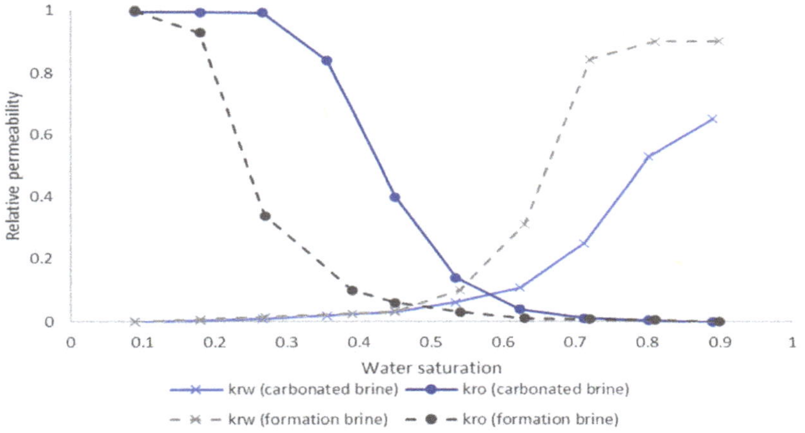

Figure 8. Imbibition relative permeability curves for water (gray) and carbonated brine (blue) injection in the oil-saturated coquina sample.

4. Discussion

The results presented in Section 3 give evidence that CO_2 dissolution (i.e., by carbonated brine injection) shifts the wettability of the aged coquina sample from intermediate to water-wet. This reduction in oil-wet behavior is in agreement with previous studies of carbonated brine injection in carbonate rocks [33,34]. Several mechanisms were proposed to support this decrease in rock hydrophobicity. Sohrabi et al. [39] related this effect with the reduction in brine pH after CO_2 dissolution. This affects surface charges at oil/water and water/solid interfaces and may destabilize polar compounds in the oil, reducing the oil-wet behavior of the rock. Chen et al. [37] studied the same problem by experimental CA measurements and surface complexation modelling. They associated pH reduction with the formation of positively charged species (i.e., $>CaOH_2^+$ for calcite surface and $-NH_2^+$ for oil surface), which causes surface repulsion, reducing the wettability toward oil. Drexler et al. [38] reported surface mineral dissolution and oil desorption to shift the wettability of oil-wet carbonate reservoir rocks. Zhu et al. [36] related the change in CA (toward water-wet) to element changes on the solid surface after introducing CO_2 in the system. A reduction in carbon content on the rock surface, which could be associated with carbonate dissolution, resulted in an increase in hydrophilic behavior of the rock surface.

The wettability alteration caused by carbonated brine is a consequence of the effect of pH reduction on the oil/brine/rock interfaces. The TAN and TBN shown in Table 2 indicate that the oil used in this work contains both acid and basic compounds. The acid compounds are in the negatively charged dissociated state at intermediate or high pH [65]. In the absence of CO_2, they interact with the positively charged surface of the carbonate rock, resulting in the intermediate-wet behavior observed for the aged sample. However, the acidic carbonated brine environment results in an increased concentration of the protonated acid components and a reduced concentration of their negatively charged dissociated species, reducing their interaction with the rock surface and, consequently, the wettability toward oil.

In addition, the increase in IFT for the carbonated brine (Section 3.3) also suggests a reduction in the interfacial activity of the polar compounds at lower pH values. Drexler et al. [43] measured the IFT for this oil and brine in the presence and absence of CO_2 in triplicate experiments, and they found a 56% increase in IFT for the systems containing carbonated brine. In addition, they reported the same trends in experiments where the pH was controlled using buffer solutions. These results give evidence that the surface activity in this oil is controlled by its acid compounds. As a result, a higher concentration of the protonated acid compounds and a lower concentration of the deprotonated species results in a diminished surface-activity. This effect was explained by Danielli [66], who measured the IFT between different organic acids in bromobenzene solutions and water, controlling the pH with buffer solutions. For all the systems containing organic acids, the IFT was reduced when the pH was high enough to favor their dissociation. Furthermore, the reduction in IFT and concentration of surface-active species at increased pH for crude oils containing acid polar compounds was also described in previous works using CO_2 dissolution [67] and acid/base buffers [68].

Besides the protonation of the acid compounds, basic compounds could also contribute to the observed wettability shift. In acidic environments, these compounds are positively charged due to protonation [69]. The presence of positively charged species at the oil/water interface induces repulsion with the positively charged surface of the carbonate rock, increasing its water-wet behavior.

Finally, chemical reaction between the rock and the carbonated brine can also play a part in the observed wettability alteration. Calcite, which is the major mineral constituent of the coquina rock samples [70], reacts with the carbonic acid formed by dissolution of CO_2. This reaction increases the concentration of Ca^{2+}, as described in the literature [33]. Furthermore, the potential determining ion Ca^{2+} was reported to weaken the interaction between positively charged surfaces of carbonates, such as chalk [71,72], and calcite [33], and the negatively charged compounds in the oil. As a result, the water-wet behavior is increased after CO_2 dissolution in the aqueous phase.

Digital rock evaluations of carbonate rocks were presented in recent publications [73–76]. However, the coquina rock considered in this work presents a wide pore size distribution ranging from nanometers to hundreds of microns, as shown in Figure 4. To the best of our knowledge, this is the first study including the impact of carbonated brine on the imbibition relative permeabilities based on numerical experiments for a highly complex carbonate rock sample with a wide range of pore sizes. The curves obtained in this work (Figure 8) indicate the positive effect of carbonated brine injection on relative permeabilities. A shift in the crossover point toward higher water saturation indicates the increase in wettability toward water for carbonated brine injection [77]. This is also supported by the reduction in the endpoint relative permeability to water observed for the carbonated brine [78]. As a result, the obtained curves for carbonate brine injection illustrate an increased flow capacity (i.e., greater relative permeability) of the oil phase for the majority of saturation values.

Relative permeability curves play a critical role in oil recovery and EOR [79]. By applying appropriate upscaling techniques and considering mobility effects, capillary limit fractional flow curves can be obtained to estimate the flow of oil and brine under no macroscopic sweeping effects [80]. The shift in wettability previously discussed indicates an increase in oil displacement by carbonated brine injection, suggesting its potential for EOR applications.

5. Conclusions

Carbonated brine injection EOR is an effective method of reducing CO_2 emissions in oil fields with high production of this gas. This work studies the effect of carbonated brine injection on the wettability and multiphase flow for a carbonate rock system analogous to Brazilian Pre-salt fields. Its main insights are summarized below:

Carbonated brine decreases the pH of the aqueous phase, favoring the protonation of acid and basic compounds in the oil. This induces an increase in the positive charge (or decrease in the negative charge) at the oil/brine interface and, consequently, increments the repulsion with the positively charged carbonate rock surface. As a result, the wettability is shifted toward water-wet.

MicroCT at 9.97 µm resolution captured 82% of the pore size distribution according to the NMR characterization, allowing the dominant system to be reproduced for pore-scale flow simulations.

Imbibition relative permeability curves obtained using pore network modelling for capillary limit immiscible flow showed changes in the curve crossover and endpoints in agreement with the wettability alteration found for carbonated brine injection. These changes contribute to increase the flow of oil, favoring its displacement for EOR processes. However, quantitative recovery calculations should also consider other effects, such as changes in mobility and brine–rock reactivity.

Author Contributions: Conceptualization, S.D. and F.H.; methodology, S.D., F.H. and W.G.; software, A.B.; validation, S.D., F.H., W.G. and A.B.; formal analysis, S.D., F.H., W.G. and A.B; investigation, S.D., F.H. and W.G.; resources, P.C.; data curation, S.D. and A.B.; writing—original draft preparation, S.D. and F.H.; writing—review and editing, S.D., F.H. and A.B.; visualization, S.D., F.H. and W.G.; supervision, S.D.; project administration, S.D. and P.C.; funding acquisition, P.C. All authors have read and agreed to the published version of the manuscript.

Funding: This research was funded by Shell Brasil, grant number ANP n° 20163-2.

Acknowledgments: This research was carried out in association with the ongoing R & D project registered as ANP n° 20163-2, "Análise Experimental da Recuperação de Petróleo para os Carbonatos do Pré-sal do Brasil através de Injeção Alternada de CO2 e Água" (UFRJ/Shell Brasil/ANP), sponsored by Shell Brasil under the ANP R & D levy as "Compromisso de Investimentos com Pesquisa e Desenvolvimento". This study was financed in part by the Coordenação de Aperfeiçoamento de Pessoal de Nível Superior-Brasil (CAPES)—Finance Code 001.

References

1. Kriegler, E.; Luderer, G.; Bauer, N.; Baumstark, L.; Fujimori, S.; Popp, A.; Rogelj, J.; Strefler, J.; van Vuuren, D.P. Pathways limiting warming to 1.5 °C: A tale of turning around in no time? *Philos. Trans. R. Soc. A Math. Phys. Eng. Sci.* **2018**, *376*, 20160457. [CrossRef] [PubMed]

2. Mac Dowell, N.; Fennell, P.S.; Shah, N.; Maitland, G.C. The role of CO_2 capture and utilization in mitigating climate change. *Nat. Clim. Chang.* **2017**, *7*, 243–249. [CrossRef]

3. Guo, J.-X.; Huang, C.; Wang, J.-L.; Meng, X.-Y. Integrated operation for the planning of CO_2 capture path in CCS–EOR project. *J. Pet. Sci. Eng.* **2020**, *186*, 106720. [CrossRef]

4. Mohsin, I.; Al-Attas, T.A.; Sumon, K.Z.; Bergerson, J.; McCoy, S.; Kibria, M.G. Economic and Environmental Assessment of Integrated Carbon Capture and Utilization. *Cell Rep. Phys. Sci.* **2020**, 100104. [CrossRef]

5. Zhang, Z.; Pan, S.-Y.; Li, H.; Cai, J.; Olabi, A.G.; Anthony, E.J.; Manovic, V. Recent advances in carbon dioxide utilization. *Renew. Sustain. Energy Rev.* **2020**, *125*, 109799. [CrossRef]

6. Berstad, D.; Nekså, P.; Gjøvåg, G.A. Low-temperature syngas separation and CO_2 capture for enhanced efficiency of IGCC power plants. *Energy Proc.* **2011**, *4*, 1260–1267. [CrossRef]

7. Aliff Radzuan, M.R.; Syarina, N.A.; Wan Rosdi, W.M.; Hussin, A.H.; Adnan, M.F. Sustainable Optimization of Natural Gas Sweetening Using A Process Simulation Approach and Sustainability Evaluator. *Mater. Today Proc.* **2019**, *19*, 1628–1637. [CrossRef]

8. Vega, F.; Baena-Moreno, F.M.; Gallego Fernández, L.M.; Portillo, E.; Navarrete, B.; Zhang, Z. Current status of CO_2 chemical absorption research applied to CCS: Towards full deployment at industrial scale. *Appl. Energy* **2020**, *260*, 114313. [CrossRef]

9.   Bui, M.; Adjiman, C.S.; Bardow, A.; Anthony, E.J.; Boston, A.; Brown, S.; Fennell, P.S.; Fuss, S.; Galindo, A.; Hackett, L.A.; et al. Carbon capture and storage (CCS): The way forward. *Energy Environ. Sci.* **2018**, *11*, 1062–1176. [CrossRef]

10.   Sanz-Pérez, E.S.; Murdock, C.R.; Didas, S.A.; Jones, C.W. Direct Capture of CO(2) from Ambient Air. *Chem. Rev.* **2016**, *116*, 11840–11876. [CrossRef]

11.   Webley, P.A. Adsorption technology for CO_2 separation and capture: A perspective. *Adsorption* **2014**, *20*, 225–231. [CrossRef]

12.   Duan, J.; Jin, W.; Kitagawa, S. Water-resistant porous coordination polymers for gas separation. *Coord. Chem. Rev.* **2017**, *332*, 48–74. [CrossRef]

13.   Ahmed, I.; Jhung, S.H. Applications of metal-organic frameworks in adsorption/separation processes via hydrogen bonding interactions. *Chem. Eng. J.* **2017**, *310*, 197–215. [CrossRef]

14.   Staciwa, P.; Narkiewicz, U.; Sibera, D.; Moszyński, D.; Wróbel, R.J.; Cormia, R.D. Carbon Spheres as CO_2 Sorbents. *Appl. Sci.* **2019**, *9*, 3349. [CrossRef]

15.   Deng, X.; Yang, W.; Li, S.; Liang, H.; Shi, Z.; Qiao, Z. Large-Scale Screening and Machine Learning to Predict the Computation-Ready, Experimental Metal-Organic Frameworks for CO_2 Capture from Air. *Appl. Sci.* **2020**, *10*, 569. [CrossRef]

16.   Brunetti, A.; Scura, F.; Barbieri, G.; Drioli, E. Membrane technologies for CO_2 separation. *J. Memb. Sci.* **2010**, *359*, 115–125. [CrossRef]

17.   Scholes, C.A.; Smith, K.H.; Kentish, S.E.; Stevens, G.W. CO_2 capture from pre-combustion processes—Strategies for membrane gas separation. *Int. J. Greenh. Gas Control* **2010**, *4*, 739–755. [CrossRef]

18.   Belaissaoui, B.; Favre, E. Membrane Separation Processes for Post-Combustion Carbon Dioxide Capture: State of the Art and Critical Overview. *Oil Gas Sci. Technol. Rev. IFP Energies Nouv.* **2014**, *69*, 1005–1020. [CrossRef]

19.   Chi, C.; Wang, X.; Peng, Y.; Qian, Y.; Hu, Z.; Dong, J.; Zhao, D. Facile Preparation of Graphene Oxide Membranes for Gas Separation. *Chem. Mater.* **2016**, *28*, 2921–2927. [CrossRef]

20.   Celia, M.A.; Bachu, S.; Nordbotten, J.M.; Bandilla, K.W. Status of CO_2 storage in deep saline aquifers with emphasis on modeling approaches and practical simulations. *Water Resour. Res.* **2015**, *51*, 6846–6892. [CrossRef]

21.   Gal, F.; Pokryszka, Z.; Labat, N.; Michel, K.; Lafortune, S.; Marblé, A. Soil-Gas Concentrations and Flux Monitoring at the Lacq-Rousse CO_2-Geological Storage Pilot Site (French Pyrenean Foreland): From Pre-Injection to Post-Injection. *Appl. Sci.* **2019**, *9*, 645. [CrossRef]

22.   Jeong, S.-M.; Ko, S.; Sean, W.-Y. Numerical Prediction of the Behavior of CO_2 Bubbles Leaked from Seafloor and Their Convection and Diffusion near Southeastern Coast of Korea. *Appl. Sci.* **2020**, *10*, 4237. [CrossRef]

23.   Pedro, J.; Araújo, A.A.; Moita, P.; Beltrame, M.; Lopes, L.; Chambel, A.; Berrezueta, E.; Carneiro, J. Mineral Carbonation of CO_2 in Mafic Plutonic Rocks, I—Screening Criteria and Application to a Case Study in Southwest Portugal. *Appl. Sci.* **2020**, *10*, 4879. [CrossRef]

24.   Gou, Y.; Hou, Z.; Liu, H.; Zhou, L.; Were, P. Numerical simulation of carbon dioxide injection for enhanced gas recovery (CO_2-EGR) in Altmark natural gas field. *Acta Geotech.* **2014**, *9*, 49–58. [CrossRef]

25.   Mazzotti, M.; Pini, R.; Storti, G. Enhanced coalbed methane recovery. *J. Supercrit. Fluids* **2009**, *47*, 619–627. [CrossRef]

26.   Iino, A.; Onishi, T.; Datta-Gupta, A. Optimizing CO_2—And Field-Gas-Injection EOR in Unconventional Reservoirs Using the Fast-Marching Method. *SPE Reserv. Eval. Eng.* **2020**, *23*, 261–281. [CrossRef]

27.   Sheng, J. *Enhanced Oil Recovery Field Case Studies*; Gulf Professional Publishing: Waltham, MA, USA, 2013; ISBN 9780123865465.

28.   Thorne, R.J.; Sundseth, K.; Bouman, E.; Czarnowska, L.; Mathisen, A.; Skagestad, R.; Stanek, W.; Pacyna, J.M.; Pacyna, E.G. Technical and environmental viability of a European CO_2 EOR system. *Int. J. Greenh. Gas Control* **2020**, *92*, 102857. [CrossRef]

29.   De Nevers, N. A Calculation Method for Carbonated Water Flooding. *Soc. Pet. Eng. J.* **1964**, *4*, 9–20. [CrossRef]

30.   Kechut, N.I.; Riazi, M.; Sohrabi, M.; Jamiolahmady, M. Tertiary Oil Recovery and CO2 Sequestration by Carbonated Water Injection (CWI). In Proceedings of the SPE International Conference on CO2 Capture, Storage, and Utilization, New Orleans, LA, USA, 10–12 November 2010; p. 12.

31. Fjelde, I.; Asen, S.M. Wettability alteration during water flooding and carbon dioxide flooding of reservoir chalk rocks. In Proceedings of the SPE EUROPEC/EAGE Annual Conference and Exhibition, Barcelona, Spain, 14–17 June 2010.

32. Al-Mutairi, S.M.; Abu-khamsin, S.A.; Hossain, M.E. A Novel Approach to Handle Continuous Wettability Alteration during Immiscible CO_2 Flooding Process. In Proceedings of the Abu Dhabi International Petroleum Conference and Exhibition, Abu Dhabi, UAE, 11–14 November 2012.

33. Seyyedi, M.; Sohrabi, M.; Farzaneh, A. Investigation of Rock Wettability Alteration by Carbonated Water through Contact Angle Measurements. *Energy Fuels* **2015**, *29*, 5544–5553. [CrossRef]

34. Ruidiaz, E.M.; Winter, A.; Trevisan, O.V. Oil recovery and wettability alteration in carbonates due to carbonate water injection. *J. Pet. Explor. Prod. Technol.* **2018**, *8*, 249–258. [CrossRef]

35. Teklu, T.W.; Alameri, W.; Kazemi, H.; Graves, R.M. Contact Angle Measurements on Conventional and Unconventional Reservoir Cores. In Proceedings of the Unconventional Resources Technology Conference, San Antonio, TX, USA, 20–22 July 2015; p. 17.

36. Zhu, Z.; Li, M.; Lin, M.; Peng, B.; Sun, L.; Chen, L. Investigation on Variations in Wettability of Reservoir Rock Induced by CO_2-Brine-Rock Interactions. In Proceedings of the SPE EUROPEC/EAGE Annual Conference and Exhibition, Vienna, Austria, 12–15 June 2011; p. 11.

37. Chen, Y.; Sari, A.; Xie, Q.; Saeedi, A. Insights into the wettability alteration of CO_2-assisted EOR in carbonate reservoirs. *J. Mol. Liq.* **2019**, *279*, 420–426. [CrossRef]

38. Drexler, S.; Silveira, T.M.G.; De Belli, G.; Couto, P. Experimental study of the effect of carbonated brine on wettability and oil displacement for EOR application in the Brazilian Pre-Salt reservoirs. *Energy Sources Part A Recover. Util. Environ. Eff.* **2019**, 1–15. [CrossRef]

39. Sohrabi, M.; Emadi, A.; Farzaneh, S.A.; Ireland, S. A Thorough Investigation of Mechanisms of Enhanced Oil Recovery by Carbonated Water Injection. In Proceedings of the SPE Annual Technical Conference and Exhibition, Houston, TX, USA, 28–30 September 2015; p. 33.

40. De Almeida, A.S.; Lima, S.D.T.C.; Rocha, P.S.; De Andrade, A.M.T.; Branco, C.C.M.; Pinto, A.C.C. CCGS opportunities in the Santos Basin pre-salt development. In Proceedings of the SPE International Conference on Health, Safety and Environment in Oil and Gas Exploration and Production 2010, Rio de Janeiro, Brazil, 12–14 April 2010; Volume 2, pp. 840–849.

41. Pepin, A.; Bize-Forest, N.; Montoya Padilla, S.; Abad, C.; Schlicht, P.; de Castro Machado, A. Pre-Salt Carbonate Reservoir Analog Selection for Stimulation Optimization. In Proceedings of the International Petroleum Technology Conference, Kuala Lumpur, Malaysia, 10–12 December 2014.

42. Pizarro, J.O.D.S.; Branco, C.C.M. Challenges in Implementing an EOR Project in the Pre-Salt Province in Deep Offshore Brasil. In Proceedings of the SPE EOR Conference at Oil and Gas West Asia, Muscat, Oman, 16–18 April 2012; p. 13.

43. Drexler, S.; Correia, E.L.; Jerdy, A.C.; Cavadas, L.A.; Couto, P. Effect of CO_2 on the dynamic and equilibrium interfacial tension between crude oil and formation brine for a deepwater Pre-salt field. *J. Pet. Sci. Eng.* **2020**, *190*, 107095. [CrossRef]

44. Campos Neto, O.P.d.A.; Lima, W.S.; Cruz, F.E.G. Bacia de Sergipe-Alagoas (Sergipe-Alagoas Basin). *Bol. Geocienc. da Petrobras* **2007**, *15*, 405–415.

45. Pettijohn, F.J. *Sedimentary Rocks*, 2nd ed.; Harper Brothers: New York, NY, USA, 1957; ISBN 978-1114189805.

46. Schäfer, W. *Ecology and Palaeoecology of Marine Environments*; The University of Chicago Press: Chicago, IL, USA, 1972.

47. Azambuja, N.C.; Arienti, L.M.; da Cruz, F.E. *Guidebook to the Rift-Drift Sergipe-Alagoas Passive Margin Basin, Brazil, Proceedings of the The 1998 AAPG International Conference and Exhibition, Rio de Janeiro, Brazil, 8–11 November 1998*; American Association of Petroleum Geologists: Tulsa, OK, USA, 1998; p. 113.

48. Tavares, A.C.; Borghi, L.; Corbett, P.; Nobre-Lopes, J.; Câmara, R. Facies and depositional environments for the coquinas of the Morro do Chaves Formation, Sergipe-Alagoas Basin, defined by taphonomic and compositional criteria. *Braz. J. Geol.* **2015**, *45*, 415–429. [CrossRef]

49. Corbett, P.W.M.; Estrella, R.; Morales Rodriguez, A.; Shoeir, A.; Borghi, L.; Tavares, A.C. Integration of Cretaceous Morro do Chaves rock properties (NE Brazil) with the Holocene Hamelin Coquina architecture (Shark Bay, Western Australia) to model effective permeability. *Pet. Geosci.* **2016**, *22*, 105–122. [CrossRef]

50. Chinelatto, G.F.; Vidal, A.C.; Kuroda, M.C.; Basilici, G. A taphofacies model for coquina sedimentation in lakes (Lower Cretaceous, Morro do Chaves Formation, NE Brazil). *Cretac. Res.* **2018**, *85*, 1–19. [CrossRef]

51. Hoerlle, F.O.; Rios, E.H.; de Azevedo Lopes da Silva, W.G.; Pontedeiro, E.M.B.D.; da Costa de Oliveira Lima, M.; Corbett, P.W.M.; Alves, J.L.D.; Couto, P. Nuclear Magnetic Resonance to Characterize the Pore System of Coquinas from Morro do Chaves Formation, Sergipe-Alagoas Basin, Brazil. *Braz. J. Geophys.* **2018**, *36*, 317–324. [CrossRef]

52. da Costa de Oliveira Lima, M.; Martins, L.P.; Rios, E.H.; Boyd, A.; Pontedeiro, E.M.B.D.; Hoerlle, F.O.; Lipovetski, T.; Neto, A.O.; Mendes, M.; Borghi, L.; et al. *Rock Typing of Coquinas from the Morro do Chaves Formation, Proceedings of the Sixteenth International Congress of the Brazilian Geophysical Society, Rio de Janeiro, Brazil, 17–22 August 2019*; Brazilian Geophysical Society: Rio de Janeiro, Brazil, 2019; pp. 1–6.

53. Coates, G.R.; Xiao, L.; Prammer, M.G. *NMR Logging: Principles and Applications*; Halliburton Energy Services: Houston, TX, USA, 1999; p. 234.

54. Hassler, G.L.; Brunner, E. Measurement of Capillary Pressures in Small Core Samples. *Trans. AIME* **1945**, *160*, 114–123. [CrossRef]

55. Albuquerque, M.R.; Eler, F.M.; Camargo, H.V.R.; Compan, A.L.M.; Cruz, D.A.; Pedreira, C.E. *Estimation of Capillary Pressure Curves from Centrifuge Measurements Using Inverse Methods, Proceedings of the Rio Oil & Gas*; IBP: Rio de Janeiro, Brazil, 24–27 September 2018.

56. Silin, D.; Patzek, T. Pore space morphology analysis using maximal inscribed spheres. *Phys. A Stat. Mech. Its Appl.* **2006**, *371*, 336–360. [CrossRef]

57. Dong, H. *Micro CT Imaging and Pore Network Extraction*; Imperial College London: London, UK, 2007.

58. Zhou, D.; Blunt, M.; Orr, F.M. Hydrocarbon Drainage along Corners of Noncircular Capillaries. *J. Colloid Interface Sci.* **1997**, *187*, 11–21. [CrossRef] [PubMed]

59. Raoof, A.; Hassanizadeh, S.M. A New Method for Generating Pore-Network Models of Porous Media. *Transp. Porous Media* **2010**, *81*, 391–407. [CrossRef]

60. Joekar-Niasar, V.; Majid Hassanizadeh, S. Effect of fluids properties on non-equilibrium capillarity effects: Dynamic pore-network modeling. *Int. J. Multiph. Flow* **2011**, *37*, 198–214. [CrossRef]

61. Blunt, M.J. Flow in porous media—Pore-network models and multiphase flow. *Curr. Opin. Colloid Interface Sci.* **2001**, *6*, 197–207. [CrossRef]

62. Buades, A.; Coll, B.; Morel, J. A non-local algorithm for image denoising. In Proceedings of the 2005 IEEE Computer Society Conference on Computer Vision and Pattern Recognition (CVPR'05), San Diego, CA, USA, 20–25 June 2005; Volume 2, pp. 60–65.

63. Kittler, J.; Illingworth, J. Minimum error thresholding. *Pattern Recognit.* **1986**, *19*, 41–47. [CrossRef]

64. Blunt, M.J. *Multiphase Flow in Permeable Media: A Pore-Scale Perspective*; Cambridge University Press: Cambridge, UK, 2017; ISBN 9781107093461.

65. Werner, J.; Persson, I.; Björneholm, O.; Kawecki, D.; Saak, C.-M.; Walz, M.-M.; Ekholm, V.; Unger, I.; Valtl, C.; Caleman, C.; et al. Shifted equilibria of organic acids and bases in the aqueous surface region. *Phys. Chem. Chem. Phys.* **2018**, *20*, 23281–23293. [CrossRef] [PubMed]

66. Danielli, J.F. The Relations between Surface pH, Ion Concentrations and Interfacial Tension. *Proc. R. Soc. Lond. Ser. B Biol. Sci.* **1937**, *122*, 155–174.

67. Lashkarbolooki, M.; Riazi, M.; Ayatollahi, S. Effect of CO_2 and natural surfactant of crude oil on the dynamic interfacial tensions during carbonated water flooding: Experimental and modeling investigation. *J. Pet. Sci. Eng.* **2017**, *159*, 58–67. [CrossRef]

68. Lashkarbolooki, M.; Ayatollahi, S. The effects of pH, acidity, asphaltene and resin fraction on crude oil/water interfacial tension. *J. Pet. Sci. Eng.* **2018**, *162*, 341–347. [CrossRef]

69. Buckley, J.S.; Takamura, K.; Morrow, N.R. Influence of Electrical Surface Charges on the Wetting Properties of Crude Oils. *SPE Reserv. Eng.* **1989**, *4*, 332–340. [CrossRef]

70. Drexler, S.; Hoerlle, F.O.; Silveira, T.M.G.; Cavadas, L.A.; Couto, P. Impact of Rock Aging Time on the Initial Wettability of Minerals and Analogue Rocks Using Pre-Salt Fluids Under Reservoir Conditions. In Proceedings of the Offshore Technology Conference Brasil, Rio de Janeiro, Brazil, 29–31 October 2019; p. 8.

71. Strand, S.; Høgnesen, E.J.; Austad, T. Wettability alteration of carbonates—Effects of potential determining ions (Ca2+ and SO42−) and temperature. *Colloids Surf. A Physicochem. Eng. Asp.* **2006**, *275*, 1–10. [CrossRef]

72. Austad, T.; Strand, S.; Madland, M.V.; Puntervold, T.; Korsnes, R.I. Seawater in Chalk: An EOR and Compaction Fluid. *SPE Reserv. Eval. Eng.* **2008**, *11*, 648–654. [CrossRef]

73. Tölke, J.; Baldwin, C.; Mu, Y.; Derzhi, N.; Fang, Q.; Grader, A.; Dvorkin, J. Computer simulations of fluid flow in sediment: From images to permeability. *Lead. Edge* **2010**, *29*, 68–74. [CrossRef]

74. Dernaika, M.R.; Sahib, M.R.; Gonzalez, D.; Mansour, B.; Al Jallad, O.; Koronfol, S.; Sinclair, G.; Kayali, A. Core Characterization and Numerical Flow Simulation in Representative Rock Types of the Raudhatain Field in Kuwait. In Proceedings of the SPE Kuwait Oil & Gas Show and Conference, Kuwait City, Kuwait, 15–18 October 2017; p. 24.

75. Zambrano, M.; Tondi, E.; Mancini, L.; Lanzafame, G.; Trias, F.X.; Arzilli, F.; Materazzi, M.; Torrieri, S. Fluid flow simulation and permeability computation in deformed porous carbonate grainstones. *Adv. Water Resour.* **2018**, *115*, 95–111. [CrossRef]

76. AlJallad, O.; Dernaika, M.; Koronfol, S.; Naseer Uddin, Y.; Mishra, P. Evaluation of Complex Carbonates from Pore-Scale to Core-Scale. In Proceedings of the International Petroleum Technology Conference, Dhahran, Kingdom of Saudi Arabia, 13–15 January 2020; p. 21.

77. Craig, F.F. *The Reservoir Engineering Aspects of Waterflooding*; Henry, L., Doherty, H.L., Eds.; Memorial Fund of AIME: Richardson, TX, USA, 1971; ISBN 9780895202024.

78. Anderson, W.G. Wettability Literature Survey Part 5: The Effects of Wettability on Relative Permeability. *J. Pet. Technol.* **1987**, *39*, 1453–1468. [CrossRef]

79. Lake, L.W. *Enhanced Oil Recovery*; Prentice Hall: Upper Saddle River, NJ, USA, 1989; ISBN 9780132816014.

80. Picchi, D.; Battiato, I. Relative Permeability Scaling From Pore-Scale Flow Regimes. *Water Resour. Res.* **2019**, *55*, 3215–3233. [CrossRef]

Article

Off-Design Operation of Conventional and Phase-Change CO$_2$ Capture Solvents and Mixtures: A Systematic Assessment Approach

Theodoros Zarogiannis [1,2]**, Athanasios I. Papadopoulos** [1,]*** and Panos Seferlis** [1,2]

[1] Chemical Process and Energy Resources Institute, Centre for Research and Technology-Hellas, 57001 Thermi, Thessaloniki, Greece; zarogiannis@certh.gr (T.Z.); seferlis@auth.gr (P.S.)
[2] Department of Mechanical Engineering, Aristotle University of Thessaloniki, 54124 Thessaloniki, Greece
* Correspondence: spapadopoulos@certh.gr

Received: 10 July 2020; Accepted: 29 July 2020; Published: 31 July 2020

Abstract: Solvent-based CO$_2$ capture technologies hold promise for future implementation but conventional solvents incur significant energy penalties and capture costs. Phase-change solvents enable a significant reduction in the regeneration energy but their performance has only been investigated under steady-state operation. In the current work, we employed a systematic approach for the evaluation of conventional solvents and mixtures, as well as phase-change solvents under the influence of disturbances. Sensitivity analysis was used to identify the impact that operating parameter variations and different solvents exert on multiple CO$_2$ capture performance indicators within a wide operating range. The resulting capture process performance was then assessed for each solvent within a multi-criteria approach, which simultaneously accounted for off-design conditions and nominal operation. The considered performance criteria included the regeneration energy, solvent mass flow rate, cost and cyclic capacity, net energy penalty from integration with an upstream power plant, and lost revenue from parasitic losses. The 10 investigated solvents included the phase-change solvents methyl-cyclohexylamine (MCA) and 2-(diethylamino)ethanol/3-(methylamino)propylamine (DEEA/MAPA). We found that the conventional mixture diethanolamine/methyldiethanolamine (DEA/MDEA) and the phase-change solvent DEEA/MAPA exhibited both resilience to disturbances and desirable nominal operation for multiple performance indicators simultaneously.

Keywords: CO$_2$ capture process; solvent-based absorption/desorption; off-design operation; phase-change solvents; sensitivity analysis

1. Introduction

The development of efficient CO$_2$ abatement systems is widely pursued as a means of mitigating the detrimental effects of global warming [1]. Currently, intense research efforts are focusing on CO$_2$ capture technologies, such as membrane processes [2], adsorption systems [3], and materials- [4,5] and solvent-based absorption/desorption processes [6]. The latter represents a technology with significant potential, which has been demonstrated even in large-scale plants [7,8]. The selection of efficient solvents or mixtures in solvent-based CO$_2$ capture plays an important role in the process configuration, the regeneration energy requirements, and the process economics [9]. While the development of new solvents has been an active research field, the ones proposed to date have not managed to reduce the regeneration energy requirements by more than 25% compared to the reference monoethanolamine (MEA) solvent, with detrimental effects on the reduction of capture costs [1]. In recent years, a new class of solvents called phase-change solvents [10] has indicated experimentally verified regeneration energy reductions of approximately 43% compared to MEA [11,12]. Unlike conventional solvents, phase-change solvents undergo a phase separation upon reaction with CO$_2$ or a change in temperature.

The resulting CO_2-lean phase may be mechanically separated and recycled to the absorber using insignificant amounts of energy, whereas the CO_2-rich phase is introduced into a desorption process. This reduction in the solvent flow rate that undergoes thermal separation, and in some cases, desorption at lower than 90 °C [6], enables the corresponding reduction in regeneration energy requirements.

The inclusion in the selection procedure of solvents or mixtures that exhibit phase-change behavior upon the absorption of CO_2 within a specific range of temperature enables the utilization of the advantages that this type of solvents bring in the reduction of the overall energy penalty for CO_2 capture. Zarogiannis et al. [13] compared phase-change solvent mixtures to conventional solvents based on their performance regarding CO_2 capture using the cyclic solvent capacity, the parasitic electricity losses, and the regeneration energy as criteria. Phase-change solvents showed improved performance in all aspects except the solvent cost due to the small industrial production for such solvents. For this purpose, a short-cut model for the absorption–desorption system has been utilized, initially developed by Kim et al. [14] but extended by Zarogiannis et al. [13] to account for phase-change solvents. However, the evaluation of the performance metrics has been accomplished by considering only the optimal design operating point at steady-state operation. Flue gas streams are usually susceptible to variabilities in their conditions, such as the flow rate, composition, concentration, and temperature, due to either production changes (e.g., variable power plant operation, fuel type change, and so forth) or other disturbances affecting the plant. The variability in the flue gas stream affects the operation of the capture plant by deviating the plant's operating conditions from the cost-effective design point with potentially detrimental effects on the achieved process and economic performance. In addition, several other factors may influence the operating efficiency of the capture plant, such as solvent degradation, solvent losses, and heat exchanger inefficiencies. It is therefore imperative to investigate the performance of candidate CO_2 capture solvents and mixtures, not only at the nominal design point but also for a wide range of off-design conditions.

To this end, the use of a control system is indispensable for addressing variability but the selection of an appropriate solvent can greatly facilitate and enhance the performance of such a control system. This is because capture solvents display different sensitivities regarding their physico-chemical properties under different operating conditions, which subsequently may affect the performance of the capture process. An unfavorable sensitivity of the key physico-chemical properties of the capture solvent mixture would eventually require greater effort from the controller to maintain a desirable process performance since the controller effort is associated with the use of resources, such as steam or an amine solvent flow rate. Consequently, the economic impact of the capture process is significant. Therefore, it is necessary to consider both nominal and off-design operation in the selection of CO_2 capture solvents and eventually select the solvent or solvent mixture that enables economically optimal operation over a wide range of operating conditions.

Existing published literature includes several contributions that investigate the off-design operation of CO_2 capture systems but only for a few conventional (non-phase-change) solvents (Table 1). Different process flowsheets and proposed control strategies are assessed according to the respective dynamic response. Table 1 presents an organized overview of published works based on the modeling approach, the investigated amine solvents, and the development of operating or control strategies.

A dynamic model may be necessary for the investigation of flexibility and the development of control strategies in a CO_2 capture process. Gaspar and Cormos [15] reported that absorption performance depends on the reaction rate and the mass transfer rate for four common amine solvents, namely monoethanolamine (MEA), diethanolamine (DEA), methyl diethanolamine (MDEA), and 2-amino-2-methyl-1-propanol (AMP). They developed a dynamic model to investigate the behavior and to evaluate the absorption capacity of solvents. Jayarantha et al. [16] considered a sensitivity analysis of an absorber for a post-combustion CO_2 capture plant. They concluded that the correlations used to describe the reactions and mass transfer are important factors that affect the efficiency of the model. The effects of the lean solvent flow rate on the CO_2 capture process were also investigated using a dynamic model with SAFT-VR (Statistical Associating Fluid Theory for potentials of Variable Range) for the determination of thermo-physical properties [17].

Table 1. Overview of the literature regarding the development of dynamic models, the employed amine solution, and their scope.

Reference	Model Used	Investigated Solvents	Purpose
[15]	Dynamic (Matlab/Simulink)	MEA, DEA, MDEA, AMP	Investigation of the dynamic process behavior of solvents
[16]	Dynamic (Matlab)	MEA	Sensitivity analysis and dynamic simulations
[17]	Dynamic (gPROMS)	MEA	Dynamic simulation
[18]	Dynamic (gPROMS)	MEA	Different dynamic simulation approaches
[19]	Dynamic (gPROMS)	MEA	Steady-state and dynamic process validation
[20]	Dynamic (Aspen)	MEA	Dynamic simulation
[21]	Dynamic (Matlab)	MEA	Dynamic model validation
[22]	Dynamic (Matlab/Simulink)	AMP	Dynamic simulation and validation
[23]	Dynamic (Aspen Dynamics)	MEA	Plantwide control
[24]	Dynamic (Aspen Plus)	MEA	Investigation of a control strategy
[25]	Mechanistic process model (gPROMS–Aspen Plus)	MEA	Investigation of decentralized control schemes
[26]	Non-linear autoregressive model with exogenous input (NLARX)	MEA	Preliminary control analysis
[27]	Dynamic, rate-based model (Matlab)	PZ, MEA	Investigation of decentralized control structure
[28]	Custom steady-state and dynamic models using OCFE	MEA, DEA, MPA	Steady-state controllability assessment and evaluation of the process dynamic response for different solvents
[29]	Dynamic (Unisim)	MEA	Investigation of operating strategy with self-optimizing control variables
[30]	Aspen HYSYS Dynamics	MEA	Investigation of control structure
[31]	Dynamic (Aspen Plus Dynamics)	MEA	Investigation of control strategy
[32]	Dynamic (gPROMS)	MEA	Control design
[33]	Dynamic, rate-based model (Aspen HYSYS)	MEA	Controllability analysis using MPC
[34]	Dynamic (Unisim)	MEA	Control strategy with alternative control structures
[35]	Dynamic (gPROMS)	MEA	Different control architectures
[36]	Dynamic (gCCS)	MEA	Multi-model modeling for advanced control design using MMPC

OCFE: Orthogonal collocation on finite elements; AMP: 2-Amino-2-methyl-1-propanol, DEA: Diethanolamine, MDEA: Methyldiethanolamine, MEA: Monoethanolamine, MPA: Monopropanolamine, PZ: Piperazine; MMPC: Multi-linear model predictive control, MPC: Model predictive control.

The use of dynamic models should be validated with either steady-state or dynamic data to assess the obtained predictions. Lawal et al. [18] validated their model using experimental data and showed that a mass-transfer-controlled model can achieve better consistency compared to an equilibrium-based model. Biliyok et al. [19] validated a dynamic model with steady-state data and dynamic data using results from a pilot plant. The validated model was used to study the impact of inlet flue gas moisture content and the effect of intercooling on the absorber. Posch and Haider [20] developed a dynamic, rate-based model in Aspen Plus® and validated their results using pilot plant data. Enaasen Flø et al. [21] performed a dynamic model validation of the post-combustion CO_2 absorption process. They claimed that changes in the flue gas and solvent flow rate affect the process more than changes in the reboiler duty. Cormos and Daraban [22] developed and validated a dynamic, rate-based model of a CO_2 capture process in an aqueous solution of AMP using experimental data.

Whereas the above studies mainly focused on the use of dynamic models to study the effects of different process parameters on the CO_2 capture process performance, there is also an increasing number of studies on the development of control strategies for absorption–desorption CO_2 capture. Lin et al. [23] investigated the plantwide control of a post-combustion CO_2 capture process using MEA. The CO_2 removal ratio was controlled by manipulating the lean solvent recycle rate. The flexibility in power plants integrated with CO_2 capture systems was also investigated by Lin et al. [24] in a commercial process simulator with two proposed control strategies. They recommended that the lean solvent flow rate and loading were considered the key variables for the process performance. Simulation studies in commercial simulators were presented by Nittaya et al. [25]. They developed a mechanistic dynamic model using MEA as a solvent and proposed three decentralized control schemes, where the first was based on the relative gain array (RGA), while the others were based on heuristics. Sensitivity analysis was performed to unveil the most suitable controlled and manipulated variables. The study showed that heuristic-based control exhibits better performance than the RGA analysis. Abdul Manaf et al. [26] employed a black-box model to investigate the transient behavior of equipment, such as the absorber, the desorber, and the lean/rich heat exchanger for the determination of the process control strategy. A preliminary analysis expressed as relative gain analysis proposed the control of the CO_2 capture rate and the regeneration energy performance through the manipulation of the lean solvent flow rate and the reboiler duty. Gaspar et al. [27] reported a decentralized control scheme using proportional-integral (PI) controllers for the piperazine (PZ) and MEA as candidate solvents for CO_2 capture. Variations in the lean solvent flow rate and reboiler duty affected the performance of the process. It was proposed that PZ is a better candidate than MEA since it compensated for the disturbances in the lean solvent flow rate, recycle flow rate, reboiler duty, and stripper's feed. Damartzis et al. [28] proposed a framework that includes a generalized process representation and an operability assessment by considering the steady-state controllability and dynamic evaluation of the process. The employed models were developed using the orthogonal collocation on finite elements (OCFE) technique in both assessment stages to efficiently manage the necessary rigorous models and to facilitate computations. Three amines were investigated, namely MEA, DEA, and monopropanolamine (MPA). Damartzis et al. [28] also illustrated that the combination of the choice of solvent and flowsheet features affects the dynamic performance of the process and thus the control scheme proposed to compensate for the possible disturbances. MPA achieved better performance under variability. Panahi and Skogestad [29] proposed a control design approach using a cost index that combines the energy utilization with a penalty for leaving CO_2 to the air by considering self-optimizing control variables for three operational regions. A dynamic simulation was reported by Mechleri et al. [30] using MEA for natural gas combined-cycle power plants. They investigated and applied an appropriate control structure to achieve flexible operation. Leonard et al. [31] proposed the control of the water balance by manipulating the temperature of flue gas in the absorber washing stage. Dynamic studies using MEA as a solvent were presented by Lawal et al. [32]. They shed light on the dynamics of a post-combustion CO_2 capture process for a 500 MWe power plant. It was shown that the power plant exhibits a faster response than the capture plant.

Model predictive control (MPC) in capture plants was the subject of the work by Sahrarei and Ricardez-Sandoval [33]. They used a dynamic, rate-based model and showed that MPC performs better than a decentralized control scheme since it maintains the controlled variables within their boundaries under step changes in the feed. Panahi and Skogestad [34] proposed alternative control structures, including MPC, that operated for three operational regions. Optimizing control architectures by comparing PID (proportional-integral-derivative) and MPC were also studied by Arce et al. [35]. Liang et al. [36] propose a multi-linear model for MEA-based CO_2 capture. The proposed approach first identifies local linear models at typical operating points. Subsequently, the nonlinearity distribution of the process was investigated for the change of the CO_2 capture rate and the change of mass flow of the flue gas. Finally, the multi-linear model was developed based on the nonlinearity distribution. The proposed linear state-space formation can be applied properly for advanced controller techniques, such as multi-linear model predictive control (MMPC).

The previous discussion highlights the increasing interest in the off-design behavior of CO_2 capture systems and also shows that this is mainly approached through the use of rigorous dynamic and/or rate-based models in dynamic mode along with a suitable control structure. Such models are indispensable for the accurate determination and evaluation of the process operation under varying conditions but often require intense effort to construct and execute. This is greatly amplified when there is a need to investigate the performance of multiple different solvents and further enhanced when such solvents include mixtures exhibiting liquid–liquid phase-change behavior. The next observation was that almost all contributions listed in Table 1 focus on MEA, except three cases that consider up to four other conventional solvents. The importance of both solvent mixtures and phase-change solvents in CO_2 capture indicates that their off-design performance needs to be investigated before selecting the most appropriate option. However, the off-design operation of neither conventional solvent mixtures nor phase-change solvents is yet to be addressed in published literature. This is mainly due to the high complexity of the models required for their simulation and the prediction of their phase-change behavior.

This work employed for the first time a systematic approach that enabled the efficient screening and selection from multi-component sets of solvents and mixtures exhibiting non-ideal, multiphase behavior. It fits into the wider need to evaluate any type of material in off-design process conditions as an additional criterion for material selection. This was first highlighted in Papadopoulos and Seferlis [37] and then in Papadopoulos et al. [38]. Papadopoulos and Seferlis [37] showed that any type of material, including solvents, used to enhance the operation of a process, may exhibit significantly different performance under the influence of variability compared to steady-state operation under nominal conditions. In the case of CO_2 capture solvents, it is necessary to select the ones that operate as close as possible to the economically desirable set-points and facilitate the minimization of resources required using the employed controller to bring the process back to its set-point under disturbances. In this work, the enhanced, shortcut model of Zarogiannis et al. [13] was exploited to identify solvents that exhibit favorable performance under off-design conditions. The model was very suitable for use with the method proposed in this work as it came with the advantages demonstrated in the case of nominal operating conditions [13]; it captured the non-ideal solvent–water–CO_2 behavior, it was much easier to develop and use than the corresponding rigorous models, and the obtained predictions were in very good agreement with the experimental data. It is therefore reasonably expected to provide valid insights regarding the off-design performance of various solvents and mixtures examined here.

More specifically, the approach of Papadopoulos et al. [38] was used within a framework that considers CO_2 capture process operability using steady-state, nonlinear sensitivity with an implicit consideration of a controller scheme [39]. A very inclusive set of controlled variables was used in the context of disturbance scenarios to assess their deviation from their desired, nominal setpoints. The employed approach could simultaneously evaluate the effects of multiple and simultaneous disturbances on the controlled and manipulated variables for a wide set of solvents and solvent mixtures. It also identified the disturbances that have major, detrimental effects on all the controlled variables for each solvent and the controlled variables that have the highest sensitivity. This is a very useful feature because fewer solvents and disturbances, as well as controlled and manipulated variables exhibiting high sensitivity, can be selected and subsequently examined in a more rigorous solvent, process, and control design problem, hence reducing the computational effort associated with it.

The proposed developments were implemented for 10 cases of solvents and solvent mixtures, including two phase-change solvents. Several indicators were used as the performance criteria of the process, such as the reboiler duty, net efficiency energy penalty points, cyclic capacity, solvent mass flow rate, solvent purchase cost, and lost revenue from parasitic electricity upon integration with power plants. Unlike previous works, for the first time, several conventional and phase-change CO_2 capture solvents were evaluated in terms of their performance under off-design conditions. The nonlinear sensitivity assessment highlights that certain economically desirable solvents may not be as attractive under off-design conditions.

2. Materials and Methods

2.1. Overview of the Controllability Assessment Framework

Papadopoulos et al. [38] proposed a systematic and generic approach for identifying materials within a framework that links optimization-based molecular design and selection (CAMD) with economic process design and controllability assessment criteria. The approach is based on a nonlinear sensitivity analysis of the solvent–process system to assess its steady-state controllability independently of the selected control algorithm [39]. The sensitivity-based investigation generates useful insights regarding the imposed control framework and the range of parameter variations, within which, the solvents–process system demonstrates optimal performance. Based on Papadopoulos et al. [38], the rationale behind this approach was that:

- A set of controlled variables associated with process performance indicators may be maintained within pre-defined levels for a set of disturbance scenarios, by calculating the necessary steady-state effort from a set of manipulated variables.
- Large variations in the steady-state position of the manipulated variables required for the compensation of relatively small disturbances indicate a limited ability by the solvent–process design configuration to address the disturbances and imply a compromise in the achieved dynamic performance by the control system.

The approach of Papadopoulos et al. [38] brings the assessment of process controllability into the material selection decision-making process early on but it is still based on the use of a rigorous, equilibrium-based model. The underlying rationale was adopted here and tailored using an enhanced short-cut process model that captures the non-ideal solvent-process behavior, as it incorporates non-linear thermodynamics for the prediction of phase compositions and stream enthalpies.

The evaluation of the off-design performance of CO_2 capture solvents and mixtures is approached through a systematic non-linear sensitivity analysis method that investigates the static operability performance of each solvent in the process under disturbance variations along a direction in which the process exhibits the greatest sensitivity. The most sensitive directions in the disturbance space are derived using the decomposition of the sensitivity matrix, which incorporates the derivatives of multiple process performance measures (e.g., reboiler duty, net energy penalty, cyclic capacity, and so forth) with respect to the operating parameters for each solvent system. The sensitivity matrix constitutes a measure of the process's operating variation under the influence of infinitesimal changes imposed on the selected parameters. The sensitivity matrix is decomposed into major directions of variability represented by the eigenvectors associated with the respective eigenvalues of the sensitivity matrix. The eigenvector direction that is associated with the largest-magnitude eigenvalue represents the dominant direction of variability for the system that causes the largest change in the performance measures. The entries in the dominant eigenvector determine the major direction of variability in the multi-parametric space and indicate the impact of each parameter in this direction. Having identified this direction, it is not necessary to explore all directions of variability (i.e., combinations of parameters) arbitrarily, hence reducing the dimensionality of the sensitivity analysis problem. The dominant eigenvector direction is then applied to the capture system for the evaluation of an aggregate performance indicator that encompasses all individual criteria for a wide variation range.

2.2. Detailed Description

The proposed approach follows the steps as presented below:

1. Define a vector E^{nom} that represents the nominal process design and operating conditions of an absorption–desorption process system that will be subjected to variability, a set D that includes the candidate solvents and mixtures, a vector X of state variables of the CO_2 capture process represented through vectors $h(X, d, E)$, $g(X, d, E)$ $\forall d \in D$ that describe the process equality and inequality constraints, and a vector Y representing the controlled variables associated to the process performance criteria.

2. Variability can be represented by considering disturbance scenarios through a vector of infinitesimal deviations dE such that each element of vector E^{nom} is represented as $\varepsilon_i^{nom} = \varepsilon_i + d\varepsilon_i$ $\forall i \in \{1, \ldots, N_\varepsilon\}$, with vector Y calculated as follows:

$$
\begin{aligned}
\underset{\forall d \in D}{\text{Calculate}} \quad & Y_1(X, d, E), \ldots, Y_N(X, d, E), \\
\text{s.t.} \quad & h(X, d, E) = 0, \\
& g(X, d, E) \leq 0, \\
& E = E^{nom} + dE, \\
& X^L \leq X \leq X^U.
\end{aligned}
\tag{1}
$$

3. The most sensitive process performance indicators in Y can then be identified by generating a local scaled sensitivity matrix P around E^{nom} as described below:

$$
P = \begin{bmatrix}
\frac{d \ln F_1}{d \ln \varepsilon_1} & \cdots & \frac{d \ln F_N}{d \ln \varepsilon_1} \\
\vdots & \ddots & \vdots \\
\frac{d \ln F_1}{d \ln \varepsilon_{N_\varepsilon}} & \cdots & \frac{d \ln F_N}{d \ln \varepsilon_{N_\varepsilon}}
\end{bmatrix}_d,
\tag{2}
$$

where $d \ln Y_i = dY_i / Y_i$ $\forall i \in \{i = 1, \ldots, N\}$ is the scaled transformation of the process performance indices, whereas $d \ln \varepsilon_i = d\varepsilon_i / \varepsilon_i$ $\forall i \in \{i = 1, \ldots, N_\varepsilon\}$ is the scaled transformation of ε_i.

4. The main directions of variability are obtained by calculating the eigenvalues of the matrix $P^T P$ and then used to rank-order the resulting eigenvectors $\{\theta_i\}$ $\forall i \in \{1, \ldots, N_\varepsilon\}$ based on the magnitude of the corresponding eigenvalues. The eigenvector corresponding to the largest-magnitude eigenvalue indicates the main direction of process variability under the influence of disturbances in the multi-parametric space defined by E^{nom}.

5. Using the dominant eigenvector, the sensitivity index $\Omega(\zeta, d)$ is calculated with the use of a relevant appropriate parameter ζ, which represents the magnitude of the disturbance magnitude along the eigenvector direction θ_1 with respect to E^{nom}, as described below:

$$
\begin{aligned}
\underset{\forall d \in D, \zeta}{\text{Calculate}} \; \Omega(\zeta, d) \; = \; & w^\Omega(\zeta) \sum_{i=1}^{N} \left| \frac{F_i(X, d, E(\zeta)) - F_i(X, d, E^{nom})}{F_i(X, d, E^{nom})} \right|, \\
\text{s.t.} \quad & h(X, d, E(\zeta)) = 0, \\
& g(X, d, E(\zeta)) \leq 0, \\
\varepsilon_i(\zeta) \; = \; & \theta_{1,i} \cdot \zeta \cdot \varepsilon_i^{nom} + \varepsilon_i^{nom} \forall i \in \{1, \cdots, N_\varepsilon\}, \\
& X^L \leq X \leq X^U.
\end{aligned}
\tag{3}
$$

where $\Omega(\zeta, d)$ is the sensitivity index under steady-state conditions, which measures the deviation of the controlled variables from their desired settings. $w^\Omega(\zeta)$ is a weight vector used to address the case of having prior knowledge regarding the importance of different controlled variables. ζ is the parameter that reflects the magnitude of change in the direction of $\theta_{1,i}$. The maximum variation along the direction is determined by the final value of ζ_{max}, which is varied within the range $[-\zeta_{max}, \zeta_{max}]$. The limits in the ζ coordinate are selected based on the variability of the process's system. It is worth noting that the nominal value of the parameter ε_i^{nom} used in sensitivity analysis is considered when ζ equals zero, $\varepsilon_i^{nom} \equiv \varepsilon_i(\zeta = 0)$. This procedure is repeated for every selected solvent or mixture of solvents. The algorithmic steps are illustrated in Figure 1.

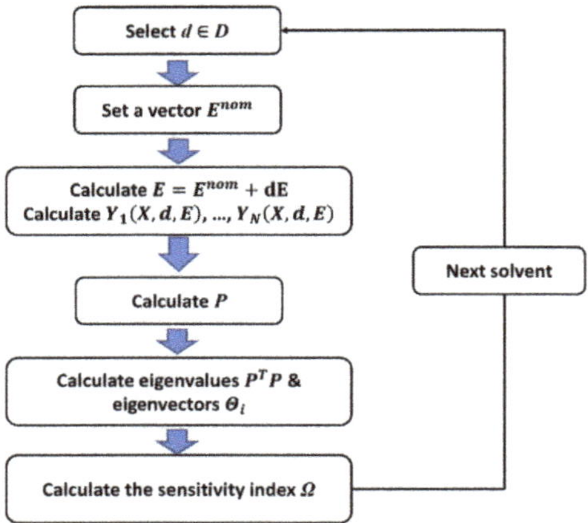

Figure 1. Logical diagram of the sensitivity analysis approach used for the process operability assessment.

After the implementation of the proposed sensitivity analysis for the assessment of the process operability, an appropriate $\zeta\prime$ is selected for all the solvents. This is used to calculate the sensitivity index $\Omega(\zeta\prime, d)$. A multi-criteria selection problem is formulated using the indices employed for the calculation of $\Omega(\zeta\prime, d)$, as described in step 6.

6. For every solvent $d \in D$, select $\Omega(\zeta\prime, d)$ at the desired point $\zeta\prime$ and develop an augmented vector such that $Y^a = [Y(d, \varepsilon(0)), Y(\zeta\prime, d)]$. Y^a is considered a combination of the optimal objective function values obtained during nominal operation and of the controllability index for each solvent. Use the elements of Y^a in a multi-criteria problem formulation that considers the solvents in D as the decision parameters to select the ones that simultaneously minimize all performance indices in Y and the sensitivity index (or indices) in Ω by generating a Pareto front as follows:

$$\begin{aligned} \text{Minimize} \quad & Y_1^a(d, \varepsilon(0)), Y_2^a(d, \zeta'), \\ \forall d \in D \\ \text{s.t.} \quad & Y_j^a(d^*) \leq Y_j^a(d) \ \forall \ j \in \{1,2\} \wedge \exists l \in \{1,2\} : Y_l^a(d^*) \leq Y_l^a(d). \end{aligned} \tag{4}$$

In step 6, the constraint of Equation (4) implies in a formal mathematical way that a solvent or mixture d in D is called a Pareto optimum or non-dominated solution if there exists no other solvent or mixture d^* in D satisfying this constraint. The constraint is illustrated for two objective functions, i.e., $Y_1^a(d, \varepsilon(0)) \equiv Y(d, \varepsilon(0))$ and $Y_2^a(d, \zeta\prime) \equiv Y(\zeta\prime, d)$. The former represents one performance index under nominal operation and the latter represents the sensitivity index for all performance indices during the variability. Multiple performance indices under nominal operation can also be considered as part of Equation (4). The Pareto optimality condition represents a minimization problem in Equation (4); however, a maximization or combinations may be similarly defined and solved by changing the direction of the inequality signs as appropriate. Note that step 6 is implemented after all calculations of Figure 1 are completed.

3. Implementation

3.1. Overview of the Process and the Amine Solvents

For completeness, a brief overview of the process and solvents is provided in this work, with all details reported in Zarogiannis et al. [13]. The flowsheets for the phase-change solvents are shown in Figure 2. In the current work, we investigated two types of phase-change solvents: a single solvent where phase separation occurs at 90 °C, hence the decanter (liquid–liquid phase separator) is placed after the intermediate heat exchanger (Figure 2a), and a phase-change solvent mixture that exhibits phase separation at 40 °C, hence the decanter is placed directly after the absorber (Figure 2b). For the conventional solvents and mixtures, the decanters are redundant.

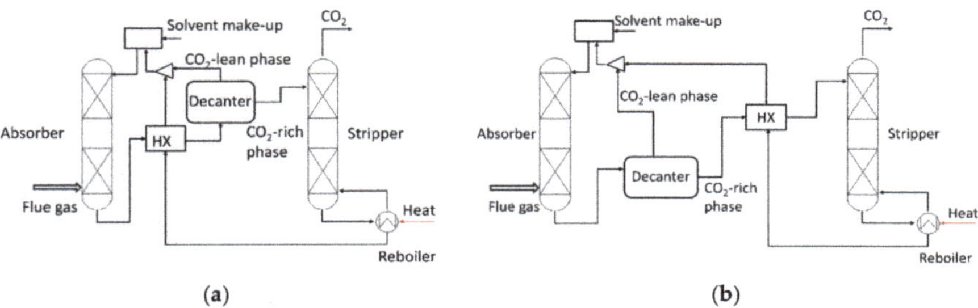

Figure 2. CO_2 capture flowsheet for phase-change solvents with phase separation (**a**) after the intermediate heat exchanger and (**b**) before the intermediate heat exchanger. The figures have been adapted from Zarogiannis et al. [13] with permission from Elsevier. HX: Heat exchanger.

The main process parameters that need to be calculated include the reboiler duty in the stripper, the mass flow rate that enters the stripper, and the stripper temperatures. For the phase-change solvents, the mass flow rate of the stream entering the stripper is calculated through mass balances that consider the phase compositions at conditions around the decanter, hence fully accounting for the liquid–liquid phase separation. For the energetic calculations, the employed shortcut model [13] focuses on two areas: the heat exchanger before the stripper and the reboiler after the stripper. The calculation of the reboiler duty takes place through the determination of the temperature of the rich stream that enters the desorber and of the lean stream that exits the stripper by implementing energy balances around the heat exchanger and the reboiler. All the details are reported in Zarogiannis et al. [13].

To illustrate the approach described in Section 2, 10 solvents and mixtures (Tables 2 and 3), including two phase-change solvents, were investigated. The single solvents included MEA, AMP, DEA, 3-(methylamino)propylamine (MAPA), and methyl-cyclohexylamine (MCA). The mixtures included MEA with methyldiethanolamine (MDEA), MPA with MDEA, DEA with MDEA, AMP with piperazine (PZ), and 2-(diethylamino)ethanol (DEEA) with MAPA. The non-phase-change amines were previously investigated as CO_2 capture solvents and were retrieved from the published literature. Phase-change solvents are characterized by the presence of a vapor–liquid–liquid equilibrium, where after the phase split, one liquid phase is rich in CO_2, while the other phase is lean in CO_2 and rich in amine. MCA is selected as a representative of phase-change solvents requiring a liquid–liquid phase separator after the intermediate heat exchanger (HX) in the absorption–desorption flowsheet (Figure 2a), whereas DEEA-MAPA requires the liquid–liquid separator before the intermediate heat exchanger (Figure 2b).

Table 2. Single solvents employed in this work. MCA is a phase-change solvent.

ID	Single Amine
MEA	 H_2N–CH$_2$CH$_2$–OH
AMP	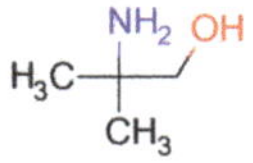 H_3C–C(CH$_3$)(NH$_2$)–CH$_2$OH
DEA	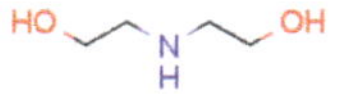 HO–CH$_2$CH$_2$–NH–CH$_2$CH$_2$–OH
MAPA	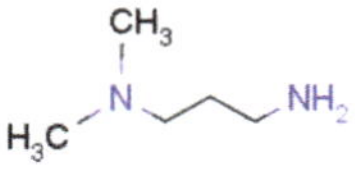 H_3C–N(CH$_3$)–CH$_2$CH$_2$CH$_2$–NH$_2$
MCA	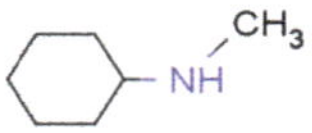 cyclohexyl–NH–CH$_3$

MAPA: 3-(Methylamino)propylamine.

Table 3. Mixtures employed in this work. 2-(Diethylamino)ethanol (DEEA)/MAPA is a phase-change mixture.

Component 1		Component 2	
MEA	 H_2N–CH$_2$CH$_2$–OH	MDEA	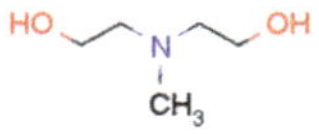 HO–CH$_2$CH$_2$–N(CH$_3$)–CH$_2$CH$_2$–OH
MPA	 HO–CH$_2$CH$_2$CH$_2$–NH$_2$	MDEA	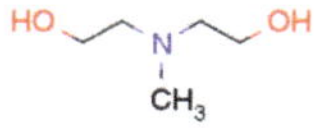 HO–CH$_2$CH$_2$–N(CH$_3$)–CH$_2$CH$_2$–OH
DEA	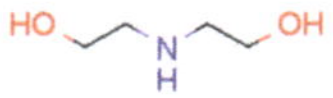 HO–CH$_2$CH$_2$–NH–CH$_2$CH$_2$–OH	MDEA	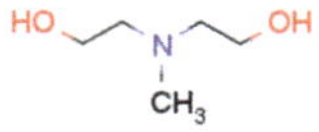 HO–CH$_2$CH$_2$–N(CH$_3$)–CH$_2$CH$_2$–OH
AMP	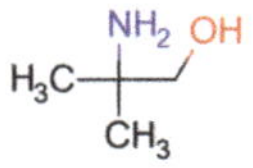 H_3C–C(CH$_3$)(NH$_2$)–CH$_2$OH	PZ	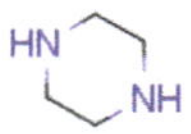 HN(CH$_2$CH$_2$)$_2$NH (piperazine)
DEEA	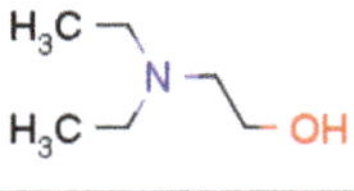 H_3C–CH$_2$–N(CH$_2$CH$_3$)–CH$_2$CH$_2$–OH	MAPA	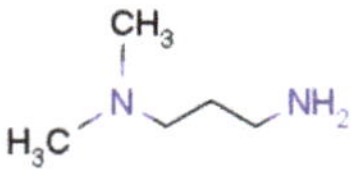 H_3C–N(CH$_3$)–CH$_2$CH$_2$CH$_2$–NH$_2$

The employed thermodynamic models include the relations of the equilibrium pressure of CO_2 and enthalpy as functions of the loading and temperature. For MEA, the models were from Oyenekan [40], whereas for AMP and AMP/PZ, they were from Oexman [41]. For DEA, MEA/MDEA, DEA/MDEA, and MPA/MDEA, the models were derived from equilibrium data obtained from gSAFT software [42]. For MPA/MDEA, the software uses an equation of state (EoS) called SAFT-γ Mie [43,44]. For the other three solvents, the model uses the SAFT-VR EoS [45]. For MAPA and DEEA/MAPA, models were derived from equilibrium data obtained from Arshad [46]. For MCA, the models were derived from equilibrium data obtained from Tzirakis et al. [47] and Jeon et al. [48].

3.2. Controlled Variables and Disturbance Scenarios

The selected controlled variables in Y included all the parameters that were used as process performance indices. These were the reboiler duty Q_{regen}, the net efficiency energy penalty NEP, the cyclic capacity $\Delta\alpha$, the solvent mass flow rate m_{am}, the solvent purchase cost C_{sol}, and the lost revenue from parasitic electricity R_{lost}. Q_{regen} indicates the energy consumed in the stripper to regenerate the solvent. NEP is also an indicator of regeneration energy but takes into account the reboiler temperature and the stripper pressure. A higher pressure facilitates solvent regeneration, hence NEP is a more inclusive indicator than Q_{regen}. $\Delta\alpha$ indicates the difference between the rich and the lean stream that exits and enters the absorber, hence it is associated with the absorber size. A higher $\Delta\alpha$ facilitates the reaction and the necessary absorber size may be smaller. m_{am} is an indicator of the solvent amount required to achieve the desired absorption. It is used as an indicator of solvent consumption, i.e., solvents of lower flow rates are desirable. On the other hand, its use assumes that all solvents would be of the same cost, hence C_{sol} is a more inclusive indicator as it accounts for the different solvent prices. Finally, R_{lost} indicates the revenue that is lost in a power plant due to the need to consume energy to capture the emitted CO_2, which would otherwise be sold to generate revenues. Parameter NEP was calculated through a correlation proposed in Zarogiannis et al. [13] for stripper pressures of 1 and 1.5 bar. The former pressure was considered for MCA due to the availability of thermodynamic data, whereas the latter pressure was used in all other solvents. Parameter R_{lost} was calculated for a 620 MW coal-fired power plant. All the other data are available in Zarogiannis et al. [13].

The parameters in E that are subjected to variations are the following:

- $\varepsilon_1 = y_{in}^{CO_2}$ is the content of CO_2 in the flue gas that enters the process. The nominal value is $\varepsilon_1(0) = 15$ vol%.
- $\varepsilon_2 = \alpha_{lean}$ is the CO_2 loading of the absorption–desorption process after the stripper. Its value depends on the selected solvents and their physico-chemical characteristics. The nominal values for each solvent were reported by Zarogiannis et al. [13]. α_{lean} can be adjusted by the reboiler duty in the stripper, which subsequently has an impact on the performance criteria.
- $\varepsilon_3 = T_1$ is the temperature of the inlet stream that enters the heat exchanger after the absorber. The nominal temperature is set to $\varepsilon_3(0) = 40\ ^\circ C$. This can be adjusted by the possible cooling effort in the absorber and the amine flow rate in the system.

The aforementioned parameters are selected because their change may affect the performance of the process. From a practical perspective, change in the CO_2 content of the flue gas may be attributed to a random disturbance or variability in the quality of the fuel or the operating regime of the power plant. Disturbance scenarios also involve a decrease or increase in α_{lean} and increase or decrease in the temperature of the inlet stream. These variations create inclinations in the operating process and help to understand potential disturbance rejection compensations that are necessary for each solvent. The starting point for every considered scenario constitutes the nominal, "desired" point, as calculated by Zarogiannis et al. [13]. The latter corresponds to operation in which there is a desired constant CO_2 capture rate at 90% with the Q_{regen} at its lowest value. These nominal points are reported in Table 4.

Table 4. Representative results of nominal operation for the investigated solvents in the performance indicators.

Solvent	Q_{regen} (GJ/ton CO_2)	NEP (%-pts.)	$\Delta\alpha$	m_{am} (ton/ton CO_2)	C_{sol} (k€/ton CO_2	R_{lost} (M€/yr)
MEA	3.72	11.2	0.3	4.7	76.3	290.1
MCA	2.12	8.8	0.22	13.83	3929.7	150.1
MAPA/DEEA	2.10	8.1	0.58	13.03	656.0	140.8
AMP	3.13	9.4	0.28	7.35	238.7	206.5
AMP/PZ	3.10	9.7	0.17	9.12	362.8	216.0
DEA	3.47	10.2	0.18	13.48	328.5	240.7
MAPA	4.56	12.6	0.49	4.15	573.3	361.3
MDEA/DEA	3.74	11.1	0.28	9.52	401.9	293.8
MEA/MDEA	4.28	11.5	0.13	17.67	746.0	302.0
MPA/MDEA	3.32	9.8	0.18	13.02	687.0	221.2

The steps and the overall range of the above parameters are represented through ζ, which is varied within the range $[-\zeta_{max}, \zeta_{max}]$. The limit in the ζ coordinate depends on the solvent or mixture of solvents; variations may result from variability in the steam availability in the reboiler, problems in the rich–lean heat exchanger, and so forth. Plotting $\Omega(\zeta, d)$ vs. ζ for different ranges can help determine the ζ_{max} value. This plot indicates that solvents with the steepest slope are the most sensitive to variability. The underlying process and thermodynamic model are non-linear, hence the $\Omega(\zeta, d)$ vs. ζ profile for every solvent may change non-linearly. The selected range for ζ values should therefore capture the plant operating patterns. With respect to the selection of the $\zeta\prime$ value, to implement step 6, a value close to $\zeta = 0$ usually provides a good indication about the solvents that are sensitive even under small variations. Inspection of the $\Omega(\zeta, d)$ vs. ζ profile for all solvents guides the selection of this value.

This work considered three different sets of performance indices:

- Set 1 included $Y_1 = Q_{regen}$, $Y_2 = m_{am}$, and $Y_3 = \Delta\alpha$.
- Set 2 included $Y_1 = NEP$, $Y_2 = m_{am}$, and $Y_3 = \Delta\alpha$.
- Set 3 included $Y_1 = R_{lost}$, $Y_2 = C_{sol}$, and $Y_3 = \Delta\alpha$.

In Set 2, Q_{regen} was replaced by NEP, which incorporates Q_{regen} together with other process parameters, such as the reboiler temperature and pressure. In Set 3, NEP was replaced by R_{lost} and m_{am} by C_{sol}, hence transforming the operating indices to economic ones. This investigation had two goals: first, to assess the sensitivity of the different solvents to disturbances, and second, to assess how different performance indices may affect the assessment of the solvents based on their sensitivity to disturbances. The latter is important because performance indices, such as the reboiler duty and mass flow rate, are often used as criteria for solvent selection. The choice of reboiler duty overlooks other important parameters, such as the reboiler temperature and pressure, which directly impact the process's economics. The use of solvent mass flow rate as a selection criterion assumes that all solvents have the same purchase value, whereas in practice, the prices of amines vary quite significantly. Each set was used for the calculation of the corresponding sensitivity indices Ω_1, Ω_2, and Ω_3, per step 5. Results are reported for all indices in Section 4.2, whereas step 6 was implemented only for Ω_3 (i.e., set 3). Pareto fronts based on two criteria were developed by considering Ω_3 and R_{lost}, C_{sol}, $\Delta\alpha$, Q_{regen}, and NEP for both a selected positive and negative ζ. This served to finally select a set of very few candidates that exhibited good trade-offs between nominal and off-design performance.

4. Results and Discussion

4.1. Influence of Parameters on the Process Operability

Tables 5–7 present the major directions of variability for each solvent capture system that corresponded to the largest in magnitude eigenvalues, as indicated by the eigenvectors of the sensitivity matrix $P^T P$. The tables show that for each solvent or solvent mixture system, a different major variability direction was derived. Table 5 illustrates the reboiler duty Q_{regen}, mass flow rate of the amine m_{am}, and cyclic capacity $\Delta \alpha$. The highest absolute values are in bold, indicating the parameter with the greatest impact on the process performance. For MEA, $y_{in}^{CO_2}$ seemed to be the most influential parameter that affected the selected performance indices. T_1, which had a negative sign in parenthesis, exhibited an opposite direction of change in comparison to $y_{in}^{CO_2}$ and a_{lean}. In the case of MEA, the eigenvector entries for a_{lean} and T_1 were very low, hence the variability in these parameters did not significantly affect the performance indices. AMP, DEA, and MAPA seemed to have the same behavior since a_{lean} was the most dominant parameter, whereas $y_{in}^{CO_2}$ and T_1 seemed to affect the process performance less. The mixture of AMP with PZ seemed to behave the same as the aforementioned solvents, apart from the inlet temperature T_1, which did not show up in the θ_1 direction. The mixtures MEA with MDEA and MPA with MDEA exhibited the same behavior, with a_{lean} being the most influential parameter in the opposite direction, as well as in the mixture DEA with MDEA. However, T_1 exhibited an opposite tendency in this case. Regarding the phase-change solvents MCA and DEEA/MAPA, a_{lean} appeared to be the most influential parameter in the same direction with the inlet temperature. In the case of DEEA/MAPA, $y_{in}^{CO_2}$ also had an impact on the performance indices. The impact of the changes in the parameter space along the eigenvector direction that corresponded to the second largest eigenvalue was not significant due to the low contribution it had on the process variability.

Table 5. Ranking of the parameters that affected Q_{regen}, m_{am}, and $\Delta \alpha$, with descending eigenvector entries from left to right. The highest absolute values are shown in bold. Underlined symbols indicate the eigenvector direction that corresponded to the second-largest eigenvalue, which is of a similar order of magnitude as the highest eigenvalues. The signs in brackets indicate the directions of change for the parameters (opposite signs indicate a change in the opposite direction).

Solvent	Order of Parameters	Solvent	Order of Parameters
MEA	$y_{in}^{CO_2}$ (+), a_{lean} (+), T_1 (−)	DEEA/MAPA	a_{lean} (−), $y_{in}^{CO_2}$ (−), T_1 (−)
AMP	a_{lean} (−), $y_{in}^{CO_2}$ (−), T_1 (0)	MEA/MDEA	a_{lean} (−), T_1 (−)
DEA	a_{lean} (−), $y_{in}^{CO_2}$ (−), T_1 (0)	MPA/MDEA	a_{lean} (−), T_1 (−)
MAPA	a_{lean} (−), $y_{in}^{CO_2}$ (−), T_1 (0)	DEA/MDEA	a_{lean} (−), T_1 (0)
MCA	a_{lean} (−), T_1 (0)	AMP/PZ	a_{lean} (−), $y_{in}^{CO_2}$ (−)

Table 6. Ranking of the parameters that affected the performance indices of NEP, m_{am}, and $\Delta \alpha$.

Solvent	Order of Parameters	Solvent	Order of Parameters
MEA	$y_{in}^{CO_2}$ (+), a_{lean} (+), T_1 (0)	DEEA/MAPA	a_{lean} (−), $y_{in}^{CO_2}$ (−), T_1 (−)
AMP	a_{lean} (−), $y_{in}^{CO_2}$ (−)	MEA/MDEA	a_{lean} (−), T_1 (−)
DEA	a_{lean} (−), $y_{in}^{CO_2}$ (−)	MPA/MDEA	a_{lean} (−), T_1 (−)
MAPA	a_{lean} (−), $y_{in}^{CO_2}$ (−), T_1 (0)	DEA/MDEA	a_{lean} (−), T_1 (0)
MCA	a_{lean} (−)	AMP/PZ	a_{lean} (−), $y_{in}^{CO_2}$ (−)

Table 7. Ranking of the parameters that affected the performance indices of R_{lost}, C_{sol}, and $\Delta\alpha$.

Solvent	Order of Parameters	Solvent	Order of Parameters
MEA	$y_{in}^{CO_2}$ (+), a_{lean} (+), T_1 (0)	DEEA/MAPA	a_{lean} (−), $y_{in}^{CO_2}$ (−), T_1 (−)
AMP	a_{lean} (−), $y_{in}^{CO_2}$ (−), T_1 (0)	MEA/MDEA	a_{lean} (−), T_1 (−)
DEA	a_{lean} (−), $y_{in}^{CO_2}$ (−)	MPA/MDEA	a_{lean} (−), T_1 (−)
MAPA	a_{lean} (−), $y_{in}^{CO_2}$ (−), T_1 (0)	DEA/MDEA	a_{lean} (−), T_1 (0)
MCA	a_{lean} (−)	AMP/PZ	a_{lean} (−), $y_{in}^{CO_2}$ (−), T_1 (−)

Table 6 illustrates the eigenvector θ_1 direction that caused the maximum variation in the considered process performance indices of NEP, m_{am}, and $\Delta\alpha$. Table 6 exhibits the same results as Table 5, except for MEA, MCA, AMP, and DEA. The most dominant parameters for this set of indices appeared to be the same as in the previous case. MCA affected the performance indices only through a_{lean}. Additionally, T_1 did not exhibit any influence on AMP and DEA.

Table 7 depicts the parameters in which the system was more sensitive to variability using the lost revenue from parasitic electricity R_{lost}, the solvent cost C_{sol}, and the cyclic capacity $\Delta\alpha$ as performance indices. Table 7 exhibits the same results as Table 6, apart from AMP and AMP/PZ.

4.2. Sensitivity Index

For each set of performance indices reported in Tables 5–7, the corresponding sensitivity indices were calculated, namely $\Omega_1(\zeta)$, $\Omega_2(\zeta)$, and $\Omega_3(\zeta)$ versus ζ. The results are shown in Figures 3–5, respectively. Solvents that exhibited mild slopes in the contours were less sensitive in variations and scored low in the sensitivity index. Solvents were more resilient to variations by keeping the performance indices close to the values at the nominal operating point and could be good candidate solvents for CO_2 capture as they exhibited the ability to alleviate the effects of disturbances. In Figures 3–5, we designated the area of positive ζ as an unfavorable scenario because disturbances caused deterioration in the performance indicators. For example, Q_{regen}, NEP, etc. increase in that direction, which was undesired. On the other hand, when a performance indicator improved due to disturbances (e.g., a reduction of Q_{regen}), this direction of variability was designated as a favorable scenario. It could be argued that in this case, solvents with a high sensitivity to variability would be desirable as they appeared to exploit the effects of disturbances. However, such a conclusion would require a more detailed analysis of other issues, such as the duration of variability and the dynamic solvent behavior. It is generally desired to avoid solvents and operating conditions that deviate significantly from the intended system operation.

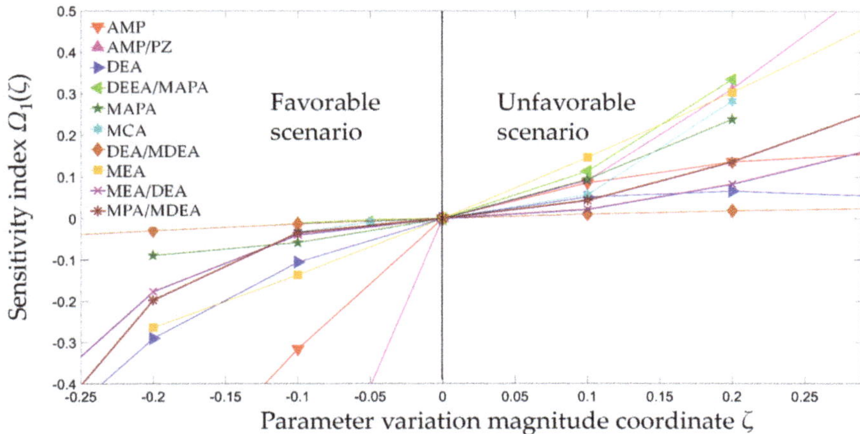

Figure 3. Sensitivity index $\Omega_1(\zeta)$ versus the parametric variation magnitude coordinate ζ.

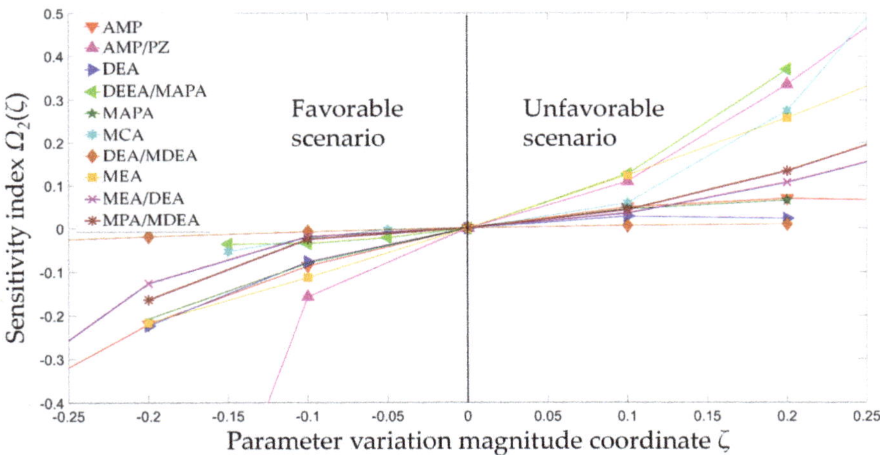

Figure 4. Sensitivity index $\Omega_2(\zeta)$ versus the parametric variation magnitude coordinate ζ.

In Figure 3, the solvent mixture DEA/MDEA appeared to be the most promising mixture through the sensitivity analysis. In the region close to $\zeta = [0.17, 0.23]$, DEA and MEA/MDEA exhibited low sensitivity, followed by MPA/MDEA and AMP. MCA, MAPA, DEEA/MAPA, and AMP/PZ seemed to be more sensitive to variability than other amines. In the region of $\zeta = [-0.05, -0.15]$ with the opposite direction, DEA/MDEA continued to be the least sensitive mixture. DEEA/MAPA also exhibited resilience against variations. MPA/MDEA, MEA/MDEA, and MCA exhibited the same behavior. DEA and MEA seemed to be more sensitive in this region. AMP and AMP/PZ seemed to have a sharp profile in this area.

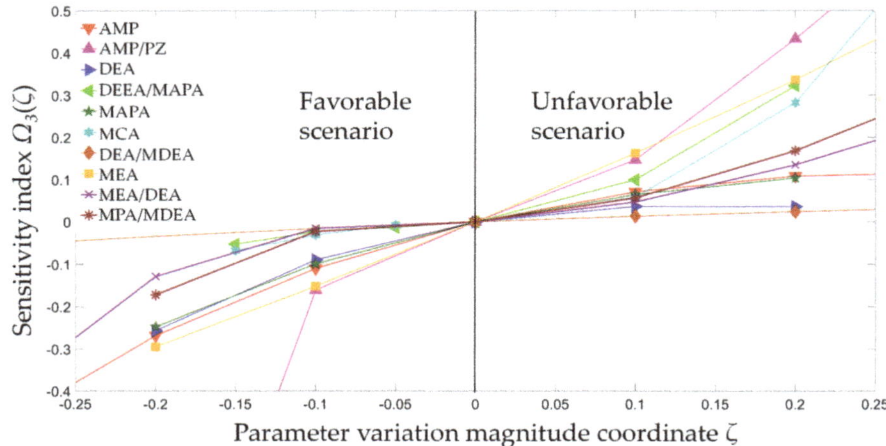

Figure 5. Sensitivity index $\Omega_3(\zeta)$ versus the parametric variation magnitude coordinate ζ.

Figure 4 shows the sensitivity index $\Omega_2(\zeta)$, which was calculated based on step 6 of the described framework. The performance indices in this case were *NEP*, m_{am}, and $\Delta\alpha$. The disturbances were the same as in the previous case. The solvent mixture DEA/MDEA continued to exhibit the least sensitivity to variability throughout the investigated range of disturbances. In the region of $\zeta = [0.15, 0.25]$, DEA seemed to be less sensitive to variations in the selected parameters. AMP, MAPA, MEA/MDEA, and MPA/MDEA seemed to be more susceptible than DEA to disturbances in this case. MEA, MCA, AMP/PZ, and DEEA/MAPA seemed to be very sensitive to variability, and such mixtures should be avoided when there are exogenous changes. In the region of $\zeta = [-0.1, -0.2]$ in the opposite direction, DEA/MDEA appeared to be the least sensitive mixture. DEEA/MAPA and MCA, followed by MPA/MDEA and MEA/MDEA, exhibited a tolerance against variations. DEA, AMP, and MAPA presented the same behavior. However, MEA and especially AMP/PZ seemed to be more sensitive in this region than the other selected solvents.

Figure 5 shows the sensitivity index $\Omega_3(\zeta)$ based on the performance indices of R_{lost}, C_{sol}, and $\Delta\alpha$. Solvents that exhibited sharp contours in the region $[0.15, 0.25]$ seemed to be sensitive in variations and exhibited high sensitivity to variability; such solvents were AMP/PZ, MEA, DEEA/MAPA, and MCA. The solvent mixture DEA/MDEA and DEA exhibited the least sensitivity to variability. MEA/MDEA, MPA/MDEA, MAPA, and AMP seemed to present the same behavior. In the region of $\zeta = [-0.1, -0.2]$, DEA/MDEA continued to be the most promising mixture. DEEA/MAPA and MCA presented the same behavior. MEA/MDEA and MPA/MDEA exhibited the same profile. DEA, AMP, and MAPA seemed to be sensitive to variations. Solvents that had a high sensitivity were MEA and AMP/PZ. It is worth noting that the slopes of the DEA/MDEA lines in either of the favorable or unfavorable scenarios were close to zero for the entire range of the considered variability in Figures 3–5. DEA was also always close to DEA/MDEA but its controllability performance was slightly worse. On the other hand, MEA had a very steep slope and a much worse controllability performance. However, its slope became less steep when combined with MDEA. The concentration of MDEA was twice the concentration of MEA, MPA, or DEA; hence, this is an indication that MDEA could exhibit desirable controllability behavior.

4.3. Selection of Solvents

To select the appropriate solvent or mixture for the post-combustion CO_2 capture process, it was beneficial to combine the sensitivity analysis with the performance indices of R_{lost}, C_{sol}, $\Delta\alpha$, Q_{regen}, and *NEP* at nominal conditions. In this context, a Pareto front of the sensitivity index $\Omega_3(\zeta\prime)$ at an appropriate $\zeta\prime$ for all the solvents with respect to the aforementioned performance criteria could be implemented (Figures 6 and 7). The selected $\zeta\prime$s were 0.2 and -0.15 to include all the proposed mixtures. The minimization of the sensitivity index with respect to the minimization of R_{lost}, C_{sol}, $\Delta\alpha$, Q_{regen}, and NEP, along with the maximization of $\Delta\alpha$, formed the respective Pareto fronts. Pareto optimal mixtures presented remarkably low sensitivity and good process performance from the nominal economic perspective.

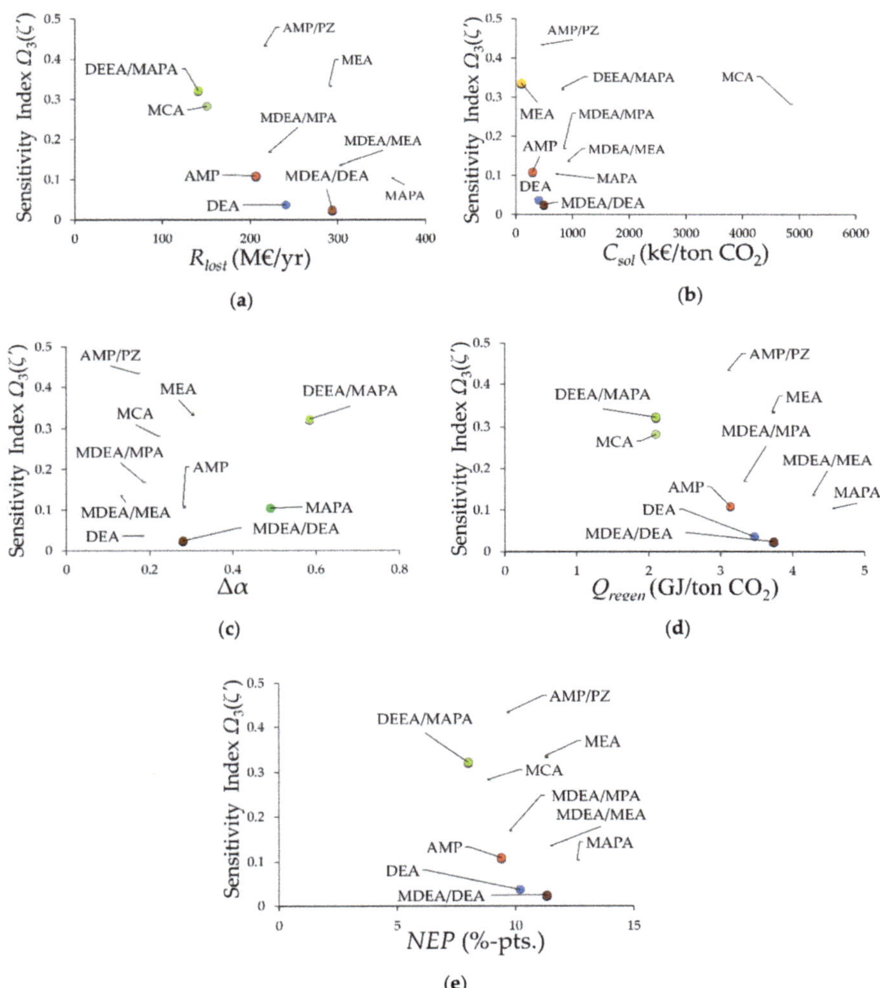

Figure 6. Pareto front between the sensitivity index $\Omega_3(\zeta' = 0.2)$ and (**a**) R_{lost}, (**b**) C_{sol}, (**c**) $\Delta\alpha$, (**d**) Q_{regen}, and (**e**) NEP. Marks other than colored circles indicate undesirable solvents.

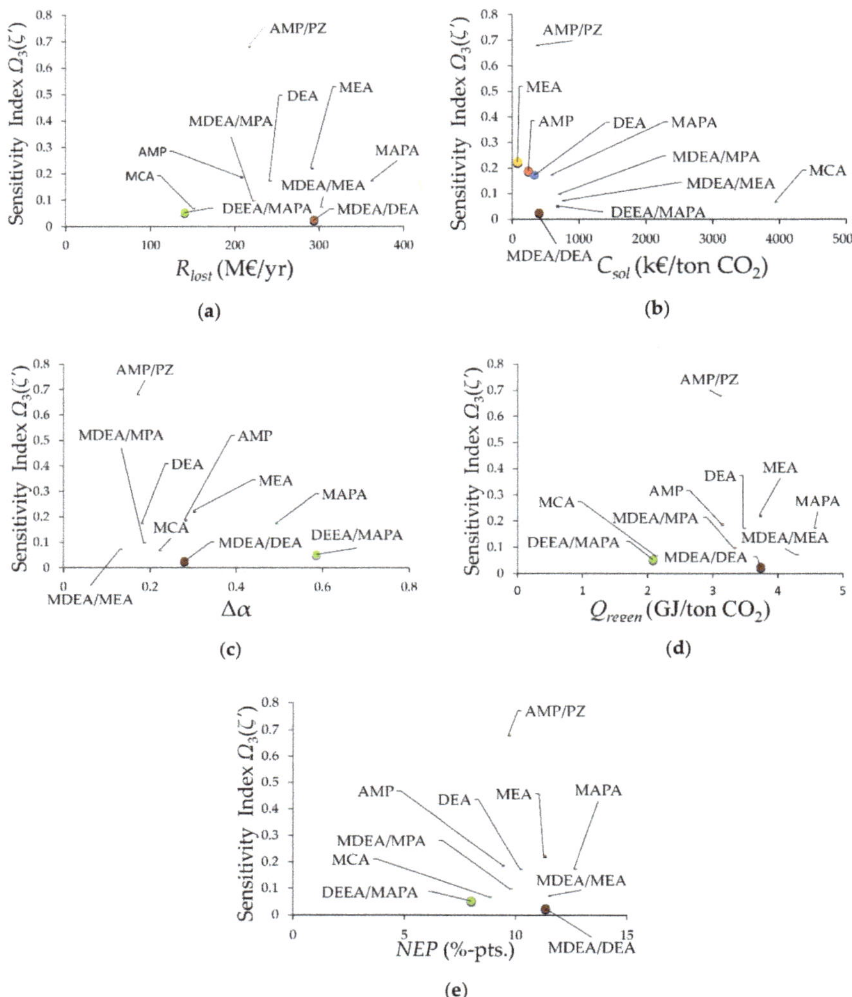

Figure 7. Pareto front between the sensitivity index $\Omega_3(\zeta' = -0.15)$ and (a) R_{lost}, (b) C_{sol}, (c) $\Delta\alpha$, (d) Q_{regen}, and (e) NEP. Marks other than colored circles indicate undesirable solvents.

Although some solvents exhibited a low R_{lost}, such as AMP/PZ and MPA/MDEA, they exhibited high sensitivity under variable conditions. These mixtures were susceptible to exogenous changes, resulting in a higher overall cost of the process. Their use may demand higher expenses, either in capital costs or investments in advanced control systems to maintain the performance of the process at a high level. The developed Pareto fronts recommend solvents that had simultaneously low R_{lost} and $\Omega_3(\zeta')$ in Figure 6a. Solvents that belonged to the Pareto front are represented with a cycle marker in a different color toward the left corner of Figure 6a. Solvents that are not part of Pareto front are represented with half-dashes of the same color since they belonged to the dominated set of solvents and should not be selected. AMP, DEA, and MDEA/DEA and both the phase-change solvents DEEA/MAPA and MCA formed the Pareto front combining both the minimization of two criteria and maintaining good nominal process performance in the corresponding index. Figure 6b illustrates the

Pareto front between the sensitivity index and the solvent cost. The minimization of both indicators is desired. The optimal solvents were MEA, AMP, DEA, and MDEA/DEA. It is worth noting that three out of the four solvents were part of the previous Pareto front. This means that they maintained the process performance since they exhibited a tolerance against variations and they were simultaneously affordable with a low lost revenue from parasitic electricity.

Figure 6c illustrates that DEEA/MAPA, MAPA, and MDEA/DEA combined flexibility in exogenous variations with high $\Delta\alpha$. Note that MDEA/DEA continued to be part of the Pareto front. Higher regeneration energy means a higher cost and thus unprofitable plant units. AMP, DEA, MDEA/DEA, and both phase-change solvents DEEA/MAPA and MCA formed the Pareto front in Figure 6d. AMP, DEA, MDEA/DEA, and DEEA/MAPA constituted the Pareto front in Figure 6e. DEA and MDEA/DEA combined flexibility in exogenous variations with high energy efficiency performance. DEEA/MAPA was the phase-change solvent that exhibited the lowest net efficiency penalty among all solvents.

In Figure 7a, MDEA/DEA and the phase-change solvent DEEA/MAPA formed the Pareto front in the case where $\zeta\prime$ was equal to -0.15, combining the minimization of both criteria. As in Figure 6b, the optimal solvents in Figure 7b were MEA, AMP, DEA, and MDEA/DEA. It is worth noting that MDEA/DEA was part of the previous Pareto front as well.

Figure 7c illustrates that DEEA/MAPA and MDEA/DEA combined flexibility in exogenous variability with a high $\Delta\alpha$. MDEA/DEA and phase-change solvent DEEA/MAPA formed the Pareto front in Figure 7d. In Figure 7e, DEEA/MAPA continued to be the phase-change solvent that exhibited the lowest net efficiency penalty among all solvents. It is clear now that although there were differences between the Pareto fronts, the mixture with the optimal trade-off between low sensitivity index and process performance was MDEA/DEA, which was part of all Pareto fronts. It is worth noting that DEEA/MAPA exhibited remarkable process performance, especially from the energy efficiency perspective and particularly in the case where $\zeta' = -0.15$.

The findings of all the above figures are summarized in Table 8. It is clear that despite some diversification between the Pareto fronts for positive and negative values of $\zeta\prime$, the solvents with the optimal trade-off between the sensitivity index Ω_3 and the nominal process performance were DEA/MDEA and DEEA/MAPA, as they appear with a very high frequency. It is also worth noting that DEA and AMP also appeared quite frequently in the Pareto fronts in the positive direction, i.e., they were quite resilient under unfavorable conditions.

Table 8. Mixtures presenting the optimal trade-off between performance and sensitivity at $\zeta\prime = 0.2$ and $\zeta\prime = -0.15$.

ζ'	R_{lost}	C_{sol}	$\Delta\alpha$	Q_{regen}	*NEP*
0.2	DEEA/MAPA DEA/MDEA MCA AMP DEA	DEA/MDEA MEA AMP DEA	DEEA/MAPA DEA/MDEA MAPA	DEEA/MAPA DEA/MDEA MCA AMP DEA	DEEA/MAPA DEA/MDEA AMP DEA
-0.15	DEEA/MAPA DEA/MDEA	DEA/MDEA MEA AMP DEA	DEEA/MAPA DEA/MDEA	DEEA/MAPA DEA/MDEA	DEEA/MAPA DEA/MDEA

Figure 8 shows the change of the performance indices compared to their value at the nominal point. It is worth noting that despite the small range of ζ, the respective performance indices changed significantly. In this respect, the performed investigation covered a very wide range of different scenarios and impacts on the performance indicators.

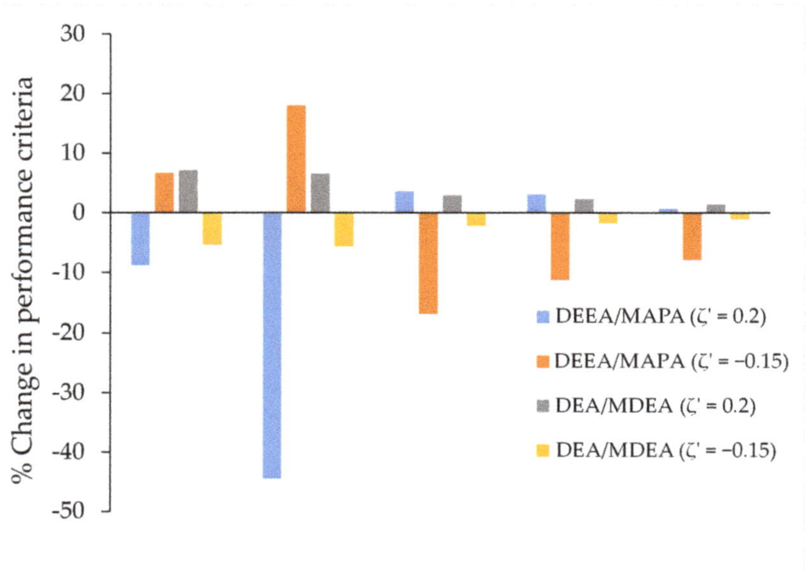

Figure 8. Change in the performance criteria for solvents DEEA/MAPA and DEA/MDEA for the investigated $\zeta\prime = 0.2$ and $\zeta\prime = -0.15$.

5. Conclusions

The present work employed a systematic framework that combined the nominal operation of a solvent-based absorption–desorption process with off-design operation as criteria for solvent selection. The framework was implemented in several analysis steps, which provided valuable insights. The proposed approach aimed to achieve a reduction in the dimensionality of the investigated parameter space and unveiled the principal directions of variability within the parameter space.

The system under study consisted of an enhanced shortcut model that was used to simulate the post-combustion CO_2 capture process and certain solvents that were subjected to variations of CO_2 content in the flue gas, lean loading, and the temperature of the inlet stream. The parameters adopted here as performance indices consisted of the reboiler duty, the net efficiency energy penalty, the cyclic capacity, the solvent mass flow rate, the solvent purchase cost, and the lost revenue from parasitic electricity. All these parameters had direct or indirect impacts on the process economics. Furthermore, the sensitivity index provided valid insights regarding the sensitivity of each solvent within a wide range. Sharp profile changes indicated that a particular solvent was unsuitable because the achieved performance would be diminished under variability. Finally, the proposed multi-criteria assessment approach identified trade-offs between the solvents, pointing to those with the highest overall performance and the minimum sensitivity to variability.

It appeared that the most important effect in the process operability was observed for $y_{in}^{CO_2}$ and a_{lean}. The former exhibited the highest eigenvalue in the case of MEA, whereas the latter exhibited the highest eigenvalue for all other solvents. This behavior was repeated regardless of the combination of the employed performance indicators. The steady-state controllability evaluation through the sensitivity analysis under variations unveiled a plethora of different responses of alternative proposed solvents. DEA/MDEA exhibited very robust behavior in either of the favorable and unfavorable scenarios investigated. MCA was also quite resilient in the case of unfavorable scenarios, while DEA was almost as resilient as DEA/MDEA under unfavorable scenarios. MPA/MDEA and MEA/MDEA appeared resilient for variations of small magnitude, whereas when variations diverted significantly from the nominal points, a process with either of these two mixtures would lose its resilience. Considering that the concentration of MDEA was twice the concentration of DEA, MPA, and MEA in the mixtures, it seemed that MDEA could be quite resilient on its own. This is because the originally steeper slopes of DEA and MEA were pulled closer to zero in the sensitivity axis when they were used with MDEA.

The Pareto optimal mixtures presented remarkably low sensitivity while maintaining operation close to the desired set-points. DEA and AMP in the positive direction, but mostly DEA/MDEA, which are part of the Pareto fronts, seemed to be both resilient to variability and underwent relatively desired operations under steady-state conditions. Although diverse trade-offs were observed among the Pareto fronts, the mixtures with the lowest sensitivity index and nominal process performance were DEA/MDEA and DEEA/MAPA. The results further indicated that, although inclusive performance indices should be used if available, simpler indices, such as those proposed in set 1, were also useful in terms of determining the performance of the solvents. The selection of solvents that exhibited low sensitivity to variability was associated with lower expenses when operating a subsequent control structure to address the reduced performance.

Even though the eigenvector directions were calculated based on local sensitivity analysis, the implemented variation along the selection eigenvector direction was quite large in magnitude (i.e., the value of coordinate ζ), which enabled the consideration of non-linear effects on the process performance criteria to be fully and explicitly explored. Obviously, the number of process parameters whose variation is worth investigating can be further expanded under the proposed framework.

Author Contributions: Software, investigation, visualization, writing—original draft preparation, T.Z.; investigation, visualization, writing—original draft preparation, writing—review and editing, supervision, project administration, funding acquisition, A.I.P.; investigation, writing—original draft preparation, writing—review and editing, supervision, project administration, funding acquisition, P.S. All authors have read and agreed to the published version of the manuscript.

Funding: This project received funding from the European Union's Horizon 2020 research and innovation program under the grant agreement 727503 – ROLINCAP – H2020-LCE-2016-2017/H2020-LCE-2016-RES-CCS-RIA.

Conflicts of Interest: The authors declare no conflict of interest.

References

1. Bui, M.; Adjiman, C.S.; Bardow, A.; Anthony, E.J.; Boston, A.; Brown, S.; Fennell, P.S.; Fuss, S.; Galindo, A.; Hackett, L.A.; et al. Carbon capture and storage (CCS): The way forward. *Energy Environ. Sci.* **2018**, *11*, 1062–1176. [CrossRef]

2. Nikolaeva, D.; Loïs, S.; Dahl, P.I.; Sandru, M.; Jaschik, J.; Tanczyk, M.; Fuoco, A.; Jansen, J.C.; Vankelecom, I.F.J. Water Vapour promotes CO_2 transport in poly(ionic liquid)/ionic liquid-based thin-film composite membranes containing zinc salt for flue gas treatment. *Appl. Sci.* **2020**, *10*, 3859. [CrossRef]

3. Manyà, J.J.; García-Morcate, D.; González, B. Adsorption performance of physically activated biochars for postcombustion CO_2 capture from dry and humid flue gas. *Appl. Sci.* **2020**, *10*, 376. [CrossRef]

4. Staciwa, P.; Narkiewicz, U.; Sibera, D.; Moszyński, D.; Wróbel, R.J.; Cormia, R.D. Carbon Spheres as CO_2 Sorbents. *Appl. Sci.* **2019**, *9*, 3349. [CrossRef]

5. Deng, X.; Yang, W.; Li, S.; Liang, H.; Shi, Z.; Qiao, Z. Large-scale screening and machine learning to predict the computation-ready, experimental metal-organic frameworks for CO_2 capture from air. *Appl. Sci.* **2020**, *10*, 569. [CrossRef]

6. Papadopoulos, A.I.; Seferlis, P. *Process Systems and Materials for CO_2 Capture: Modelling, Design, Control and Integration*; Papadopoulos, A.I., Seferlis, P., Eds.; John Wiley & Sons: Chichester, UK, 2017.

7. Stéphenne, K. Start-up of world's first commercial post-combustion coal fired ccs project: Contribution of shell cansolv to saskpower boundary dam iccs project. *Energy Procedia* **2014**, *63*, 6106–6110.

8. Armpriester, A. *WA Parish Post Combustion CO2 Capture and Sequestration Project Final Public Design Report*; Petra Nova Parish Holdings: Washington, DC, USA, 2017.

9. Damartzis, T.; Papadopoulos, A.I.; Seferlis, P. Process flowsheet design optimization for various amine-based solvents in post-combustion CO_2 capture plants. *J. Clean. Prod.* **2016**, *111*, 204–216. [CrossRef]

10. Jessop, P.G.; Heldebrant, D.J.; Li, X.; Eckertt, C.A.; Liotta, C.L. Green chemistry: Reversible nonpolar-to-polar solvent. *Nature* **2005**, *436*, 1102. [CrossRef]

11. Zhang, J. Study on CO_2 Capture Using Thermomorphic Biphasic Solvents with Energy-Efficient Regeneration. Ph.D. Thesis, TU-Dortmund University, Dortmund, Germany, 2013.

12. Pinto, D.D.D.; Knuutila, H.; Fytianos, G.; Haugen, G.; Mejdell, T.; Svendsen, H.F. CO_2 post combustion capture with a phase change solvent. Pilot plant campaign. *Int. J. Greenh. Gas Control* **2014**, *31*, 153–164. [CrossRef]

13. Zarogiannis, T.; Papadopoulos, A.I.; Seferlis, P. Efficient Selection of Conventional and Phase-Change CO_2 Capture Solvents and Mixtures Based on Process Economic and Operating Criteria. *J. Clean. Prod.* **2020**, 122764. [CrossRef]

14. Kim, H.; Hwang, S.J.; Lee, K.S. Novel shortcut estimation method for regeneration energy of amine solvents in an absorption-based carbon capture process. *Environ. Sci. Technol.* **2015**, *49*, 1478–1485. [CrossRef] [PubMed]

15. Gaspar, J.; Cormos, A.-M. Dynamic modeling and absorption capacity assessment of CO_2 capture process. *Int. J. Greenh. Gas Control* **2012**, *8*, 45–55. [CrossRef]

16. Jayarathna, S.A.; Lie, B.; Melaaen, M.C. Dynamic modelling of the absorber of a post-combustion CO_2 capture plant: Modelling and simulations. *Comput. Chem. Eng.* **2013**, *53*, 178–189. [CrossRef]

17. Mac Dowell, N.; Samsatli, N.J.; Shah, N. Dynamic modelling and analysis of an amine-based post-combustion CO_2 capture absorption column. *Int. J. Greenh. Gas Control* **2013**, *12*, 247–258. [CrossRef]

18. Lawal, A.; Wang, M.; Stephenson, P.; Yeung, H. Dynamic modelling of CO_2 absorption for post combustion capture in coal-fired power plants. *Fuel* **2009**, *88*, 2455–2462. [CrossRef]

19. Biliyok, C.; Lawal, A.; Wang, M.; Seibert, F. Dynamic modelling, validation and analysis of post-combustion chemical absorption CO_2 capture plant. *Int. J. Greenh. Gas Control* **2012**, *9*, 428–445. [CrossRef]

20. Posch, S.; Haider, M. Dynamic modeling of CO_2 absorption from coal-fired power plants into an aqueous monoethanolamine solution. *Chem. Eng. Res. Des.* **2013**, *91*, 977–987. [CrossRef]

21. Enaasen Flø, N.; Knuutila, H.; Kvamsdal, H.M.; Hillestad, M. Dynamic model validation of the post-combustion CO_2 absorption process. *Int. J. Greenh. Gas Control* **2015**, *41*, 127–141. [CrossRef]

22. Cormos, A.-M.; Daraban, I.M. Dynamic modeling and validation of amine-based CO_2 capture plant. *Appl. Therm. Eng.* **2015**, *74*, 202–209. [CrossRef]

23. Lin, Y.-J.; Pan, T.-H.; Wong, D.S.-H.; Jang, S.-S.; Chi, Y.-W.; Yeh, C.-H. Plantwide Control of CO_2 Capture by Absorption and Stripping Using Monoethanolamine Solution. *Ind. Eng. Chem. Res.* **2011**, *50*, 1338–1345. [CrossRef]

24. Lin, Y.-J.; Wong, D.S.-H.; Jang, S.-S.; Ou, J.-J. Control strategies for flexible operation of power plant with CO_2 capture plant. *AIChE J.* **2012**, *58*, 2697–2704. [CrossRef]

25. Nittaya, T.; Douglas, P.L.; Croiset, E.; Ricardez-Sandoval, L.A. Dynamic modelling and control of MEA absorption processes for CO_2 capture from power plants. *Fuel* **2014**, *116*, 672–691. [CrossRef]

26. Abdul Manaf, N.; Cousins, A.; Feron, P.; Abbas, A. Dynamic modelling, identification and preliminary control analysis of an amine-based post-combustion CO_2 capture pilot plant. *J. Clean. Prod.* **2016**, *113*, 635–653. [CrossRef]

27. Gaspar, J.; Ricardez-Sandoval, L.; Jørgensen, J.B.; Fosbøl, P.L. Controllability and flexibility analysis of CO_2 post-combustion capture using piperazine and MEA. *Int. J. Greenh. Gas Control* **2016**, *51*, 276–289. [CrossRef]

28. Damartzis, T.; Papadopoulos, A.I.; Seferlis, P. Solvent effects on design with operability considerations in post-combustion CO_2 capture plants. *Chem. Eng. Res. Des.* **2018**, *131*, 414–429. [CrossRef]

29. Panahi, M.; Skogestad, S. Economically efficient operation of CO_2 capturing process. Part II. Design of control layer. *Chem. Eng. Process. Process Intensif.* **2012**, *52*, 112–124. [CrossRef]

30. Mechleri, E.D.; Biliyok, C.; Thornhill, N.F. Dynamic Simulation and Control of Post-combustion CO_2 Capture with MEA in a Gas Fired Power Plant. In *Computer Aided Chemical Engineering*; Elsevier: Amsterdam, The Netherlands, 2014; Volume 33, pp. 619–624. ISBN 9780444634344.

31. Léonard, G.; Mogador, B.C.; Belletante, S.; Heyen, G. Dynamic modelling and control of a pilot plant for post-combustion CO_2 capture. In *Computer Aided Chemical Engineering*; Elsevier: Amsterdam, The Netherlands, 2013; Volume 31, pp. 451–456.

32. Lawal, A.; Wang, M.; Stephenson, P.; Obi, O. Demonstrating full-scale post-combustion CO_2 capture for coal-fired power plants through dynamic modelling and simulation. *Fuel* **2012**, *101*, 115–128. [CrossRef]

33. Hossein Sahraei, M.; Ricardez-Sandoval, L.A. Controllability and optimal scheduling of a CO_2 capture plant using model predictive control. *Int. J. Greenh. Gas Control* **2014**, *30*, 58–71. [CrossRef]

34. Panahi, M.; Skogestad, S. Economically efficient operation of CO_2 capturing process part I: Self-optimizing procedure for selecting the best controlled variables. *Chem. Eng. Process. Process Intensif.* **2011**, *50*, 247–253. [CrossRef]
35. Arce, A.; Mac Dowell, N.; Shah, N.; Vega, L.F. Flexible operation of solvent regeneration systems for CO_2 capture processes using advanced control techniques: Towards operational cost minimisation. *Int. J. Greenh. Gas Control* **2012**, *11*, 236–250. [CrossRef]
36. Liang, X.; Li, Y.; Wu, X.; Shen, J.; Lee, K.Y. Nonlinearity analysis and multi-model modeling of an MEA-based post-combustion CO_2 capture process for advanced control design. *Appl. Sci.* **2018**, *8*, 1053. [CrossRef]
37. Papadopoulos, A.I.; Seferlis, P. A Framework for Solvent Selection Based on Optimal Separation Process Design and Controllability Properties. *Comput. Aided Chem. Eng.* **2009**, *26*, 177–181. [CrossRef]
38. Papadopoulos, A.I.; Seferlis, P.; Linke, P. A framework for the integration of solvent and process design with controllability assessment. *Chem. Eng. Sci.* **2017**, *159*, 154–176. [CrossRef]
39. Seferlis, P.; Grievink, J. Process design and control structure evaluation and screening using nonlinear sensitivity analysis. *Comput. Chem. Eng.* **2004**, *17*, 326–351.
40. Oyenekan, B.A. Modeling of Strippers for CO_2 Capture by Aqueous Amines. Ph.D. Thesis, University of Texas at Austin, Austin, TX, USA, 2007.
41. Oexman, J. Post Combustion CO_2 Capture: Energetic Evaluation of Chemical Absorption Processes in Coal—Fired Steam Power Plants. Ph.D. Thesis, Technischen Universität Hamburg, Hamburg, Germany, 2011.
42. PSE gSAFT. 2018. Available online: www.psenterprise.com/products/gsaft (accessed on 9 May 2020).
43. Papaioannou, V.; Lafitte, T.; Avendaño, C.; Adjiman, C.S.; Jackson, G.; Müller, E.A.; Galindo, A. Group contribution methodology based on the statistical associating fluid theory for heteronuclear molecules formed from Mie segments. *J. Chem. Phys.* **2014**, *140*, 054107. [CrossRef]
44. Dufal, S.; Lafitte, T.; Haslam, A.J.; Galindo, A.; Clark, G.N.I.; Vega, C.; Jackson, G. The A in SAFT: Developing the contribution of association to the Helmholtz free energy within a Wertheim TPT1 treatment of generic Mie fluids. *Mol. Phys.* **2015**, *113*, 948–984. [CrossRef]
45. Gil-Villegas, A.; Galindo, A.; Whitehead, P.J.; Mills, S.J.; Jackson, G.; Burgess, A.N. Statistical associating fluid theory for chain molecules with attractive potentials of variable range. *J. Chem. Phys.* **1997**, *106*, 4168–4186. [CrossRef]
46. Arshad, M.W. Measuring and Thermodynamic Modeling of De-Mixing CO_2 Capture Systems. Ph.D. Thesis, Technical University of Denmark, Lyngby, Denmark, June 2014.
47. Tzirakis, F.; Tsivintzelis, I.; Papadopoulos, A.I.; Seferlis, P. Experimental measurement of Vapour-liquid-liquid equilibria of phase-change solvents. In Proceedings of the 26th Thermodynamics Conference, Punta Umbría, Spain, 26–28 June 2019.
48. Jeon, S.B.; Cho, S.W.; Lee, S.S.; Jang, S.Y.; Oh, K. Absorption characteristics of carbon dioxide into an O/W emulsion absorbent containing *N*-methylcyclohexylamine/2,6 dimethylpiperidine. *J. Taiwan Inst. Chem. Eng.* **2014**, *45*, 2673–2680. [CrossRef]

Article

Mineral Carbonation of CO_2 in Mafic Plutonic Rocks, II—Laboratory Experiments on Early-Phase Supercritical CO_2-Brine-Rock Interactions

Patrícia Moita [1,2], Edgar Berrezueta [3], Halidi Abdoulghafour [4], Massimo Beltrame [2], Jorge Pedro [1,4], José Mirão [1,2], Catarina Miguel [2], Cristina Galacho [2,5], Fabio Sitzia [2], Pedro Barrulas [2] and Júlio Carneiro [1,4,*]

[1] Departamento de Geociências, Escola de Ciências e Tecnologia, Universidade de Évora, Rua Romão Ramalho 59, 7000-671 Évora, Portugal; pmoita@uevora.pt (P.M.); jpedro@uevora.pt (J.P.); jmirao@uevora.pt (J.M.)

[2] Laboratório HERCULES, Universidade de Évora, Largo Marquês de Marialva 8, 7000-809 Évora, Portugal; massimo@uevora.pt (M.B.); cpm@uevora.pt (C.M.); pcg@uevora.pt (C.G.); fsitzia@uevora.pt (F.S.); pbarrulas@uevora.pt (P.B.)

[3] Instituto Geológico y Minero de España, C/Matemático Pedrayes 25, 33005 Oviedo, Spain; e.berrezueta@igme.es

[4] Instituto de Ciências da Terra, Universidade de Évora, Rua Romão Ramalho 59, 7000-671 Évora, Portugal; halidi@uevora.pt

[5] Departamento de Química, Escola de Ciências e Tecnologia, Universidade de Évora, Rua Romão Ramalho 59, 7000-671 Évora, Portugal

[*] Correspondence: jcarneiro@uevora.pt; Tel.: + 351-912-857-538

Received: 16 June 2020; Accepted: 21 July 2020; Published: 23 July 2020

Abstract: The potential for mineral carbonation of CO_2 in plutonic mafic rocks is addressed through a set of laboratory experiments on cumulate gabbro and gabbro-diorite specimens from the Sines Massif (Portugal). The experiments were conducted in an autoclave, for a maximum of 64 days, using a CO_2 supersaturated brine under pressure and temperature conditions similar to those expected around an injection well during early-phase CO_2 injection. Multiple techniques for mineralogical and geochemical characterization were applied ante- and post-carbonation experiments. New mineralogical phases (smectite, halite and gypsum), roughness increase and material loss were observed after exposure to the CO_2 supersaturated brine. The chemical analysis shows consistent changes in the brine and rock specimens: (i) increases in iron (Fe) and magnesium (Mg) in the aqueous phase and decreases in Fe_2O_3 and MgO in the specimens; (ii) a decrease in aqueous calcium (Ca) and an increase in CaO in the cumulate gabbro, whereas in the gabbro-diorite aqueous Ca increased and afterwards remained constant, whereas CaO decreased. The geochemical model using the CrunchFlow code was able to reproduce the experimental observations and simulate the chemical behavior for longer times. Overall, the study indicates that the early-stage CO_2 injection conditions adopted induce mainly a dissolution phase with mineralogical/textural readjustments on the external area of the samples studied.

Keywords: CO_2 storage; supercritical CO_2; mafic plutonic rocks; experimental test

---

1. Introduction

The International Energy Agency, in its flagship report "Energy Technology Perspectives", has demonstrated that CO_2 capture, utilization and storage (CCUS) is a key technology for CO_2 emissions reduction, essential to achieve the targets set in the Paris Agreement [1].

In the CCUS chain of technologies, CO_2 is captured in large stationary sources and transported to a utilization or permanent storage in deep geological formations [2,3]. Adequate geological environments for CO_2 storage are provided by depleted hydrocarbon reservoirs, uneconomic coal seams, deep saline aquifers or mafic and ultramafic rocks [4–8]. The latter relies on the effectiveness of mineral carbonation, in which the CO_2 reacts with the enriched calcium (Ca), magnesium (Mg) and iron (Fe) minerals, to precipitate as carbonate minerals, thus ensuring safe and permanent sequestration of the CO_2 in solid phases.

Most subsurface carbon storage projects to date have injected CO_2 into sedimentary formations, either deep saline aquifers or hydrocarbon fields, but the exciting results obtained at two pilot sites where CO_2 is injected in basalts, the Carbfix and Wallula projects [9,10], have raised the profile of mafic and ultramafic rocks as suitable candidates for in situ mineral carbonation.

Although at a less developed research stage than other CO_2 storage environments, the possibility of using mafic and ultramafic rock massifs should be considered when they occur near major CO_2 emission sources. That is the case of the main industrial clusters in Sines and Setúbal. As described in an accompanying article [11], mafic and ultramafic rock massifs in southern Portugal may present a valid alternative for mineral carbonation (be it in situ, ex situ, or even enhanced weathering) of CO_2 captured in that cluster. The interested reader is directed to that paper for details on the rationale for selecting the rock massifs and their characterization.

CO_2–brine–rock interaction experiments are a well-established method to understand and explore the mechanisms and processes of geological storage [12,13]. However, unlike most previous experiments, this article focuses on mafic plutonic rocks. These have seldom been considered for in situ mineral carbonation, given the low porosity and permeability, but previous experiments on ex situ or enhanced weathering applications have been published (e.g., [14,15]).

This study deals with mineral carbonation experiments, in the laboratory, to assess the rate of reaction between CO_2 and rock samples from the Sines Massif, a subvolcanic massif mainly composed of gabbros, diorites and subordinated syenites. The laboratory experiments were designed to replicate the early stages of interaction between the mineral phases and a brine supersaturated in CO_2, as one would expect around an injection well. Indeed, we attempted to understand the initial dissolution of the rock minerals that should provide the cations to react with the dissolved CO_2.

The possible mineralogical-textural changes of the rock after interaction with brine-CO_2 was studied by a multi-analytical approach (optical microscopy (OM); scanning electron microscopy with X-ray detector (SEM-EDS); X-ray diffraction (XRD); and infrared Fourier transform spectroscopy (FTIR)) and the chemical compositional evolution of the brine and whole rock by comparative analyses (inductively coupled plasma optical emission spectrometry (ICPM-OES) and X-ray fluorescence (XRF)) before and after the experiment. Finally, numerical geochemical computation (CrunchFlow code) was used to interpret, replicate and predict the system's behavior for periods of time longer than 64 days, the maximum duration of the experiments.

The article is organized as follows: a brief background on mineral carbonation experiments and tests is presented, followed by a description of the methodology applied to characterize the rock samples and conduct the laboratory experiments. The results are then interpreted and discussed in terms of the chemical and mineral changes observed in the brine and rock surface, and according to a numerical model that reproduces the laboratory experiments and extrapolates them to longer times.

2. Background on Mineral Carbonation Experiments and Tests

During mineral carbonation, the conversion of CO_2 to stable minerals starts with the CO_2 dissolution in the aqueous phase, a function of the fluid ionic strength, pressure and temperature [16]. The resulting carbonic acid (H_2CO_3) decomposition releases hydrogen protons (H^+), lowering the pH (Equation (1)). Subsequent consumption of the H^+ due to reaction with the Mg-Ca-Fe-rich minerals releases the metallic cations and increases the pH of the solution (Equations (2) to (6)). At suitable pH and saturation conditions, cations combine with hydrogen carbonate and form (Ca, Mg, Fe) CO_3.

$$CO_2 \text{ (g)} + H_2O \text{ (aq)} = H_2CO_3 \text{ (aq)} = H^+ \text{ (aq)} + HCO_3 \text{ (aq)} \tag{1}$$

$$\text{(Forsterite) } Mg_2SiO_4 \text{ (s)} + 4H^+ \text{ (aq)} = 2Mg^{2+} \text{ (aq)} + SiO_2\text{(aq)} + 2H_2O \text{ (l)} \tag{2}$$

$$\text{(Ca-plagioclase) } CaAl_2Si_2O_8 \text{ (s)} + 8H^+\text{(aq)} = Ca^{2+} \text{ (aq)} + 2Al^{3+} \text{ (aq)} + 2SiO_2\text{(aq)} + 4H_2O \text{ (l)} \tag{3}$$

$$\text{(Clinopyroxene) } CaMgSi_2O_6 \text{ (s)} + 4H^+ \text{ (aq)} = Ca^{2+} \text{ (aq)} + Mg^{2+} \text{ (aq)} + 2SiO_2\text{(aq)} + 2H_2O \text{ (l)} \tag{4}$$

$$Mg^{2+} \text{ (aq)} + HCO_3^- \text{ (aq)} = MgCO_3 \text{ (s)} + H^+ \text{ (aq)} \tag{5}$$

$$Mg^{2+} \text{ (aq)} + Ca^{2+} \text{ (aq)} + 2HCO_3^- \text{ (aq)} = (Ca,Mg)CO_3 \text{ (s)} + 2H^+ \text{ (aq)} \tag{6}$$

Numerous laboratory experiments have been performed to study the reactivity of olivine, serpentine, pyroxene, amphibole and plagioclase mineral groups, and glass basalt lavas with CO_2-enriched solutions [17–20]. Most of these experiments were conducted in batch conditions at controlled pressure, temperature and P_{CO_2}, using crushed rock. Sodium hydrogen carbonate ($NaHCO_3$) is often added in the reactor to increase the hydrogen carbonate concentration in solution and buffer the pH up to 7.7 and 8.0 [21]. This pH range, temperatures from 155 °C to 185 °C and total pressures from 11.5 MPa to 19 MPa [21–23] provide the optimal conditions for carbonation enhancement.

In general, the parameters that affect the rate of carbonate minerals' precipitation are brine composition, temperature, pressure and, principally, pH [24]. Mineral carbonation is favored over a higher pH—for instance, above 6.5 pH for Ca-Mg carbonates [25] or above 9.0 pH for $CaCO_3$ [9].

In the CarbFix project, the water in the Hellisheidi carbon injection site has a temperature ranging from 15 to 35 °C and the in situ pH ranges from 8.4 to 9.8 [26]. The injected water (with dissolved CO_2) has a temperature of 25 °C and a pH of 3.7 to 4.0 [27]. According to. [28], during the injection phase, the water with CO_2 will create porosity in the near vicinity of the injection by dissolving primary and secondary minerals. Furthermore, away from the injection well, secondary minerals will precipitate due to the reaction with the Ca-Mg-Fe-rich reservoir rocks.

Other authors [14,29,30] observed a decrease in the permeability of the carbonated rock. Inter- and intragranular pores (10 μm pore throats) were filled with magnesite and characterized using μRaman and SEM-TEM images. SEM-TEM observations [14] showed a magnesite layer separated from the olivine by submicron siderite grains, and poorly crystallized phyllosilicates. Iron oxide and amorphous silica are also observed as secondary phases in batch and flow-through experiments [14,22,29]. Phyllosilicates and chalcedony affect both the porosity and permeability, causing a decrease in the carbonation rate due to the creation of an exfoliated passivation layer rich in silica.

Carbonation experiments have been performed by [31] on continental flood basalt (CFB) samples (10.3 MPa and 90 °C) from eastern Washington, and by Schaef and McGrail [32] in basalt samples representing formations from North America, India and Africa.

Both studies noted that calcite was the first mineral to precipitate. For instance, Schaef and McGrail [32] observed small calcite nodules precipitating after 86 days. Post-reacted samples from Deccan basalts [32] displayed larger and opaque precipitates (after 280 days) and more reddish-brown grains containing a large calcite component (66.0–82.0 wt. %) and minor magnesite (9.1–22.0 wt. %) or siderite (3.1–15.0 wt. %) components. The basalt grains representing Southern Africa (Karoo) were insignificantly reactive, displaying very few carbonate nodules with high Mg (31.0 wt. %) and Fe (29.0 wt. %) contents. More recent basalt carbonation experiments have been performed by [33] on recent Auckland basalt (0.3 Ma). After 140-day experiments (100 °C and 5.5 MPa), they observed ankerite and aluminosilicates as secondary precipitation and an increase in both porosity and permeability. This observation contrasts with the major reported carbonation studies [29,30], where the permeability was observed to decrease due to carbonate growth. Theoretical and experimental studies of rock-CO_2 interactions in wet conditions and low temperature (25–90 °C) and first phases (0.5–100 days) indicate the presence of chemical reactions and, consequently, textural-mineralogical-chemical changes [25]. However, according to [25], in order to optimize low-temperature mineral carbonations, an "equilibrium" between initial CO_2 (acid supply), rock to water ratio and temperature needs to be adjusted in order to effectively mineralize CO_2 within a reasonable time scale.

3. Methodology

To evaluate the potential for mineral carbonation in plutonic mafic rocks, a six-step methodology was followed (Figure 1):

1. Selection of representative samples for study and definition of conceptual conditions (rock-brine-CO_2) to be studied (Figures 1 and 2a).
2. Mineralogical, textural and chemical characterization of the specimens before exposure to brine and supercritical CO_2 (SC CO_2) (Figures 1 and 2b).
3. Exposure of the specimens to CO_2 supersaturated brine at selected conditions (supercritical CO_2: 8 MPa and 40 °C) in the autoclave (Figures 1 and 2c): (a) Stage 1—CO_2 pressurized injection (3 h); (b) Stage 2—CO_2 pressurized stabilization (1, 4, 16 and 64 days) and (c) Stage 3 CO_2—pressure release (3 h).
4. Upon conclusion of the laboratory experiments in Step 3, mineralogical, textural and chemical characterization of specimens and brine chemical analysis were conducted.
5. Geochemical modelling of the mineral carbonation experiments using CrunchFlow.
6. Interpretation of results and correlation of experimental and modelling data.

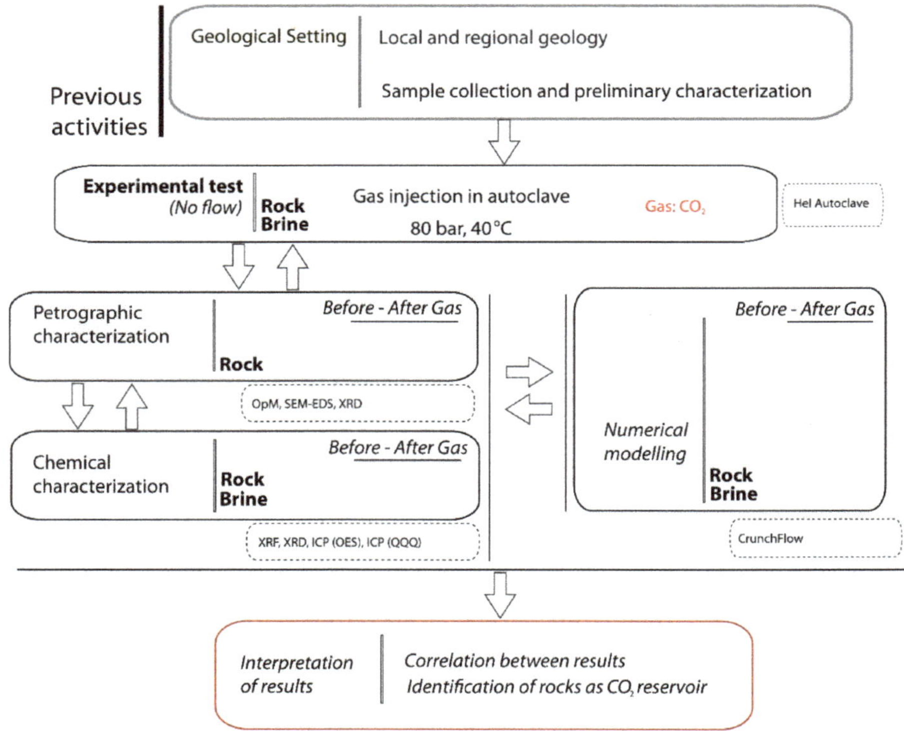

Figure 1. Schematic representation of the work sequence followed in this study.

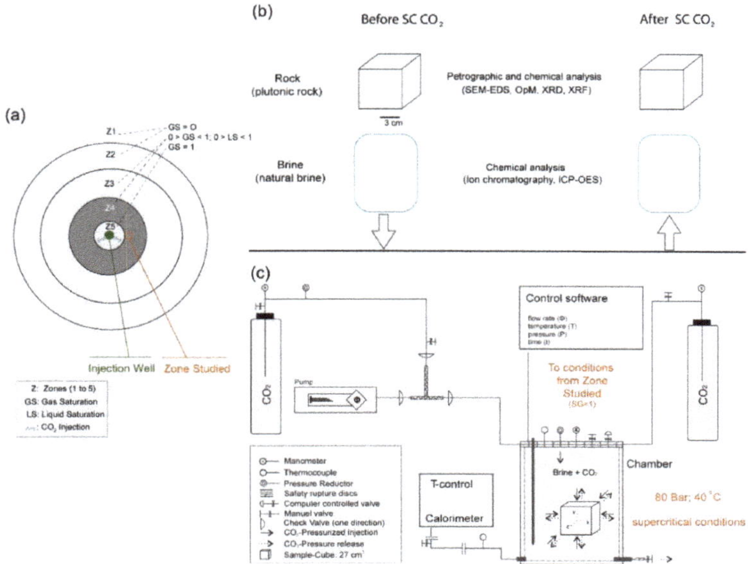

Figure 2. (a) Conceptual diagram of the reactive zones (Z1, Z2, Z3, Z4 and Z5) around the injection well according to [34] and [35]. (b) Sample preparation for mineralogical and geochemical analyses before and after SC CO_2 exposition. (c) Layout of the experimental setup. Reactor system used for the pressurized CO_2 injection (modified from. [36]).

3.1. Materials

Two different lithologies of igneous rock from the Sines Massif, Portugal were sampled and used in the experiments: (i) Experiment 1 with a cumulate gabbro (CG) from a cliff near Praia do Norte and (ii) Experiment 2 with a gabbro-diorite (GD) sampled at quarry Monte Chãos [11].

The CG displays a medium to coarse cumulate texture and is formed of clinopyroxene (45–55%), olivine (15–20%), brown amphibole (10–15%), plagioclase (5–10%) and primary ilmenite (5%); it occasionally shows accessory alteration products (e.g., chlorite, actinolite, serpentine). The GD exhibits a layered medium to coarse texture and is composed of plagioclase (50–60%), clinopyroxene (20–25%), subordinate olivine (5–10%), biotite (10–15%) and ilmenite (5–10%). Despite some fractures with chlorite and incipient sericitization, the sample has no significant alteration.

The coarse-grained CG sample was cut into 40 subsample cubes of 27 cm^3 each and four of 1 cm^3. They were divided in four run sets and one reference set (0-day set). For Experiment 1, each set of specimens had seven cubes of 27 cm^3, two parallelepipeds of 27/2 cm^3 and one of 1 cm^3 (Table 1). The coarse-grained gabbro-diorite sample was cut into 80 subsamples cubes of 27 cm^3 each and four of 1 cm^3. For Experiment 2, each set of specimens had 15 cubes of 27 cm^3, two parallelepipeds of 27/2 cm^3 and one of 1 cm^3 (Table 1).

Table 1. Experimental conditions of rock-brine-CO_2 exposure in autoclave.

Experiment	Sample	Samples of 27 cm^3	Samples of 27/2 cm^3	Samples of 1 cm^3	Brine (cm^3)	CO_2 (cm^3)	Run (Days)
1	Cumulate gabbro (CG)	7	2	1	600	1184	1, 4, 16, 64
2	Gabbro-diorite (GD)	15	2	1	1350	218	1, 4, 16, 64

The brine used in the experiments is a natural brine sampled from an old borehole in a saline aquifer (see Section 4.2).

3.2. Experimental Procedure (Autoclave)

The experimental setup of the autoclave employed in this experiment (Figure 2c) is based on similar systems described by [37] and [38]. Specific initial conditions in the autoclave were considered due to the planned target: sample material (rock-type and representative sample size), geological environment (pressure, temperature and salinity) and technical equipment (materials for chamber, software, pumps, etc.) for the final arrangement of the experimental device and run conditions. The autoclave (Figure 2c) [36,39] has two CO_2 cylinders (standard industrial CO_2 at 4.5 MPa) that are linked to the other elements of the system by steel connectors (diameter: 5 mm). The first CO_2 cylinder is directly connected with the chamber. The second CO_2 cylinder is connected to a piston pump that operates with a flow of 0.01 g/s. In case of gas leakage in the chamber during the experiment, this pump maintains the experimental pressure defined for the test. The inside of the chamber has a capacity of 2 dm^3. This is coated with polytetrafluoroethylene (PTFE) to protect the material against corrosion. At the bottom of the chamber, a thermostat controls the internal temperature. The calorimeter and pump are linked to the chamber with pressure and temperature sensors and are connected to a computer.

In detail, the experiment consisted of exposure of mafic rocks to CO_2-supersaturated brine (i.e., SC CO_2-rich brine) in the autoclave to a pressure (P) of 8 MPa and to a temperature (T) of 40 °C without flow. The P and T conditions were selected to exceed the CO_2 supercritical (SC CO_2) point [40,41] and to simulate the conditions of injection and storage of CO_2 [2,42]. These conditions are representative to a depth of approx. 800 m. The selected exposure time (1, 4, 16 and 64 days) was chosen to identify possible changes in the rock (dissolutions and/or precipitation) during the first injection phases.

The experiments began with the immersion of specimen rocks within natural brine in the chamber. Once the rock is introduced and fully immersed in the brine, CO_2 is injected (Table 1).

The experimental runs comprised: (a) a pressurized CO_2 injection (3 h, from 4.5 MPa and 20 °C conditions to the SC condition); (b) a pressurized stabilization (period of test, no CO_2 flow inside the chamber) and (c) CO_2 pressure release (3 h, from supercritical conditions to ambient conditions). The final volume of CO_2 in chamber to 80 bar and 40 °C was (i) 1184 cm^3 in experiment with CG and (ii) 218 cm^3 in experiment with gabbro-diorite. The times of filling and emptying the chamber with SC CO_2 were the same (3 h, from ambient conditions to supercritical conditions and supercritical conditions to ambient conditions, respectively), following the chamber manufacturer's recommendations. This is the time required to reach the target pressure and temperature values from the initial ambient conditions.

3.3. Material Characterization

To obtain a precise characterization of the rock specimens and to evaluate the changes after SC CO_2 exposure, in mineralogy, texture and chemistry, a set of complementary analytical techniques was repeatedly applied.

Petrography on thin-sections, by optical microscopy (OM), was performed by a Leica bright-field microscope (LEICA DM 2500P, Wetzlar, Germany).

The X-ray powder and in situ diffractograms were produced using a Bruker AXS-D8 Advance (Bruker Corp., Billerica, MA, USA), with Cu-Kα radiation (λ = 0.1540598 nm), under the following conditions: scanning between 3° and 75° (2θ), scanning velocity of 0.05° 2θ/s, accelerating voltage of 40 kV, and current of 40 mA. In order to evaluate the mineralogical composition of the specimen surface, in situ grazing incidence geometry experiments were conducted, with incidence of 1.5° and 2θ scanning from 8 to 60°.

A Hitachi S-3700N SEM (Hitachi High Technologies, Berlin, Germany), coupled with a Bruker XFlash 5010 SDD detector (Bruker Corp, Billerica, MA, USA), was used for the surface sample chemical analysis. The analysis was performed under a low vacuum at 40 Pa, with a current of 20 kV. An infrared spectrometer Bruker Hyperion 3000 equipped with a single-point MCT detector cooled with liquid nitrogen and a 20 × ATR objective with a Ge crystal of 80 µm diameter was used.

An infrared spectrometer Brüker Hyperion 3000 equipped with a single-point MCT detector cooled with liquid nitrogen and a 20 × an attenuated total reflectance (ATR) objective with a Ge crystal of 100 μm diameter was used. The infrared spectra were acquired with a spectral resolution of 4 cm^{-1}, 32 scans, in the 4000-650 cm^{-1} region. In order to ensure the representativeness of the data, each cube was analysed in nine different spots, screening the most exposed surface, and each spot was analysed three times.

The whole-rock geochemistry analysis was provided by XRF, which allows for the quantification of major oxides (SiO_2, TiO_2, Al_2O_3, Na_2O, K_2O, CaO, MgO, MnO, FeO, P_2O_5), sulfur and some minor elements (Rb, Sr, Y, Zr, Nb, Th, Cr, Co, Ni, Cu, Zn, Ga, As, Pb, Sn, V, U, Cl). Analyses were performed with an S2 Puma energy-dispersive X-ray spectrometer (Bruker), using a methodology similar to that adopted by [43]. A description of the standard reference materials (SRM) utilized in the calibration method can be found elsewhere [44]. After the determination of loss on ignition (LOI), samples were fused on a Claisse LeNeo heating chamber, using a flux (Li-tetraborate) to prepare fused beads (ratio sample/flux = 1/10). The software utilized for acquisition and data processing was Spectra Elements 2.0, which reported the final oxide/element concentrations and the instrumental statistical error (Stat. error) associated to the measurement.

OM, XRD, XRF, FTIR and SEM-EDS analyses were performed at HERCULES Laboratory (University of Évora, Portugal).

Linear roughness was measured by a homemade microprofilometer consisting of 55 parallel needles put in contact. The spacing between the needles' center-line is 500 μm for a total measurable length of 3.5 mm. After the profile measure, a photo of the needles' position was taken. Profiles of the samples were detected by drawing a line joining all the extremities of the needles, with the help of Adobe Photoshop CC 2018. The profile line in JPG format was uploaded on online software (WebPlotDigitizer, https://apps.automeris.io/wpd/) that allowed us to identify the X and Y values of the line according to the pixels composing the JPG photo. Pixel values were later converted to microns according to a scale.

The brine analyses were performed at IGME (Madrid, Spain) by ion chromatography (Dionex 600 de Vertex) and ICP-OES (Varian Vista MPX) before and after the experiment on each run. Iron and magnesium brine content determination, from Experiment 1, was performed at HERCULES Laboratory (Évora, Portugal) with an Agilent 8800 ICP Triple Quad (ICP-QQQ), operating with an RF power of 1550 W, RF matching of 1.7 V, a sample depth of 10 mm, carrier gas (Ar) of 1.1 L/min and plasma gas (Ar) of 15 L/min. Prior to the analysis, the equipment was calibrated with a tuning solution from Agilent, and the sensitivity and resolution were optimized and the doubly charged ions ($< 1.84\%$) and oxides ($< 1.10\%$) were minimized.

3.4. Geochemical Modelling

3.4.1. Code Descriptions and Capabilities

The simulations were performed using CrunchFlow [45,46], a multicomponent reactive flow and transport code for studying fluid rock interactions in porous media. The reactive transport code has many advantages such as a complete equilibrium thermodynamic treatment, flexible kinetic rate law formulations for each mineral depending on the reaction mechanism, and a range of 3D flow capabilities. The reactive transport code [47] numerically solves the mass balance of solutes, as shown in Equation (7):

$$\frac{\partial(\phi C_j)}{\partial t} = \nabla \cdot \left(D\nabla C_j\right) - \nabla\left(qC_j\right) + R_j \ (j = 1, 2, 3 \dots n) \tag{7}$$

where ϕ is the updated porosity, C_j is the concentration of component j (mol m^{-3}), q is the Darcy velocity (m s^{-1}), R_j is the total reaction rate affecting component j (mol m^{-3} s^{-1}) and D is the combined dispersion-diffusion coefficient ($m^2 s^{-1}$). The carbonation experiments were conducted in closed batch conditions, that is, without flow, so that Equation (7) can be simplified as Equation (8):

$$\frac{\partial\left(\phi C_j\right)}{\partial t} = R_j \; (j = 1,\, 2,\, 3\ldots n) \tag{8}$$

The reaction rate is only described as function of time. The mineral dissolution to aqueous phases and precipitation of secondary phases are kinetically controlled, thus a reaction rate is given in term of primary species. The reaction kinetic is treated based on the rate law types including the transition state theory (TST), irreversible, monod, dissolution only and precipitation only [46]. In this work, a common theoretical framework provided by the TST rate law [48–50] is adopted. The dissolution and precipitation are treated as reversible at equilibrium, by explicitly including a dependence on Gibbs energy or saturation state (Equation (9)):

$$R_m = \mp A_m k_m \left(\prod a_n\right)\left|\left(\frac{Q_m}{K_{eq}}\right)^n - 1\right|^m \tag{9}$$

where R_m is the rate of precipitation (rate > 0) or dissolution (rate < 0) of mineral m in mol $L^{-1}\,s^{-1}$, A_m is the reactive surface area and k_m the kinetic constant of mineral m (in $mol.m^{-2}.s^{-1}$). The sign of the saturation index $\Omega = \log\left(\frac{Q_m}{K_{eq}}\right)$ determines the sign of the reaction rate. Negative means dissolution and positive means precipitation. The kinetic constant at given temperature T (K) is calculated from Equation (10):

$$k = k_{25}\mathrm{Exp}\left[\frac{E_a}{R}\left(\frac{1}{T} + \frac{1}{298.15}\right)\right] \tag{10}$$

where k_{25} is the kinetic constant at 25 °C, E_a is the apparent activation energy (KJ mol^{-1}) and R is the gas constant (J $mol^{-1}K^{-1}$). The change in initial porosity (ϕ_i) and mineral bulk surface area (A_m) owing to dissolution is computed from Equation (11):

$$A_m = A^{\text{initial}}\left(\frac{\phi_m}{\phi_{(i)m}}\right)^{\frac{2}{3}}\left(\frac{\phi}{\phi_i}\right) \tag{11}$$

and the change due to precipitation is computed from Equation (12):

$$A_m = A^{\text{initial}}\left(\frac{\phi}{\phi_i}\right). \tag{12}$$

3.4.2. Input Conditions for Rock and Fluid Composition

The numerical simulations presented in this article only describe the CG experiments; modelling of the gabbro-diorite experiments is ongoing and will be described elsewhere. The initial mineral composition of the rock was derived from Canilho [51] and adjusted with petrography and mineral chemistry, obtained through SEM-EDS [11]. For the numerical simulations, it was assumed that CG is mainly composed of clinopyroxene and olivine, associated with minor amounts of amphibole and calcic plagioclase. Iron oxides and small retrogradation mineral phases were not considered. As in the autoclave experiments, the modelling considered a closed system in which the solid rock takes 20% of the volume and the remaining space is taken up by the brine. As a requirement of the code database, the CG mineral phases are described with the endmembers of the solid solutions, as represented in Table 2. The reactive surface area for each mineral was calculated based on a random variable of the specific surface areas (SSA), defined in the literature as ranging from 10 to 250 cm^2/g (e.g., applying Brunauer-Emmett-Teller (BET) measurements) for mafic and ultramafic rock [25,52–56] from Equation (13):

$$A_{bulk} = \frac{\varphi_m SSAM_w}{V_m} + SA. \tag{13}$$

Table 2. Considered modal composition assumed for cumulate gabbro modelling and reactive surface area considered for the calculations.

Rock Composition	Vol. Fraction (%)	10 cm²/g	75 cm²/g	120 cm²/g	170 cm²/g	220 cm²/g
		Reactive Surface Area (m²/m³) (or $A_{initial}$)				
Albite, $NaAlSi_3O_{8)}$	0.35	9.15	68.65	1.09×10^2	1.55×10^2	2.01×10^2
Anorthite, $CaAl_2Si_2O_8$	3.25	88.32	6.62×10^2	1.06×10^3	1.5×10^3	1.94×10^3
Diopside, $CaMgSi_2O_6$	10.8	3.53×10^2	2.64×10^3	4.23×10^3	5.9×10^3	7.76×10^3
Forsterite, Mg_2SiO_4	2.42	77.1	5.78×10^2	9.25×10^2	1.31×10^3	1.69×10^3
Fayalite, Fe_2SiO_4	2.21	96.63	7.24×10^2	1.16×10^2	1.64×10^3	2.12×10^3
Enstatite, $MgSiO_3$	0.56	19.97	1.34×10^2	2.15×10^2	3.05×10^2	3.95×10^2
Ferrosilite, $FeSiO_3$	0.41	16.0	1.2×10^2	1.92×10^2	2.72×10^2	3.52×10^2
Φ (brine fraction)	80					
Total	100					

The term SA (geometric surface area = 0.67 cm²/g) is the ratio between the rock area and the rock volume = 1.68 cm²/cm³. Because the dissolution reactions depend on the minerals' surface areas, we adjusted the SA with the first term in Equation (13), in order to reproduce the ideal conditions for CO_2 mineralization. In fact, mineral surface area is the most uncertain and complex kinetic parameter and the evolution (especially for multimineral systems) is not quantitatively understood at present [53]. The definition of this parameter requires random values as the measured values using BET overestimates the reaction rates [53,55,57].

The kinetic constants for the reacting minerals were taken from the literature [52,53]. The fit of the model to the experimental data (aqueous calcium, silica, magnesium and iron) concentrations and pH) was performed based on Table 2. The initial concentration of dissolved CO_2 was calculated to be 5.5×10^{-1} mol kgw^{-1} according to the Duan and Sun [16] model, with an imposed total pressure of 80 bars. From the thermodynamic and kinetic point of view, 73 aqueous species were considered in the simulations. The equilibrium constants were taken from the EQ3/6 thermodynamic database [58] included in CrunchFlow. The activity coefficients were calculated using the extended Debye-Hückel formulation (b-dot model) [59], with parameters from the same database. The brine composition in the simulations is similar to that of the brine used in the experiments, with pH = 6.85 (before CO_2 addition). The brine is enriched in Ca and Mg (5.12×10^{-2} and 2.62×10^{-2} mol.kgw^{-1}) and contains important amounts of sulfur and sodium chloride (Table 3).

Table 3. Chemical composition of solution used as input in the simulation.

Components	(mg kgw^{-1})	(mol kgw^{-1})
Ca^{2+}	2050	5.12×10^{-2}
Mg^{2+}	560	2.33×10^{-2}
Fe^{2+}	5.16	1.01×10^{-4}
$SiO_2(aq)$	11.5	1.55×10^{-4}
K^+	260	6.64×10^{-3}
Na^+	85,450	3.7
SO_4^{2-}	5400	5.62×10^{-2}
Cl^{-1}	133,500	3.56

4. Results

4.1. Petrographic and Chemical Characterisation of Rocks before and after SC CO_2 Exposure

The two lithologies (CG and GD) used in experiments and tested as described above show similar textural and mineralogical results. After interaction of the rock specimens with the SC CO_2

supersaturated brine in the autoclave chamber, an increase in the dissolution of specimens can be perceived; after the one-day run there is a noticeable surface roughness, which increases for the longer runtimes and eventually leads to material fragmentation for 16-day and 64-day runs (Figure 3). The mean linear roughness (Ra) is higher on CG from 14.80 μm to 21.89 μm for 16 and 64 days, respectively, whereas for GD the Ra increases from 1.01 μm to 1.85 μm for 16 and 64 days, respectively.

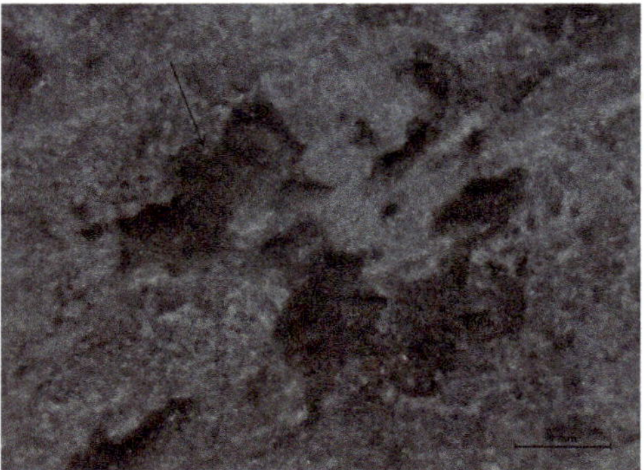

Figure 3. Stereo-zoom image that displays the effect of minerals dissolution (arrow) on the surface of a gabbro-diorite specimen after 64 days within brine.

The petrographic analyses, through transmitted light microscopy, did not show significant mineralogical or textural differences after the experiments. It is important to note that even if there were any changes, they should be located on the surface and the thin section production process could have eliminated them. The powder-XRD performed on global fraction before immersion [11] reflects the modal igneous composition given by clinopyroxene, olivine, amphibole, plagioclase and magnetite for CG and plagioclase, clinopyroxene, olivine and mica for the GD (Table 4). Clinochlore has been identified on both rock samples and probably derived from the alteration of the mafic mineralogical phases. After the 1-, 4-, 16- and 64-day runs, the obtained powder diffractograms (Table 4) shows besides the igneous mineralogy the new presence of talc, vermiculite, and halite on both CG and GD in most specimens. Moreover, XRD performed on the surface of CG and GD 64-days specimens, through grazing incidence (Figure 4), additionally reveals the presence of smectite and gypsum, which were not detected in 0-day specimens.

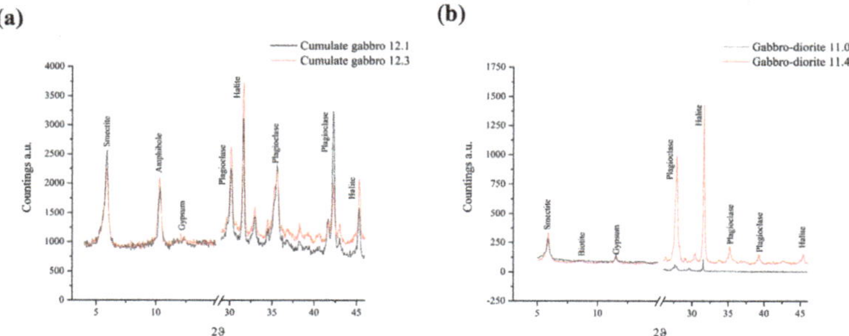

Figure 4. Diffractograms obtained through grazing apparatus on specimens after 64 days within brine: (**a**) cumulate gabbro; and (**b**) gabbro-diorite. For each we present diffractograms obtained at two distances, where the highest value is nearest the surface of the specimen.

Table 4. Mineralogical composition obtained through powder XRD. Tr: trace amount.

	Sample/Days	Clinopyroxene	Amphibole	Plagioclase	Clinochlore	Vermiculite	Talc	Mica	Olivine	Halite	Rutile	Ilmenite	Magnetite
Experiment 1	CG_0	32	24	13	4			2	5				2
	CG_1	38	28	17	4		Tr	3	8	1			2
	CG_4	26	29	24	4	Tr	Tr	3	10	1			2
	CG_16	37	21	21	3	Tr	Tr	2	12	1			2
	CG_64	34	24	26	3	Tr	Tr	2	8	1			2
Experiment 2	GD_0	16	4	69	Tr			6	3			3	
	GD_1	12	4	73	1		Tr	5	1	1		2	
	GD_4	12	4	73	1	Tr	Tr	5	1	1		2	
	GD_16	10	5	73	1	Tr	Tr	5	1	1	1	1	
	GD_64	16	4	69	2		Tr	4	2	1		1	

As described for the CG [60], the SEM images obtained before and after each experiment on GD (Figure 5a,b) essentially show an increase in ferromagnesian minerals' dissolution, reflected in the higher surface roughness, as reported by other authors [61]. The added phases show different reflectance on backscattered electron (BSE) images and, according to EDS elemental map distribution (Figure 5c,d), correspond to significant enrichment in chlorine (Cl), sodium (Na), sulfur (S) and carbon (C). The precipitation of salts in the form of efflorescence covers a large part of the surface of the specimens and the cavities generated by the dissolution.

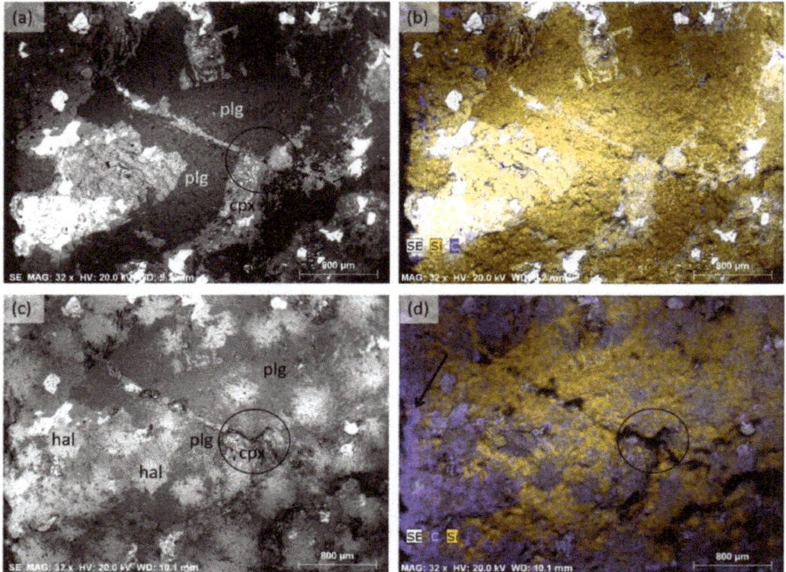

Figure 5. SEM images of a gabbro-diorite specimen before (upper) and after (lower) SC CO_2 brine: (**a**) backscattered electron (BSE) images at 0 days; (**b**) BSE and EDS elemental maps distribution for silicon and carbon at 0 days; (**c**) BSE images after 64 days; and (**d**) BSE and EDS elemental maps distribution for silicon and carbon after 64 days, with preferential dissolution of clinopyroxene (circle) and carbon enrichment (organic accumulation) on the edge of specimen (arrow). plg—plagioclase, cpx—clinopyroxene, hal—halite.

The carbon enrichment observed on the edge of 64-day specimens (CG and GD) and associated with salts (Figures 5d and 6a) corresponds to organic material, which accumulates due to runoff during the drying process at 40 °C. The origin of this material remains to be identified, but is probably related to the presence of organic material (hydrocarbon) within the original brine.

In fact, the carbon particles observed at the surface (Figure 6a) are not related to the crystallization of carbonates or any other mineral phase [60]. The ATR-FTIR spectra of these carbon-enriched areas present infrared fingerprints pointing to the presence of triglyceride-enriched areas whose origin is still to be determined [58]. In addition, the analysis by ATR-FTIR of CG and GD after 16 days (Figure 6b) of interaction with CO_2 revealed a decrease in the intensity of the absorption band

Whole-rock geochemistry data for CG and GD before and after 64 days within a SC CO_2-brine solution are shown in Table 5. For CG the most significant variations after a six-day run were an increase in CaO (+0.4 wt. %) and a decrease in Fe_2O_3 (−0.4 wt. %), whereas in GD CaO and Fe_2O_3 decreased ((−0.42 wt. % and (−0.3 wt. %, respectively). The enrichment in sodium was more evident in GD specimens after 64 days within SC CO_2-brine (+0.13 wt. % vs. 0.04 wt. % for CG). Sulfur increased in both samples after 64 days (+0.05 wt. % and +0.08 wt. % for GB and GD, respectively).

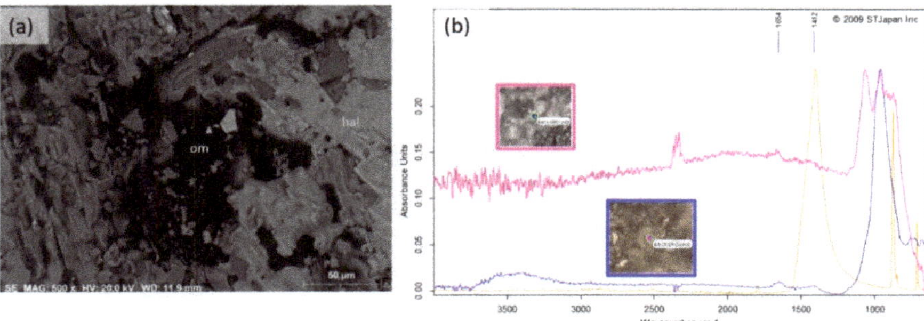

Figure 6. (**a**) Backscattered electron image of a detail of carbon-rich particles (om—organic matter) associated with halite (hal); (**b**) ATR-FTIR spectra of gabbro-diorite (GD) at 0 days (blue) and 16 days (pink). The calcite reference spectrum is also shown for comparison (yellow; STJapan Inc. database). The insets are the spots of analysis.

Table 5. Whole-rock geochemistry before and after runs of 64 days for cumulate gabbro (CG) and gabbro-diorite (GD).

	CG 0 (wt. %)	Stat. Error	CG 64 (wt. %)	Stat. Error	GD 0 (wt. %)	Stat. Error	GD 64 (wt. %)	Stat. Error
SiO_2	42.30	± 0.0344	42.20	± 0.0345	49.00	± 0.0356	49.50	± 0.0356
TiO_2	3.34	± 0.0175	3.20	± 0.0175	3.26	± 0.0176	3.16	± 0.0176
Al_2O_3	9.40	± 0.0295	9.50	± 0.0300	16.20	± 0.0368	16.40	± 0.0369
Fe_2O3	15.50	± 0.0133	15.10	± 0.0135	11.20	± 0.0115	10.90	± 0.0115
P_2O_5	0.28	± 0.00427	0.34	± 0.00439	0.85	± 0.00506	0.74	± 0.00494
MnO	0.40	± 0.005	0.40	± 0.005	0.32	± 0.005	0.31	± 0.005
MgO	12.90	± 0.0506	12.80	± 0.0503	4.48	± 0.0349	4.27	± 0.0344
CaO	12.70	± 0.0428	13.10	± 0.0444	8.33	± 0.0375	7.91	± 0.0371
BaO	0.23	± 0.013	0.20	± 0.013	0.28	± 0.013	0.27	± 0.013
Na_2O	0.84	± 0.0519	0.88	± 0.0512	3.46	± 0.0607	3.59	± 0.0612
K_2O	0.19	± 0.0345	0.23	± 0.0356	1.42	± 0.0395	1.42	± 0.0396
S	0.19%	± 0.00172	0.24%	± 0.00178	0.15%	± 0.00162	0.23%	± 0.00174
LOI	0.89		0.95%		0.03%		0.25%	
total	99.17		99.14%		98.98%		98.95%	
	(ppm)		**(ppm)**		**(ppm)**		**(ppm)**	
Rb	9	± 2.08	10	± 2.14	40	± 2.24	39	± 2.27
Sr	286	± 2.57	313	± 2.67	748	± 3.09	763	± 3.13
Y	15	± 2.30	15	± 2.37	34	± 2.46	35	± 2.49
Zr	80	± 2.82	74	± 2.90	199	± 3.14	208	± 3.19
Nb	20	± 2.52	14	± 2.60	62	± 2.65	65	± 2.68
Th	9	± 2.94	13	± 3.02	10	± 3.10	10	± 3.15
Cr	478	± 29.0	526	± 30.5	20	± 25.4	65	± 25.9
Co	198	± 5.65	198	± 5.76	139	± 5.02	134	± 5.00
Ni	117	± 4.12	135	± 4.29	7	± 3.42	14	± 3.54
Cu	62	± 4.82	64	± 4.91	42	± 4.88	49	± 5.01
Zn	103	± 6.07	100	± 6.19	108	± 6.37	101	± 6.26
Ga	14	± 4.30	15	± 4.41	25	± 4.67	19	± 4.70
As	7	± 4.29	2	± 4.39	8	± 4.53	10	± 4.61
Pb	0	± 0	2	± 16.7	12	± 17.2	0	± 0
Sn	7	± 27.3	0	± 0	0	± 0	13	± 28.0
V	479	± 68.0	505	± 68.9	323	± 68.0	249	± 67.8
U	0	± 0.209	1	± 0.215	2	± 0.221	2	± 0.225
Cl	43	± 0.364	54	± 0.371	53	± 0.359	70	± 0.383

4.2. Brine Evolution

Before starting the carbonation experiments, the composition of the brine was analysed using a combination of techniques (see Tables 6 and 7, "Pure Brine" column). The brine used in each run of experiments was recovered immediately after the end of the run for a comparative chemical analysis.

Table 6. Chemical analysis of pure brine and brine taken from the reaction chambers (one, four, 16 and 64 days) for Experiment 1 with CG (Figure S1 as supplementary material).

	Pure Brine (mg/L)	Brine Post Test (mg/L)	Brine Post Test (mg/L)	Brine Post Test (mg/L)	Brine Post Test (mg/L)
	0 Days	1 Day	4 Days	16 Days	64 Days
Na^+	85.45×10^3 $\pm 11.96 \times 10^3$	67.11×10^3 $\pm 93.96 \times 10^3$	86.02×10^3 $\pm 12.04 \times 10^3$	84.05×10^3 $\pm 11.77 \times 10^3$	79.68×10^3 $\pm 11.16 \times 10^3$
K^+	260 ± 31	260 ± 31	305 ± 37	310 ± 37	415 ± 50
Mg^{2+}	560 ± 100	590 ± 106	590 ± 106	610 ± 109	680 ± 122
Ca^{2+}	2050 ± 205	1900 ± 190	1890 ± 189	1860 ± 186	1650 ± 165
SO_4^{2-}	5400 ± 756	5400 ± 756	5600 ± 784	5700 ± 798	5300 ± 742
Cl^-	13.35×10^4 $\pm 1.6 \times 10^4$	10.40×10^4 $\pm 1.3 \times 10^4$	$13.30 10^4$ $\pm 1.6 \times 10^4$	12.40×10^4 $\pm 1.5 \times 10^4$	11.90×10^4 $\pm 1.4 \times 10^4$
HCO^-_3	40 ± 4.0	35 ± 3.5	38 ± 3.8	35 ± 3.5	33 ± 3.3
Fe (total)	5.2 ± 0.3	6 ± 0.3	11.3 ± 0.6	18.9 ± 0.9	26.4 ± 1.3
NO^-_3	0	0	0	0	0
SiO_2	11.5 ± 1.2	20 ± 2.8	24.9 ± 3.5	39 ± 5.5	37 ± 5.2
pH (pH Unit.)	6.9 ± 0.2	4.5 ± 0.1	4.9 ± 0.1	5.1 ± 0.2	5.5 ± 0.2
Cond (mS/cm)	80,000	85,000	80,000	75,000	75,000

Table 7. Chemical analysis of pure brine and brine taken from the reaction chambers (1, 4, 16 and 64 days) with GD (Figure S2 as supplementary material).

	Pure Brine (mg/L)	Brine Post Test (mg/L)	Brine Post Test (mg/L)	Brine Post Test (mg/L)	Brine Post Test (mg/L)
	0 Days	1 Day	4 Days	16 Days	64 Days
Na^+	87.61×10^3 $\pm 12.26 \times 10^3$	10.25×10^3 $\pm 14.35 \times 10^3$	99.72×10^3 $\pm 13.96 \times 10^3$	$10.70 \times 10^3 \pm$ 14.98×10^3	$78.54 \times 10^3 \pm$ 11.00×10^3
K^+	235 ± 28.2	250 ± 30	240 ± 28.8	270 ± 32.4	275 ± 33
Mg^{2+}	580 ± 116	560 ± 112	560 ± 112	580 ± 116	600 ± 120
Ca^{2+}	1520 ± 152	1640 ± 164	1680 ± 168	1660 ± 166	1650 ± 165
SO_4^{2-}	6900 ± 966	7200 ± 1008	7300 ± 1022	7600 ± 1064	6400 ± 896
Cl^-	9.70×10^4 $\pm 1.2 \times 10^4$	11.70×10^4 $\pm 1.4 \times 10^4$	12.70×10^4 $\pm 1.5 \times 10^4$	11.20×10^4 $\pm 1.3 \times 10^4$	11.10×10^4 $\pm 1.3 \times 10^4$
HCO^-_3	40 ± 4	40 ± 4	50 ± 5	110 ± 11	100 ± 10
NO^-_3	0	0	0	0	0
SiO_2	8 ± 0.8	10 ± 1	39 ± 3.9	59 ± 5.9	19 ± 1.9
pH (pH Unit.)	7 ± 0.2	5 ± 0.2	5.2 ± 0.2	5.4 ± 0.2	5.6 ± 0.2
Cond. (mS/cm)	80,000	77,500	77,000	77,000	80,000

The variation in the brine chemistry with time for both experiments—Experiment 1 with CG and Experiment 2 with GD—is shown in Tables 6 and 7. The pH, measured just after concluding each run, is more acidic for Experiment 1 (with CG). In fact, the initial pH of 6.85 dropped to 4.5, 4.87, 5.1 and 5.5, after one, four, 16 and 64 days, respectively, in the brine-rock-CO_2 system (Table 1) whereas for Experiment 2 (with GD) the pH decreased from 7 (brine without CO_2) to 5.81, 5.61, 6.38 and 6.61. In Experiment 2, the number of specimens was doubled (higher rock/brine ratio), but there was not a clear increase in the concentration of ions in solution, as suggested by similar values of conductivity for both experiments.

For Experiment 1 (CG), in general, the behavior of cations in solution was variable, despite the tendency of Ca^{2+} decreasing, whereas K^+, Mg^{2+}, SiO_2 and total Fe increased.

For Experiment 2 (GD), besides Na^+, all other analysed cations had lower concentrations in solution when compared to Experiment 1. Furthermore, the analysed elements did not always show the same behavior as described for Experiment 1. After an increase until the four-day run,

Ca^{2+} tended to stabilize. K^+ showed an increase, whereas Mg^{2+} concentrations tended to stay the same or increase slightly. SiO_2 clearly increased, but the 64-day run saw a decrease in silica.

In both experiments, SO_4^{2-} had similar behavior: a smooth enrichment, followed by a decrease for the 64-day run; Na^+ and Cl^- had more unpredictable behavior, displaying either enrichment or impoverishment trends along the four runs.

4.3. Geochemical Modelling of Experiments with Cumulate Gabbro

4.3.1. Outlet Solution Composition

The results of the experimental and simulated variations and tendencies in the output concentration over time are shown in Figure 7. The measured and simulated pH in brine supersaturated with CO_2 are presented as a function of the reactive surface area (Figure 7a) and compared to the variation of dissolved CO_2 (Figure 7b).

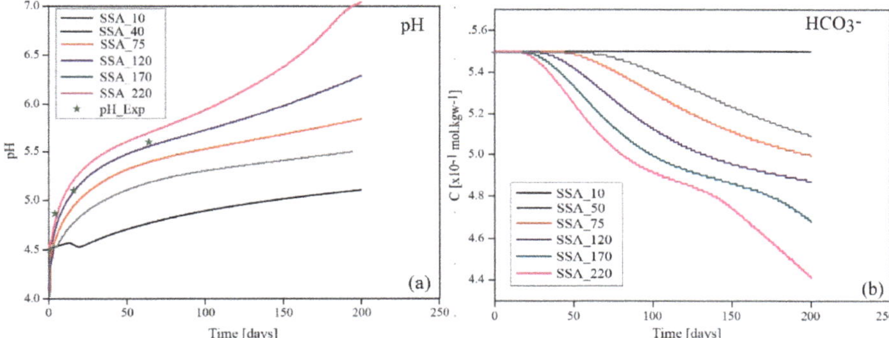

Figure 7. (a) Simulated pH variations of the outlet fluid compared with the measured pH; (b) simulated dissolved CO_2 (as HCO_3^-) variations as a function of time computed with different specific surfaces. The measured HCO_3^- values are obtained after opening the chamber and are not relevant for comparison.

Initially, the fluid pH at the onset of the modelling was 4.0 and increased rapidly to 4.5, and then gradually to 7.0, in the range of the measured values. Clearly, over the simulated time (0 to 200 days), pH ranged from 4.0 to 5.1 for the low SSA, increasing up to 7 for high SSA (170–220 $cm^2.g^{-1}$). The interaction between the rock and acidified fluid is a function of the dissolved CO_2 and the reactive surface area, since the decrease in HCO_3^- is inversely correlated to the increase in pH. Hence, the increase in pH above 5.5 was correlated to the consumption of the dissolved CO_2. As a result, at t < 50 days for SSA < 75 $cm^2.g^{-1}$, and t < 10 days, for SSA = 170–220 $cm^2.g^{-1}$, the pH remains lower than 5.0 and the dissolved CO_2 concentration remained constant. The constant concentration of HCO_3^- at a low pH denotes a dominant dissolution phase, where the concentration of the primary aqueous species (Ca, Mg, Fe, SiO_2) increases with time (Figure 8). However, at moderate and high pH values, CO_2 consumption is inferred from the decline in HCO_3^- concentration (below 0.48 $mol.kgw^{-1}$), triggering CO_2 mineralization into carbonates.

Simulated aqueous silica, Ca, Mg and Fe were observed to initially increase and then level off or decrease over increasing time. For instance, a minor increase in Ca concentration with respect to the initial concentration (ΔC_{Ca} = 0.14 $mol·kgw^{-1}$) was observed for t < 40 days before a continuous decrease began. The pattern of increase and decrease match perfectly with the evolution of dissolved CO_2, and are in agreement with the observations in previous studies [14,25,29,55].

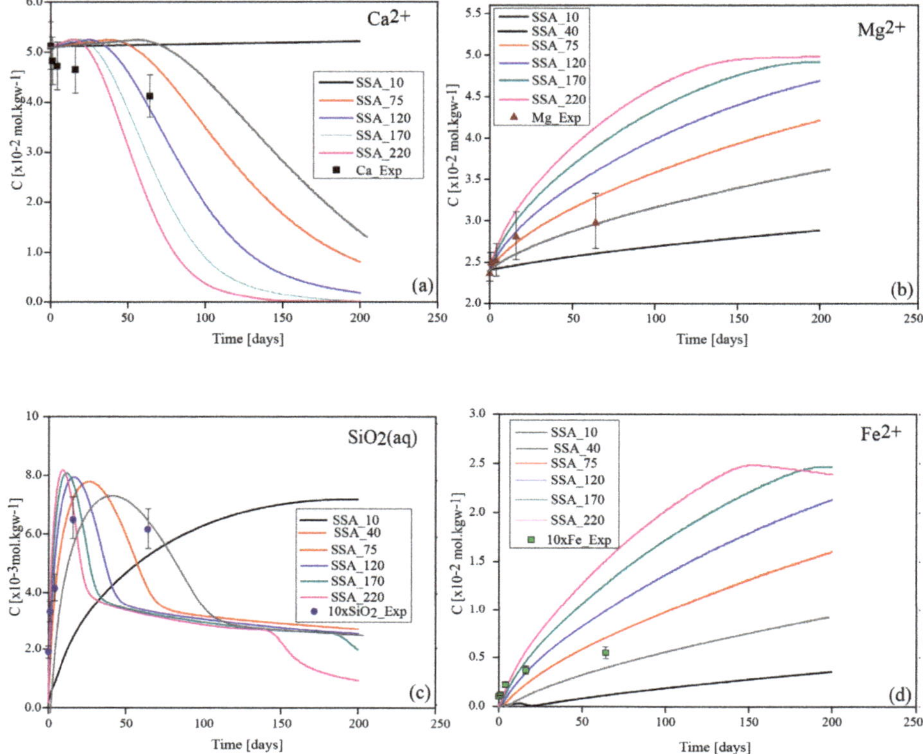

Figure 8. Variation in the solution composition with time compared to the experimental data. (**a**) Ca^{2+}; (**b**) Mg^{2+}; (**c**) $SiO_2(aq)$; (**d**) Fe^{2+}.

The simulated magnesium concentration doubled before levelling off (after 100 days). Moreover, the $SiO_2(aq)$ concentration stabilized shortly before a sharp and progressive decrease, indicating quantitatively important secondary silicate phases' precipitation. A steady increase in iron concentration was observed; the simulation indicates levelling off occurred later, after 150 days, and was attributed to fayalite dissolution. Aluminum values, initially 1.0×10^{-4} mol/kg, dropped sharply by five orders of magnitude, following the albite, Ca zeolite and kaolinite precipitation observed in the first simulation.

Because of the assumptions made in the initial mineralogy, the simulation could not reproduce the potassium and $SO4^-$ behaviors.

4.3.2. Minerals' Dissolution and Precipitation

Simulated primary and secondary minerals saturation indices (SI) are shown in Figure 9a,b. All primary minerals (except albite) are seen to dissolve (negative saturation index and rate, Table 3). SI of albite remains positive over time, implying that the modest decrease in Na^+ (from 0.371 to 3.69 mol.kgw^{-1}) is a consequence of albite volume increase. The dissolution rates (Table 8) indicate that olivine, plagioclase and diopsidic minerals are the most reactive phases.

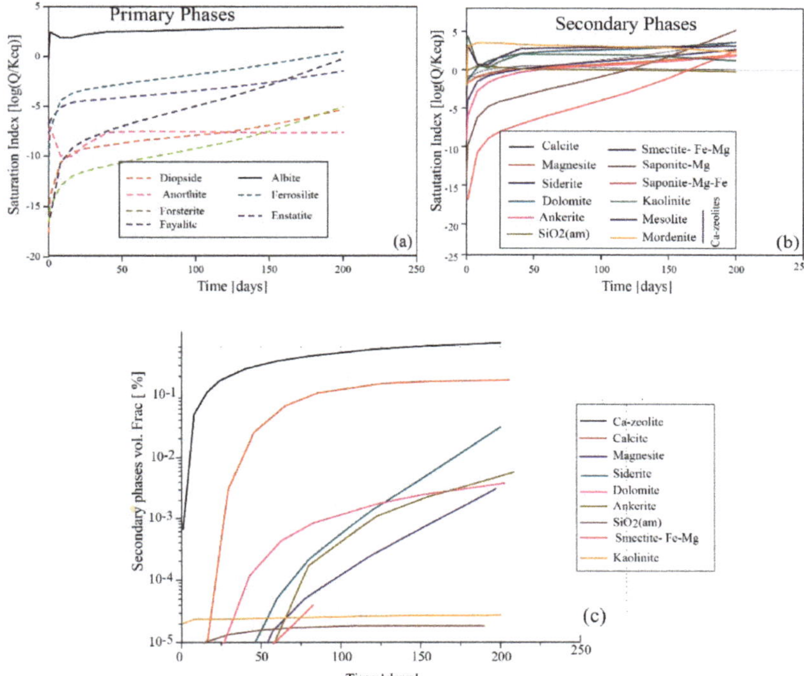

Figure 9. Evolution of the simulated saturation index: for the primary (**a**) and secondary minerals (**b**) and the variation of minerals' volume fraction (in m^3 mineral/m^3 rock) vs. time (**c**).

Table 8. Simulated dissolution and precipitation rate at $t = 200$ days (simulation with specific surface areas (SSA) = 120 $cm^2 \cdot g^{-1}$).

Primary Minerals	Reaction Rate ($mol \cdot L^{-1} \cdot s^{-1}$)	Secondary Minerals	Reaction Rate ($mol \cdot L^{-1} \cdot s^{-1}$)
Albite	2.77×10^{-11}	Siderite	4.13×10^{-9}
Anorthite	-1.08×10^{-9}	Calcite	5.01×10^{-10}
Fayalite	-1.24×10^{-10}	Magnesite	4.01×10^{-10}
Forsterite	-1.64×10^{-10}	Ca_zeolites	2.41×10^{-10}
Enstatite	-1.73×10^{-10}	Ankerite	8.58×10^{-11}
Diopside	-6.48×10^{-11}	Dolomite	4.84×10^{-11}
Ferrosilite	-2.57×10^{-14}	Smectite	1.31×10^{-12}
		Kaolinite	5.37×10^{-15}
		SiO$_2$	2.14×10^{-16}

Two mineral assemblages resulting from the CG alteration are predicted by the geochemical model. A Ca and Fe-Mg carbonate assemblage and another composed of Ca-Mg-Fe silicates (and aluminum) such as calcium-rich zeolites, clay minerals and amorphous silica. Ca-zeolites (mesotile and mordenite) and kaolinite are the first secondary minerals to form simultaneously with albite volume increase, followed by the amorphous silica (Figure 9c). The Si-Al-rich assemblage formation results in a sudden decrease in aluminum and silicon from the outlet solution. The carbonates' mineral phases correspond to calcite, dolomite, magnesite, siderite and ankerite. Calcite and dolomite appear in the simulations after 16 days, followed by siderite and magnesite (after 40 days). Ankerite was the last carbonate phase to precipitate, concomitant with Fe-Mg-smectite.

5. Discussion

5.1. Experimental Results

The petrographic study of the two samples, with SEM-EDS, did not reveal any carbonate phase precipitation. However, an increase in textural roughness in the samples was detected with the number of days of exposure into brine SC CO_2, as also mentioned in [61]. The diffractograms on global powdered fractions for the two experimental specimens are not considerably different between 0 and 64 days. Nevertheless, they reflect trace amounts of talc and vermiculite, as alteration products, after most experimental runs of mafic silicates. Also present is halite. The grazing incidence diffraction of CG and GD specimens, which helps to detect the mineralogy overlaid on the face specimens, additionally indicated the presence of gypsum and smectite (Table 9). Considering that talc and vermiculite were not detected by this technique within 64 days, those mineral phases should already be present before the experiments, resulting from the replacement of iron-magnesium igneous phases. Their trace amounts may justify the fact that they are not systematically detected in powder XRD. The absence of significant differences in the global mineralogical and chemical compositions of samples from the two experiments allows us to conclude that the reactions between minerals and fluids were not significant. On the other hand, the mineralogical and physical changes are limited to external areas of the CG and GD specimens exposed to CO_2-rich brine. The changes could be due to local precipitations/dissolutions at the specimen surface and could represent the early effects of influence of the CO_2-rich brine on the rock.

Table 9. Comparative summary of modelled and experimental mineral phases from cumulate gabbro (GXRD: grazing incidence geometry XRD; phases expressed as vol %).

	0 days	1 Day	4 Days	16 Days	64 Days
	Primary Phases			**Secondary Phases**	
Modelling	Clinopyroxene (52) Orthopyroxene (1.2) Olivine (10) Plagioclase (11)	Kaolinite (2.07×10^{-5}) Ca-zeolites (2.82×10^{-3})	SiO_2 am. (2.13×10^{-6}) Kaolinite (2.4×10^{-5}) Ca-zeolites (4.16×10^{-3})	Calcite (4.8×10^{-5}) Dolomite (5.28×10^{-6}) SiO_2 am (5.92×10^{-6}) Kaolinite (2.42×10^{-5}) Ca-zeolites (4.36×10^{-3})	Smectite (2.13×10^{-6}) Calcite (8.85×10^{-2}) Dolomite (8.32×10^{-4}) Magnesite (5.92×10^{-5}) Siderite (2.17×10^{-4}) Ankerite (3.95×10^{-4}) Ca-zeolites (5.54×10^{-2}) SiO_2 am. (1.31×10^{-5}) Kaolinite (2.75×10^{-5})
Experimental	Clinopyroxene (45–55) Olivine (15–20) Amphibole (10–15) Plagioclase (5–10) Ilmenite (5)	Gypsum (SEM-EDS) Halite (XRD + SEM-EDS)	Gypsum (SEM-EDS) Halite (SEM-EDS)	Gypsum (SEM-EDS) Halite (SEM-EDS)	Smectite (GXRD) Gypsum (GXRD + SEM-EDS) Halite (GXRD + SEM-EDS)

The measure of pH values in the brine is conditioned by the release of CO_2 during depressurization after chamber opening, and thus the acidity during the experiments should be even lower than that measured when the chamber is decompressed (Tables 6 and 7). However, the increase in pH correlates positively with the time of rock-brine interaction (Tables 6 and 7). The trends of the pH measurements correspond to those found in previous experimental works carried out in an autoclave [25,37–39].

The variability of the concentration of elements in the brine (Tables 6 and 7) reflects the reaction dynamics imposed by the composition of the high-salinity brine, the rock composition (Table 5) and the experimental conditions. In general, the concentrations of the experimental solutions, with higher Na and lower Ca and Mg concentrations for GD, agree with the less mafic composition of this lithotype when compared to the CG. The correlation of the measurements of Ca^{2+}, Mg^{2+} and total Fe in the brine and CaO, MgO and Fe_2O_3 in the rock is as follows:

- In the case of CG, the Ca^{2+} concentration measured in the brine shows a decrease after each run. This decrease in Ca^{2+} is not described elsewhere [14,25]. On the contrary, and in line with the behavior of Mg^{2+} and total Fe, what is generally described is an increase in its concentration. This was also observed with the GD experiments.

- The Ca^{2+} (aqueous) decrease can be explained by the high content of this ion in the initial brine. Under the experimental conditions, Ca^{2+} should be consumed when forming secondary minerals (Ca-Al-Si; calcium aluminum silicates), as predicted by the modelling. The hypothesis that precipitation of gypsum within the autoclave could account for the Ca^{2+} decrease should be discarded, since the sulphate ion concentration is not sufficiently high. Modelling results indicate that for gypsum to precipitate the sulphate concentration should be at least double.

- The 0.4 wt. % increase in CaO in the total rock composition after a 64-day run seems to reflect not only its incorporation in the structure of Al-silicates but also the crystallization of gypsum during the drying process, after concluding the run.

- Increases of iron and magnesium in solution are in accordance with the decrease, albeit reduced, of these elements (0.4 wt. % FeO and 0.1 wt. % MgO) in the whole rock composition after 64 days. These observations are also in agreement with the textural observations, where the Fe-Mg mineral phases are the first to react with acidified brine, releasing these components into the solution.

Regarding the experiment with GD, the variation of Ca^{2+} and Mg^{2+} concentrations in brine shows a very small increase, or a tendency to remain constant. The increase is in line with what is described in the literature [25]. Such an increase in solution is reflected in the decreasing concentrations of CaO (0.42 wt. %) and MgO (0.21 wt. %) in total rock for the 64-day run specimens. FeO shows a similar trend, decreasing by 0.3 wt. % in the same specimens.

5.2. Numerical Simulation and Correlation with Experimental Results

Observations on fluid chemistry and pH changes over time indicate that, under similar conditions of P_{CO_2} and temperature, the simulation with SSA = 120 cm^2/g shows a relative agreement with the experimental data (Figures 7 and 8). The considered SSA is in the range of the specific area of olivine, plagioclase and pyroxene used in [53]. The predicted mobility of calcium is higher than for other components, with a slight increase in concentration for the first 24 days, before being incorporated within the secondary phases. As mentioned above, this increase is not reflected in Experiment 1.

The model also shows that Ca^{2+} has been consumed in Ca-zeolites and/or carbonates; for instance, mesolite and mordenite are the first phases to consume calcium at a low pH < 5.5. Moreover, the gradual increase in Mg concentration is justified by the progressive dissolution of diopside and forsterite, accompanied by a moderate precipitation of dolomite and later of magnesite. The results of the newly formed mineralogy (Table 9) indicate the presence of early neoformation phases (e.g., smectite) that can capture calcium within its structure. Nevertheless, no carbonates were observed or detected during the experiment. In CG specimens, most likely, the secondary phases might be local and covered by the considerable amount of salts lining the rock surface.

The simulation also indicates that the Δ(Mg/Si) and Δ(Ca/Si) ratios are 1.45 and 0.35, respectively, and suggests that olivine, diopside and Ca-plagioclase (the main CG minerals) dissolve stoichiometrically [25]. Therefore, the high ratio between silicon in the simulation and the experiment (Si/Si* = 10) testify to the nonstoichiometric dissolution of CG in the experiment. Furthermore, (Δ(Mg*/Si*) = 4) is significantly higher and denotes that magnesium dissolves faster than silica (incongruent dissolution). At pH > 5.5, Ca-Mg carbonates (namely, calcite and dolomite) start precipitating and continue up to a neutral/alkaline pH, consistent with the observations from other authors [25,55]. Calcite remains the dominant carbonate phase at a high pH, indicating its stability under alkaline conditions. Iron displayed a similar trend to Mg^{2+}, showing a high release rate. Consequently, a tiny fraction of siderite and Fe-Mg-rich smectite was precipitated, concomitant with the dissolution of the primary Fe-rich minerals (fayalite) and followed by ankerite (Ca (Fe, Mg) $(CO_3)_2$). The calculated silica and iron concentrations are significantly higher compared to Experiment 1.

Besides the incongruent dissolution observed in the experiments, CG rocks are less reactive than basaltic rocks. As a matter of comparison, Gysi and Stefánsson [25] measured elevated P_{CO_2}, ion and Si values in glass basalts, in the same range of the simulations (2 mmol/kgw). Note that the brine used in Experiment 1 has a CO_2 concentration 1.5 times higher than that used previously by Gysi and Stefánsson [55], and thus, according to the those authors' results, further secondary phases are expected, since dissolved CO_2 is available during the time of the experiment. The limitation in secondary phases (in CG) is due to the rock reactivity. The experiment's duration is another factor influencing the secondary phases' precipitation.

The total volume of secondary minerals predicted is 0.8 vol %, dominated mostly by zeolites (0.61% in rock volume), carbonates, especially calcite (0.172% in rock volume), and insignificant amount of clays. The small amounts of secondary phases render their identification in SEM very difficult, as reported in previous studies [62]. After 50 days, the $\Delta(Ca) < 0$ corresponds to a stage where zeolites and carbonates competed with a dominant volume of Ca-zeolites. The Mg/Si and Fe/Si ratios increased over time, resulting from the continuous dissolution of clinopyroxene and olivine and precipitating trace amounts of Mg-Fe-rich carbonates, as well as clay minerals.

SEM and XRD observations showed a significant amount of salt crystals lining the samples' surfaces. However, small amounts of Na_2O were measured in the rock, contrasting with the noticeable quantity of crystals observed in SEM images. The Na and Cl concentrations also showed a cyclic decrease and increase in concentration, not consistent with an unlikely continuous evaporation of the brine during the experiments. We believe that halite and gypsum crystallized after opening the chamber and degassing, or during drying.

As discussed in Section 2, other authors with similar experimental approaches have observed mineral carbonation occurring in time frames comparable to those applied in our experiments. Still, the lack of noticeable precipitation of carbonate minerals during our experiments was not unexpected, since the experiment was designed to look at early-phase dissolution conditions. The proportion of SC CO_2/brine/rock, was defined to ensure that the brine would always be supersaturated CO_2 during the experiments, as is expected in early-phase injections when dissolution is the dominant process. Such proportions imply that, during the whole duration of the experiments, the pH remained very low in the autoclave, and carbonates could not precipitate. Exposure of the samples to the CO_2-supersaturated brine for longer periods would probably result in consumption of the CO_2, pH increase and precipitation of carbonate minerals. Given the time constraints of the project, for the second run of experiments, with the gabbro-diorite sample, the duration of the runs was kept constant, but the CO_2/brine/rock proportion was changed. The volume of CO_2 was decreased and the mass of rock increased, to see if mineral carbonation would be observed during at least the 64-day runs. The results of this second round of experiments have not yet been fully explored as geochemical modelling is still being done, but they will be the basis for defining the physical conditions and duration of the next set of experiments.

6. Conclusions

The results obtained reveal, in the first analysis, the complexity of the systems considered: not only the mineralogy observed in the CG and GD but also the composition of the highly saline brine used in the experiments. Thus, a full fit between modelling and experimental results was not always verified.

The preliminary results of the numerical modelling indicate that carbonate precipitation is possible in a batch reactor at a low temperature and pressure (40 °C and 80 kbar). However, the amount of carbonated minerals is very small. In all the tests conducted, Ca-Mg-Fe types of carbonates (calcite, dolomite, magnesite, siderite and ankerite) are predicted, mostly after 64 days (about 0.18% in volume). At this time, carbonates are coupled with the secondary silicates (0.62 vol %) namely zeolites, clays and amorphous silica. Zeolites and amorphous silica were the first predicted phases, forming just after two days.

After 64 days, the experimental data are in line with the modelling as far as the Ca-Al-Si mineral phases. The presence of smectite was verified by grazing XRD. However, calcite, or any other carbonate also predicted by the model in small quantities, was not recorded by any of the analysis techniques (SEM-EDS, XRD, FTIR). Both experiments detected a large amount of salt (gypsum and halite) lining on specimens' surfaces that was not reproduced by the model. Those should have precipitated after opening the chamber/during drying. Some vestigial secondary phases not detected by XRD can be hidden under the salts and not observed by SEM-EDS.

Experiment 2 with GD was not modelled, but the experimental results are in line with the observations from Experiment 1: textural features, given enhanced dissolution of ferromagnesian minerals and increased roughness with run time, and secondary mineralogy precipitation (smectite, halite and gypsum).

The measured complex variations in brine composition and total rock compositions reflect the dissolution under SC CO_2 conditions but also the contribution of the high-salinity brine.

The sensitive analysis conducted on the reactive surface area demonstrated that, for the cumulate gabbros, a specific surface area of 120 cm^2/g is required to fit the geochemical model instead of the 250 cm^2/g reported for basalt glass. Alternatively, the BET method can be applied. This preliminary model's results will provide interesting benchmarking, and other alternatives are being considered to adequately match the laboratory results.

The experimental and modelling results for 64 days essentially reflect a dissolution mechanism. For higher pH values, and for 64-day runs, the prevailing mechanism changes and significant carbonation occurs. Although in small amounts, the carbonate precipitation would be significant if upscaled to a reservoir scale. Subsequent experiments within the INCARBON project will attempt to replicate the conditions under which mineral precipitation will be the prevailing process, in order to establish a clear picture of the mineral carbonation potential in the target rocks.

Supplementary Materials: The following are available online at http://www.mdpi.com/2076-3417/10/15/5083/s1, Figure S1: Variations of brine SC-CO_2 composition, during experiment with cumulate-gabbro: (a) Ca^{2+}, (b) Mg^{2+} and K^+; (c) total Fe and SiO_2; (d) SO_4^{2-}; (e) Na^+; (f) Cl^-. Figure S2: Variations of brine SC-CO_2 composition, during experiment with gabbro-diorite: (a) Ca^{2+}, (b) Mg^{2+} and K^+; (c) SiO_2; (d) SO_4^{2-}; (e) Na^+; (f) Cl^-.

Author Contributions: E.B., P.M., J.P. and J.C. were responsible for the experimental design applied in this study; H.A. was responsible for the numerical modelling characterization; E.B. was responsible for the carbonation experiments. P.M., J.P. and M.B. obtained and interpreted the SEM-EDS data; J.M., M.B., P.M. and C.G. obtained and interpreted the XRD data; C.M. obtained and interpreted the ATR-FTIR data; P.B. obtained and interpreted brine data with ICP-MS. F.S. determined the roughness; E.B., H.B., P.M., J.P. and J.C. interpreted the data and planned this paper. All the authors discussed the results and edited the manuscript. All authors have read and agreed to the published version of the manuscript.

Funding: This work was supported by national funds through the Portuguese FCT—Fundação Para a Ciência e Tecnologia—in the scope of project INCARBON (contract PTDC/CTA-GEO/31853/2017).

Acknowledgments: APS—Ports of Sines and the Algarve Authority, S.A.—is gratefully acknowledged for cooperation on the collection of rock samples from the Monte Chãos quarry. The two anonymous reviewers of the submitted manuscript are acknowledgment for their very constructive comments and suggestions that greatly contributed to the improvement of the final version of this article.

References

1. IEA. *Energy Technology Perspectives 2017: Catalysing Energy Technology Transformations*; IEA/OECD, Ed.; IEA: Paris, France, 2017; p. 443. [CrossRef]

2. Bachu, S. Sequestration of CO_2 in geological media: Criteria and approach for site selection in response to climate change. *Energy Convers. Manag.* **2000**, *41*, 953–970. [CrossRef]

3. Benson, S.M.; Cole, D.R. CO_2 Sequestration in Deep Sedimentary Formations. *Elements* **2008**, *4*, 325–331. [CrossRef]

4. Sayegh, S.; Krause, F.; Girard, M.; DeBree, C. Rock/Fluid Interactions of Carbonated Brines in a Sandstone Reservoir: Pembina Cardium, Alberta, Canada. *SPE Form. Eval.* **1990**, *5*, 399–405. [CrossRef]

5. Izgec, O.; Demiral, B.; Bertin, H.; Akin, S. CO_2 injection into saline carbonate aquifer formations I: Laboratory investigation. *Transp. Porous Media* **2008**, *72*, 1–24. [CrossRef]

6. Gaus, I. Role and impact of CO_2-rock interactions during CO_2 storage in sedimentary rocks. *Int. J. Greenh. Gas Control* **2010**, *4*, 73–89. [CrossRef]

7. Saeedi, A.; Rezaee, R.; Evans, B.; Clennell, B. Multiphase flow behaviour during CO_2 geo-sequestration: Emphasis on the effect of cyclic CO_2–brine flooding. *J. Pet. Sci. Eng.* **2011**, *79*, 65–85. [CrossRef]

8. Klein, F.; McCollom, T. From serpentinization to carbonation: New insights from a CO_2 injection experiment. *Earth Planet. Sci. Lett.* **2013**, *379*, 137–145. [CrossRef]

9. Druckenmiller, M.L.; Maroto-Valer, M.M. Carbon sequestration using brine of adjusted pH to form mineral carbonates. *Fuel Process. Technol.* **2005**, *86*, 1599–1614. [CrossRef]

10. McGrail, B.P.; Spane, F.A.; Amonette, J.E.; Thompson, C.R.; Brown, C.F. Injection and Monitoring at the Wallula Basalt Pilot Project. *Energy Procedia* **2014**, *63*, 2939–2948. [CrossRef]

11. Pedro, J.; Araújo, A.A.; Moita, P.; Beltrame, M.; Lopes, L.; Chambel, A.; Berrezueta, E.; Carneiro, J. Mineral carbonation of CO_2 in Mafic Plutonic Rocks, I – Screening criteria and application to a case study in SW Portugal. *Appl. Sci.* **2020**, *10*, 4879. [CrossRef]

12. Rosenbauer, R.; Koksalan, T.; Palandri, J. Experimental investigation of CO_2-brine-rock interactions at elevated temperature and pressure: Implications for CO_2 sequestration in deep-saline aquifers. *Fuel Process. Technol.* **2005**, *86*, 1581–1597. [CrossRef]

13. Liu, Q.; Maroto-Valer, M. Investigation of the pH effect of a typical host rock and buffer solution on CO2 sequestration in synthetic brines. *Fuel Process. Technol.* **2010**, *91*, 1321–1329. [CrossRef]

14. Andreani, M.; Luquot, L.; Gouze, P.; Godard, M.; Hoisé, E.; Gibert, B. Experimental Study of Carbon Sequestration Reactions Controlled by the Percolation of CO_2-Rich Brine through Peridotites. *Environ. Sci. Technol.* **2009**, *43*, 1226–1231. [CrossRef]

15. Romão, I.S.; Gando-Ferreira, L.M.; da Silva, M.M.V.G.; Zevenhoven, R. CO_2 sequestration with serpentinite and metaperidotite from Northeast Portugal. *Miner. Eng.* **2016**, *94*, 104–114. [CrossRef]

16. Duan, Z.; Sun, R. An improved model calculating CO_2 solubility in pure water and aqueous NaCl solutions from 273 to 533 K and from 0 to 2000 bar. *Chem. Geol.* **2003**, *193*, 257–271. [CrossRef]

17. Chizmeshya, A.V.G.; McKelvy, M.J.; Squires, K.; Carpenter, R.W.; Bearat, H. *A Novel Approach to Mineral Carbonation: Enhancing Carbonation While Avoiding Mineral Pretreatment Process Cost*; Arizona State Univ: Tempe, AZ, USA, 2007. [CrossRef]

18. Daval, D.; Martinez, I.; Corvisier, J.; Findling, N.; Goffé, B.; Guyot, F. Carbonation of Ca-bearing silicates, the case of wollastonite: Experimental investigations and kinetic modeling. *Chem. Geol.* **2009**, *265*, 63–78. [CrossRef]

19. Daval, D.; Sissmann, O.; Menguy, N.; Saldi, G.D.; Guyot, F.; Martinez, I.; Corvisier, J.; Garcia, B.; Machouk, I.; Knauss, K.G.; et al. Influence of amorphous silica layer formation on the dissolution rate of olivine at 90 °C and elevated pCO_2. *Chem. Geol.* **2011**, *284*, 193–209. [CrossRef]

20. Sissmann, O.; Daval, D.; Brunet, F.; Guyot, F.; Verlaguet, A.; Pinquier, Y.; Findling, N.; Martinez, I. The deleterious effect of secondary phases on olivine carbonation yield: Insight from time-resolved aqueous-fluid sampling and FIB-TEM characterization. *Chem. Geol.* **2013**, *357*, 186–202. [CrossRef]

21. O.´Connor, W.; Dahlin, D.; Nilsen, D.; Rush, G.E.; Walters, R.; Turner, P. Carbon Dioxide Sequestration by Direct Mineral Carbonation: Results from Recent Studies and Current Status. In Proceedings of the First National Conference on Carbon Sequestration, Washington, DC, USA, 14–17 May 2001.

22. Béarat, H.; McKelvy, M.J.; Chizmeshya, A.V.G.; Gormley, D.; Nunez, R.; Carpenter, R.W.; Squires, K.; Wolf, G.H. Carbon Sequestration via Aqueous Olivine Mineral Carbonation: Role of Passivating Layer Formation. *Environ. Sci. Technol.* **2006**, *40*, 4802–4808. [CrossRef]

23. Gerdemann, S.J.; O'Connor, W.K.; Dahlin, D.C.; Penner, L.R.; Rush, H. Ex Situ Aqueous Mineral Carbonation. *Environ. Sci. Technol.* **2007**, *41*, 2587–2593. [CrossRef]

24. Soong, Y.; Fauth, D.L.; Howard, B.H.; Jones, J.R.; Harrison, D.K.; Goodman, A.L.; Gray, M.L.; Frommell, E.A. CO_2 sequestration with brine solution and fly ashes. *Energy Convers. Manag.* **2006**, *47*, 1676–1685. [CrossRef]

25. Gysi, A.P.; Stefánsson, A. CO_2-water-basalt interaction. Low temperature experiments and implications for CO_2 sequestration into basalts. *Geochim. Cosmochim. Acta* **2012**, *81*, 129–152. [CrossRef]

26. Alfredsson, H.A.; Oelkers, E.H.; Hardarsson, B.S.; Franzson, H.; Gunnlaugsson, E.; Gislason, S.R. The geology and water chemistry of the Hellisheidi, SW-Iceland carbon storage site. *Int. J. Greenh. Gas Control* **2013**, *12*, 399–418. [CrossRef]

27. Matter, J.M.; Stute, M.; Snaebjornsdottir, S.O.; Oelkers, E.H.; Gislason, S.R.; Aradottir, E.S.; Sigfusson, B.; Gunnarsson, I.; Sigurdardottir, H.; Gunnlaugsson, E.; et al. Rapid carbon mineralization for permanent disposal of anthropogenic carbon dioxide emissions. *Science* **2016**, *352*, 1312–1314. [CrossRef]

28. Kaszuba, J.; Janecky, D.; Snow, M. Carbon dioxide reaction processes in a model brine aquifer at 200°C and 200 bars: Implications for geologic sequestration of carbon. *Appl. Geochem.* **2003**, *18*, 1065–1080. [CrossRef]

29. Peuble, S.; Godard, M.; Luquot, L.; Andreani, M.; Martinez, I.; Gouze, P. CO_2 geological storage in olivine rich basaltic aquifers: New insights from reactive-percolation experiments. *Appl. Geochem.* **2015**, *52*, 174–190. [CrossRef]

30. Peuble, S.; Godard, M.; Gouze, P.; Leprovost, R.; Martinez, I.; Shilobreeva, S. Control of CO_2 on flow and reaction paths in olivine-dominated basements: An experimental study. *Geochim. Cosmochim. Acta* **2019**, *252*, 16–38. [CrossRef]

31. McGrail, B.P.; Schaef, H.T.; Ho, A.M.; Chien, Y.J.; Dooley, J.J.; Davidson, C.L. Potential for carbon dioxide sequestration in flood basalts. *J. Geophys. Res. Solid Earth* **2006**, *111*, B12. [CrossRef]

32. Schaef, H.T.; McGrail, B.P. Dissolution of Columbia River Basalt under mildly acidic conditions as a function of temperature: Experimental results relevant to the geological sequestration of carbon dioxide. *Appl. Geochem.* **2009**, *24*, 980–987. [CrossRef]

33. Kanakiya, S.; Adam, L.; Esteban, L.; Rowe, M.C.; Shane, P. Dissolution and secondary mineral precipitation in basalts due to reactions with carbonic acid. *J. Geophys. Res. Solid Earth* **2017**, *122*, 4312–4327. [CrossRef]

34. André, L.; Audigane, P.; Azaroual, M.; Menjoz, A. Numerical modeling of fluid–rock chemical interactions at the supercritical CO_2–liquid interface during CO_2 injection into a carbonate reservoir, the Dogger aquifer (Paris Basin, France). *Energy Convers. Manag.* **2007**, *48*, 1782–1797. [CrossRef]

35. Gaus, I.; Audigane, P.; Andre, L.; Lions, J.; Jacquemet, N.; Dutst, P.; Czernichowski-Lauriol, I.; Azaroual, M. Geochemical and solute transport modelling for CO_2 storage, what to expect from it? *Int. J. Greenh. Gas Control* **2008**, *2*, 605–625. [CrossRef]

36. Berrezueta, E.; González-Menéndez, L.; Breitner, D.; Luquot, L. Pore system changes during experimental CO_2 injection into detritic rocks: Studies of potential storage rocks from some sedimentary basins of Spain. *Int. J. Greenh. Gas Control* **2013**, *17*, 411–422. [CrossRef]

37. Luquot, L.; Gouze, P. Experimental determination of porosity and permeability changes induced by injection of CO_2 into carbonate rocks. *Chem. Geol.* **2009**, *265*, 148–159. [CrossRef]

38. Tarkowski, R.; Wdowin, M.; Manecki, M. Petrophysical examination of CO_2-brine-rock interactions—results of the first stage of long-term experiments in the potential Zaosie Anticline reservoir (central Poland) for CO_2 storage. *Environ. Monit. Assess.* **2014**, *187*, 4215. [CrossRef]

39. Berrezueta, E.; Ordoñez Casado, B.; Quintana, L. Qualitative and quantitative changes in detrital reservoir rocks caused by CO_2 – brine – rock interactions during first injection phases (Utrillas sandstones, northern Spain). *Solid Earth* **2016**, *7*, 37–53. [CrossRef]

40. Lake, L.W. *Enhanced Oil Recovery*; Prentice Hall: Englewood Cliffs, NJ, USA, 1989.

41. Span, R.; Wagner, W. A New Equation of State for Carbon Dioxide Covering the Fluid Region from the Triple-Point Temperature to 1100 K at Pressures up to 800 MPa. *J. Phys. Chem. Ref. Data* **1996**, *25*, 1509–1596. [CrossRef]

42. Holloway, S. An overview of the underground disposal of carbon dioxide. *Energy Convers. Manag.* **1997**, *38*, S193–S198. [CrossRef]

43. Georgiou, C.D.; Sun, H.J.; McKay, C.P.; Grintzalis, K.; Papapostolou, I.; Zisimopoulos, D.; Panagiotidis, K.; Zhang, G.; Koutsopoulou, E.; Christidis, G.E.; et al. Evidence for photochemical production of reactive oxygen species in desert soils. *Nat. Commun.* **2015**, *6*, 7100. [CrossRef]

44. Beltrame, M.; Liberato, M.; Mirão, J.; Santos, H.; Barrulas, P.; Branco, F.; Gonçalves, L.; Candeias, A.; Schiavon, N. Islamic and post Islamic ceramics from the town of Santarém (Portugal): The continuity of ceramic technology in a transforming society. *J. Archaeol. Sci. Rep.* **2019**, *23*, 910–928. [CrossRef]

45. Steefel, C.; Lasaga, A. A Coupled Model for Transport of Multiple Chemical-Species and Kinetic Precipitation Dissolution Reactions with Application to Reactive Flow in Single-Phase Hydrothermal Systems. *Am. J. Sci.* **1994**, *294*, 529–592. [CrossRef]

46. Steefel, C.I. *CrunchFlow Software for Modeling Multicomponent Reactive Flow and Transport*; User's Manual; Earth Sciences Division, Lawrence Berkeley, National Laboratory: Berkeley, CA, USA, 2009; pp. 12–91.

47. Steefel, C.; Appelo, T.; Arora, B.; Jacques, D.; Kalbacher, T.; Kolditz, O.; Lagneau, V.; Lichtner, P.; Mayer, K.; Meeussen, J.; et al. Reactive transport codes for subsurface environmental simulation. *Comput. Geosci* **2014**, 19. [CrossRef]

48. Lasaga, A.C. Rate laws of chemical reactions. In *kinetics of geochemical processes*; Lasaga, A., Kirkpatrick, R., Eds.; Mineralogical Society of America: Washington DC, USA, 1981; Volume 8, pp. 1–68.

49. Lasaga, A. Chemical Kinetics of Water-Rock Interaction. *J. Geophys. Res.* **1984**, *89*, 4009–4025. [CrossRef]

50. Aagaard, P.; Helgeson, H.C. Thermodynamic and kinetic constraints on reaction rates among minerals and aqueous solutions; I, Theoretical considerations. *Am. J. Sci.* **1982**, *282*, 237–285. [CrossRef]

51. Canilho, M.H. *Elementos de geoquímica do maciço ígneo de Sines*; Ciências da Terra UNL: Caparica, Portugal, 1989; pp. 65–80.

52. Palandri, J.; Kharaka, Y. *A Compilation of Rate Parameters of Water-Mineral. Interaction Kinetics for Application to Geochemical Modeling*; Geological Survey: Reston, VA, USA, 2004; p. 71.

53. Xu, T.; Apps, J.A.; Pruess, K. Numerical simulation of CO_2 disposal by mineral trapping in deep aquifers. *Appl. Geochem.* **2004**, *19*, 917–936. [CrossRef]

54. Geloni, C.; Giorgis, T.; Battistelli, A. Modeling of Rocks and Cement Alteration due to CO_2 Injection in an Exploited Gas Reservoir. *Transp. Porous Media* **2011**, *90*, 183–200. [CrossRef]

55. Gysi, A.P.; Stefánsson, A. CO_2–water–basalt interaction. Numerical simulation of low temperature CO_2 sequestration into basalts. *Geochim. Cosmochim. Acta* **2011**, *75*, 4728–4751. [CrossRef]

56. Geloni, C.; Giorgis, T.; Biagi, S.; Previde Massara, E.; Battistelli, A. Modeling of Well Bore Cement Alteration as a Consequence of CO_2 Injection in an Exploited Gas Reservoir. In Proceedings of the 5th Meeting of the Wellbore Integrity Network, Calgary, AB, Canada, 12–13 May 2009.

57. Koenen, M.; Wasch, L. Reactive Surface Area in Geochemical Models-Lessons Learned from a Natural Analogue. In Proceedings of the Second EAGE Sustainable Earth Sciences (SES) Conference and Exhibition, Pau, France, 30 September–4 October 2013.

58. Wolery, T.J.; Jackson, K.J.; Bourcier, W.L.; Bruton, C.J.; Viani, B.E.; Knauss, K.G.; Delany, J.M. Current Status of the EQ3/6 Software Package for Geochemical Modeling. In *Chemical Modeling of Aqueous Systems II*; American Chemical Society: Washington, DC, USA, 1990; Volume 416, pp. 104–116.

59. Helgeson, H.C.; Garrels, R.M.; MacKenzie, F.T. Evaluation of irreversible reactions in geochemical processes involving minerals and aqueous solutions—II. Applications. *Geochim. Cosmochim. Acta* **1969**, *33*, 455–481. [CrossRef]

60. Moita, P.; Berrezueta, E.; Pedro, J.; Miguel, C.; Beltrame, M.; Galacho, C.; Mirão, J.; Araújo, A.; Lopes, L.; Carneiro, J. Experiments on mineral carbonation of CO_2 in gabbro's from the Sines massif-first results from project InCarbon. *Comunicações Geológicas* **2020**, *107*, 91–96.

61. Wells, R.K.; Xiong, W.; Giammar, D.; Skemer, P. Dissolution and surface roughening of Columbia River flood basalt at geologic carbon sequestration conditions. *Chem. Geol.* **2017**, *467*, 100–109. [CrossRef]

62. Dávila, G.; Luquot, L.; Soler, J.M.; Cama, J. 2D reactive transport modeling of the interaction between a marl and a CO_2-rich sulfate solution under supercritical CO_2 conditions. *Int. J. Greenh. Gas Control* **2016**, *54*, 145–159. [CrossRef]

 *applied sciences*

Article

Evolutions of CO_2 Adsorption and Nanopore Development Characteristics during Coal Structure Deformation

Linlin Wang [1,*] and Zhengjiang Long [1,2]

[1] Jiangsu Key Laboratory of Coal-Based Greenhouse Gas Control and Utilization,
   Low Carbon Energy Institute, China University of Mining and Technology, Xuzhou 221008, China;
   TS17010048A3@cumt.edu.cn
[2] Ministry of Education Key Laboratory of Coalbed Methane Resource & Reservoir Formation Process,
   School of Resource and Geoscience, China University of Mining and Technology, Xuzhou 221008, China
* Correspondence: linwang@cumt.edu.cn

Received: 30 June 2020; Accepted: 17 July 2020; Published: 21 July 2020

Abstract: The coal structure deformation attributed to actions of tectonic stresses can change characteristics of nanopore structure of coals, affecting their CO_2 adsorption. Three tectonically deformed coals and one undeformed coal were chosen as the research objects. The isotherm adsorption experiments of four coal specimens were carried out at the temperature of 35 °C and the pressure of 0 to 7 MPa. Nanopore structures were characterized using the liquid nitrogen adsorption method. The results show that there exist maximum values of excess and absolute adsorption capacity, which increase with increasing coal deformation degree. As the degree of coal deformation increases, the pore volume and specific surface area present an obvious increasing trend in the case of micropores, exhibiting an increase at first (cataclastic coal and ganulitic coal) and then stabilization (crumple coal), in the case of mesopores, and showing a gradual decrease in the case of macropores. The mesopores are the key factor of CO_2 adsorption of tectonically deformed coals, followed by the micropores and the limited effect of macropores at the strong coal deformation stage.

Keywords: CO_2 adsorption; nanopore; coal structure deformation; tectonically deformed coal

---

1. Introduction

In China, the exploration, development, and use of coalbed methane has been paid more and more attention to in recent years. Although coalbed methane resources are very rich in China, the output of coalbed methane shows a steady downward trend [1]. A great amount of coalbed methane is stored in the coal seam damaged by action of tectonic stress. The effect of tectonic stress causes the damage and transformation of coal body structure, and thereby induces the tectonically deformed coals [2–4]. However, "two high and two low" characteristics of tectonically deformed coals, i.e., high gas content, high gas desorption rate, low permeability, and low coal strength, cause the dangerous areas of gas outburst in the tectonically deformed coal development areas, and thereby restrict the efficient exploitation of coalbed methane [5,6]. Since the coal has larger adsorption capacity of CO_2 compared to CH_4, the replacement effect by injecting CO_2 into the coal seam can enhance the recovery rate of coalbed methane (CO_2-ECBM) and reduce greenhouse gas emissions effectively [7]. Therefore, CO_2 adsorption characteristics of tectonically deformed coal is one of key problems during the coalbed methane exploitation by CO_2 injection.

The gas in coal mainly at adsorption state occurs on the inner surface of coal matrix particles. Therefore, the pore structure is an important factor affecting the gas adsorption characteristics of coal. Many previous studies have indicated the relationship between pore structure and gas adsorption of coal [8,9]. Meanwhile, they have proposed that the coal adsorption is affected by coal pore volume, pore area, pore size, pore shape, and pore complexity, and revealed that developmental characteristics of micropores or nanopores are main factors affecting coal adsorption capacity to a large extent. For example, Castello et al. [10] found that the adsorption capacity of coal was proportional to the micropore volume. Chen et al. [11] believed that the specific surface area of nanopores determined the adsorption capacity of coal. Jian et al. [12] studied the relationships between the pore structure and adsorption characteristics of low rank coal, and found that developmental characteristics of microporous pore volume determined change laws of gas adsorption characteristics.

At present, there have been some studies on tectonically deformed coal adsorption, and some basic laws have been instituted. It has been found that the adsorption capacity of tectonically deformed coal is significantly greater than that of primary coal at the same coal rank, which is attributed to significant changes in pore volume and the specific surface area of micropores, as well as the developmental degree and connectivity of pore fractures of coal suffering from tectonic stress [13,14]. However, there are still many problems to be investigated related to the CO_2 adsorption characteristics of tectonically deformed coal. For example, what are basic laws of CO_2 adsorption characteristics of tectonically deformed coal? How do vary the CO_2 adsorption capacity with increasing coal deformation degree? What are the evolutionary characteristics of nanopores of coal with coal deformation degrees, and their effects on the CO_2 adsorption?

Therefore, tectonically deformed coals with different deformation degrees and deformation types are chosen as the research object. Isothermal adsorption experiments and pore structure tests and analysis were carried out to reveal evolution laws of CO_2 adsorption and nanopores with coal deformation degree. Finally, the control mechanisms of nanopore structure evolution on CO_2 adsorption of tectonically deformed coals were investigated.

2. Specimens and Experimental Methods

2.1. Coal Specimens and Their Characteristics

2.1.1. Coal Specimen Collection

In the present study, the sampling site is located at the Zhuxianzhuang coal mine in the Suxian mining area of Huaibei coalfield (Figure 1). The main coal bearing strata are the Permian Shihezi and Shanxi formations. The Shanxi formation, with a thickness of 96 m to 143 m and 120 m on average, is widely distributed throughout the whole area. As one of the main coal bearing sections in the area, the Shanxi formation presents well-developed number 10 and 11 coal seams. The stable number 10 coal seam has a cumulative thickness varying from 2.0 m to 3.5 m, and is the main minable coal seam in the Suxian mining area. The regional control structure of the Suxian mining area is composed of the Sudong syncline in the overlying system, and the Sunan and Nanping synclines in the underlying system of the Xishipo thrust fault. The Zhuxianzhuang coal mine is located in the overlying system of the Xisipo thrust fault, with very large compression and shear stress and intense tectonic deformation. Therefore, complete types of tectonically deformed coals have developed there.

The underground sampling is located near the newly exposed structure of the number 10 coal seam. Tectonically deformed coal samples of different deformation degrees are collected from different distances to the fault plane (Figure 1). The methods of collection and preservation of coal specimens are based on the "Sampling of coal seams" (GB/T 482-2008) and "Sampling of coal petrology" (GB/T 19222-2003). During sampling, the original structure of the coal body should be maintained as fully as possible. Three tectonically deformed coal specimens and one undeformed coal specimen are chosen as the research objects.

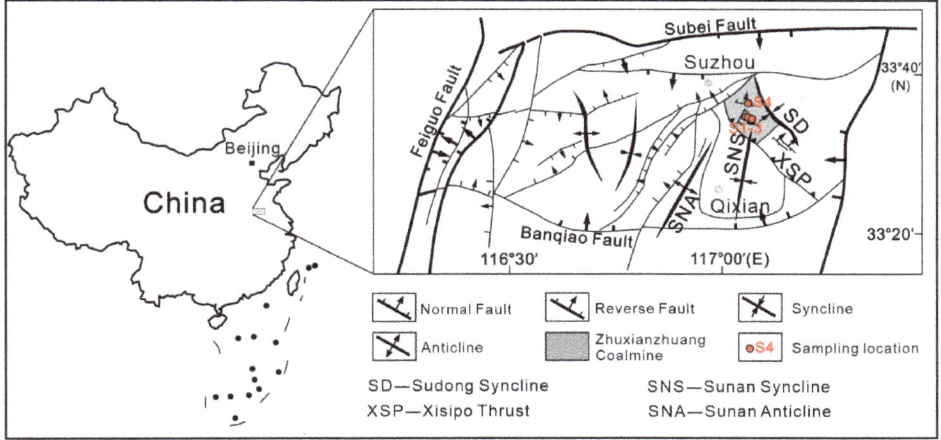

Figure 1. Regional structure of the study area and sampling locations.

2.1.2. Coal Specimen Properties

The basic properties of the coal specimens are shown in Table 1. Vitrinite reflectance and macerals were carried on a microspectrophotometer (AXIO Imager M1m, Carl Zeiss, Germany). According to proximate analysis of coal (GB/T 212-2008), the proximate analyses of coal specimens were carried out by employing the Self Employed Industrial Instrument and Analyzer (SDTGA8000a, Hunan Sundy Technology Co., Ltd., China). The ultimate analyses were carried out by an element analyzer (PerkinElmer 2400 II). Table 1 shows that the maximum vitrinite reflectance varies from 0.9% to 1.17%, and the volatile content (wt.%) changes from 30.27% to 34.06%. Therefore, the collected specimens are highly volatile bituminous coal A.

Table 1. Properties of coal specimens.

Coal Specimen	Type	$R_{o,max}$/%	Proximate Analysis (wt.%)			Maceral Analysis (vol.%)			Ultimate Analysis (wt.%) [d]				
			Mad	Ad	VMdaf	V	E	I	C	H	O	N	S
S1	Undeformed coal	0.90	2.21	11.47	34.06	83.21	2.37	8.82	77.25	4.28	16.88	1.31	0.28
S2	Cataclastic coal	0.96	1.82	8.92	33.14	82.33	2.56	7.62	77.61	4.32	16.42	1.26	0.39
S3	Granulitic coal	1.17	2.04	10.16	31.68	82.75	2.36	7.14	76.39	4.23	17.94	1.24	0.20
S4	Crumple coal	1.05	2.36	9.54	30.27	85.94	4.10	5.19	75.62	4.30	18.64	1.29	0.15

Note: Mad, moisture as air-dry basis; Ad, ash as dry basis; VMdaf, volatile matter as dry, ash-free basis; V, vitrine; E, exinite; I, inertinite; [d], as dry basis.

Macro- and microstructure deformation characteristics of four coal specimens were observed, including coal body structure, macro- and microfractures and folds, and etc. Then the three types of technically deformed coal specimens were recognized, i.e., undeformed coal, cataclastic coal, and crumpl coal; meanwhile, their deformation degree and deformation series were determined (Table 2 and Figure 2).

Table 2. Macro and micro deformation characteristics of coal specimens.

Specimen No.	Tectonic Coal Type	Deformation Degree	Deformation Series	Macro Characteristics	Micro Characteristics
S1	Undeformed coal	No or weak deformation		Complete coal body structure, well preserved primary banded structure, and easily distinguished petrographic constituent (Figure 2a)	Rare structural fractures (Figure 2b)
S2	Cataclastic coal	Weak deformation	Brittle deformation	Damaged structure, while visible primary banded structure, and developed fractures (Figure 2c)	A set of structural fractures developed (Figure 2d)
S3	Granulatic coal	Medium deformation		Completely destroyed primary structure; broken coal body with the particle size ranging from millimeter to centimeter scale (Figure 2e)	Extremely developed micro fractures, difficult tracing of fractures with undirectionality (Figure 2f)
S4	Crumple coal	Strong deformation	Ductile deformation	Disappeared primary structure of the coal body, presence of irregular arc or twisting shapes, similar to fold shape, and abnormally developed fractures (Figure 2g)	Microfold shape and abnormally developed microfracture (Figure 2h)

Figure 2. Macro and micro photos of coal specimens. (**a**) Undeformed coal (macroscope); (**b**) undeformed coal (microscope); (**c**) cataclastic coal (macroscope); (**d**) cataclastic coal (microscope); (**e**) granulitic coal (macroscope); (**f**) granulitic coal (microscope); (**g**) crumple coal (macroscope); (**h**) crumple coal (microscope).

2.2. Experimental

2.2.1. Adsorption Experiments

The coal specimens were ground, and the particle size was in the range of 180–250 μm (60–80 mesh). About 100 g for each coal specimen were weighed to put into a thermostat with supersaturated K_2SO_4 solution for 48 h, in order to make coal specimens containing equilibrium moisture. Then, the isotherm adsorption experiments on the coal specimens were carried out by an ISO-300 isotherm adsorption instrument with the temperature of 35 °C and pressure varying from 0 to 7 MPa. The experimental principle is shown in Figure 3.

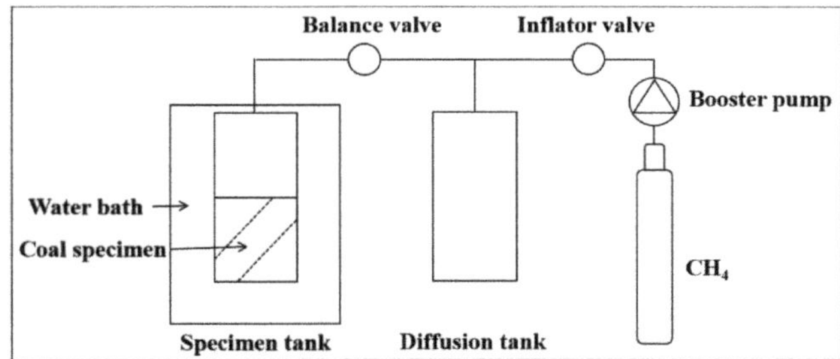

Figure 3. Schematic diagram of isothermal adsorption experiment apparatus.

The detailed experiment procedure is as follows:

(1) Put the coal specimens in the specimen tank and check the air tightness.
(2) The free space volume of specimen tank and reference tank was measured.
(3) Open the water bath heating system and set the experimental temperature to 35 °C.
(4) Open the booster pump and inflation valve and close the balance valve to charge CH_4 into the reference tank, and after the pressure is stable, open the balance valve and close the inflation valve. Then, the isothermal adsorption experiment begins.
(5) After adsorption equilibrium, record the experimental equilibrium pressure and adsorption data, and repeat step (4) to obtain the adsorption data of other pressure points. Seven pressure points were set in this experiment, and the pressure balance time was 72 h.
(6) After the completion of the adsorption experiment, open the balance valve and vent valve, and adjust the exhaust speed to make the maximum experimental pressure drop to standard atmospheric pressure within 2 h.
(7) After the experiment, close the booster pump, cylinder valve, and water bath heating system, and take out the coal specimens.

According to the experimental method of high-pressure isothermal adsorption to coal (capacity method) (GB/T 19560-2008), and based on the dynamic change of gas pressure before and after adsorption, the gas adsorption increment is calculated using the material balance equation, and thereby the gas adsorption amount in the coal under different pressures can be obtained. The detailed calculation process is as follows.

Attributed to the conservation of mass during gas adsorption, the difference between the sum of the gas amount in the reference tank after the *i*th inflation, the gas amount in the specimen tank, and the total gas amount in both tanks after adsorption balance is regarded as the adsorption increment of the adsorption balance pressure point. The *i*th adsorption increment is given by

$$\Delta\eta_i^{ex} = \frac{1}{RT}\left(\frac{P_{i-1}V_s}{Z_{i-1}} + \frac{P_{i1}V_r}{Z_{i1}} - \frac{P_{i2}V_s}{Z_{i2}} - \frac{P_{i3}V_r}{Z_{i3}}\right) \tag{1}$$

Since the gas pressure in the reference tank is equal to that in the specimen tank at the adsorption equilibrium state, i.e., $P_{i2} = P_{i3}$, the equation is written as

$$\Delta\eta_i^{ex} = \frac{1}{RT}\left[V_r\left(\frac{P_{i1}}{Z_{i1}} - \frac{P_{i2}}{Z_{i2}}\right) - V_s\left(\frac{P_{i2}}{Z_{i2}} - \frac{P_{i-1}}{Z_{i-1}}\right)\right] \tag{2}$$

The adsorption amount after the *i*th adsorption equilibrium is described as

$$\eta_i^{ex} = \sum_{x=1}^{i} \Delta\eta_x^{ex} \tag{3}$$

Therefore, the adsorption capacity can be calculated by

$$V_i = 22.4 \times 10^3 \frac{\eta_i^{ex}}{m} \tag{4}$$

where $\Delta\eta_i^{ex}$ is the adsorption increment after the ith adsorption equilibrium (mol); V_i is the adsorption capacity after the ith adsorption equilibrium (cm^3·g^{-1}); m is the mass of coal specimen (g); T is the experimental temperature (K); R is universal constant ($R = 8.314$); V_s is the free space volume of the experimental tank (cm^3); V_r is the free space volume of the reference tank (cm^3); P_{i-1} is the gas pressure in the experimental tank (MPa); P_{i1} is the gas pressure in the reference tank before the ith adsorption (MPa); P_{i2} is the gas pressure in the experimental tank after the ith adsorption equilibrium (MPa); P_{i3} is the gas pressure in the reference tank after the ith adsorption equilibrium (MPa); and Z_{i-1}, Z_{i1}, Z_{i2}, and Z_{i3} are compression factors corresponding to the pressures of P_{i-1}, P_{i1}, P_{i2}, and P_{i3}, respectively.

2.2.2. Liquid Nitrogen Adsorption

The liquid nitrogen adsorption method reflects the pore volume, pore specific surface area, and pore size distribution of a solid through the adsorption law of nitrogen on the solid surface. The specimens with particle sizes in the range of 60–80 mesh were dried for 4 h at 80 °C. Liquid nitrogen absorption tests were carried out by the surface area and porosimetry system (ASAP-2020, Micromeritics Instrument Corp., Norcross, GA, USA) at 77.3 K. We can obtain the isotherm adsorption curve of adsorption amount versus relative pressure. The specific surface area and volume of nanopores (from 2 to 200 nm) of coal specimens were analyzed by the Barrett-Joyner-Halenda (BJH) method.

3. Results and Discussion

3.1. Adsorption Characteristics' Evolution during Coal Structure Deformation

3.1.1. Excess Adsorption Curve

The experimental measurement of the adsorption capacity is known as Gibbs excess adsorption capacity. Figure 4 shows that isotherms of four coal specimens are not monotonous at the pressure of 0–7 MPa. The excess adsorption capacity of CO_2 presents an increase at first, and then a decrease with increasing pressure, and shows the single peak curve, which coincides with type I of Gibbs adsorption isotherms proposed by Donohue and Aranovich [15]. An increase of coal deformation degree causes the overall increase in the excess adsorption capacity. According to granulitic coal and crumple coal with strongly structural deformation, the adsorption isotherms present crossing points; in the low pressure zone of 0–3.5 MPa, the adsorption capacity of granulitic coal is greater than that of crumple coal; as the pressure exceeds 3.5 MPa, the adsorption capacity of crumple coal is greater than that of granulitic coal. The maximum values of excess adsorption capacity of undeformed coal, cataclastic coal, granulitic coal, and crumple coal were 9.48 m^3·g^{-1}, 12.04 m^3·g^{-1}, 15.83 m^3·g^{-1}, and 16.74 m^3·g^{-1}, respectively. The maximum excess adsorption capacity increased with increasing degrees of coal deformation.

The maximum excess adsorption capacity is present at pressure ranging from 4 MPa to 5 MPa, and with increasing coal deformation degree the pressure corresponding to maximum excess adsorption amount shifts to the high pressure. As shown in Figure 4, the maximum adsorption amount is located at the pressure of 4 MPa in cases of undeformed coal and cataclastic coal, and at the pressure larger than 5 MPa in cases of strongly deformed granulitic coal and crumple coal. More researches will be carried out in the future to validate the results attributed to the limited number of specimens in the present study.

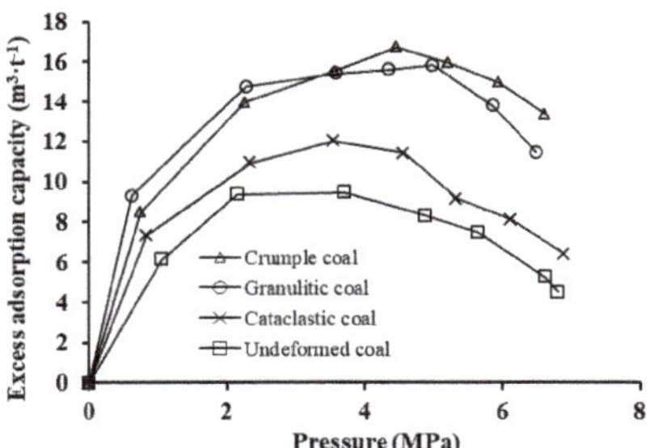

Figure 4. Isotherms for excess adsorption of CO_2 on coal.

3.1.2. Absolute Adsorption Curve

As mentioned above, isotherms of four coal specimens exhibited an increase at first and then a decrease with increasing pressure, and showed a single peak curve rather than monotonous characteristics. That indicates that with increasing pressure, CO_2 gradually approaches a critical state and exhibits supercritical fluid quality, especially with pressure larger than 5 MPa. Therefore, the adsorption capacity measured by isothermal adsorption experiment is the excess adsorption capacity, also known as apparent adsorption capacity; the corresponding absolute adsorption capacity is the real adsorption capacity. According to the relationship between the excess adsorption capacity and the real adsorption capacity, as described in Equation (5) [16,17], the difference between them is mainly attributed to the densities of the adsorbed phase and bulk gas phase.

$$V_{ab} = V_{ex}/(1 - \rho_g/\rho_{ad}) \tag{5}$$

where V_{ab} and V_{ex} denotes absolute and excess adsorption capacity, respectively ($m^3 \cdot t^{-1}$); and ρ_{ad} and ρ_g denote the density of the adsorption phase and gas phase, respectively ($kg\ m^{-3}$). When supercritical CO_2 is in the adsorption phase, it is often regarded as the gas with ultimate compression, which his attributed to its non-liquefying property. The volume occupied by a CO_2 molecule is its intrinsic volume, and reduces the free space of the molecule, which is consistent with the meaning of volume correction term b in a van der Waals gas state equation. Therefore, the volume correction term b value given by Equation (6) [18] is employed to calculated the density of CO_2 adsorption phase, i.e., the ultimate compression density of CO_2 equal to 1028 $kg \cdot cm^{-3}$. The gas phase density of CO_2 is chosen according to values determined by U.S. National Institute of standards and Technology (NIST):

$$\rho_{ad} = 1/b \times M \tag{6}$$

where ρ_{ad} denotes the density of adsorption phase ($kg \cdot m^{-3}$) and M represents the gas molar mass ($g \cdot mol^{-1}$). Therefore, according to Equations (5) and (6), the real adsorption capacity can be obtained using the apparent adsorption capacity under the same temperature and pressure conditions (Figure 5).

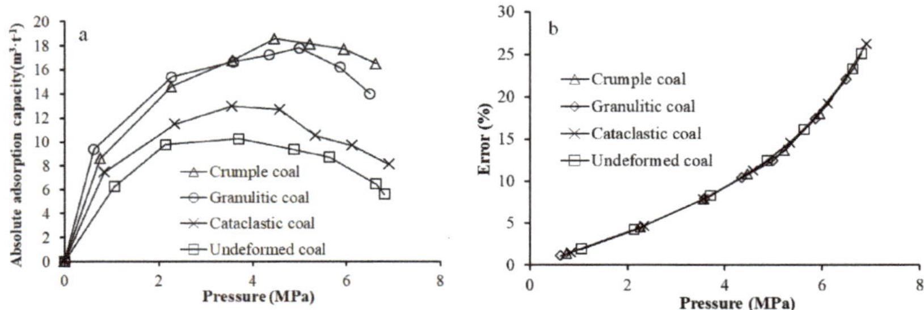

Figure 5. Absolute adsorption capacity of coal (**a**), and its error with excess adsorption capacity (**b**).

Figure 5a shows that the absolute adsorption capacity of tectonically deformed coals presents similar evolution laws with excess adsorption capacity. An increase of pressure at first causes an increase, and then a decrease of the isotherm curve, which indicates type I isotherm characteristics. An increase of coal deformation degree induces a gradual increase in the absolute adsorption capacity. The maximum adsorption value shifts to the high pressure point. In order to compare and analyze the relationship between absolute adsorption capacity and excess adsorption capacity with pressure, the error between excess adsorption capacity and absolute adsorption capacity at each pressure point is calculated, as shown in Equation (7):

$$Error = (V_{ex} - V_{ab})/V_{ex} \tag{7}$$

The error analysis indicates that an increase of pressure causes the rapid increase in the error between excess adsorption and absolute adsorption shown in Figure 5b. Meanwhile, error curves of four specimens almost coincide, which indicates that the error has no relationship with the degree of coal deformation.

It should be noted that although the adsorption phase is taken into account in the absolute adsorption correction, the downward trend of the adsorption curve has not been eliminated. According to this downward trend, if the pressure continues to increase, the adsorption capacity may be negative, as indicated by Krooss et al. [19] and Hall et al. [20]. During CO_2 adsorption, an increase of pressure causes the large amount of CO_2 adsorbed on the surface of the coal matrix as the adsorption phase and the coal matrix to expand, which results in the change of free space volume of specimen tank, and finally causes the deviation of theoretical calculation results of Gibbs excess adsorption.

3.2. Nanopore Structural Evolutions during Coal Structure Deformation

Figure 6 shows that the relationship between the pore volume/pore-specific surface area of coal and the pore diameter presents a three-stage change, with boundaries of pore diameters of 10 nm and 50 nm, respectively. Therefore, nanopores of tectonically deformed coals are divided into the micropore (2–10 nm), mesopore (10–50 nm), and macropore (50–200 nm). The nanopore structures of tectonically deformed coals are obviously affected by the tectonic deformation of coal. An increase of degree of coal deformation, i.e., in the order of undeformed coal, cataclastic coal, granulitic coal, and crumple coal, causes obvious increases in the total pore volume and total specific surface area of tested pore size range (2–200 nm); the total pore volume increases from 4.371×10^{-3} $cm^3 \cdot g^{-1}$ to 14.746×10^{-3} $cm^3 \cdot g^{-1}$, and the total specific surface area increases from 0.517 m^2/g to 7.211 $m^2 \cdot g^{-1}$ (Figure 7a,e).

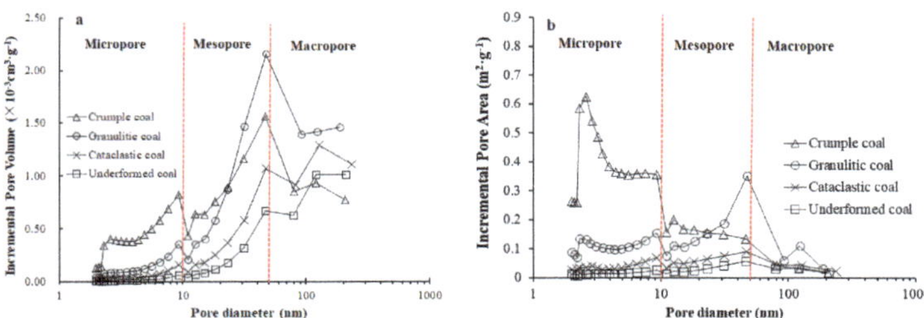

Figure 6. Relationships between incremental pore volume (**a**), incremental pore specific surface area, (**b**) and average pore diameter.

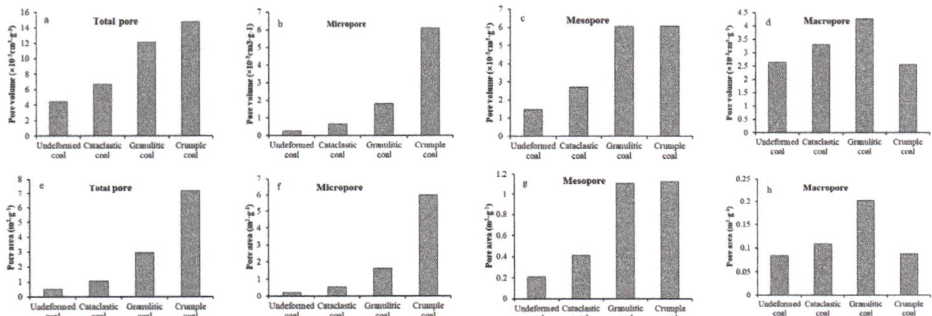

Figure 7. Various types of pore volume (**a–d**) and specific surface area (**e–h**): (**a**) Pore volume of total pore; (**b**) Pore volume of micropore; (**c**) Pore volume of mesropore; (**d**) Pore volume of macropore; (**e**) Pore area of total pore; (**f**) Pore area of micropore; (**g**) Pore area of mesopore; (**h**) Pore area of macro pore.

It is clearly seen that an increase of degree of coal deformation causes the relatively consistent change trends of various types of pore volume and specific surface area. Specifically, as the degree of coal deformation increases, the pore volume and specific surface area present an obvious increasing trend in the case of micropores (Figure 7b,f), an increase at first (cataclastic coal and granulitic coal) and then stabilization (crumple coal) in the case of mesopores (Figure 7c,g), and a gradual decrease in the case of macropores (Figure 7d,h).

3.3. Relationships between CO_2 Adsorption Capacity and Nanopore Structure

It is shown in Figure 8 that the effects of pore volume and specific surface area on the absolute adsorption capacity of tectonically deformed coals is relatively consistent at the pressure stage of 0–7 MPa. In the order of undeformed coal, cataclastic coal, and granulitic coal, the absolute adsorption capacity of coal specimens presents obviously linear positive correlation with the pore volume and specific surface area of macropores (black lines in Figure 8a,e); as it changes from granulitic coal to crumple coal, the pore volume and specific surface area of macropores decreases obviously, and the absolute adsorption capacity increases slightly (red lines in Figure 8a,e). The absolute adsorption capacity of tectonically deformed coals is obviously controlled by mesopores; the volume and specific surface area of mesopores both present good linear positive correlation with the absolute adsorption capacity (Figure 8b,f).

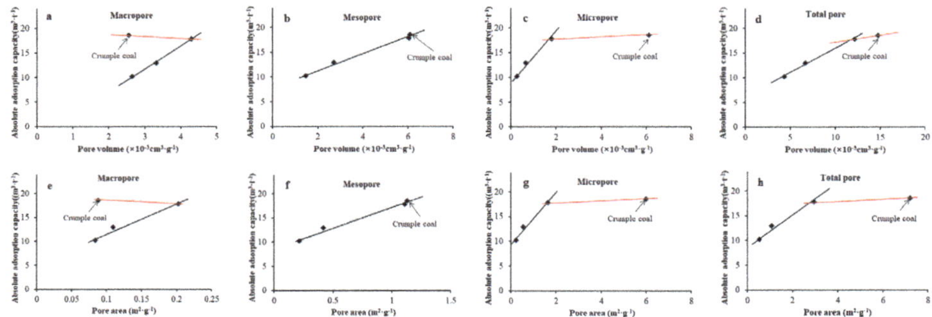

Figure 8. Relationships between absolute adsorption capacity and pore volume (**a–d**) and specific surface area of coals (**e–h**): (**a**) Absolute adsorption capacity versus pore volume of macropore; (**b**) Absolute adsorption capacity versus pore volume of mesopore; (**c**) Absolute adsorption capacity versus pore volume of micropore; (**d**) Absolute adsorption capacity versus pore volume of total pore; (**e**) Absolute adsorption capacity versus pore area of macropore; (**f**) Absolute adsorption capacity versus pore area of mesopore; (**g**) Absolute adsorption capacity versus pore area of micropore; (**h**) Absolute adsorption capacity versus pore area of total pore.

Increases in micropores mainly contribute to increases in the total pore volume and specific surface area. Therefore, the relationships between total pore volume and specific surface area of tectonically deformed coals and absolute adsorption capacity are consistent with those of micropores. In the order of undeformed coal, cataclastic coal, and granulitic coal, the absolute adsorption capacity of the coal specimens presents obviously linear positive relationships with pore volume, specific surface area of micropores, total pore volume, and specific surface area of the coals (black lines in Figure 8c,d,g,h). As it changes from granulitic to crumple coal, there are obvious increases in the pore volume and specific surface area of micropores and total pores; however, the absolute adsorption capacity increases more gently (red lines in Figure 8c,d,g,h).

Therefore, during coal deformation, tectonic stress causes damage of the coal matrix block from the undeformed coal to cataclastic coal and granulitic coal. The volume and specific surface area of all types of pores exhibit obvious increases, especially for macropores and mesopores, which causes the increase in the CO_2 adsorption capacity of the coal. As the tectonic deformation increases gradually (crumple coal), the coal structure is more broken; the tectonic stress has affected the smaller nanopores with newly formed, large numbers of micropores, as well as micropores formed due to damages of existing macropores and mesopores [21]. Therefore, in the case of crumple coal, an increase of tectonic deformation causes obvious decreases in the pore volume and specific surface area of macropores, slight increases in the pore volume and specific surface area of mesopores, and sharp increases in pore volume and specific surface area of micropores. However, the absolute adsorption capacity of crumple coal presents a slight increase, which coincides with the change trend of mesopores, and does not increase or decrease obviously with changes of macropores and micropores. The above analyses indicate that the mesopores are the key factor of CO_2 adsorption of tectonically deformed coals at the pressure stage of 0–7 MPa, followed by the micropores and the limited effect of macropores at the strong coal deformation stage. Those also reveal that curves of excess adsorption and absolute adsorption of granulitic coal and crumple coal present intersection points. As indicated in Figure 9, when the pressure is smaller than 3.5 MPa, CO_2 is mainly adsorbed in larger pores (23–200 nm) of coals, and the pore volume and specific surface area of larger pores (23–200 nm) of granulitic coal are larger than those of crumple coal, which indicates the larger adsorption capacity of granulitic coal as compared to crumple coal. As the pressure increases (3.5–7 MPa stage), CO_2 is adsorbed in smaller pores of coals (2–23 nm); at this stage, the pore volume and specific surface area of crumple coal are larger than those of crumple coal, which indicates a larger adsorption capacity of crumple coal.

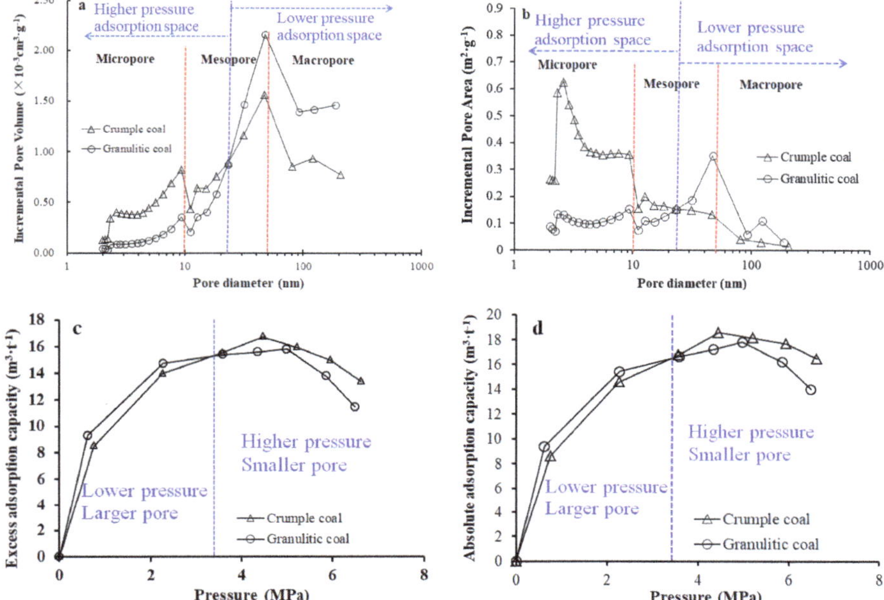

Figure 9. Adsorption space distributions at higher and lower pressure stages: (**a**) Division of higher and lower pressure adsorption space of pore volume distribution curve; (**b**) Division of higher and lower pressure adsorption space of pore area distribution curve; (**c**) Division of higher pressure and lower pressure stage of isothermal excess adsorption curve; (**d**) Division of higher pressure and lower pressure stage of isothermal absolute adsorption curve.

4. Conclusions

This study presents evolutions of CO_2 adsorption and nanopore structure characteristics with coal structure deformation, and reveals the effects of nanopore structure on CO_2 adsorption in tectonically deformed coals.

Isotherms of excess and absolute adsorption of four coal specimens exhibit an increase at first, and then a decrease with increasing pressure. There exist maximum values of excess and absolute adsorption capacity increasing with increasing coal deformation degree. Maximum values of excess adsorption capacity of undeformed coal, cataclastic coal, granulitic coal, and crumple coal are 9.48 $m^3 \cdot g^{-1}$, 12.04 $m^3 \cdot g^{-1}$, 15.83 $m^3 \cdot g^{-1}$, and 16.74 $m^3 \cdot g^{-1}$, respectively. The maximum excess adsorption capacity is present at a pressure ranging from 4 MPa to 5 MPa, and with increasing coal deformation degree the pressure corresponding to the maximum excess adsorption amount shifts to the higher pressure. The error analysis indicates that an increase of pressure causes the rapid increase in the error between excess adsorption and absolute adsorption.

In addition, as the degree of coal deformation increases, the pore volume and specific surface area present an obviously increasing trend in the case of micropores, exhibiting an increase at first (cataclastic coal and granulitic coal) and then stabilization (crumple coal), in the case of mesopores, and showing the gradual decrease in the case of macropores. The effects of pore volume and specific surface area on the absolute adsorption capacity of tectonically deformed coals is relatively consistent at the pressure stage of 0–7 MPa. The mesopores are the key factor of CO_2 adsorption of tectonically deformed coals at the pressure stage of 0–7 MPa, followed by the micropores and the limited effect of macropores at the strong coal deformation stage. The volume and specific surface area of mesopores both present good linear positive correlation with the absolute adsorption capacity. Differences in mesopore development result in the intersection of curves of excess adsorption and absolute adsorption in granulitic coal and crumple coal.

Author Contributions: Specimen preparation, macro- and microstructure analysis, Z.L.; isotherm adsorption experiment and liquid nitrogen adsorption, L.W. All authors discussed the results and contributed to the final manuscript. All authors have read and agreed to the published version of the manuscript.

Funding: This research was funded by National Natural Science Foundation of China, grant numbers 41702169 and 41572138, and China Postdoctoral Foundation, grant number 2019M662013.

Conflicts of Interest: The authors declare no conflict of interest.

References

1. Hou, S.Y. Development status of China coalbed methane industry in recent years. *China Coalbed Methane* **2018**, *15*, 42–45.
2. Gou, D.Y. Research on Structural Physics of Coal and Gas Outburst. Ph.D. Thesis, China University of Mining and Technology, Beijing, China, 1996.
3. Zhang, H.R. Discussion on the origns of structural coal in Xiahuayuan coal mine. *Min. Saf. Environ. Prot.* **1999**, *5*, 31–34.
4. Ju, Y.W.; Jiang, B.; Hou, Q.L.; Wang, G.L. The new structure-genetic classification system in tectonically deformed coals and its geological significance. *J. China Coal Soc.* **2004**, *29*, 513–517.
5. Jiang, B.; Qin, Y.; Ju, Y.W.; Wang, J.L.; Li, M. The coupling mechanism of the evolution of chemical structure with the characteristics of gas of tectonic coals. *Earth Sci. Front.* **2009**, *16*, 262–271.
6. Zhang, Y.G. Evolution of Deformed Coal and Process of Coal Mechanochemistry. Ph.D. Thesis, Taiyuan University of Technology, Taiyuan, China, 2006.
7. Xing, W.L. Study on Characteristics of Adsorption/Desorption and Diffusion for CO_2, CH_4, N_2, and Their Mixtures in Coal. Ph.D. Thesis, Dalian University of Technology, Dalian, China, 2016.
8. Chen, R.; Qin, Y.; Zhang, P.F.; Wang, Y.Y. Changes in pore structure of coal caused by CS_2 treatment and its methane adsorption response. *Geofluids* **2018**, *2018*, 1–11. [CrossRef]
9. Zhao, J.C.; Qin, Y.; Shen, J.; Zhou, B.Y.; Li, C.; Li, G. Effects of pore structures of different maceral compositions on methane adsorption and diffusion in anthracite. *Appl. Sci.* **2019**, *9*, 5130. [CrossRef]
10. Castello, D.L. Advances in the study of methane storage in a porous carbon aceous materials. *Fuel* **2002**, *81*, 1777–1803. [CrossRef]
11. Chen, X.J.; Liu, J.; Wang, L.; Qi, L.L. Influence of pore size distribution of different metamorphic grade of coal on adsorption constant. *J. China Coal Soc.* **2013**, *39*, 294–300.
12. Jian, K.; Fu, X.H.; Ding, Y.M.; Wang, H.D.; Li, T. Characteristics of pores and methane adsorption of low-rank coal in China. *J. Nat. Gas Sci. Eng.* **2015**, *27*, 207–218. [CrossRef]
13. Wang, X.H.; Wang, Y.B.; Gao, S.S.; Hong, P.F.; Zhang, M.J. Differences in pore structures and absorptivity between tectonically deformed and undeformed coals. *Geol. J. China Univ.* **2012**, *18*, 528–532. [CrossRef]
14. Li, Y.K. Research on the Methane Adsorption/Desorption/Diffusion Characteristics of Tectonic Coal. Master's Thesis, Henan Polytechnic University, Jiaozuo, China, 2015.
15. Donohue, M.D.; Aranovich, G.L. A new classification of isotherms for Gibbs adsorption of gases on solids. *Fluid Phase Equil.* **1999**, *158*, 557–563. [CrossRef]
16. Moffat, D.H.; Weale, K.E. Sorption by coal of methane at high pressures. *Fuel* **1955**, *34*, 449–462.
17. Kim, H.J.; Shi, Y.; He, J.W.; Lee, H.H.; Lee, C.H. Adsorption characteristic of CO_2 and CH_4 on dry and wet coal from subcritical to supercritical conditions. *Chem. Eng. J.* **2011**, *171*, 45–53. [CrossRef]
18. Cui, Y.J. *Adosrption of CH_4, N_2, CO_2, Single and Multicomponent Gas on Coal*; Xian Branch, China Coal Researeh Institute: Xian, China, 2003.
19. Kroos, B.M.; Bergen, F.V.; Gensterblum, Y.; Siemons, N.; Pagnier, H.J.M.; David, P. High-pressure methane and carbon dioxide adsorption on dry and moisture-equilibrated pennsylvanian coals. *Int. J. Coal Geol.* **2002**, *51*, 69–92. [CrossRef]

20.   Hall, F.E.; Chunhe, Z.; Gasem, K.A.M.; Robinson, R.L.; Yee, D. Adsorption of pure methane, nitrogen, and carbon dioxide and their binary mixtures on wet Fruitland coal. In Proceedings of the SPE Eastern Regional Conference and Exhibition, Charleston, WV, USA, 8–10 November 1994; pp. 329–344.
21.   Pan, J.N.; Zhu, H.T.; Hou, Q.L.; Wang, H.C.; Wang, S. Macromolecular and pore structures of Chinese tectonically deformed coal studied by atomic force microscopy. *Fuel* **2015**, *139*, 94–101. [CrossRef]

 *applied sciences*

Article

Mineral Carbonation of CO_2 in Mafic Plutonic Rocks, I—Screening Criteria and Application to a Case Study in Southwest Portugal

Jorge Pedro [1,2], António A. Araújo [1,2], Patrícia Moita [1,3], Massimo Beltrame [3], Luis Lopes [1,2], António Chambel [1,2], Edgar Berrezueta [4] and Júlio Carneiro [1,2,*]

[1] Departamento de Geociências, Escola de Ciências e Tecnologia, Universidade de Évora, Rua Romão Ramalho, 59, 7000-671 Évora, Portugal; jpedro@uevora.pt (J.P.); aaraujo@uevora.pt (A.A.A.); pmoita@uevora.pt (P.M.); lopes@uevora.pt (L.L.); achambel@uevora.pt (A.C.)

[2] Instituto de Ciências da Terra, Universidade de Évora, Rua Romão Ramalho, 59, 7000-671 Évora, Portugal

[3] Laboratório Hercules, Universidade de Évora, Largo Marquês de Marialva, 8, 7000-809 Évora, Portugal; massimo@uevora.pt

[4] Instituto Geológico y Minero de España, C/Matemático Pedrayes 25, 33005 Oviedo, Spain; e.berrezueta@igme.es

[*] Correspondence: jcarneiro@uevora.pt; Tel.: +351-266745301

Received: 16 June 2020; Accepted: 9 July 2020; Published: 16 July 2020

Abstract: This article describes the screening, ranking and characterization of ultramafic and mafic rocks in southern Portugal for mineral carbonation as an alternative to conventional CO_2 storage in sedimentary rocks. A set of criteria including mineralogy, structure, surface area, distance to CO_2 sources, expected volume, and socioeconomic conditions was applied to screen ultramafic and mafic rock massifs in the Alentejo region, southern Portugal. Ranking of the massifs indicated that the plutonic massifs of Sines and of Torrão-Odivelas were the most promising. A characterization was made of the Sines massif, a subvolcanic massif composed mostly of gabbros and diorites, located immediately adjacent to the CO_2 sources and outcropping along 300 km^2 onshore and offshore. These studies confirmed that these rock samples exhibited the appropriate mineralogical and geochemical features, but also indicated that the secondary porosity provided by the fracture patterns was very small.

Keywords: CO_2 capture; utilization and storage; mafic plutonic rocks; mineral carbonation; screening and ranking; Sines massif; Portugal

---

1. Introduction

According to the International Energy Agency [1], achieving the Paris Climate Agreement targets of global warming below 2 °C by 2100 while developing efforts to limit temperature rising to 1.5 °C [2] implies wide scale deployment of CO_2 capture, utilization, and storage (CCUS). The scenarios developed by the IEA by the 2060 CCUS should contribute to 14% of the required reduction in CO_2 emissions to achieve a scenario where temperatures are less than 2 °C, or 32% to achieve the 1.5 °C limit scenario.

Portugal has been on the forefront of policies to reduce CO_2 emissions, having established a roadmap to reach carbon neutrality by 2050 [3]. This policy has been reinforced by the Green Deal announced in December 2019 by the European Commission, which sets the same carbon neutrality target by 2050 for the European Union [4].

The roadmap does not explicitly consider CCUS as one of the technologies that should contribute to carbon neutrality in Portugal, although earlier studies have demonstrated that CCUS will be required in some industry sectors (e.g., cement) as early as 2030 [5,6].

According to the European Union carbon market database (EU ETS), in 2018, the Portuguese CO_2 emissions from the industrial and power sectors amounted to 26.2 Mt CO_2 [7], with the most important emissions cluster located in southern Portugal, extending from Setúbal to the Sines industrial area, being responsible for 44% of all emissions in the country (Figure 1a). Although the main point source in the region, the Sines Coal Power Plant, is to be decommissioned in 2023, the remaining sources will account for more than 30% of all emissions from industrial sources (Figure 1b). Furthermore, the socioeconomic relevance of the industry to the region imposes the need to seek alternatives to decarbonization in the region.

The possibility of capturing CO_2 at the Sines and Setúbal sources and storing it in offshore deep saline aquifers was addressed previously [8,9], but was decided against due to low storage capacity and high storage costs. The possibility of CO_2 transport to other parts of the country with more favorable geological storage conditions was studied in Seixas et al. [6] and is being further pursued in the STRATEGY CCUS project [10], funded by the European Union's Horizon 2020 to study the development of CCUS in Southern and Eastern Europe, including a promising region in Portugal.

Nonetheless, the multiple ultramafic and mafic rock massifs in Alentejo could provide an alternative for CO_2 storage by mineral carbonation. The potential of ultramafic and mafic rocks for CO_2 storage rests in their ability to stabilize CO_2 via mineral carbonation (e.g., [11–14]). These rocks are enriched in $Ca^{2+,}$ Mg^{2+}, and Fe^{2+} silicate minerals, which react with CO_2 precipitate carbonate minerals [15], leading to the trapping of the CO_2 in a solid phase at a much faster rate than can be expected in sedimentary silicate rocks. The feasibility of this process was demonstrated in the CarbFix (Iceland) project [16] and Wallula (USA) project [17], in a process designated as in situ mineral carbonation, in which CO_2 is dissolved in water and injected into basalts. However, extensive studies have also addressed an ex situ process, in which the ultramafic and mafic rocks are mined and the mineral carbonation process takes place in plants under optimized pressure and temperature conditions [18–20]. A third possibility that has been put forward as a geo-engineering technique is the utilization of ultramafic rocks to sequester CO_2 through enhanced weathering, in which the mined and crushed rock is spread in large areas under atmospheric conditions to react with CO_2, promoting direct CO_2 capture from the air [21,22].

Plutonic rocks, with a mineralogical and geochemical composition similar to basalts have, to the best of our knowledge, not been considered for in situ mineral carbonation, since the low porosity is seen as an impediment to industrial-scale implementation. However, if fracture patterns provide enough secondary permeability and the CO_2 volume to inject is not high, mafic plutonic rocks may prove a valid environment for in situ mineral carbonation. That possibility, together with potential utilization in ex situ mineral carbonation or enhanced weathering, indicates that ultramafic and mafic rocks in the Alentejo region should be studied as an alternative to reduce CO_2 emissions in the region.

This article is the first of two on the mineral carbonation potential in Alentejo. This first article describes the procedure for selecting the ultramafic and mafic rock massifs; sets a screening and ranking procedure for defining them with the highest potential; and describes the methods, techniques, and results obtained in the geological, petrographic, and geochemical characterization of the Sines massif, which was ranked with the highest potential. A second article [23] describes the experiments carried out in a laboratory to characterize the mineral carbonation potential of rock samples from the Sines Massif.

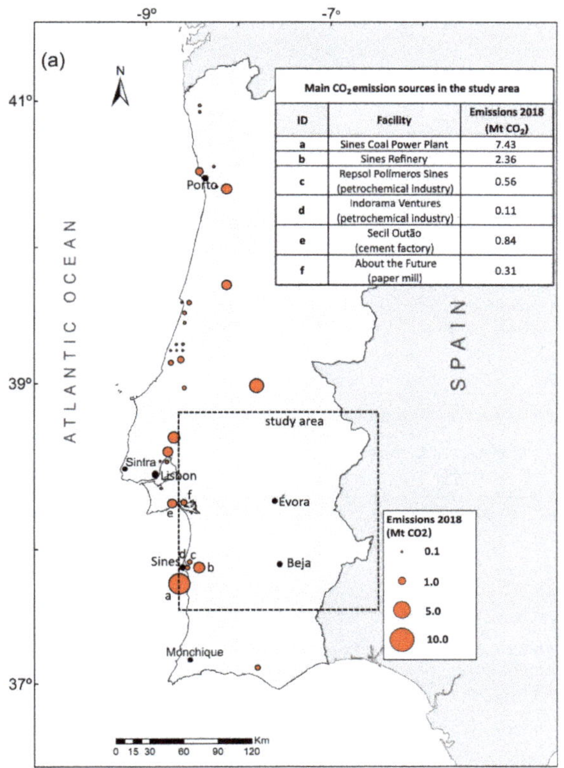

ID	Facility	Emissions 2018 (Mt CO$_2$)
a	Sines Coal Power Plant	7.43
b	Sines Refinery	2.36
c	Repsol Polímeros Sines (petrochemical industry)	0.56
d	Indorama Ventures (petrochemical industry)	0.11
e	Secil Outão (cement factory)	0.84
f	About the Future (paper mill)	0.31

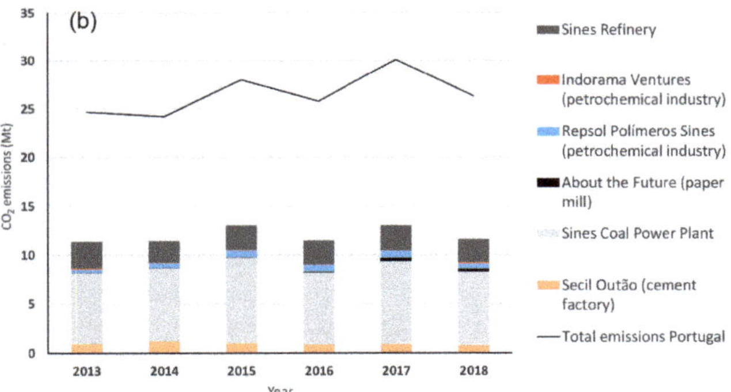

Figure 1. (**a**) Distribution of main CO$_2$ emissions sources in Portugal reported to the European Union carbon market database (EU ETS). The location of ultramafic and mafic rock massifs in the study area is shown in Figure 3. (**b**) Evolution of CO$_2$ emissions in the study area and total national emissions from 2013 to 2018, as reported to the EU ETS. The sources in the target area were accountable for 44% of the CO$_2$ emissions from power and industrial sources in the country in 2018. The Sines coal power plant is to be decommissioned in 2023, but the remaining sources in Sines and Setúbal will still be responsible for more than 30% of the emissions from the industrial sector in the country.

2. Methodology

Figure 2 depicts the overall approach to define and characterize the rock massifs, the techniques applied, and the laboratory experiments, which are described in the second article [23] to establish the potential for mineral carbonation.

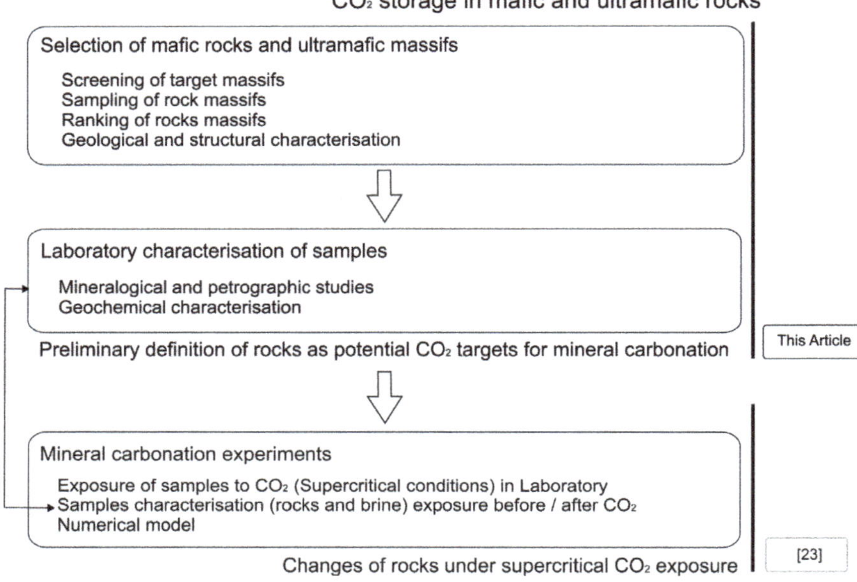

Figure 2. Diagram illustrating the methodology employed to screen, rank, and characterize the rock massifs and rock samples. The samples were subsequently submitted to mineral carbonation experiments in the laboratory, the results of which are the subject of a second article [23].

A first screening of mafic and ultramafic geological formations located in the study area, the Alentejo region (Figure 1a), was made, using the geological map of Portugal (scales: 1/50,000, 1/200,000 and 1/500,000), an analysis of published information, and publicly available maps and reports.

A first field trip was made to outcrops of the screened rock massifs and samples were collected for laboratory characterization, comprising thin section production, mineral identification, textural description, and evaluation of the porosity by optical and conventional microscopy.

Next we used the mineralogical characterization of those samples, as well as other data resulting from desk studies, to rank the different ultramafic and mafic rock massifs in terms of their theoretical suitability for mineral carbonation. The criteria adopted in the ranking procedures are further described in Section 2.1.

For the two highest-ranking rock massifs, we conducted detailed fieldwork and experiments to identify the relevant fracture patterns at outcrop, along scanlines, and in boreholes. Representative samples were submitted to a battery of petrographic, mineral chemistry, and geochemistry techniques to fully characterize the samples. Those same techniques were applied to study the changes imposed by a set of experimental runs in which the rock samples were put in contact with a brine saturated in CO_2 under pressure and temperature conditions that simulated the supercritical injection of CO_2. The techniques utilized for the characterization of the rock samples are described in Section 2.2, while the results of the mineral carbonation experiments are addressed in Moita et al. [23].

2.1. Screening and Ranking Criteria

The selection of geological reservoirs for CO_2 storage in deep saline aquifers, depleted hydrocarbon fields, or even uneconomic coal beds has well-established screening and ranking criteria [24–27], but the selection of targets for mineral carbonation has yet to have sound and widely accepted procedures.

Nonetheless, given the number of ultramafic and mafic rock massif outcroppings in the study area, we found a systematic selection procedure to be necessary, and we defined nine criteria for screening and ranking of these formations. The criteria included geological conditions and socioeconomic and environmental constraints. Each criterion was subdivided into two to three classes, to which relative weights were assigned using integer values. In fact, the range of values assigned to each criterion was debatable, but the chosen values reflected the relative importance that we attached to each criterion, ranging from a maximum of nine in the criteria considered most relevant, down to a maximum of three in what we considered to be secondary. The criteria to which we assigned only two values were related to the existence, or not, of a certain condition. Finally, the studied geological units were hierarchized on the basis of the sum of the obtained values. For some criteria, the most unfavorable situation was considered eliminatory.

The criteria applied are listed in Table 1 and were as follows:

- Geological conditions

1. Lithological composition—the mineralogical composition of the rocks is a fundamental aspect to address in order to achieve the desired objectives. An expedited modal classification based on the relative percentage of mafic minerals (i.e., olivine, pyroxene, amphibole, and biotite) observed in thin sections was used to classify each massif into three classes, favoring those targets with the highest percentage of minerals enriched in calcium, magnesium, and iron.

2. Area—since data on subsurface geology in the study area are quite rare, the outcrop area is a relevant indicator for a first assessment of the size of the targets under study. In most cases, the mapped area corresponds to the outcrop area and can be estimated directly, but in cases where the unit under study is partially covered by more recent sediments, its determination is more difficult. Three classes were defined, and in situations of uncertainty, we always chose a conservative assessment.

3. Expected volume—the volume of rock masses was estimated by considering the previous criterion (area); the shape of each geological unit (stratiform or batholith); and, in some cases, any available geophysical information. As was the case for the previous criterion, in situations of uncertainty, a conservative assessment was always adopted.

4. Existence of a seal unit—the existence of an impermeable layer overlaying the target unit represents a particularly favorable structural situation, as this cover will act as a barrier to CO_2 leakage for in situ mineral carbonation. The Carbfix project developed an injection method in which CO_2 is injected dissolved in water, and thus CO_2 buoyancy will not occur and the existence of a seal is not strictly necessary [11,28]. In Alentejo, the basal levels of tertiary deposits overlaying the Paleozoic and Mesozoic massifs generally correspond to impermeable clayey sediments. In the most favorable situations, where this tertiary coverage exists, a weight of 3 was assigned. In the remaining situations, where the formations crop out without any cover, or are covered by sands with Miocene age or later, a zero weight was assigned.

5. Fracture density—the main constraints when injecting fluids into plutonic rocks are low permeability and porosity. Fracture density controls the permeability and porosity of rock masses; a higher fracture density facilitates fluid circulation and thus in situ carbonation. For each geological formation, the fracture density was assessed by fracture pattern studies in outcrops and quarries using a scanline approach with measurement of fracture frequency. The massifs were divided into three categories: more than 10 fractures/m, 3–10 fractures/m, and fewer than 3 fractures/m, to which indices 9, 6, and 1 were assigned, respectively.

- Socioeconomic and environmental constraints

6. Distance to CO_2 sources—transport of CO_2 over long distances is a highly penalizing factor. Thus, for this criterion, three categories were defined as a function of the distance to which values 9, 6, and 1 were assigned, respectively, for distances of less than 10 km, between 10 and 100 km, and over 100 km.

7. Social and demographic situation—building an industry for in situ carbonation will not be well accepted in urban areas. For this criterion, two categories were defined, wherein an index of 3 was assigned to rural regions and zero was assigned to urban areas.

8. Existence of productive aquifers—groundwater is a value that must be preserved. Thus, geological units that correspond to productive aquifers were completely excluded on the basis of this criterion.

9. Environmental restrictions—this type of project is prohibited in natural parks and other protected areas. This criterion was considered, but no geological formations favorable to in situ carbonation were identified in protected areas.

Table 1. Screening and ranking criteria. The shaded bars represent the relative importance of the class in each criterion. The final ranking results from the sum of the weight obtained in each criterion. If any massif was assigned the class "eliminatory criterion" it was not further considered in the screening and ranking procedure.

Criteria	Classes	Weight	
Geological conditions			
C1—Lithological composition	Ultramafic—more than 90% of mafic minerals	9	▬
	Mafic—40–90% mafic minerals	6	▬
	Intermediate—10–39% of mafic minerals	1	▮
C2—Outcropping area	Over 20 km^2	3	▪
	From 10 to 20 km^2	2	▪
	Less than 10 km^2	1	▮
C3—Expected volume	Over 20 km^3	6	▬
	From 10 to 20 km^3	3	▪
	Less than 10 km^3	Eliminatory criterion	
C4—Existence of a seal unit	Existent	3	▪
	Not known	0	
C5—Fracture density	More than 10 fractures/m	9	▬
	3–10 fractures/m	6	▬
	Fewer than 3 fractures/m	1	▮
Socioeconomic constraints			
C6—Distance to CO_2 sources	Less than 10 km	9	▬
	From 10 to 100 km	6	▬
	Over 100 km	1	▮
C7—Social and demographic situation	Urban area	3	▪
	Rural	0	
C8—Existence of productive aquifers	No	3	▪
	Yes	Eliminatory criterion	
C9—Environmental restrictions	No restrictions	0	
	Protected areas	Eliminatory criterion	

The final ranking of the rock massifs was obtained by adding the weights assigned to each criterion.

2.2. Petrographic, Mineral Chemistry, and Geochemical Techniques

The samples collected in the first field trip were processed in thin sections for petrography and mineral chemistry analysis and powdered for X-ray diffraction (XRD) and X-ray fluorescence (XRF) analysis (Figure 3).

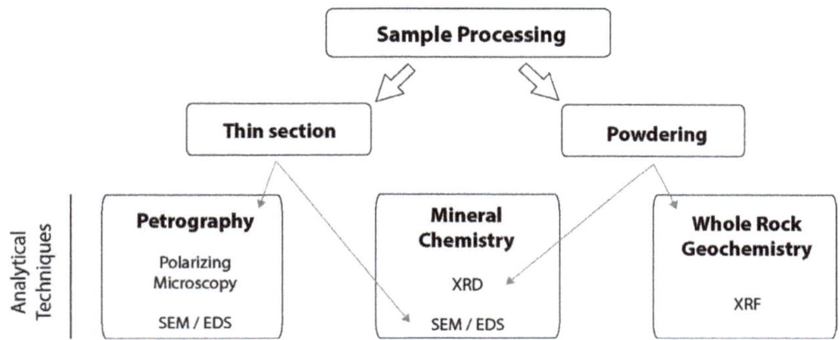

Figure 3. Diagram illustrating the sample processing and the analytical techniques used.

The petrography included conventional petrographic techniques by transmitted and reflected light polarizing microscopy, complemented with scanning electron microscopy with energy-dispersive spectroscopy (SEM–EDS), which also provided a valuable semi-quantitative analysis of the mineral chemistry using a Hitachi S3700N interfaced with a Quanta EDS microanalysis system. The Quanta system was equipped with a Bruker AXS XFlash Silicon Drift Detector (129 eV Spectral Resolution at FWHM/MnKα). Standardless PB/ZAF (self-calibrating EDS spectrum procedure) quantitative elemental analysis was performed using Bruker Espirit software. The operating conditions for EDS analysis were 20 kV accelerating voltage, 10 mm working distance, and 120 μA emission current. The detection limits for major elements (>Na) were on the order of 0.1 wt %.

The petrographic analysis was also complemented by detailed identification of crystalline phases through XRD analysis using a Bruker AXS D8 Discover diffractometer with a Cu-Kα source, operating at 40 kV and 40 mA with a Lynxeye one-dimensional detector. Scans were performed from 3 to 75° 2θ, with a 0.05 2θ step and 1 s/step measuring time by point. Diffract-Eva Bruker software with PDF-2 database (Powder Diffraction File by the International Centre for Diffraction Data) was utilized to interpret all XRD patterns.

The whole rock geochemistry analysis was conducted by XRF, which allows for the quantification of major oxides (SiO_2, TiO_2, Al_2O_3, Na_2O, K_2O, CaO, MgO, MnO, FeO, and P_2O_5), sulfur, and some minor elements (Rb, Sr, Y, Zr, Nb, Th, Cr, Co, Ni, Cu, Zn, Ga, As, Pb, Sn, V, U, and Cl). Analyses were performed with an Energy-Dispersive X-Ray Spectrometer S2 Puma (Bruker) using a methodology similar to that adopted by Georgiou [29]. A description of the standard reference materials (SRM) utilized in the calibration method can be found elsewhere [30]. Samples were fused on a Claisse LeNeo heating chamber using a flux (Li-tetraborate) to prepare fused beads (ratio sample/flux = 1/10). The software utilized for acquisition and data processing was Spectra Elements 2.0, which reported the final oxide/element concentrations and the instrumental statistical error associated with the measurement.

3. Results

A first selection of several geological formations located in the Alentejo region (Figure 4) was made, which was considered potentially interesting due to their dimensions and lithological nature using the geological map of Portugal (scales 1/50,000, 1/200,000, and 1/500,000) and other general bibliographic elements.

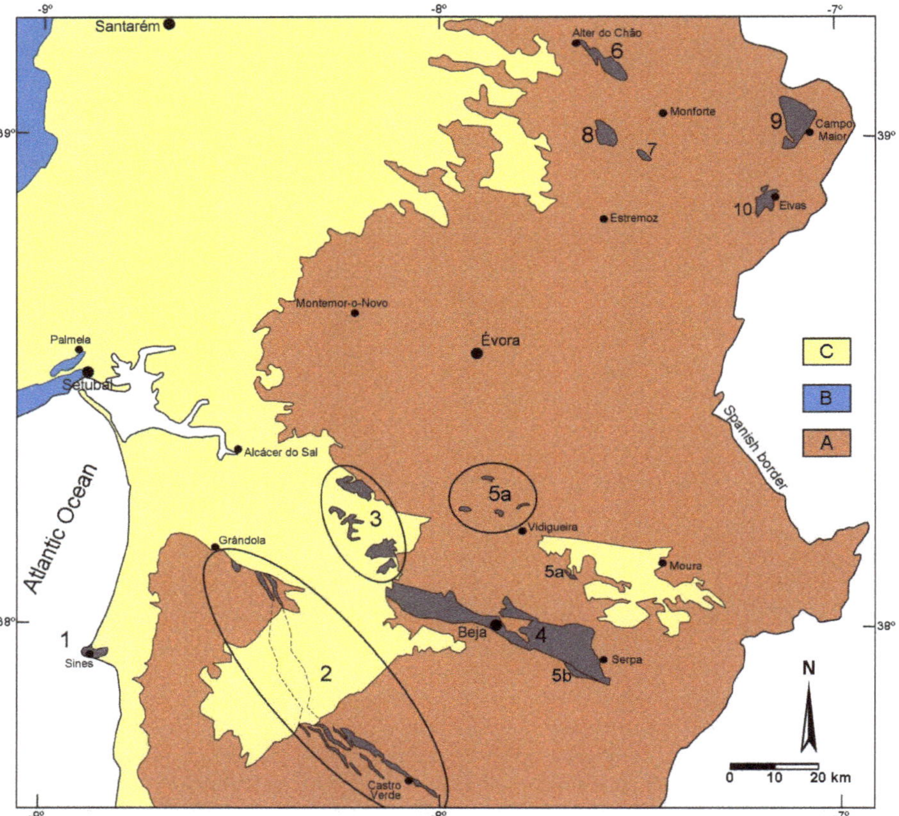

Figure 4. Location of the studied units: (**A**) Variscan basement; (**B**) Mesozoic cover; (**C**) Cenozoic cover; (**1**) Sines massif; (**2**) diabases of the Iberian Pyrite Belt; (**3**) gabbros of the Torrão-Odivelas region; (**4**) Beja gabbros; (**5**) ophiolitic sequences: (**a**) internal ophiolitic sequences; (**b**) Beja-Acebuches ophiolitic complex; (**6**) Alter do Chão/Cabeço de Vide massif; (**7**) Veiros massif; (**8**) Vale de Maceiras massif; (**9**) Campo Maior massif; (**10**) Elvas massif.

Most of these units correspond to igneous, ultramafic, or mafic intrusive rocks of Paleozoic age (gabbros and diorites of Torrão-Odivelas; Beja gabbros and igneous massifs of Alter do Chão/Cabeço de Vide, Veiros, Vale de Maceiras, Campo Maior, and Elvas). The Sines massif is also mostly composed by mafic rocks, but, having a Cretaceous age, is considerably more recent. The diabases from the Iberian Pyrite Belt are part of a Lower Carbonic volcano sedimentary complex and the ophiolitic sequences are mostly ultramafic and mafic rocks associated with the obduction of oceanic lithosphere during the Variscan orogeny.

For each of the 10 geological units represented in Figure 4, we determined the values corresponding to the nine criteria described. The sum of these values was used to rank the studied geological formations. Table 2 presents the result of these calculations and the hierarchy of the units in terms of their potentialities.

Table 2. Ranking of ultrabasic and basic geological formations in the Alentejo region. A massif with an ideal condition would have a total score of 45.

Criterion Geological Formation	C1	C2	C3	C4	C5	C6	C7	C8	C9	Σ
1—Sines massif	6	3	6	0	6	9	0	0	0	30
2—Diabases from Iberian Pyrite Belt	1	2	3	3	9	6	3	0	0	27
3—Gabbros of Torrão-Odivelas	6	3	6	3	1	6	3	0	0	28
4—Beja gabbros	6	3	6	0	1	1	3	Elim	0	—
5—Ophiolitic sequences	9	1	Elim	0	9	6	3	0	0	—
6—Alter do Chão/Cabeço Vide massif	9	3	6	0	6	1	3	Elim	0	—
7—Veiros massif	6	1	Elim	0	1	1	3	0	0	—
8—Vale de Maceiras massif	6	2	3	0	1	1	3	0	0	16
9—Campo Maior massif	6	3	6	0	6	1	3	0	0	25
10—Elvas massif	6	2	3	0	6	1	0	0	0	18

Elim—eliminatory criterion.

Among the studied units, two were eliminated due to lack of size (the Veiros massif and ophiolitic sequences because they are very fragmented), and two others were eliminated because they corresponded to productive aquifers (Beja gabbros and Alter do Chão/Cabeço de Vide massif).

Regarding the remaining six units, we emphasize the excellent rating achieved by the Sines massif, gabbros of Torrão-Odivelas, and diabases from the Iberian Pyrite Belt. The Sines massif, which was ranked first and was selected for the first experiments, has the disadvantage of being located in an urban area, but the great advantage of its location near the CO_2-producing sources and its extension for some kilometers into the continental shelf.

Further research is ongoing for the Sines and Torrão-Odivelas massifs. The remainder of this article, as well as the article by Moita et al. [23] in this issue, address the works already conducted at the Sines massif.

3.1. The Sines Massif

3.1.1. Geological Setting

The Sines massif (Figure 5) is located on the west Portuguese coast (southwest Iberia). Onshore it has a relatively small area ($\approx$10 km^2), largely covered by Plio-Quaternary sediments, mainly outcropping along the seacoast and in the Montes de Chãos quarry (Figure 6). It exhibits an elliptical shape that is elongated into the continental shelf and has been interpreted by Teixeira [31] and Ribeiro et al. [32] as a subvolcanic ring structure. Carvalho et al. [33], resorting to geophysical data (magnetic, gravimetric, and seismic reflection), were able to model the offshore of the massif, along a NE–SW trend, and covering an area of $\approx$300 km^2.

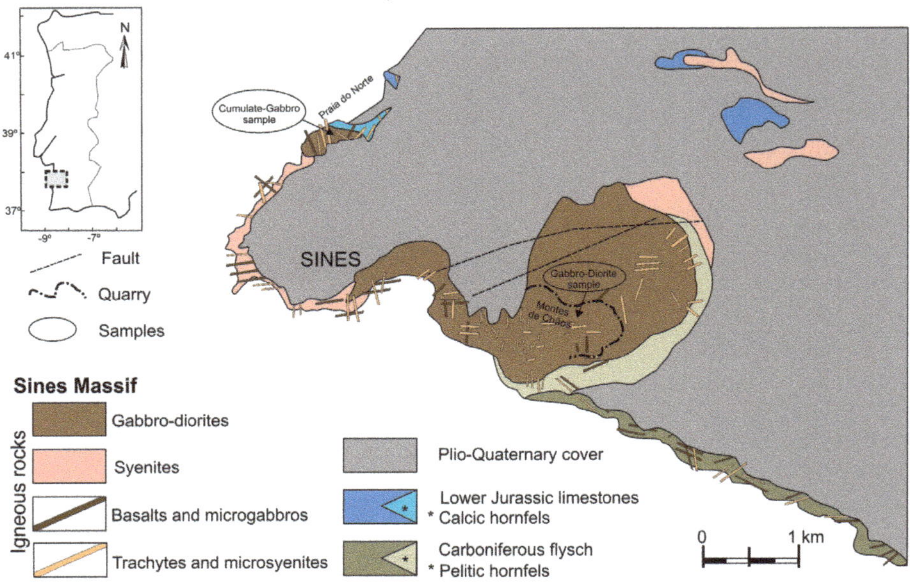

Figure 5. Geological map of Sines massif (adapted from Inverno et al. [34]) with sampling localities.

Figure 6. Mesoscopic features of Sines massif at seacoast near Praia do Norte (**a**) and Montes de Chãos quarry (**b**). Detail of sampled cumulate gabbro at a cliff near Praia do Norte (**c**) and gabbro-diorite at Montes de Chãos quarry (**d**).

Together with the Sintra and Monchique massifs, the volcanic complex of Lisbon, and several other minor intrusions, the Sines Massif is part of a Late Cretaceous alkaline magmatic cycle dated to ca. 22 Ma (94–72 Ma) [35].

The Sines massif is mainly formed of gabbros, diorites, and syenites, with a profusion of dykes of variable composition (basalts, microgabbros, microdiorites, trachybasalts, lamprophyres, trachytes, and microsyenites [34,36]). Despite the Cenozoic sedimentary cover, the geological map of Portugal at 1/50,000 scale (sheet 42-C, Santiago de Cacém [34]) shows that these intrusive rocks cut the Carboniferous flysch at the south (Mira Formation) and Lower Jurassic carbonate rocks at the north. A model of multiphase emplacement is accepted for the genesis of the Sines massif, with the intrusion of gabbro-diorite, followed by the intrusion of syenite, and finally the net of dykes that cut the intrusive massif rocks and the surrounding country rocks.

3.1.2. Fracture Characterization

In the field, the observed rocks of the Sines massif displayed very low primary porosity, as is expected for plutonic rocks. Therefore, the characterization of the fracture patterns is fundamental for an indirect assessment of the permeability of the massif and, consequently, for the viability of the gabbro and diorite being used for in situ mineral carbonation. The exposure at the walls of Montes de Chãos quarry is the best place to analyze the fracture patterns of the Sines massif. The obtained fracturing density diagram poles is shown in Figure 7.

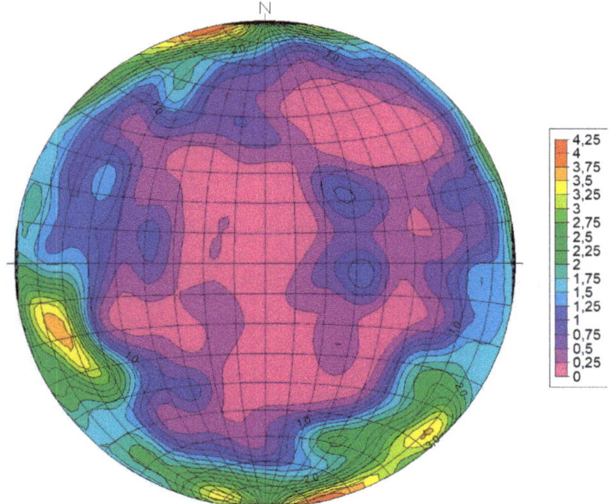

Figure 7. Fracture density diagram poles for the Monte Chãos quarry, showing a great dispersion in the orientation of discontinuities. The projection uses the Schmidt equal area net, and the number of discontinuities represented is 425. The values in the scale represent the percentage of occurrence.

For a more detailed fracture analysis, we performed several scanlines. This method consists of stretching a line along the walls of the quarry and measuring the intersection of the line with any discontinuity. To quantify the rock mass rating, we freely adapted the RQD (Rock Quality Designation) index, applied routinely to evaluate the fracture frequency in drill cores, which was calculated as the percentage of unfractured rock segments longer than 10 cm along the scanlines. That is, RQD was used here to quantify, along the scanlines, the percentage of consecutive fractures that were more than 10 cm apart. The scan lines and RQD results are listed in Table 3.

Table 3. Scanline data, RQD (Rock Quality Designation) index, and scanline statistics. Latitude and longitude refer to the starting point of each scanline. Scanline 2* corresponds to the continuity of resuming observations along the same azimuth of scanline 2 after following 25 m of no collected data due to outcrop inaccessibility.

		Scanline 1	Scanline 2	Scanline 2*	Scanline 3
	Length (cm)	1275	1250	590	2630
	Latitude	37.94978°	37.9514°		37.94924°
Data	Longitude	−8.85078°	−8.84634°		−8.84187°
	Azimuth	70°	60°		135°
	Number of discontinuities	63	62	66	153
	Discontinuities per m	4.9	5.0	11.2	5.8
	Average	24.5	24.5	10.7	10.7
Length of section between two consecutives fractures (cm)	Standard deviation	20.7	23.2	11.0	11.0
	Maximum value	103	120	57	57
	Minimum value	4	3	1	1
	Sum of sections > 10 cm	1167	1150	373	373
	RQD (%)	91.5	92.0	63.2	63.2

The data show a variety of orientations, which means it is not possible to establish a geometrically well-defined fracture pattern. The average spacing between fractures (10.7–24.5 cm) revealed a heterogeneous pattern with medium-density fractures, suggesting some secondary porosity, which increases the potential surface area available for reaction.

These results are consistent with those recently obtained by drilling cores in an internal report requested by the Administration of the Port of Sines [37], which states that the gabbro-diorite massif in general has good to excellent geotechnical quality, occasionally reasonable, especially in the most superficial areas of the massif. During the surveys carried out, the extraction of continuous rock cores without fractures was frequent, revealing the excellent geotechnical quality of the gabbro-diorite massif.

3.1.3. Petrography, Mineral Chemistry, and Geochemistry

Two coarse-grained ultramafic to mafic samples from the Sines massif were collected for laboratory characterization and mineral carbonation experiments—a melanocratic cumulate gabbro at a cliff near Praia do Norte (Figures 5 and 6c) and a gabbro-diorite at Montes de Chãos quarry (Figures 5 and 6d).

The cumulate gabbro sample displayed a medium to coarse igneous texture that preserved the ferromagnesian mineral phases. The mineral composition was found to be clinopyroxene (45–55%), olivine (15–20%), brown-amphibole (10–15%), plagioclase (5–10%), and primary ilmenite (5%) (Figure 8), although occasionally showed accessory alteration products (e.g., chlorite, actinolite, and serpentine). These secondary mineral phases occurred around the olivine, as well as around some pyroxene crystals. The enrichment in ferromagnesian minerals and their textural relations revealed cumulated textures, with amphibole crystals developing a poikilitic texture.

The mineral chemistry (Table 4) indicated the presence of magnesium olivines (Fo = 0.52–0.67), Ca–Fe–Mg clinopyroxenes with diopside–augite compositions, Ti-rich amphiboles (tschermakite horneblende), and calcic plagioclases with bytownite–labradorite compositions (An = 0.67–0.79). Except for ilmenite, which was not detected in the XRD diffractogram where the presence of magnetite was revealed, suggesting equilibrium at low temperatures in the oxide phases, we confirmed all primary mineralogy by XRD analysis (Figure 9). The rock geochemistry data (Table 5) showed very low SiO_2 contents (42.3 wt %), compatible with an ultrabasic composition, and high MgO (12.90 wt %), CaO (12.7 wt %), Fe_2O_3 (15.15 wt %), Cr (478 ppm), and Ni (117 ppm). This compositional spectrum, coupled with a high value of TiO_2 (3.34 wt %), agrees with the observed enrichment in ferromagnesian mineral and denotes magmatic alkaline features.

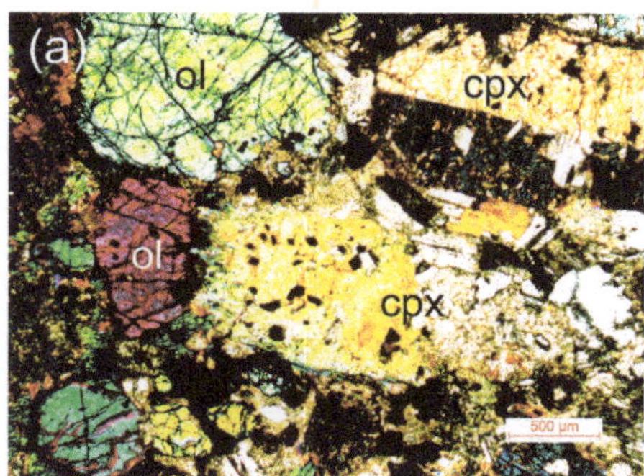

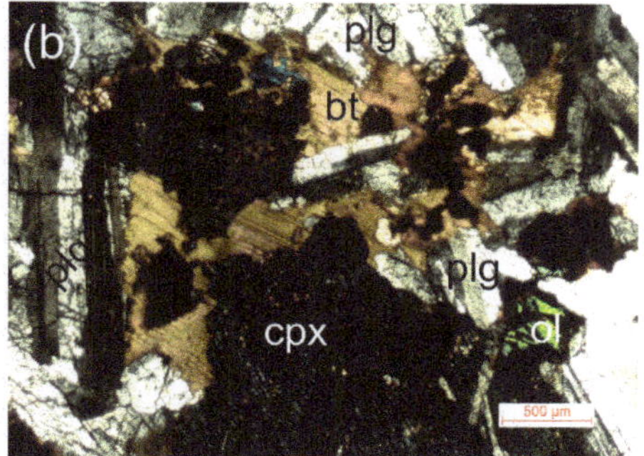

Figure 8. Microphotographs of (**a**) cumulate gabbro and (**b**) gabbro-diorite under polarized light microscope. ol—olivine, cpx—clinopyroxene, plg—plagioclase, bt—biotite.

Table 4. Mineral chemistry by scanning electron microscopy with energy-dispersive spectroscopy (SEM–EDS) in atomic percent (at %).

at %	Olivine $n = 5$		Pyroxene $n = 3$		Amphibole $n = 4$		Plagioclase $n = 4$		Ilmenite $n = 2$	
Cumulate Gabbro										
	Min	Max	Min	Max	Min	Max	Min	Max	Min	Max
O	61.567	63.646	60.440	61.007	59.764	60.797	62.562	65.073	61.711	61.529
Si	10.764	11.855	15.941	16.566	12.313	12.986	13.904	14.814	0.804	0.600
Ti	n.d.	n.d.	0.350	0.439	1.637	1.771	n.d.	n.d.	15.833	16.344
Al	0.100	0.618	1.744	1.974	5.204	5.540	12.922	14.320	0.534	0.840
Mg	12.844	17.564	8.382	8.833	7.560	7.860	n.d.	n.d.	1.321	2.518
Fe	8.741	11.914	2.565	2.849	3.726	3.938	n.d.	n.d.	19.115	17.567
Ca	0.207	0.366	7.920	8.869	5.190	5.470	5.161	6.562	0.275	0.215
Mn	0.000	0.222	0.000	0.133	0.000	0.129	n.d.	n.d.	1.321	0.383
Na	n.d.	n.d.	1.056	1.122	2.124	2.390	1.661	2.558	n.d.	n.d.
K	n.d.	n.d.	n.d.	n.d.	0.524	0.584	0.000	0.295	n.d.	n.d.
Ca/(Ca + Na)							0.67	0.79		
Mg/(Mg + Fe)	0.52	0.67								

at %	Olivine $n = 3$		Pyroxene $n = 4$		Biotite $n = 3$		Plagioclase $n = 4$		Ilmenite $n = 2$	
Gabbro-Diorite										
	Min	Max	Min	Max	Min	Max	Min	Max	Min	Max
O	57.796	58.844	58.346	58.900	55.762	58.686	59.420	61.240	58.265	60.052
Si	11.846	12.429	16.456	17.472	12.624	14.348	18.240	19.420	2.412	1.765
Ti	n.d.	n.d.	0.450	0.571	1.801	2.563	n.d.	n.d.	8.199	16.736
Al	0.416	1.429	1.614	2.020	6.975	7.953	11.050	12.140	1.611	0.656
Mg	11.298	12.612	8.717	9.051	6.689	8.348	n.d.	n.d.	2.007	1.685
Fe	14.933	16.937	3.237	3.617	5.626	6.764	n.d.	n.d.	26.594	18.319
Ca	0.213	0.426	8.111	9.361	n.d.	n.d.	3.960	5.220	0.598	0.272
Mn	0.375	0.446	0.130	0.160	n.d.	n.d.	n.d.	n.d.	0.314	0.514
Na	n.d.	n.d.	0.982	1.132	1.005	1.639	1.633	2.530	n.d.	n.d.
K	n.d.	n.d.	n.d.	n.d.	4.523	4.547	0.270	0.390	n.d.	n.d.
Ca/(Ca + Na)							0.61	0.72		
Mg/(Mg + Fe)	0.40	0.46								

n—number of analyses; n.d.—not determined.

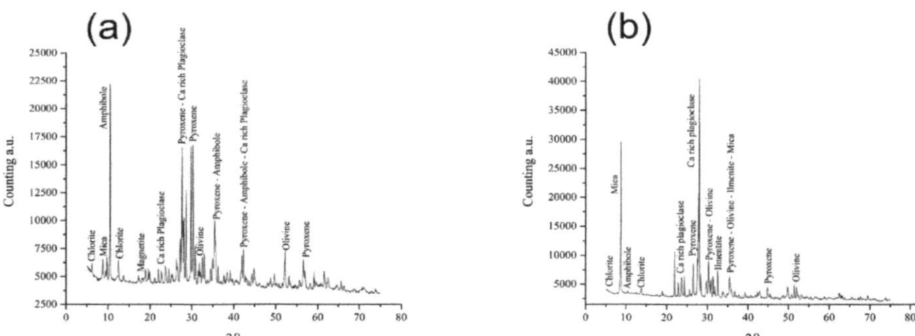

Figure 9. XRD diffractograms of (**a**) cumulate gabbro and (**b**) gabbro-diorite.

Table 5. Geochemical data in weight percent (wt %) and parts per million (ppm).

	Cumulate Gabbro		Gabbro-Diorite	
	wt %	Stat Error	wt %	Stat Error
SiO$_2$	42.30	±0.0344	49.00	±0.0356
TiO$_2$	3.34	±0.0175	3.26	±0.0176
Al$_2$O$_3$	9.40	±0.0295	16.20	±0.0368
Fe$_2$O3	15.50	±0.0133	11.20	±0.0115
P$_2$O$_5$	0.28	±0.00427	0.85	±0.00506
MnO	0.40	±0.005	0.32	±0.005
MgO	12.90	±0.0506	4.48	±0.0349
CaO	12.70	±0.0428	8.33	±0.0375
BaO	0.23	±0.013	0.28	±0.013
Na$_2$O	0.84	±0.0519	3.46	±0.0607
K$_2$O	0.19	±0.0345	1.42	±0.0395
LOI	0.89	-	0.03	-
Total	98.97	-	98.83	-
	ppm		ppm	
S	1900	±17.2	1500	±16.2
Rb	9	±2.08	40	±2.24
Sr	286	±2.57	748	±3.09
Y	15	±2.30	34	±2.46
Zr	80	±2.82	199	±3.14
Nb	20	±2.52	62	±2.65
Th	9	±2.94	10	±3.10
Cr	478	±29.0	20	±25.4
Co	198	±5.65	139	±5.02
Ni	117	±4.12	7	±3.42
Cu	62	±4.82	42	±4.88
Zn	103	±6.07	108	±6.37
Ga	14	±4.30	25	±4.67
As	7	±4.29	8	±4.53
Pb	0	±0	12	±17.2
Sn	7	±27.3	0	±0
V	479	±68.0	323	±68.0
U	0	±0.209	2	±0.221
Cl	43	±0.364	53	±0.359

The gabbro-diorite sample exhibited a hypidiomorphic texture and was found to be composed of plagioclase (50–60%) and clinopyroxene (20–25%), and, to a lesser extent, olivine (5–10%), biotite (10–15%), and ilmenite (5–10%). The sample was heterogeneous, with plagioclase-enriched layers alternated with ferromagnesian-enriched layers. Despite some fractures with chlorite and incipient sericitization, the sample had no significant alteration.

The mineral chemistry data (Table 4) indicated a more evolved composition with ferrous olivines (Fo = 0.40–0.46), diopside–augite compositions for clinopyroxenes, slightly fewer calcium plagioclases (An = 0.61–0.72), ferromagnesium micas, and Fe-rich ilmenite. The XRD (Figure 9) and mineral chemistry (Table 4) data support the previous observed and analyzed mineralogy. The geochemical data revealed (Table 5) a basic composition (SiO_2 = 49.00 wt %) and displayed lower MgO (4.48 wt %), CaO (8.33 wt %), and Fe_2O_3 (11.20 wt %), enhanced by low Cr (20 ppm) and Ni (7 ppm) and high Sr (748 ppm) values, which were directly related to the modal abundance of plagioclase, clinopyroxene, and olivine. The established basic composition and the TiO_2 value (3.26 wt %) obtained for the gabbro-diorite agree with the values reported by Canilho [38].

4. Discussion

In general, according to Schaef et al. [39], Rosenbauer et al. [40], and Alfredsson et al. [41], the injection of carbon dioxide into basaltic rocks has several advantages in comparison with sedimentary basins if the storage of CO_2 is through its mineral carbonation. Effective mineral carbonation of CO_2 is best achieved with ultrabasic or basic volcanic rocks due to the high contents of Mg^{2+}, Fe^{2+}, and Ca^{2+} in ferromagnesian mineral phases, which can react with CO_2 and precipitate as carbonates (e.g., magnesite, siderite, or calcite).

Since basalt carbonation could therefore become an important carbon storage solution [15], we selected 10 ultramafic and mafic rock areas in this study as a potential area for CO_2 storage. In a preliminary phase of site characterization for CO_2 geological storage in ultrabasic and basic volcanic rocks, we selected one geological formation with the potential to be considered as adequate for CO_2 storage. Comparing the 10 studied areas, we found that the Sines massif had more favorable characteristics (Table 3) than the other areas with respect to all the relevant parameters evaluated (Table 2). Its location, only a few kilometers from the main CO_2 sources in the region, which would avoid excessive transport costs, coupled with the large expected volume of cumulate gabbro with a high percentage of mafic minerals, provided the best prospect in the study area for the implementation of mineral carbonation projects.

Despite the subvolcanic (intrusive) nature of the Sines massif, the study of mafic lithologies, in particular the cumulate gabbros with more than 60% of ferromagnesian minerals, presented the chemical and mineralogical features theoretically required for mineral carbonation. Frequently, the boundaries between the cumulate gabbros and diorites were unclear, and thus it became necessary to test not only the more mafic lithologies but also the dioritic rocks in order to assess the mineral carbonation potential of the rock massif.

As such, samples of both rock types were tested in laboratory experiments to quantify their reactivity with a natural brine highly saturated with CO_2, under supercritical conditions, to understand the chemical reactions that are expected to occur in the area immediately surrounding an injection well. These laboratory experiments were designed to assess whether the CO_2–water–rock interaction would dissolve the cations of cumulate gabbros and gabbro-diorites necessary to react the CO_2 and promote fast mineral carbonation.

The Sines massif is ideally located close to the CO_2 sources in the region, which are hundreds of meters to a few kilometers away from the cumulate gabbros outcrops. A further advantage is that the subvolcanic massif extends from those outcrops to the offshore, where it encompasses a much larger area, most of which is covered by recent sediments. A scenario of CO_2 injection in the cumulate gabbros in the near offshore could be used to assess whether the fracture permeability is sufficient to allow for injection. Geophysical surveys and reinterpretation of previous data were necessary to clarify the offshore area, volume, and fracture pattern. The studies of fracture density did not show a consistent and pervasive fracture pattern, and it is expected that fracture density decreases with depth, which may make an in situ carbonation process unviable.

Mineral carbonation can also be achieved ex situ by resorting to the mining of the cumulate gabbros in quarries and promoting its reaction with CO_2 in specifically built facilities. A large quarry already exists in the Sines massif. Enhanced weathering is another possibility. The biggest contributor to the natural cycle of removing carbon dioxide from the atmosphere is the chemical weathering of certain types of rocks. This process is obviously very slow, but some studies (e.g., [22–42]) have demonstrated that, under the right conditions, the utilization of crushed mafic rocks could result in a net removal of CO_2 from the atmosphere. The mineral carbonation experiments being conducted should indicate whether the cumulate gabbros and gabbro-diorites from the Sines massif have some potential for these applications.

5. Conclusions

Several outcropping mafic rock massifs in the south of Portugal may present an opportunity for mineral carbonation from the CO_2 sources located in the region, which make up the largest industrial cluster in the country. The conditions of these rock massifs, composed primarily of plutonic rocks, are far from ideal in terms of porosity and permeability, and thus we adopted a set of ranking criteria to select the rock massifs that may present the highest potential. The adopted ranking criteria considered geological factors and socioeconomic and environmental constrains, with the most relevant being the percentage of mafic minerals, the fracture density, and the proximity to CO_2 sources to minimize transport costs. Criteria such as exceptionally low expected volume, the existence of productive aquifers, and environmentally protected areas were eliminatory factors.

Application of the screening and ranking procedure allowed us to select, from the identified 10 ultramafic and mafic rock massifs in the study area, the two with the best conditions for mineral carbonation—the Sines massif and the gabbros from Torrão-Odivelas massif.

The petrographic, mineral chemistry, and geochemical study of samples from the Sines massif demonstrated that the mineralogy of the cumulate gabbros is, theoretically, very favorable for mineral carbonation, with a higher percentage of ferromagnesian minerals than was previously documented. This has prompted mineral carbonation experiments in which the rock samples are submerged in a brine supersaturated in CO_2, under pressure and temperature conditions enough to ensure supercritical CO_2 conditions, in order to understand the CO_2–brine–CO_2 reactions. These experiments and their results are described in a second article [23].

Author Contributions: J.P., A.A.A., P.M., and J.C. were responsible for the selection of mafic and ultramafic massif rocks. J.P., A.A.A., P.M., M.B., L.L., and J.C. sampled the massif rocks. A.A.A. performed the ranking of massif rocks. J.P., A.A.A., P.M., L.L., A.C., and J.C. collected the geological and structural field data. J.P., P.M., and M.B. were responsible for sample preparation, petrography, mineral chemistry, and geochemical sample characterization. J.P., A.A.A., P.M., E.B., and J.C. interpreted the data and planned this paper. All the authors discussed the results and edited the manuscript. All authors have read and agreed to the published version of the manuscript.

Funding: This work was supported by national funds through the Portuguese FCT—Fundação Para a Ciência e Tecnologia—in the scope of project INCARBON, contract PTDC/CTA-GEO/31853/2017.

Acknowledgments: APS—Ports of Sines and the Algarve Authority, S.A. is gratefully acknowledged for cooperation on the collection of rock samples from the Monte Chãos quarry. We also would like to thank two anonymous reviewers for their constructive comments.

Conflicts of Interest: The authors declare no conflict of interest. The funders had no role in the design of the study; in the collection, analyses, or interpretation of data; in the writing of the manuscript; or in the decision to publish the results.

References

1. IEA. *Energy Technology Perspectives 2017: Catalysing Energy Technology Transformations*; IEA/OECD: Paris, France, 2017. [CrossRef]
2. UNFCCC. *Adoption of the Paris Agreement—Proposal by the President*; UNFCCC—United Nations Framework Convention on Climate Change: Paris, France, 2015; p. 31.

3. APA. *Roadmap for Carbon Neutrality 2050—Long-Term Strategy for Carbon Neutrality of the Portuguese Economy by 2050*; Agência Portuguesa do Ambiente: Lisbon, Portugal, 2019; p. 102.

4. European Commission. *The European Green Deal. Communication from the Commission to the European Parliament, the European Council, the Council, the European Economic and Social Committee and the Committee of the Regions*; European Comission: Brussels, Belgium, 2019; p. 24.

5. Boavida, D.; Carneiro, J.; Martinez, R.; van den Broek, M.; Ramirez, A.; Rimi, A.; Tosato, G.; Gastine, M. Planning CCS Development in the West Mediterranean. *Energy Procedia* **2013**, *37*, 3212–3220. [CrossRef]

6. Seixas, J.; Fortes, P.; Dias, L.; Carneiro, J.; Boavida, D.; Aguiar, R.; Marques, F.; Fernandes, V.; Helseth, J.; Ciesielska, J.; et al. *CO_2 Capture and Storage in Portugal: A Bridge to a Low Carbon Economy*; FCT-UNL: Lisbon, Portugal, 2015; p. 42.

7. EU ETS. European Union Emission Trading System—European Union Transaction Log. Available online: https://ec.europa.eu/clima/ets/oha.do (accessed on 2 December 2019).

8. Carneiro, J.; Martinez, R.; Suaréz, I.; Zarhloule, Y.; Rimi, A. Injection rates and cost estimates for CO_2 storage in the west Mediterranean region. *Environ. Earth Sci.* **2015**. [CrossRef]

9. Van den Broek, M.; Boavida, D.; Cabal, H.; Carneiro, J.; Fortes, P.; Gouveia, J.; Labriet, M.; Lechón, Y.; Martinez, R.; Mesquita, P.; et al. *Report with Selection of Most Promising CCS Infrastructure Options. COMET Technical Note TN6.4*; University of Utrech: Utrecht, The Netherlands, 2013; p. 173.

10. Veloso, F.M.L. STRATEGY CCUS-Strategic planning of Regions and Territories in Europe for low-carbon energy and industry through CCUS Coordination and Support Action (CSA). In Proceedings of the CO2GEONET Open Forum, Venice, Italy, 6–9 May 2019.

11. Matter, J.M.; Stute, M.; Snaebjornsdottir, S.O.; Oelkers, E.H.; Gislason, S.R.; Aradottir, E.S.; Sigfusson, B.; Gunnarsson, I.; Sigurdardottir, H.; Gunnlaugsson, E.; et al. Rapid carbon mineralization for permanent disposal of anthropogenic carbon dioxide emissions. *Science* **2016**, *352*, 1312–1314. [CrossRef] [PubMed]

12. Zevenhoven, R.; Eloneva, S.; Teir, S. Chemical fixation of CO_2 in carbonates: Routes to valuable products and long-term storage. *Catal. Today* **2006**, *115*, 73–79. [CrossRef]

13. Goldberg, D.S.; Takahashi, T.; Slagle, A.L. Carbon dioxide sequestration in deep-sea basalt. *Proc. Natl. Acad. Sci. USA* **2008**, *105*, 9920–9925. [CrossRef]

14. Andreani, M.; Luquot, L.; Gouze, P.; Godard, M.; Hoisé, E.; Gibert, B. Experimental Study of Carbon Sequestration Reactions Controlled by the Percolation of CO_2-Rich Brine through Peridotites. *Environ. Sci. Technol.* **2009**, *43*, 1226–1231. [CrossRef]

15. Gislason, S.R.; Oelkers, E.H. Carbon Storage in Basalt. *Science* **2014**, *344*, 373–374. [CrossRef]

16. Druckenmiller, M.L.; Maroto-Valer, M.M. Carbon sequestration using brine of adjusted pH to form mineral carbonates. *Fuel Process. Technol.* **2005**, *86*, 1599–1614. [CrossRef]

17. McGrail, B.P.; Spane, F.A.; Amonette, J.E.; Thompson, C.R.; Brown, C.F. Injection and Monitoring at the Wallula Basalt Pilot Project. *Energy Procedia* **2014**, *63*, 2939–2948. [CrossRef]

18. Hanchen, M.; Prigiobbe, V.; Baciocchi, R.; Mazzotti, M. Precipitation in the Mg-carbonate system—Effects of temperature and CO_2 pressure. *Chem. Eng. Sci.* **2008**, *63*, 1012–1028. [CrossRef]

19. Romão, I.S.; Gando-Ferreira, L.M.; da Silva, M.M.V.G.; Zevenhoven, R. CO_2 sequestration with serpentinite and metaperidotite from Northeast Portugal. *Miner. Eng.* **2016**, *94*, 104–114. [CrossRef]

20. Sanna, A.; Uibu, M.; Caramanna, G.; Kuusik, R.; Maroto-Valer, M.M. A review of mineral carbonation technologies to sequester CO_2. *Chem. Soc. Rev.* **2014**, *43*, 8049–8080. [CrossRef] [PubMed]

21. Beerling, D.J.; Leake, J.R.; Long, S.P.; Scholes, J.D.; Ton, J.; Nelson, P.N.; Bird, M.; Kantzas, E.; Taylor, L.L.; Sarkar, B.; et al. Farming with crops and rocks to address global climate, food and soil security. *Nat. Plants* **2018**, *4*, 138–147. [CrossRef] [PubMed]

22. Wu, J.C.S.; Sheen, J.-D.; Chen, S.-Y.; Fan, Y.-C. Feasibility of CO_2 Fixation via Artificial Rock Weathering. *Ind. Eng. Chem. Res.* **2001**, *40*, 3902–3905. [CrossRef]

23. Moita, P.; Berrezueta, E.; Abdoulghafour, H.; Beltrame, M.; Pedro, J.; Mirão, J.; Miguel, C.; Galacho, C.; Sitzia, F.; Barrulas, P.; et al. Mineral carbonation of CO_2 in mafic plutonic rocks. II: Early-phase SC CO_2-brine-rock interaction. *Appl. Sci.* **2020**, *10*, 5083.

24. Bachu, S. Sequestration of CO_2 in geological media: Criteria and approach for site selection in response to climate change. *Energy Convers. Manag.* **2000**, *41*, 953–970. [CrossRef]

25. CO2CRC. *Storage Capacity Estimation, Site Selection and Characterisation for CO_2 Storage Projects*; RPT08-1001; Cooperative Research Centre for Greenhouse Gas Technologies: Canberra, Australia, 2008; p. 52.

26. NETL. *Site Screening, Selection, and Initial Characterization for Storage of CO$_2$ in Deep Geologic Formations. 2013 Revised Edition*; DOE/NETL-2013/1605; National Energy Technology Laboratory: Pittsburgh, PA, USA, 2013; p. 110.

27. Oldenburg, C. *Health, Safety, and Environmental Screening and Ranking Framework for Geologic CO$_2$ Storage Site Selection*; LNBL: Sacramento, CA, USA, 2005; p. 22.

28. Ragnheidardottir, E.; Sigurdardottir, H.; Kristjansdottir, H.; Harvey, W. Opportunities and challenges for CarbFix: An evaluation of capacities and costs for the pilot scale mineralization sequestration project at Hellisheidi, Iceland and beyond. *Int. J. Greenh. Gas Control* **2011**, *5*, 1065–1072. [CrossRef]

29. Georgiou, C.D.; Sun, H.J.; McKay, C.P.; Grintzalis, K.; Papapostolou, I.; Zisimopoulos, D.; Panagiotidis, K.; Zhang, G.; Koutsopoulou, E.; Christidis, G.E.; et al. Evidence for photochemical production of reactive oxygen species in desert soils. *Nat. Commun.* **2015**, *6*, 7100. [CrossRef]

30. Beltrame, M.; Liberato, M.; Mirão, J.; Santos, H.; Barrulas, P.; Branco, F.; Gonçalves, L.; Candeias, A.; Schiavon, N. Islamic and post Islamic ceramics from the town of Santarém (Portugal): The continuity of ceramic technology in a transforming society. *J. Archaeol. Sci. Rep.* **2019**, *23*, 910–928. [CrossRef]

31. Teixeira, C. La structure annulaire subvolcanique des massifs éruptifs de Sintra, Sines et Monchique. In *Est. Cient. Oferecidos em Homenagem ao Prof. Carrington da Costa*; Junta de Investigação do Ultramar: Lisbon, Portugal, 1962; pp. 41–49.

32. Ribeiro, A.; Antunes, M.T.; Ferreira, M.P.; Rocha, R.B.; Soares, A.F.; Zbyszewski, G.; Moitinho de Almeida, F.; Carvalho, D.D.; Monteiro, J.H.; Serviços Geológicos de, P. *Introduction à la Géologie Générale du Portugal*; Serviços Geológicos de Portugal: Lisbon, Portugal, 1979.

33. Carvalho, J.P.G.; Torres, L.M.; Afilhado, A. Delimitação do maciço sub-vulcânico de Sines offshore a partir de dados geofísicos. *Comum. Serv. Geol. Port.* **1998**, D57–D60.

34. Inverno, C.; Manuppella, G.; Zbyszewski, G.; Pais, J.; Ribeiro, L. *Notícia Explicativa da Folha 42-C Santiago do Cacém. Carta Geológica de Portugal de 1/50 000*; Serviços Geológicos de Portugal: Lisbon, Portugal, 1993.

35. Miranda, R.; Valadares, V.; Terrinha, P.; Mata, J.; Azevedo, M.d.R.; Gaspar, M.; Kullberg, J.C.; Ribeiro, C. Age constraints on the Late Cretaceous alkaline magmatism on the West Iberian Margin. *Cretac. Res.* **2009**, *30*, 575–586. [CrossRef]

36. Canilho, M.H. Estudo geológico-petrográfico do maciço eruptivo de Sines. *Bol. Mus. Lab. Min. FCUL* **1972**, *12*, 77–161.

37. GeoAlgar. *Pedreira Monte Chãos Estudo Geológico-Geotécnico. Internal Report to APS (Administração do Porto de Sines)*; GeoAlgar: Sines, Portugal, 2019; p. 38.

38. Canilho, M.H. Elementos de geoquímica do maciço ígneo de Sines. *Ciências Da Terra UNL* **1989**, *10*, 65–80.

39. Schaef, H.T.; McGrail, B.P.; Owen, A.T. Carbonate mineralization of volcanic province basalts. *Int. J. Greenh. Gas Control* **2010**, *4*, 249–261. [CrossRef]

40. Rosenbauer, R.J.; Thomas, B.; Bischoff, J.L.; Palandri, J. Carbon sequestration via reaction with basaltic rocks: Geochemical modeling and experimental results. *Geochim. Cosmochim. Acta* **2012**, *89*, 116–133. [CrossRef]

41. Alfredsson, H.A.; Oelkers, E.H.; Hardarsson, B.S.; Franzson, H.; Gunnlaugsson, E.; Gislason, S.R. The geology and water chemistry of the Hellisheidi, SW-Iceland carbon storage site. *Int. J. Greenh. Gas Control* **2013**, *12*, 399–418. [CrossRef]

42. Luckow, P.; Wise, M.A.; Dooley, J.J.; Kim, S.H. Large-scale utilization of biomass energy and carbon dioxide capture and storage in the transport and electricity sectors under stringent CO$_2$ concentration limit scenarios. *Int. J. Greenh. Gas Control* **2010**, *4*, 865–877. [CrossRef]

Article

Valorization of Slags Produced by Smelting of Metallurgical Dusts and Lateritic Ore Fines in Manufacturing of Slag Cements

Theofani Tzevelekou [1], Paraskevi Lampropoulou [2,*], Panagiota P. Giannakopoulou [2], Aikaterini Rogkala [2], Petros Koutsovitis [2], Nikolaos Koukouzas [3] and Petros Petrounias [2]

[1] Laboratory of Metallurgy (METLAB), Department of Chemical Engineering, University of Patras, 26504 Patras, Greece; ftzevelekou@elkeme.vionet.gr

[2] Section of Earth Materials, Department of Geology, University of Patras, 26504 Patras, Greece; peny_giannakopoulou@windowslive.com (P.P.G.); krogkala@upatras.gr (A.R.); pkoutsovitis@upatras.gr (P.K.); Geo.plan@outlook.com (P.P.)

[3] Chemical Process & Energy Resources Institute, Centre for Research & Technology Hellas (CERTH), 15125 Maroussi, Greece; koukouzas@certh.gr

* Correspondence: p.lampropoulou@upatras.gr

Received: 2 June 2020; Accepted: 4 July 2020; Published: 7 July 2020

Featured Application: The smelting process can be applied for the direct recycling of the nickel ferrous dust collected in the gas cleaning systems of the rotary kilns (R/Ks) during the pre-reduction of laterite ore in the course of ferronickel production by the pyrometallurgical route. The method can also be applied for smelting of ore fines. Contained Ni is recovered in the liquid metal resulting in the production of high purity low Ni-alloyed steel grades following secondary metallurgical treatment. The composition of the produced slags by this method can be efficiently adjusted so that the slags could be used in the manufacturing of special slag type composite cements substituting clinker. Present findings indicate that there is also potentially space for improvement in conventional EAF steel slags composition to allow for their wider use in building sector. In general, the developed zero residues recycling process concept can be applied for the recycling of metallurgical wastes/byproducts (dusts, fines etc.) by "one-stage" reduction smelting in DC-arc furnaces without any pretreatment (agglomeration, briquetting, sintering) recovering the contained valuable metals in the liquid bath, while producing slags suitable for further utilization in the manufacturing of high added value environmental friendly building materials (cements) accountable for less CO_2 footprint; relative to the amount of clinker being substituted.

Abstract: A pyrometallurgical process was developed for the recycling of Ni bearing dusts and laterite ore fines by direct reduction smelting in DC (direct current) arc furnace. In the course of the performed industrial trials, besides the Ni-recovery in the liquid bath, slag composition was deliberately adjusted in order to produce a series of metallurgical slags with different chemical and mineralogical composition. The aim of this study was to investigate their suitability as clinker substitute in cement manufacturing. Examined parameters were slag FeO_x content, basicity and applied cooling media (air, water cooling). A series of composite Portland and slag cements were manufactured in laboratory scale incorporating 20% and 40% of each slag, respectively; the rest being clinker of OPC (ordinary Portland cement) and 5% gypsum. The extended mineralogical analysis and microstructural properties of the produced slags were examined and correlated with the properties of the produced cements. The physical and mechanical characteristics of all examined cement products were found to meet the requirements of the regulation set for cements. The present research revealed that the most critical parameter in the compressive strength development of the slag cements is the mineralogical composition of the slag. Even in cases where rapid cooling to obtain glassy matrix is not feasible, adjustment of slag analysis to obtain mineralogical phases similar to those met in clinker of OPC, even at higher FeO contents (up to ~21wt.%), can result in production of slag with considerable latent hydraulic properties. These results indicate that there is potentially space for adjustments in conventional EAF (electric arc furnace) steel slags composition to allow for their wider use in cement manufacturing with significant environmental and economic benefits resulting from the reduction of energy requirements, CO_2 emissions and natural raw materials consumption.

Keywords: slag valorization; metallurgical dusts; slag cement; CO_2 emissions; EAF slag; zero waste

---

1. Introduction

Electric arc furnace (EAF) steelmaking has been proven a promising recycling route for a variety of industrial, mining and metallurgical wastes enabling the recovery of contained metals in the produced steel products, while exploiting their chemical and/or energy content [1–7]. Technological advancements in auxiliaries, such as hollow graphite electrode in direct current (DC)-arc furnaces, special tuyères and manipulator lances have enabled further the efficient recycling of powder waste materials, as well as the feeding of fluxes and carbonaceous reductants along with carrier and process gases, allowing overall high recovery yield of valuable metals in the liquid bath [4,8,9]. In modern steelmaking furnaces these "redox-tools" also provide the ability to adjust the slag chemistry at request, so that, besides favoring the metallurgical yield of the process, it may well become appropriate for further utilization, as a secondary raw material in the production of high added value products. Slag valorization capability depends on its chemical and mineralogical composition resulting from the originating production process and applied cooling methodology. Generally, knowledge of the chemical, mineralogical, and morphologic characteristics of ferrous slags is essential, because their cementitious and mechanical properties, which play the key role in their utilization, are closely related to these features. For example, the properties in abrasion and attrition of steel slag are influenced by its morphology and mineralogy. Similarly, the volumetric stability of steel slag is a function of its chemistry and mineralogy. The most well established valorization case of ferrous slags is the utilization of Ground-Granulated Blast-Furnace slag (GGBS or GGBFS) as a substitution of clinker in cement manufacturing. Besides the conservation of natural resources, this substitution results in significant reduction of CO_2 emissions per ton of produced cement, relative to the amount of clinker being substituted. In general, the production of 1.0 t of Portland cement causes 0.8–1 t CO_2, while indicatively 1.0 t of slag cement containing 75% of GGBS causes emission of only 0.3 t CO_2 [10,11]. These data include the calcination process, emissions from fossil fuel burning and the use of electricity.

For the case of EAF slag, its properties render it a good candidate for many engineering applications, including as a (heavyweight) aggregate for concrete, as radiation-shielding concrete, as an environmentally friendly substitute of natural aggregates such as barite, hematite or limonite [12]. Nevertheless, although the approach of EAF slag utilization in the final stage of cement manufacturing as clinker substitute similar to the exploitation of blast furnace slag has been limited researched [13–15] it has not been usually practiced, mostly due to its weak hydraulic properties [16]. Further, steel slag is not currently included in the European Norm in force for the types of cements [17]. Other hindering reasons, also constraining its direct wide use in concrete, are volume instability and potential swelling attributed to hydroxylation and subsequent carbonation of their free CaO and MgO contents [18–20], long-term oxidation of metallic iron from divalent to trivalent valence, as well as the transformation of β-C_2S ($2CaO \cdot SiO_2$) to γ-C_2S [12]. Finally, an environmental factor that has been discouraging in using some steel slags in constructions is the potential leaching of undesired metals over the life time of each construction application; however, this risk is particularly relevant for slags from stainless steel production [12]. Hot stage slag engineering aiming to influence slags chemical composition to enable higher value-added applications and energy recuperation has been gaining attention in the vision for sustainable metallurgy [21–23]. The improvement of the EAF slag cementitious properties by adjusting its chemical and mineralogical composition during the smelting and cooling periods, while maintaining a positive impact on the quality of produced metal, could potentially widen its use in construction sector, overcoming the aforementioned obstacles. In this direction, to tackle leachability issues, additions of bauxite, Al_2O_3 containing residues and aluminum metal as a method to increase the stability of stainless steel EAF slags and to stabilize chrome (Cr) was suggested by Mudersbach et al. [24], while Primavera et al. [25] has described the addition of a stabilizer oxide in the treatment of electric arc furnace (EAF) slags. However, studies for optimized EAF-slag, the primary targeted use remains as an aggregate in construction applications i.e., road bed. This use, even it is quite beneficial, does not maximize the value added benefits for the steel manufacturer. Preferably, its successful use in cement manufacturing as clinker substitute should upgrade its added value in the circular economy chain, yield important environmental and economic benefits for both steel and cement industry, contribute to the conservation of natural resources, as well as to further reduction of CO_2 emissions in cement industry, by expanding the cycle of alternatives to clinker secondary raw materials [19,26].

A reduction smelting process was developed for the efficient recycling of metallurgical Ni-bearing dust by injection through the hollow electrode in DC—EAF [4,27,28]. This dust is collected in the gas cleaning systems of the rotary kilns (R/Ks) during the pre-reduction of laterite ore in the course of ferronickel production by the pyro-metallurgical route. This method can also be applied for the smelting of ore fines; their collection and treatment prior to their feeding in the R/Ks will considerably reduce the amount of generated dust. A significant advantage of the developed process in the DC-arc furnace is that it allows the direct smelting of the fine materials without any pretreatment (agglomeration, briquetting, sintering) [4,8]. During the industrial testing of the process, apart from the effective dust and ore fines recycling and nickel recovery in the metal, the interest was also focused on the production of different steel slag types in terms of chemical and mineralogical composition. The objective of the present research was to investigate their suitability as clinker substitute in common cement manufacturing and further the correlation of the obtained results to the microstructural characteristics of the used slags, to establish the development of a zero residues production melting process for metallurgical dusts and ore fines recycling as presented to the flow sheet in the corresponded supplementary file. Examined parameters were slag FeO_x content, basicity and applied cooling media (air, water cooling). A series of cements incorporating 20% and 40% of each slag respectively were manufactured in laboratory scale. Since the current European Norm [17] for common cements refers only to the utilization of ground granulated blast furnace slag (GGBS) in cement production, the physical and mechanical characteristics of the manufactured cements in the present study have been compared to the requirements for composite Portland cements with blast furnace slag (CPCs) and slag cements (SCs) for the mixtures with 20% and 40% slag content, respectively. The main difference is that blast furnace slag accepted in cement production according to EN 197-1 [17], contains at least two-thirds by mass of glassy slag and has minimum iron oxide content.

2. Materials and Methods

2.1. Tests in Raw Materials

Industrial reduction smelting trials for the recycling of Ni-bearing dust and laterite fines have been conducted in DC–EAF furnace with very high Ni-recovery in the produced metal, as described in [4], where the furnace installation, the data of conducted industrial trials and their metallurgical evaluation have been thoroughly analyzed. The liquid metal following secondary metallurgical treatment resulted in the production of low nickel alloyed steel grades [4]. Typical chemical analysis of the most important constituents of the dust and ore fines is presented in Table 1, while their detailed characterization including mineralogical and microstructural analysis by XRD, SEM/EDS and grain size distribution can be found in [4,28]. Slag chemistry adjustment was carried out during the trials in order to produce slags with desired range of chemical composition. The slag analysis was deliberately influenced by proper fluxes and reductants additions during the smelting campaigns, as described in [4,26,28]. The selected slag parameters examined were the FeO_x content, the basicity of the slags expressed as CaO/SiO_2 (wt.%/wt.%) and the effect of slag cooling procedure (air or water cooling).The objective was to investigate the effect of these parameters on the slags utilization as clinker substitute in the production of slag containing cements. This was achieved by pre-calculated additions of reducing agents and fluxes; namely, carbon and metallurgical lime, respectively. Having the targeted use in mind, addition of dolomitic lime was avoided in order not to increase slags MgO content and also for promoting FeO_x reduction. Two different means of slag cooling were also applied: (a) air by tapping liquid slag in a pit in the plant yard and leave it to cool and (b) water by purring liquid slag in a basket filled with water. In total, nine different slags were collected.

Table 1. Typical chemical analysis (wt.%) of Ni-bearing dust and lateritic ore fines [4].

	Dust	Ore Fines
FeO_x	30.44	36.65
SiO_2	35.00	34.85
MnO	0.50	0.87
Cr_2O_3	3.00	5.82
Al_2O_3	11.55	4.31
CaO	4.10	4.18
MgO	5.27	5.92
NiO	1.80	1.37
S	0.44	0.20
C_{tot}	3.60	0.94
Moisture	0.24	1.82

The chemical analysis of the slags was determined by X-ray fluorescence, XRF. The mineralogical analysis of the slags were determined by powder X-ray diffraction (XRD), using a Bruker D8 Advance diffractometer (Bruker, Billerica, MA, USA), with Ni-filtered CuKα radiation. The scanning area for bulk mineralogy of the samples covered the 2θ interval 20–70°, with a scanning angle step size of 0.015° and a time step of 0.1 s. The mineral phases were determined using the DIFFRAC plus EVA 12® software (Bruker-AXS, Billerica, MA, USA) based on the ICDD powder diffraction file of PDF-2 2006. Study of the slags microstructure was carried out by Scanning Electron Microscopy (SEM) (JEOL JSM-6300), while several microanalyses were performed with Energy Dispersive X-ray Spectroscopy (SEM/EDS).

2.2. Tests in Cement Production

A series of composite Portland and slag cements were manufactured in laboratory scale by co-grinding 20% and 40% of each slag, respectively with clinker of ordinary Portland cement (OPC). In all cement mixtures 5% gypsum was used and the remaining mass was clinker of OPC. Reference OPC (95% clinker, 5% gypsum) mixtures were also manufactured for comparison purposes. The experimental procedure involved separate crushing of the slag and the clinker, so that 95% is below 5 mm. The crushed materials were mixed with gypsum in the aforementioned ratios and subsequently co-ground in a 5-kg ball mill until targeted specific surface (Blaine) of 4500 cm^2/g was achieved. Indicatively two cements incorporating 20% slag but targeting to lower Blaine of 3650 cm^2/g were also comparatively manufactured along with corresponding reference OPC. Once grinding was completed, proper specimens were prepared for the determination of the experimental cement's technological properties according to the corresponding standards specified in EN 197-1 [17], the standard in force for cement products. Initial and final setting time, expansion and early, late and long-term compressive strengths up to 90 days of hydration were measured.

In the final step of this study, co-evaluation of the results was performed and correlations between the mineralogical analysis and microstructural properties of the produced steel slags with the physical and mechanical characteristics of the produced cements were acknowledged. Since the European Norm in force for the types of cements [17] refers only to the utilization of GGBS in cement production, experimental cements classification was made indicatively following the corresponding ranges described in [17] for blast furnace slags. The obtained results of this research are hereafter presented and discussed.

3. Results and Discussion

3.1. Slags Characterization

3.1.1. Chemical Composition

The chemical composition of the so produced slags in wt.% used for cements manufacturing and the mean of their cooling are presented in Table 2. Due to the applied experimental conditions of the trials [4,28], iron is expected to be mainly in divalent state in slags 3 and 4, while higher trivalent iron content is present in slag 5. Accordingly, higher fraction of Cr should be expected to be divalent than trivalent form due to the highly reducing conditions of the smelting process [28,29]. The lowest FeO content in slags 4-2 was achieved by aluminum addition during smelting via the aluminothermic reduction reaction $3(FeO) + 2\,[Al] = Al_2O_3 + 3\,[Fe]$, increasing also the Al_2O_3 content of the slags.

Table 2. Chemical composition of the produced slags used in cements manufacturing.

No. [*]	Bas.	FeO_x (wt.%)	Al_2O_3 (wt.%)	MgO (wt.%)	MnO (wt.%)	Cr_{tot} (wt.%)	S (wt.%)
3-1a, b	1.68	10.7	7.72	10.8	2.48	0.53	0.29
3-2a, b	1.32	5.32	8.57	11.5	2.51	0.60	0.30
4-1a, b	1.50	7.30	11.6	8.93	3.12	0.96	0.36
4-2a, b	2.61	3.85	12.00	7.54	1.68	0.19	0.39
5a	2.15	21.1	7.36	4.36	2.36	0.88	0.20

a—air-cooled, b—water cooled slags, bas—basicity (wt.% CaO/wt.% SiO_2), (*)—industrial trials traceability index.

3.1.2. Mineralogical Characterization and Microstructure

The mineralogical analyses of the slags are summarized in Tables 3 and 4 for the air and water cooled samples, respectively. The high number of phases detected in each XRD pattern (Figure 1), therefore high overlapping of peaks, made their distinguishing quite difficult and SEM/EDS analysis assisted substantially towards this direction (Table 5). Special care was taken to present selected microanalyses that are either stoichiometric or closely approximate the stoichiometry, since the fractured analyzed surfaces of samples prevented accurate quantitative analyses. Moreover, characteristic microphotographs of their structure obtained by optical microscopy and SEM are presented in Figures 2–5.

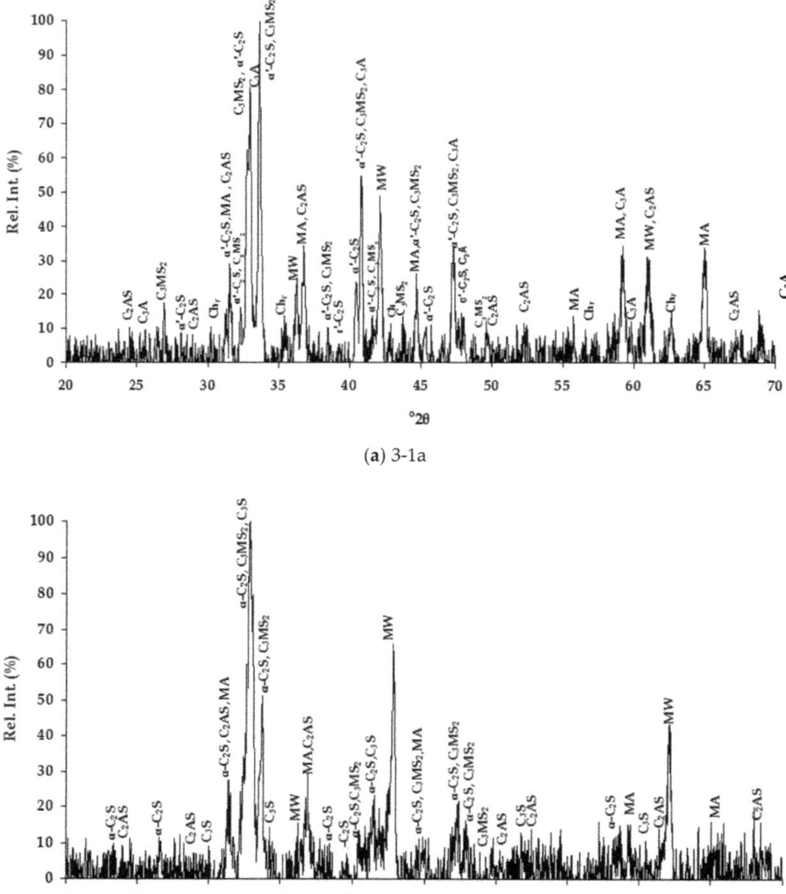

(**a**) 3-1a

(**b**) 3-1b

Figure 1. *Cont.*

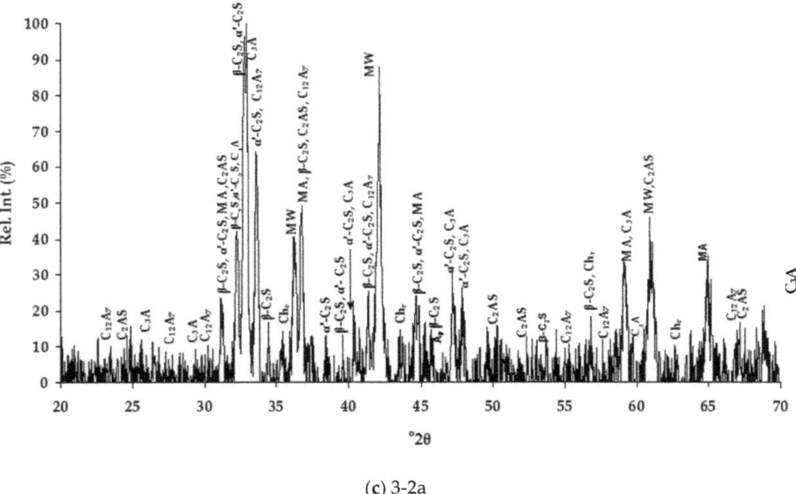

(**c**) 3-2a

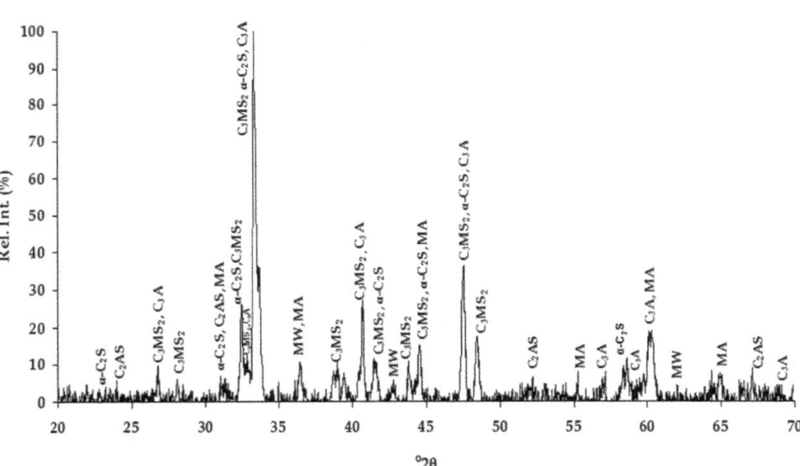

(**d**) 3-2b

Figure 1. *Cont.*

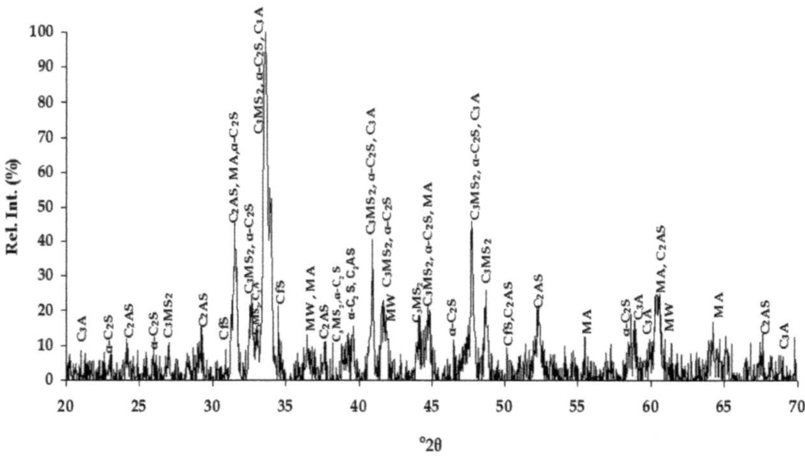

(**e**) 4-1a

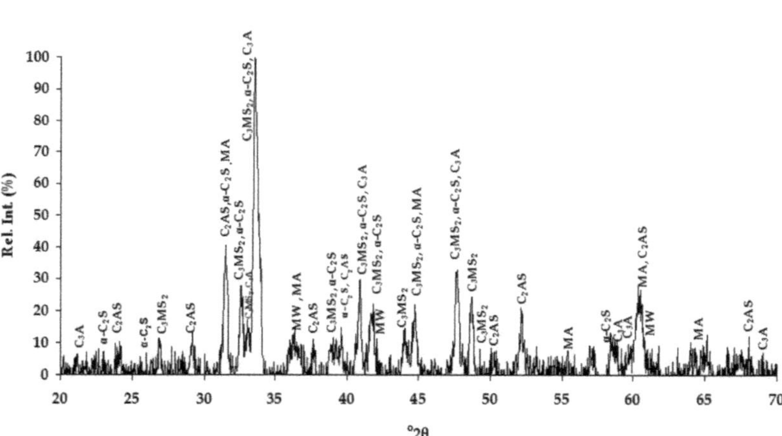

(**f**) 4-1b

Figure 1. *Cont.*

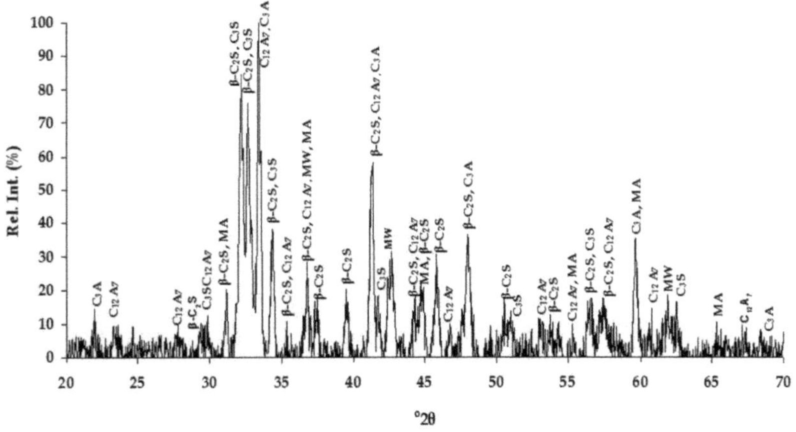

(**g**) 4-2g

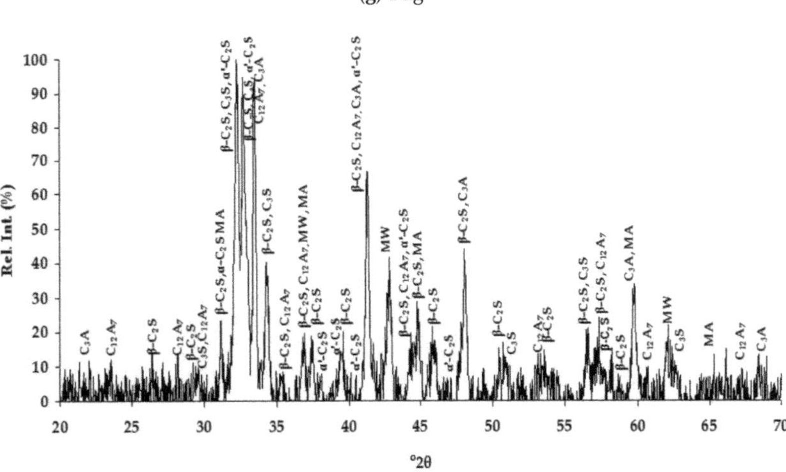

(**h**) 4-2b

Figure 1. *Cont.*

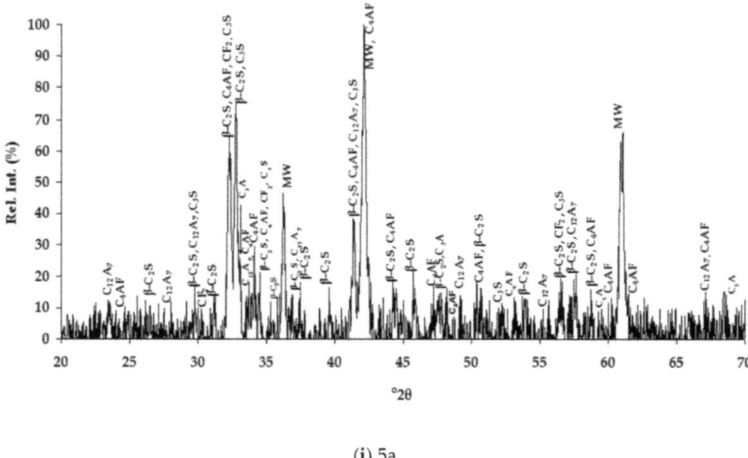

(i) 5a

Figure 1. X-ray diffraction patterns of the studied slags: (**a**) slag 3-1a; (**b**) 3-1b; (**c**) 3-2a; (**d**) 3-2b; (**e**) 4-1a; (**f**) 4-1b; (**g**) 4-2a; (**h**) 4-2b; (**i**) 5a. Abbreviations—α'-C_2S (α'-$2CaOSiO_2$-$Ca_{14}Mg_2(SiO_4)_8$)—α'-dicalcium silicate-bredigite; α-C_2S (α-$2CaOSiO_2$)—a-dicalcium silicate; β-C_2S (β-$2CaOSiO_2$)—belite, β-dicalcium silicate; C_3S ($3CaOSiO_2$)—alite, tricalcium silicate; C_3A ($3CaOAl_2O_3$)—tricalcium aluminate; C_3MS_2 ($3CaOMgO_2SiO_2$)—merwinite; MW ($Mg_{1-x}Fe_xO$)—magnesiowüstite; C_2AS ($2CaOAl_2O_3SiO_2$)—gehlenite; MA ($MgOAl_2O_3$)—spinel; Chr ($FeOCr_2O_3$)—chromite; CfS ($CaOFeOSiO_2$)—kirschteinite; $C_{12}A_7$ ($12CaO_7Al_2O_3$)—mayenite; C_4AF ($4CaOAl_2O_3Fe_2O_3$)—brownmillerite; CF_2 ($CaO_2Fe_2O_3$)—calcium iron oxide.

Table 3. Mineralogical phases of air-cooled slags.

	3-1a	3-2a	4-1a	4-2a	5a
α-C_2S			+		
α'-C_2S	+	+	+		
β-C_2S		+		+	+
C_3S				+	+
C_3MS_2	+		+		
C_3A	+	+	+	+	+
$C_{12}A_7$		+		+	+
C_2AS	+	+	+		
$Mg_{1-x}Fe_xO$	+	+	+	+	+
Spinel	+	+	+	+	
Chromite	+	+			
C_4AF					+
CF_2					+
$CaFeSiO_4$			+		

Table 4. Mineralogical phases of water-cooled slags.

	3-1b	3-2b	4-1b	4-2b
α-C_2S	+	+	+	
α'-C_2S				+
β-C_2S				+
C_3S	+			+
C_3MS_2	+	+	+	
C_3A		+	+	+
$C_{12}A_7$				+

Table 4. *Cont.*

	3-1b	3-2b	4-1b	4-2b
C_2AS	+	+	+	
$Mg_{1-x}Fe_xO$	+	+	+	+
Spinel	+	+	+	+
Ettringite	+			
Monosulfate				+

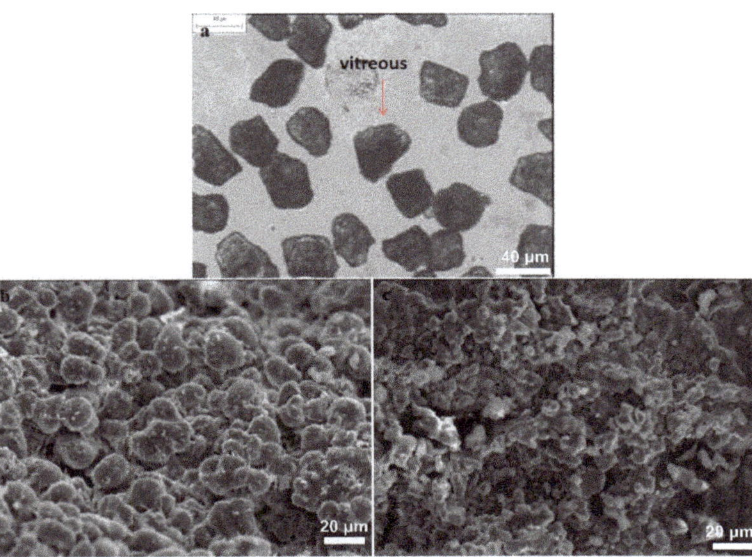

Figure 2. (**a**) Image of 35–45-µm fraction of water-cooled slag under polarized light; (**b**,**c**) SEM (SEI) micrographs of round-shaped C_2S crystals development.

Figure 3. (**a**) SEM (SEI) micrograph of MW crystals development in the cubic system; (**b**) micrograph of C_4AF plate crystals development; (**c**) micrograph of hexagonal C_3S crystal in the slag matrix; (**d**) micrograph of longwise merwinite crystals development in the slag matrix.

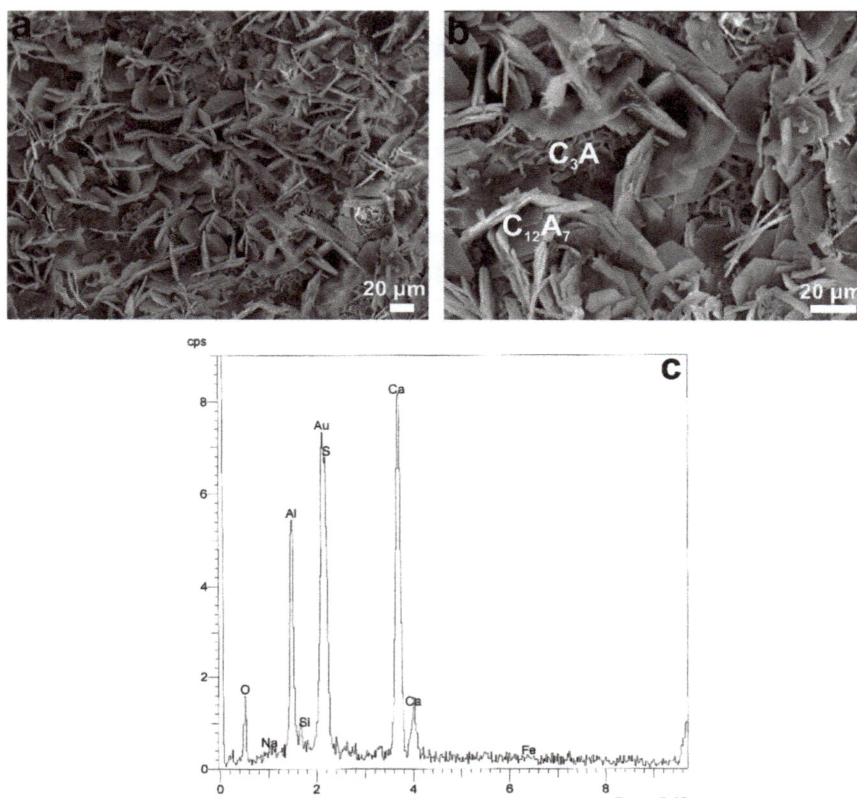

Figure 4. SEM (SEI) micrographs of calcium aluminates and hydrated calcium sulfo-aluminate phases in 4-2b slag (**a**) rich in calcium aluminates micro-area; (**b**) microstructure of calcium aluminate phases at higher magnification; (**c**) SEM/EDS spectrum of thin hexagonal plate (monosulfate).

According to the obtained results, well-formed crystals were also detected in the slags cooled in water as shown in Figure 1. The constrained amorphous (glass) content was also verified by the optical observation under polarized light of the water-cooled 35–45 µm slags fraction as indicatively shown in Figure 2a. In order to obtain vitreous granulated slag, as in GGBS, high pressure water jets are required [19]. The main differences in the mineralogical phases identified between air and water-cooled slags of the same chemical composition were the polymorphs of dicalcium silicate attributed to the different rates of cooling. The environmental conditions during the smelting trials and thus slags tapping (temperature <0 °C, rainfall) and formation of microcrystalline material prevented the formation of γ-C_2S (γ-2CaO·SiO$_2$) and the disintegration of the cooled slag phase due to $\beta \rightarrow \gamma$-C_2S transformation even in the air-cooled slags. The later takes place in cases of slow cooling rate and its presence is not desired in cementitious materials, since it exhibits slight hydraulic properties not contributing to strength development. The co-existence of α and/or α' and β-C_2S phases in the slag samples indicates that the theoretical thermodynamic transformation was not completed due to rapid temperature drop mainly in the initial phase of the cooling process [30]. The incorporation of Mg in C_2S could lead to the formation of bredigite, $Ca_{14}Mg_2(SiO_4)_8$—once the solubility limits in α and α'-C_2S are exceeded. Usually, its composition varies between $Ca_{1.75}Mg_{0.25}$ and $Ca_{1.7}Mg_{0.3}$ at ~1300 °C per SiO$_4$ [31] and its presence has been reported in cases of EAF steelmaking slags [32]. However,

α'-C_2S and bredigite have different structures, it has been confirmed [31,33] that the similarity in their XRD patterns makes distinguishing them very difficult (Figure 1). In this work, the nomenclature of α'-C_2S has been currently attributed to a probable combination of both phases. The semiquantitative analysis obtained by SEM-EDS verified solubility of Mg in C_2S, as well as the existence of purer C_2S micro-areas, while Al, Mn and Fe incorporated in the crystal lattice was also identified (Table 5, Figure 2). The incorporation of foreign ions in the structure of dicalcium silicates, as well as deviation from the stoichiometric composition, favor the stabilization of the α' and β phases [34].

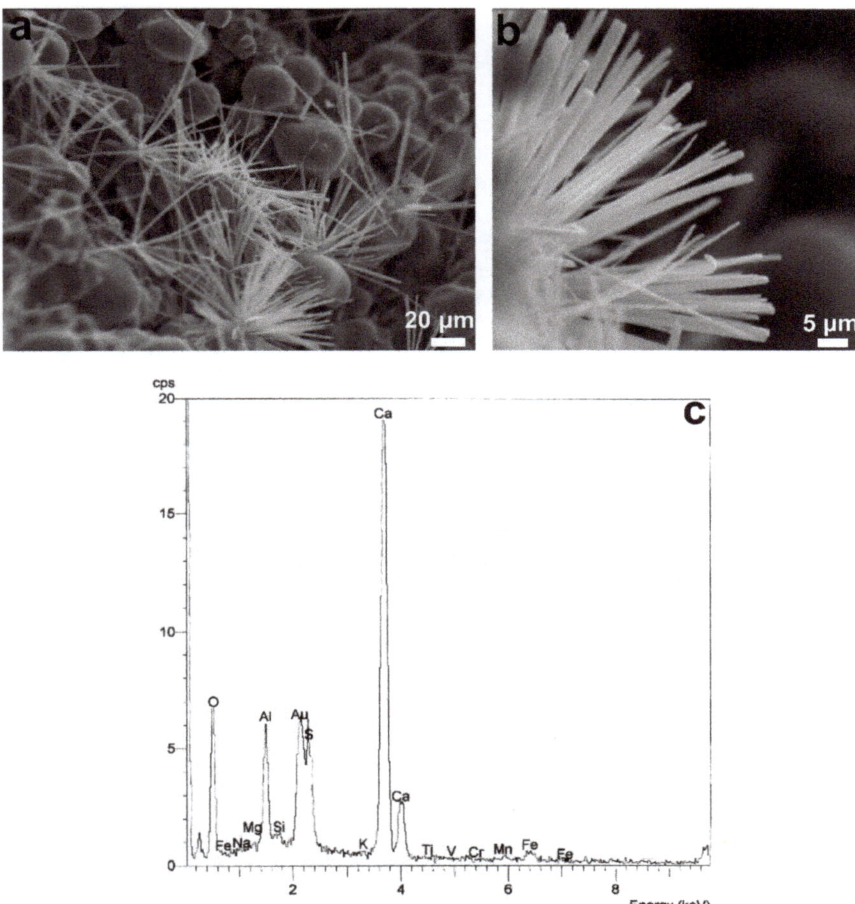

Figure 5. Ettringite needle like crystals formation in 3-1b water-cooled slag (**a**) microstructure of ettringite needles development in the slag; (**b**) magnified micrograph and; (**c**) SEM/EDS spectrum of ettringite needle-shaped crystals.

In all examined samples, magnesiowüstite phase was identified (Table 5, Figure 3a). Magnesiowüstite comprises a solid solution between MgO and FeO and it is commonly found in the mineralogical composition of steelmaking slags [19,21,35,36]. It is reflected in the XRD patterns by displaced peaks located between pure wüstite (41.890 2θ strongest line) and periclase ones (42.917 2θ strongest line), depending on the Mg/Fe^{2+} ratio and it has been assigned as $Mg_{1-x}Fe_xO$ (MW). Investigations have shown that the composition of these types of solid solutions in the slag depends on slag chemistry and on the cooling rate. Slower cooling rate will promote the enrichment of FeO into the structure, while MgO predominates by increasing the cooling rate [21,36]. This may explain the difference observed when comparing the microanalysis of MW phase in slags, 3-1a versus 3-1b (Table 5). Nevertheless, it may be noticed that the strongly reduced %FeO in the chemical analysis of 4-2 slag led to rich Mg-MW even in the air-cooled slag sample.

Incorporation of Manganese, Calcium and Chromium (most possibly divalent) in magnesiowüstite solid solutions was also detected in the course of SEM-EDS analysis in the slag samples. Chromium was also identified in XRD analysis in chromite phase in two of the air-cooled slag samples (3-1,2a). Increase of cooling rate prevents the thermodynamic chromite spinel formation from the MW solid solution with temperature drop, preserving Chromium in the MW phase. In high basicity EAF slags, primary crystallization of Chromium occurs in magnesiowüstite [21,36]. This is illustrated in the microanalyses of the mageniowüstite solid solution phases in the higher basicity air-cooled slag samples 4 and 5, (Table 5). Increase of FeO amount in these solid solutions has also been reported to decrease its leachability from the slag, which is an advantageous property for its further envisaged use [21,36].

Overall, iron in the examined slags was mainly in the divalent state bound in the MW phase as described above. In slag 5, where limited reduction of iron oxides was targeted through the applied experimental conditions, brownmillerite, $Ca_4Al_2Fe_2O_{10}$ (C_4AF) and calcium ferrite ($CaO.2Fe_2O_3$) were additionally identified, also indicating presence of trivalent iron in the slag. Brownmillerite is commonly met in the clinker of Portland cement, while calcium ferrite phase in contrast to MW has been reported as phase of hydraulic nature, too [18]. Micro-areas rich in C_4AF crystals were detected by SEM/EDS examinations in slag 5 sample (Table 5, Figure 3b).

Tricalcium silicate $3CaO.SiO_2$ (C_3S), being the most important phase in cement for early strength development among the calcium silicates, was mainly detected in the slag products, 4-2a, b and 5, where the basicity was higher, while some traces were also found in slag 3-1b. Typical micrograph and microanalyses of identified C_3S hexagonal crystal in the slags microstructure is shown in Figure 3c and Table 5, respectively. Presence of merwinite, $3CaO·MgO·2SiO_2$ (C_3MS_2), common constituent of EAF steelmaking slags was also identified in the XRD and SEM/EDS analysis of the slags. Characteristic micro-area of solidified merwinite crystals is shown in Figure 3d and representative C_3MS_2 microanalyses are provided in Table 5. According to [15] low grain size merwinite in EAFs contributes to slags cementitious properties exhibiting a hydration behavior similar to that of di- and tri-calcium silicates.

Calcium aluminate phases, such as $3CaO·Al_2O_3$ (C_3A) and $12CaO·7Al_2O_3$ ($C_{12}A_7$), which also contribute to strength development upon hydration, participated in the mineralogical composition of the slags, being more abundant in slags 4-2(Table 5, Figure 4), which had the highest content of Al_2O_3. In addition, gehlenite C_2AS ($2CaOAl_2O_3SiO_2$) was also often detected (Figure 1, Table 5). The retention of some of the water-cooled slags in water during their cooling resulted in the fast hydration of part of the calcium aluminate phases. The presence of sulfur in these slags led to the formation of monosulfate, C_4ASH_{12} ($4CaO·Al_2O_3·SO_3·12H_2O$), found as thin hexagonal plates, (Figure 4c) or ettringite$C_6AS_3H_{32}$ ($6CaO·Al_2O_3·3SO_3·32H_2O$) detected in the form of needles (Figure 5). These hydrated phases were only detected locally in SEM backscattered images. It has also been reported that in most cases it is not possible to detect ettringite by XRD [37]. These phases are usually encountered during the hydration of Portland cements due to the reaction of gypsum with calcium aluminates according to the following reactions:

$$C_3A + 3C\overline{S}H_2 + 26H = C_6A\overline{S}_3H_{32} \text{ (ettringite formation)}$$

$$2C_3A + C_6A\overline{S}_3H_{32} + 4H = 3C_4A\overline{S}H_{12} \text{ (monosulfate)}$$

During cement hydration process, if added gypsum is consumed before the complete hydration of C_3A, the remaining C_3A reacts with ettringite producing monosulfate. The higher calcium aluminate content, due to higher Al_2O_3 percentage in the case of slag 4-2b has possibly allowed monosulfate formation by a similar mechanism.

3.2. Properties of Composite Portland Cements (CPCs) Manufactured by 20% Slag

The current European Norm [17] for common cements refers only to the utilization of ground granulated blast furnace slag (GGBS) in cement production. Thus, the physical and mechanical characteristics of all the composite cement products, manufactured using the air-cooled and water-cooled slags in the present study, are compared to the requirements for composite Portland cements (CPCs) with blast furnace slags, (CEM II). The target of this study is to evaluate the feasibility of the produced slags utilization in common cement production and further the correlation of the obtained results to the microstructural characteristics of the used slags. The main difference is that blast furnace slag accepted in cement production according to EN 197-1 [17], contains at least two-thirds by mass of glassy slag and has minimum iron oxide content.

3.2.1. Compressive Strength

The early, late and long term compressive strengths development of the CPCs with 20wt.% slag are presented in Table 6 and Figure 6a,b for the air and water-cooled slags, respectively. In addition, the results are compared to the OPC (95% clinker and 5% gypsum). All investigated mixtures were found to be within the values of EN 197-1 [17]. The early strengths (2 days of curing) were found higher than 2 Pa and 28-day curing strengths were in the range of 48.2–59.4 MPa. The preservation of relatively high early strengths observed in all the composite cements despite their limited and/or no C_3S content is most possibly attributed to the contained calcium aluminates.

Concerning the effect of iron oxide content, in general, for high FeO_x containing slags, the reduction of the % FeO_x is expected to enhance its cementitious properties, as this increases mainly the mass of residual calcium and silicon oxides, the combination of which leads to the formation of desired hydraulic phases. According to the results of this study, the %FeO_x of the slags, within the studied concentration limits, 3.85 wt.%–21.1 wt.%, did not present significant influence on the compressive strengths of the produced cements. The highest values were obtained using slag 4-2, with 3.85% FeO_x content both for the air (a) and water-cooled slag-product (b). In fact, the CPC prepared by 4-2b slag acquired slightly higher compressive strengths than those of the blank OPC. This can be attributed not only to the lower FeO_x content, but also to most the hydraulic phases formed in these slags as presented above. These slags were abundant in calcium silicates (C_2S, C_3S) and calcium aluminates (C_3A, $C_{12}A_7$). The strengths of the rest of the composite cements lay at similar levels, despite the difference in their FeO_x content. Even the cement prepared with the slag having the highest FeO_x content (~21%), presented considerable compressive strength values. This can be correlated not only with the calcium silicate compounds (C_3S, C_2S) detected in this slag, but also with the fact that part of the iron was in trivalent state resulting in the formation of brownmillerite, C_4AF, which exhibits good and intermediate contribution in the early in the final cement compressive strengths, respectively. This observation demonstrates that the state of Fe in the slag influences the morphology and its latent hydraulic properties in agreement with the findings of [18]. It is also noticed that the main peaks of the XRD pattern of this slag were attributed to magnesiowüstite. However, its negative effect on strengths development was counterbalanced by the presence of the other active phases in the slag mass. These results manifest that the strength development of the manufactured composite cement is highly affected by the mineralogical composition of the used slag. According to these results, in case of slags with crystalline structure, the most important feature for positive contribution to the compressive strength of the cement is the formation of mineralogical phases similar to those met in clinker.

Table 5. Representative semiquantitative analysis SEM/EDS of identified mineralogical phases in the slags; dicalcium silicate (C_2S), magnesiowüstite (MW) solid solution phases, brownmillerite (C_4AF), tricalcium silicate (C_3S), merwinite (C_3MS_2) and calcium aluminates (C_3A, $C_{12}A_7$, C_2AS). —below detection limit.

Phase	C_2S						MW			C_4AF	C_3S	C_3MS_2	C_3A	$C_{12}A_7$	C_2AS
Sample No/(wt.%)	4-1a	4-1a	4-2a	4-2a	3-1b	3-1a	3-1b	4-2a	5a	5a	5a	3-1a	4-2b	4-2b	3-1a
SiO_2	35.99	39.70	35.83	33.34	36.17	–	2.22	–	0.75	1.98	26.57	37.18	–	–	19.75
Al_2O_3	–	–	1.87	–	1.08	–	1.46	–	0.89	22.59	–	–	41.18	51.97	24.59
FeO_x	–	1.27	–	–	1.49	60.87	15.99	11.63	62.82	31.53	–	3.04	–	–	9.00
MnO	–	–	–	–	0.68	14.02	3.39	11.24	6.44	–	–	1.14	–	–	2.10
MgO	1.27	1.26	0.89	–	2.84	25.09	69.49	75.17	22.57	–	–	10.35	–	–	0.90
CaO	62.74	57.77	61.40	66.66	57.74	–	3.31	–	4.98	43.88	73.43	48.29	58.82	48.03	43.65
CrO_x	–	–	–	–	–	–	4.15	1.96	1.55	–	–	–	–	–	–
Nr of ions in	4 O						1 O			10 O	5 O	8 O	6 O	33 O	7 O
Si	1.020	1.100	1.008	0.964	1.016	–	0.016	–	0.007	0.170	1.010	2.048	–	–	0.987
Al	–	–	0.064	–	0.036	–	0.013	–	0.010	2.230	–	–	2.142	14.091	1.442
Fe	–	0.028	–	–	0.036	0.508	0.099	0.073	0.516	2.070	–	0.144	–	–	0.378
Mn	–	–	–	–	0.016	0.119	0.021	0.071	0.054	–	–	0.056	–	–	0.091
Mg	0.052	0.052	0.036	–	0.12	0.373	0.766	0.839	0.330	–	–	0.848	–	–	0.070
Ca	1.904	1.716	1.852	2.068	1.740	–	0.026	–	0.052	4.100	2.985	2.856	2.784	11.847	2.331
Cr	–	–	–	–	–	–	0.024	0.012	0.012	–	–	–	–	–	–

Table 6. Compressive strength and grinding time of manufactured CPCs (20% slag). Blaine 4500 cm^2/g, w/c: 0.5.

Slag	Grinding Time (min)	Compressive Strength (MPa)					
		1 Day	2 Days	7 Days	28 Days	60 Days	90 Days
–	75	17.3	28.3	41.6	57.9	63.1	64.4
3-1a	82	14.5	23.8	34.3	50.0	53.6	56.0
3-2a	111	19.5	25.2	37.1	52.5	53.8	56.5
4-1a	81	16.6	27.1	39.9	48.2	55.0	57.9
4-2a	106	16.7	25.7	39.6	57.0	60.1	64.2
5a	85	16.2	22.9	34.6	52.3	55.2	57.3
3-1b	89	14.1	23.0	35.1	50.4	59.0	60.3
3-2b	95	13.5	22.3	32.6	48.6	58.6	58.5
4-1b	90	15.9	25.3	38.2	51.3	56.2	58.0
4-2b	100	19.1	28.1	41.0	59.4	64.6	67.1

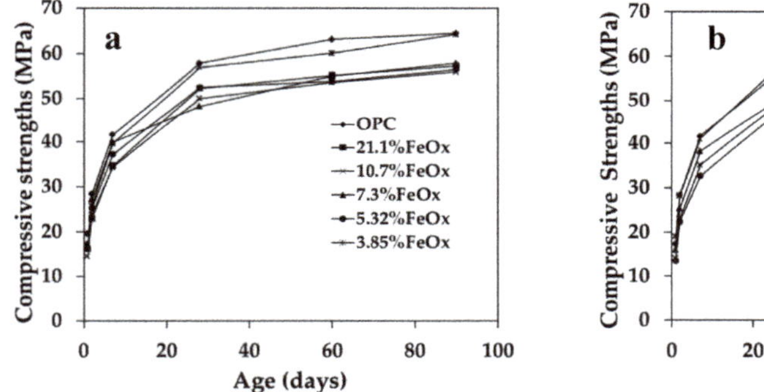

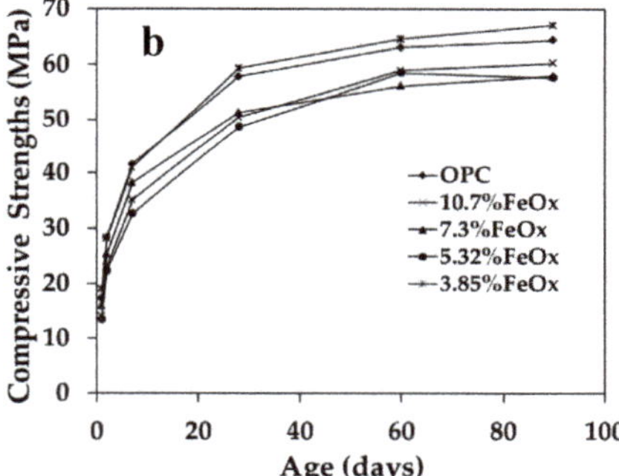

Figure 6. Compressive strengths of CPCs (20% slag) manufactured with: (a) air-cooled slags; (b) water-cooled slags, Blaine: 4500 cm^2/g.

The beneficial effect of slag basicity on the 28-days of curing achieved compressive strengths of the CPCs for the air and water-cooled slags is presented in Figure 7. In general, the higher the slag basicity in the range of 0.9–2.7, the higher are the latent hydraulic properties of the slag expected to be, since slag's basicity is directly related to the formation of the hydraulic calcium silicate phases [19,38,39]. Deviation from absolute linearity in the obtained results is related to the fact that strength development does not solely depend on calcium silicates, thus on the CaO/SiO$_2$ ratio, but also on the other slag's constituents, which are not at the same levels in the slags. In example, the concentration of alumina, which participates in the formation of cementitious phases, also has an effect.

As illustrated in Figure 8, the means of cooling, given the way it was employed, did not result in significant differences upon the obtained strength results. The slightly higher long term values in case of water-cooled slags may be possibly correlated to some amorphous content, as well as the presence of more active C$_2$S polymorphs [40]. Probably, cooling with high pressure water jets and formation of vitreous slags of these compositions could favor even higher strengths.

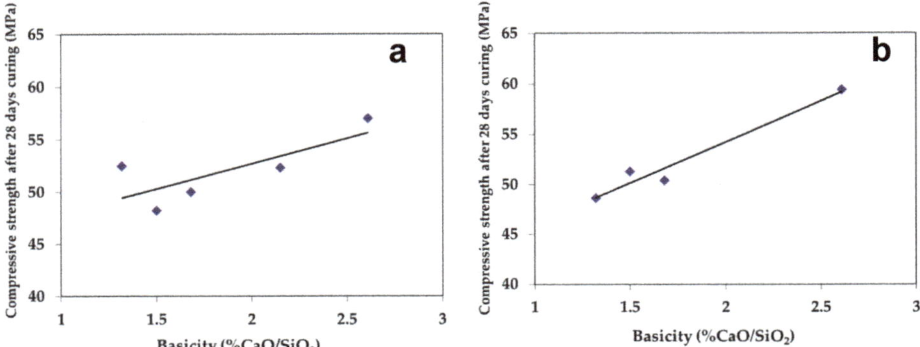

Figure 7. Effect of slag basicity on the compressive strengths after 28 days of curing of CPCs (20% slag): (**a**) air-cooled slags; (**b**) water-cooled slag.

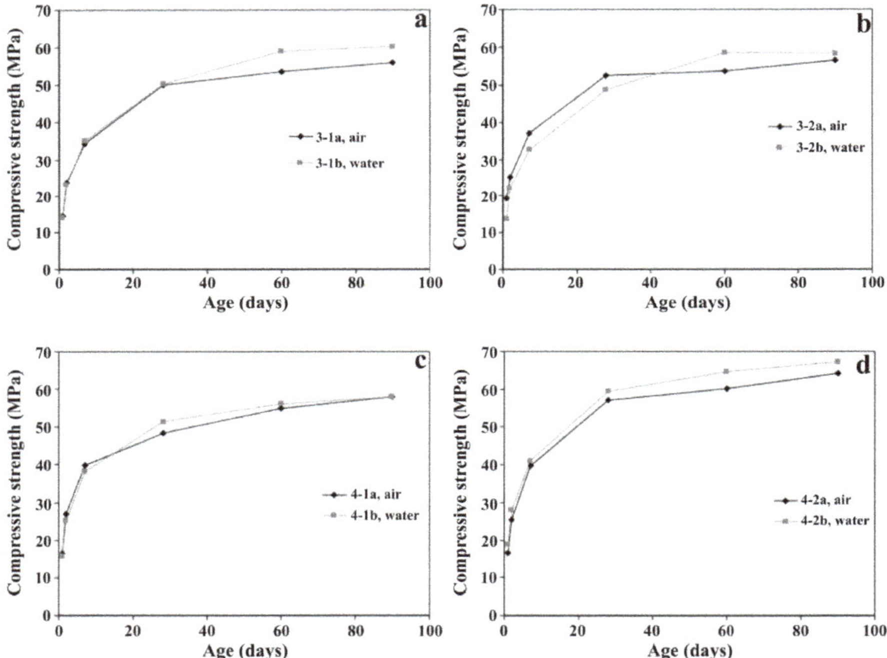

Figure 8. (**a–d**) Effect of cooling media on the development of compressive strengths in CPCs (20% slag) manufactured with air and water-cooled slags of the same chemical analysis.

3.2.2. Setting Time

The initial and final setting times of the manufactured CPS are presented in Table 7. For all the examined slags, the initial setting time was higher than the minimum one specified in EN 197-1 [17] for all cement types. Indicatively, according to the standard [17], 75 min is the highest initial setting time required for the 32.5 N/R cements, while the required time is reduced for higher strength cement types, i.e., for 52.5 N/R cements the corresponding minimum requirement is 45 min. Setting and consequently hardening of the water–cement mixture is the result of complex physicochemical changes, among which hydration reactions play a key role. Overall, cement setting is directly linked with its concentration in calcium aluminates and the percentage of added calcium sulfate. The high setting rate of pure cement comes from C_3A, which reacts instantly with water. Calcium sulfate addition aims at the modification of C_3A hydration reaction through the formation of an intermediate salt, which has the ability to significantly increase initial setting time. Depending on the concentration of aluminum and sulfate ions in solution, the crystalline precipitate can be either ettringite in the form of needle shape crystals or monosulfate with hexagonal crystal structure. Ettringite crystals, being the first to form, contribute in gradual setting of cement paste and finally in initial strength development. Different setting processes take place depending on the reactivity of calcium aluminates and their mass ratio to available sulfate ions [28]. Due to the increased rates of the aforementioned phenomena in case cement is exposed to humidity during its storage period, its subsequent setting behavior changes. Absorption of humidity results in reduction of C_3A reactivity and setting retardation. Furthermore, cement exposure to humidity in any case results in reduced compressive strengths. In some of the water-cooled slags of the present study, formation of such salts was detected in their mineralogical composition as presented above, possibly as a result of their contents in calcium aluminate phases and sulfur, their residence time in water during cooling and relative humidity during collection and storage. According to the results of Table 7, all the prepared cement specimens present relatively comparable setting times to OPC. Overall, the highest initial setting time was measured in the cement containing 3-1b slag, in which ettringite was identified resulting in setting retardation. For the cements manufactured with air-cooled slags, cement with slag 3-1a also presented the highest setting time. These results indicate that most possibly C_3A content was relatively low in 3-1 slags, which may partially explain why the hydration reaction in 3-1b stopped to ettringite and did not proceed to monosulfate formation. For the rest of the cements, the shortest times were measured in those specimens, where slags abundant in calcium aluminate phases was used. Moreover, $C_{12}A_7$ and C_4AF phases exhibit similar behavior to C_3A, concerning fast hydration rate and reaction to sulfates. This is why the shortest setting time was measured in the cement manufactured by slag 5, containing all three crystalline phases. Furthermore, in all cases the cements prepared with the water-cooled slags exhibited a delay in setting initiation compared to the cements, where the corresponding air-cooled slags was used. This can be attributed to the reduction of calcium aluminates reactivity due to humidity. The slowest time was recorded for 4-2b slag, where hexagonal monosulfate crystals was detected, indicating excess of contained calcium aluminates. In any case, cement setting time can be adjusted by the proper calcium sulfate addition depending on the technological requirements of the cement provided that the chemical specification regarding SO_3 concentration, 3.5% for CEM II Portland cements is also met. This value corresponds to 1.4% S. Sulfur range 0.2 wt.%–0.39 wt.% in the studied slag lays well below this limit value.

Table 7. Initial and final setting time of manufactured CPCs (20% slag).

Slag	Initial Setting Time (min)	Final Setting Time (min)
–	120	170
3-1a	140	190
3-2a	120	165
4-1a	110	160
4-2a	115	145
5a	100	150
3-1b	165	230
3-2b	145	185
4-1b	140	200
4-2b	120	155

3.2.3. Soundness (Expansion)

All manufactured CPCs exhibited good behavior in terms of volume stability as shown in the expansion values determined by Le Chatelier soundness test (Table 8), which are well over than the upper limit of 10 mm, indicating that free CaO and MgO, the crystalline phases well known to cause expansion issues during the hydration of cement, if present, are in very low concentrations. It is noted that no traces of free lime were detected either in the XRD analysis or in the SEM-EDS examination of the slag samples. Furthermore, in all cases pure periclase was not identified. Instead formation of MW solid solutions with different FeO/MgO ratios was always spotted, accompanied by incorporation of mainly Mn and less Ca. These formations seem to inhibit transformation to hydrated compounds causing the expansion. According to Qian et al. [35], the higher the FeO/MgO ratio of such solutions, the lower the potential of brucite transformation reaction, which is accountable for the expansion in case of pure periclase.

Table 8. Expansion of manufactured CPCs (20% slag).

Slag	Expansion (mm)
–	0.0
3-1a	0.5
3-2a	0.5
4-1a	0.5
4-2a	0.5
5a	0.5
3-1b	0.5
3-2b	0.5
4-1b	1.5
4-2b	1.5

3.2.4. Effect of Blaine

The higher grinding times that were required in case of CPCs manufacturing compared to OPC for Blaine 4500 cm^2/g, Table 6, indicate higher energy requirements. In view of economic and environmental considerations, grinding targeting to lower Blaine level was indicatively examined. For this purpose, two CPCs with 20% slag were selectively prepared at 3650 cm^2/g blaine using slags 4-2a and 5 having the lowest and highest iron oxide content, respectively. According to the obtained results (Table 9, Figure 9), although the compressive strengths of the produced cements are lower than the corresponding ones at 4500 cm^2/g, as expected, they still fall inside the EN-197-1 [17] cement specifications. In fact, cement with 4-2a slag conforms to the limits of the high strength cement category 52.5 N even at the lower Blaine. These results point out that even at lower Blaine the produced slags could be used up to 20% in cement mass replacing clinker. The reduced grinding time employed in case of the lower compared to the higher Blaine, corresponds to reduced manufacturing energy requirements. In case of commercial use, cement Blaine and thus grinding time depends on the requirements of the construction application for which it is intended, namely the strength specs. Nevertheless, we should not forget that despite the additional grinding time required for the preparation of composite cements, the economic benefit to the cement producer is still significant, due to the substitution of 21% of the mass of clinker by slag, which is an energy upgraded material and does not need to go through the high energy demanding firing stage in the rotary kiln during the production of cement.

Table 9. Compressive strength and grinding time of manufactured CPCs (20% slag) at Blaine 3650 cm^2/g, w/c: 0.5.

Slag	Grinding Time	Compressive Strength (MPa)			
		1 Day	2 Days	7 Days	28 Days
–	55	15.60	23.80	37.10	53.00
4-2a	69	16.50	24.00	36.80	53.75
5a	60	13.80	21.50	32.00	46.30

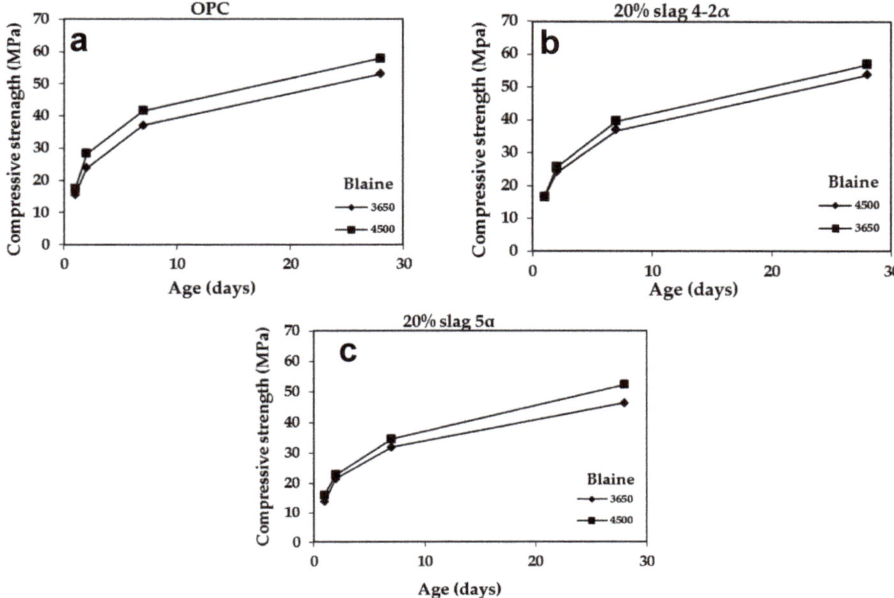

Figure 9. Effect of Blaine on strength development of: (**a**) OPC; (**b**) slag 4-2a; (**c**) slag 5a.

3.3. Properties of Slag Cements (SCs) manufactured by 40% Slag

The physical and mechanical characteristics of all the slag cement products, manufactured using the air-cooled and water-cooled slags in the present study, are evaluated comparatively to the requirements of EN 197-1 [17] for slag cements with blast furnace slags, (CEM III). Slag cement with 4-1a slag was not manufactured because the available slag mass was not sufficient. Further, in slag cement made with slag 4-1b different clinker batch was used due to inadequate clinker mass. New reference OPC was also prepared by the new clinker for comparison reasons noted by (*) in the corresponding results tables thereafter.

3.3.1. Compressive Strength

The early, late and long term compressive strengths of the slag cements (SCs) with 40wt.% slag are presented in Table 10. The obtained compressive strengths of all the slag-cements ranging between 35.4 and 49.2 MPa are within the limits of EN 197-1 [17]. The results indicate that the compressive strengths of the slag cements are not solely affected by the FeO content of used slags in agreement with the findings of CPCs. The highest strengths for SCs with 40% slag, for both air-cooled and water-cooled slags, are obtained for the case of slag 4-2 in agreement with the results of the CPCs with 20% slag. In fact, as it is illustrated in Figure 10, the slag cement manufactured with 4-2b slag exhibits such a considerable increase in the long term compressive strength development that after 90 days, the values attained by OPC are reached.

Table 10. Compressive strength and grinding time of manufactured SCs (40% slag). Blaine 4500 cm^2/g, w/c: 0.5.

Slag	Grinding Time (min)	Compressive Strength MPa					
		1 Day	2 Days	7 Days	28 Days	60 Days	90 Days
–	75	17.3	28.3	41.6	57.9	63.1	64.4
–*	95	21.0	32.7	45.7	58.7	61.4	64.5
3-1a	102	12.0	18.5	28.2	36.7	43.0	46.2
3-2a	95	12.2	16.8	24.4	37.5	41.2	43.4
4-2a	90	9.3	16.5	25.5	39.7	49.0	51.1
5a	90	10.8	15.0	23.3	36.8	41.5	45.8
3-1b	100	10.6	15.4	24.2	36.6	43.6	47.7
3-2b	100	10.3	15.2	23.9	35.4	40.0	39.4
4-1b*	101	11.4	18.4	28.5	38.5	42.0	45.0
4-2b	102	11.2	17.8	29.8	49.2	58.4	65.0

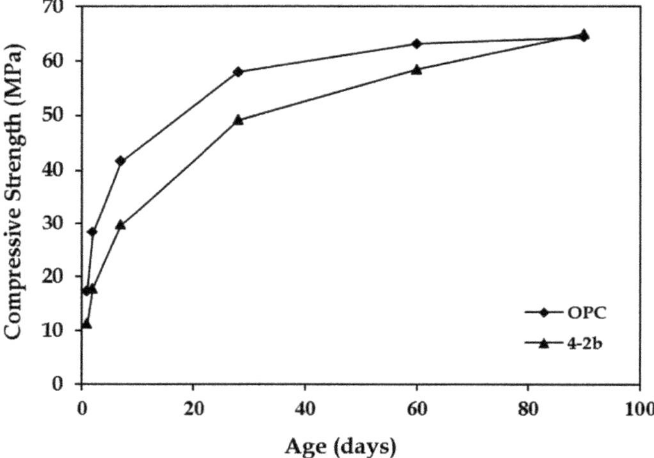

Figure 10. Comparative compressive strength development of OPC and slag cement manufactured with 40% slag 4-2b, Blaine: 4500 cm^2/g.

In Figures 11 and 12 the normalized late (28 days) and long term (90 days) compressive strengths of the prepared cements in relation to their slag content are presented for the mixtures produced with the air and water-cooled slags, respectively. As normalized strength the following strength ratio was defined: Compressive strength obtained by the mixture where slag was used/ Compressive strength obtained by the corresponding OPC. Overall, it can be observed that the increase in the slag content results in the decrease of the strength. Nevertheless, the contained C$_2$S of the slags results in a smoother decrease in case of long term strengths.

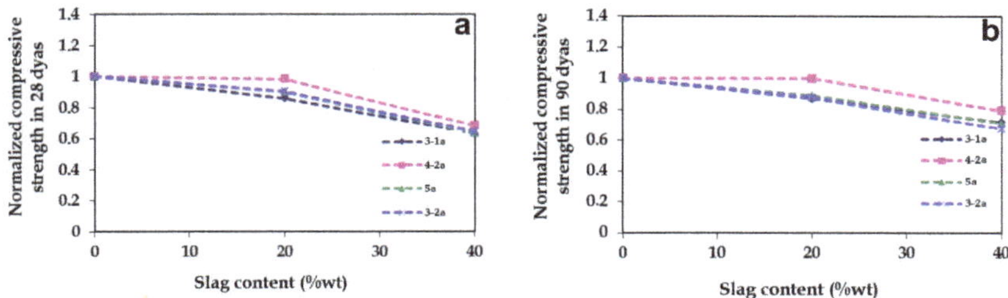

Figure 11. Normalized compressive strength of cements manufactured with air-cooled slags in relation to their slag content (**a**) after 28 days of curing and (**b**) after 90 days of curing.

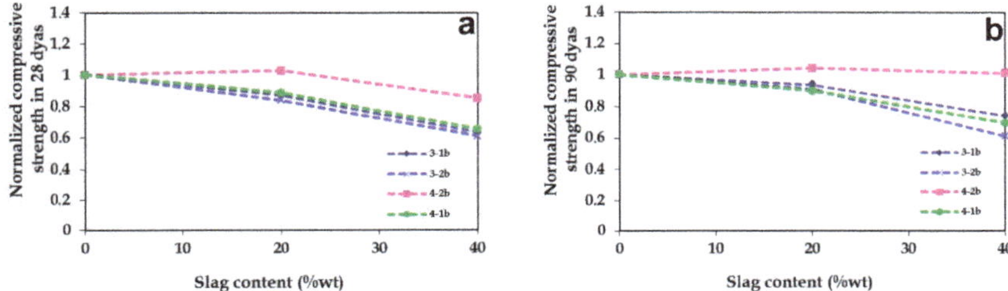

Figure 12. Normalized compressive strength of cements manufactured with water-cooled slags in relation to their slag content (**a**) after 28 days of curing and (**b**) after 90 days of curing.

3.3.2. Setting Time

The initial and final setting times of the manufactured slag cements is presented in Table 11. Initial setting times of cements with 40% slag are lower than the corresponding ones with 20% and in some cases lay marginally within the limits of EN 197-1 [17] for 32.5 typical strength cements. The higher initial setting time was measured for the mixture with slag 3-1a, in agreement to the results obtained in case of CPCs with 20% slag. Overall, the relatively fast onset setting of the slag cement mixtures is mainly correlated to their calcium aluminates content of the used slags as described above. However, the obtained results cannot totally be/gr explained by the mineralogical composition of the slags. For example, for the water-cooled slags 4-1b and 3-2b, although the use of 20% in cement resulted in increase of the initial setting time compared to OPC, increasing slag content to 40%, resulted in decreased initial setting time. Further, unexpected fast onset of setting was also measured for cement with slag 3-1b, in which the ettringite was detected. These results are most probably attributed to the "false set" phenomenon, which is usually experienced in the case of partially hydrated cements deriving from improper storage conditions. It is noted that the used clinker was stored for a period of almost one and a half year in plant's warehouse for the elaboration of all the lab scale experiments.

Table 11. Initial and final setting time of manufactured SCs (40% slag).

Slag	Initial Setting Time (min)	Final Setting Time (min)
–	120	170
–*	135	185
3-1a	125	185
3-2a	90	170
4-2a	75	145
5a	75	125
3-1b	75	155
3-2b	80	140
4-1b*	90	170
4-2b	85	145

3.3.3. Soundness (Expansion)

The expansion values measured for the slag cements are presented Table 12. However, these values are higher than those of cements with 20% slag, they still remain well below the standard limit value of 10 mm. These encouraging results support that 40% slag content in the cement substituting clinker can take place without causing volume stability issues.

Table 12. Expansion of manufactured SCs (40% slag).

Slag	Expansion (mm)
–	0.0
–*	0.0
3-1a	0.5
3-2a	1.3
4-2a	2.0
5a	0.6
3-1b	1.6
3-2b	1.0
4-1b	2.0
4-2b	1.5

3.3.4. Grinding Time

Concerning the grinding time of the slag cements, it was generally observed to be higher than that one required for the OPC and the corresponding CPCs. Exceptions are the mixtures of slag 3-2a and 4-2a, for which reduction in grinding time was observed in case of 40% slag content. It is noticeable that these slags exhibited the highest grinding time among the cements with 20% slag. In addition, for 4-2b slag, which presented a relatively high grinding time in case of 20%, in the case of 40% the difference is only two min. The combination of these results indicates that when the content of slag of relatively high hardness increases in the mixture, a "self-grinding" phenomenon takes place, where the slag itself contributes to its grinding, thereby reducing the time required in order for the corresponding cement mixture to achieve the same Blaine. This observation is particularly important in terms of energy consumption.

3.4. Suppressing CO_2 Emissions by Use of Slag in Cement Manufacturing

International policy frameworks developed by the Paris Agreement Sustainable Development Goals (SDG) calls for immediate actions to be taken to suppress CO_2 emissions. According to the World cement Association, cement producers have developed plans to contribute towards this direction, by outlining pathways for low-carbon cement production. Despite the profound benefits that cement offers in shaping our modern built environment, it is undoubtedly a massive source of carbon dioxide to the atmosphere. According to estimates from the International Energy Agency, it accounts for about 7 percent of all global carbon emissions, meaning that it is the second-largest single industrial emitter, second only to the iron and steel industry. Further to this, it is expected that cement production will vastly increase by within the next 20 years due to the increasing demand in cement from developing countries. Thus, solutions must be implemented to reduce CO_2 emissions because this it may likely put the Paris Agreement's global climate targets in jeopardy (e.g., [41,42]). The production of Portland cement leaves a notable carbon footprint due to the use of fossil fuels to create the necessary intense heating but more important, from the chemical decomposition of limestone that releases large amounts of CO_2 as a byproduct. Proposed solutions include the use of alternative fuels, the implementation of innovative carbon capture technologies, as well as the use of recycled byproducts from other industries, such as slag, fly ash and clay material.

Scientists and cement industries are combining forces and are expected to succeed in gradually reducing CO_2 emissions, although there are still many barriers to be dealt with [43]. Blast furnace slag production can partially replace clinker in cement by even up to 50% or directly in to be used in concrete. Moving one step forward, the results of the present research demonstrate that the allowable main cement constituents list of the current cement norm [17] could potentially be expanded to include other kind of slags beyond the blast furnace slags. The testing principles applied in this study demonstrate that the selection criteria should not rule out slags just based on the chemical analysis and amorphous slag content as currently in place for the GGBFS, but consider whether the properties of the corresponding obtained cements meet the required technical specifications for their application. Increasing the potential volume of slag cement would result in significant environmental and economic benefits, since its production requires less than a fifth of the energy compared to the production of Portland cement and releases less than a fifteenth of the carbon dioxide emissions. Additional sustainable benefits are that manufacture of slag cement does not require the quarrying of virgin limestone materials, but also the fact that the slag is considered as a key link to achieve a sustainable circular economy, to move our society from a linear economic system by using slag as an eco-friendly byproduct.

4. Conclusions

The examined slags produced in the course of the industrially tested Ni-dust and laterite ore fines recycling smelting process could be used in the cement industry for the manufacturing of special type composite Portland or slag cements up to 40% by mass with significant environmental and economic benefits, since the physical and mechanical characteristics of all examined cement products were found to meet the requirements of the regulation set for cements. For the 4-2a, b slags, exhibiting the highest strength results, even higher addition could be most possibly feasible. The present research revealed that the most critical parameter in the compressive strength development of the slag cements is the mineralogical composition of the slag. Increasing slag basicity enhances formation of desired hydraulic calcium silicates, as well as calcium aluminates in presence of adequate alumina. Formation of magnesiowüstite solid solutions and iron valence also affect the obtained physical and mechanical properties of the corresponding slag cements, respectively, suppressing potential expansion issues. Rapid cooling enhances slag's cementitious properties. However, the results showed that even in cases where rapid cooling to obtain glassy matrix is not feasible, adjustment of slag analysis to obtain mineralogical phases similar to those met in clinker of OPC, even at higher FeO contents (up to 21wt.%), can result in production of slag with considerable latent hydraulic properties. These findings indicate that there is also potentially space for improvement in conventional EAF steel slags chemical composition to allow for their wider use in building sector. Overall, this research demonstrated that the recycling of metallurgical wastes, through the reduction smelting route in DC–EAF, provides perspectives for development of zero residues metallurgical recycling processes, since besides the recovery of valuable metals in the liquid bath, produced slag analysis can be adjusted, on one hand in favor of the metallurgical yield of the process and on the other to allow its further utilization in the manufacturing of high added value environmental friendly building materials accountable for less CO_2 footprint; relative to the amount of clinker being substituted.

Supplementary Materials: The following are available online at http://www.mdpi.com/2076-3417/10/13/4670/s1, Figure S1: Flow sheet of the zero residues production smelting process for metallurgical dusts and ore fines recycling [4].

Author Contributions: T.T. coordinated the research, designed and carried out the testing, the interpretation of the results and the writing of the manuscript; P.L. participated in the interpretation of the results and contributed to the manuscript writing; P.P.G. participated in the interpretation of the results and contributed to the manuscript writing; A.R. participated in the interpretation of the results and contributed to the manuscript writing; P.K. participated to the manuscript writing; N.K. participated to the manuscript writing; P.P. participated in the interpretation of the results and contributed to the manuscript writing. All authors have read and agreed to the published version of the manuscript.

Funding: The industrial Ni-dust and laterite ore fines smelting trials, in the course of which, the slags were produced were jointly funded by the government of Niedersachsen and Georgsmarienhütte Gmbh steelmaking company.

Acknowledgments: The present announcement is dedicated to the memory of Ing. D.C. Papamantellos for his overall guidance and contribution. The industrial Ni-dust smelting trials and slags production took place in the DC–EAF of Georgsmarienhütte Gmbh steel making company. LARCO GMMS.A. provided the Ni-dust and laterite ore-fines. Manufacturing and testing of the slag cements was performed in the R&D center of TITAN S.A. Finally, special thanks are due to Editor and four anonymous reviewers for their precious comments on improving the manuscript.

Conflicts of Interest: The authors declare no conflicts of interest.

References

1. Yunos, N.F.M.; Zaharia, M.; Ismail, A.N.; Idris, M.A. Transforming Waste Materials as Resources for EAF Steel making. *Int. J. Mater. Eng.* **2014**, *4*, 167–170.
2. Jones, R.T.; Hayman, D.A.; Denton, G.M. Recovery of cobalt, nickel and copper from slags using DC-arc furnace technology, International Symposium on Challenges of Process Intensification. In Proceedings of the 35th Annual Conference of Metallurgists, Montreal, QC, Canada, 24–29 August 1996.
3. Burström, G.Y.; Kuhn, M.; Piret, J. Reduction of Steelmaking Slags for Recovery of Valuable Metals and Oxide Materials. In Proceedings of the 6th International Conference on Molten Slags, Fluxes and Salts, Stockholm, Sweden-Helsinki, Finland, 12–17 June 2000.
4. Tzevelekou, T.V.; Geck, H.G.; Hüllen, P.; Höfer, F.; Papamantellos, D.C. Direct Smelting of Metallurgical Dusts and Ore Fines in a 125t DC-HEP Furnace. *Steel Res. Int.* **2004**, *75*, 382–392. [CrossRef]
5. Memoli, F.; Mapelli, C.; Guzzon, M. Recycling of Ladle Slag in the EAF: A Way to Improve Environmental Conditions and Reduce Variable Costs in Steel Plants. *Iron Steel Technol.* **2007**, *4*, 68–76.
6. Heo, J.H.; Kim, T.S.; Sahajwalla, V.; Park, J.H. Observations of FeO Reduction in Electric Arc Furnace Slag by Aluminum Black Dross: Effect of CaO Fluxing on Slag Morphology. *Metal Mater. Trans. B* **2020**, *51*, 1201–1210. [CrossRef]
7. Schliephake, H.; Algermissen, D. How far away is the steel industry from the target now waste? Proceedings EUROSLAG. In Proceedings of the 10th European Slag Conference Slag based products–best practices for Circular Economy, Thessaloniki, Greece, 8–11 October 2019.
8. Alexopoulou, I.E.; Angelopoulos, G.N.; Papamantellos, D.C.; Rentizelas, G. The dust recycling problem of the FeNi production at Larymna/Greece—Part II: New perspectives and proposals. *Erzmetall* **1994**, *47*, 651–657.
9. Kotze, J. Pilot plant production of ferronickel from nickel oxide ores and dusts in a DC arc furnace. *Miner. Eng.* **2002**, *15*, 1017–1022. [CrossRef]
10. Osmanović, Z.; Haračić, N.; Zelić, J.; Lihić, A. *Slag Cement Production in Terms of Emission Reducing*; ISEM: Adiyaman, Turkey, 2014.
11. Schorcht, F.; Kourti, I.; Scalet, B.M.; Roudier, S.; Delgado Sancho, L. *Best Available Techniques (BAT) Reference Document for the Production of Cement, Lime and Magnesium Oxide: Industrial Emissions Directive 2010/75/EU: (Integrated Pollution Prevention and Control)*; Luxembourg Publications Office of the European Union: Brussels, Belgium, 2013.
12. Pellegrino, C.; Faleschini, F.; Meyer, C. *Developments in the Formulation and Reinforcement of Concrete*, 2nd ed.; Woodhead Publishing Series in Civil and Structural Engineering: Vancouver, BC, Canada, 2019.
13. Kim, H.S.; Kim, K.S.; Jung, S.S.; Hwang, J.I.; Choi, J.S.; Sohn, I. Valorization of electric arc furnace primary steelmaking slags for cement applications. *Waste Manag.* **2015**, *41*, 85–93. [CrossRef] [PubMed]
14. Zhang, L.; Drelich, J.W.; Neelameggham, N.R.; Guillen, D.P.; Haque, N.; Zhu, J.; Sun, Z.; Wang, T.; Howarter, J.A.; Tesfaye, F.; et al. (Eds.) *Energy Technol 2017—Carbon Dioxide Management and Other Technologies, The Minerals, Metals & Materials Series, 1, XII, 499*; Springer: Berlin, Germany, 2017. [CrossRef]
15. Luckman, M.; Vitta, S.; Venkateswaran, D. Cementitious and pozzolanic behavior of electric arc furnace steel slags. *Cem. Concr Res.* **2009**, *39*, 102–109.
16. Branca, T.A.; Colla, V.; Algermissen, D.; Granbom, H.; Martini, U.; Morillon, A.; Pietruck, R.; Rosendahl, S. Reuse and Recycling of By-Products in the Steel Sector: Recent Achievements Paving the Way to Circular Economy and Industrial Symbiosis in Europe. *Metals* **2020**, *10*, 345. [CrossRef]

17. British Standard EN 197 Part 1. *Cement-Composition, Specifications and Conformity Criteria for Common Cements*; BSI: London, UK, 2011.

18. Murphy, N.; Meadwcroft, T.R.; Barr, P.V. Enhancement of the cementitious properties of steelmaking slag. *Can. Metal. Q.* **1997**, *36*, 315–331. [CrossRef]

19. Shi, C.; Qian, J. High performance cementing materials from industrial slags–a review. *Resour. Conserv. Recycl.* **2000**, *29*, 195–207. [CrossRef]

20. Rojas, M.F.; de Rojas, M.I.S. Chemical assessment of the electric arc furnace slag as construction material: Expansive formations. *Cem. Concr. Res.* **2004**, *34*, 1881–1888. [CrossRef]

21. Engstrom, F.; Pontikes, Y.; Geysen, D.; Jones, P.T.; Bjorkman, B.; Blanpain, B. Review: Hot stage engineering to improve slag valorisation options. In Proceedings of the 2nd International Slag Valorization Symposium, The transition of sustainable materials management, Leuven, Belgium, 18–20 April 2011.

22. Pontikes, Y.; Malfliet, A. Slag Valorisation as a Contribution to Zero-Waste Metallurgy. *J. Sustain. Metall.* **2016**, *2*, 1–2. [CrossRef]

23. Li, Y.; Gang, D.; Guo, Z. Convert hot slag into value-added materials by modification methods. In Proceedings of the 5th International Slag Valorization Symposium, Leuven, Belgium, 3–5 April 2017.

24. Mudersbach, D.; Kühn, M.; Geisler, J.; Koch, K. Chrome Immobilisation in EAF-Slags from High-Alloy Steelmaking: Tests at FEhS Institute and Development of an Operational Slag Treatment Process. In Proceedings of the First International Slag Valorisation Symposium, Leuven, Belgium, 6–7 April 2009.

25. Primavera, A.; Pontoni, L.; Mombelli, D.; Barella, S.; Mapelli, C. EAF Slag Treatment for Inert Materials' Production. *J. Sustain. Metall.* **2016**, *2*, 3–12. [CrossRef]

26. Tzevelekou, T.; Stivanakis, V.; Papamantellos, D.C.; Hoefer, F.; Haniotakis, E.; Fragoulis, D. Properties of slags and Portland-slag-cements produced by smelting of metallurgical dusts and ore fines in steelmaking DC-HEP furnace. In Proceedings of the Advances in Mineral Resources Management and Environmental Geotechnology, Chania, Greece, 7–9 June 2004.

27. Tzevelekou, T.V.; Geck, H.G.; Hüllen, P.V.; Höfer, F.; Poulakis, S.; Papamantellos, D.C. Recycling of nickel ferrous rotary kiln dust by injection through hollow electrode in a 125t DC-HEP furnace. In Proceedings of the International Symposium on Metals and Energy Recovery Proc, Skellefteå, Sweden, 25–26 June 2003.

28. Tzevelekou, T.V. Development of a Process for Recycling and Producing New Materials by Reduction Smelting of Das Cleaning Systems Dusts from the Ferronickel Industry. Ph. D Thesis, Department of Chemical Engineering, University of Patras, Patras, Greece, 2004.

29. Papamantellos, D.C. *UntersuchungenzurEntchromung von Roheisen, Doktor-Arbeit, FakultätfürBergbau und Hüttenwesen der RWTH*; Aachen University: Aachen Germany, 1965.

30. Hong, H.; Fu, Z.; Min, X. Effect of cooling performance on the mineralogical character of Portland cement clinker. *Cem. Concr. Res.* **2001**, *31*, 287–290. [CrossRef]

31. Moseley, D.; Glasser, F.P. Identity, Composition and Stability of bredigite-structure phase T. *Cem. Concr. Res.* **1981**, *11*, 559–565. [CrossRef]

32. Luxán, M.P.; Sotolongo, R.; Dorrego, F.; Herrero, E. Characteristics of the slags produced in the fusion of scrap steel by electric arc furnace. *Cem. Concr. Res.* **2000**, *30*, 517–519. [CrossRef]

33. Bai, J.; Chaipanich, A.; Kinuthia, J.M.; O'Farrell, M.; Sabir, B.B.; Wild, S.; Lewis, M.H. Compressive strength and hydration of wastepaper sludge ash-ground granulated blast furnace slag blended pastes. *Cem. Concr. Res.* **2003**, *33*, 1189–1202. [CrossRef]

34. Hofmänner, F. *Microstructure of Portland Cement Clinker*; «Holder bank» Management & Consulting Ltd.: Holderbank, Switzerland, 1975.

35. Qian, G.R.; Sun, D.D.; Tay, J.H.; Lai, Z.Y. Hydrothermal reaction and autoclave stability of Mg bearing RO phase in steel slag. *Br. Ceram. Trans.* **2002**, *101*, 159–164. [CrossRef]

36. Engström, F. Mineralogical Influence on Leaching Behaviour of Steelmaking Slags. Ph.D. Thesis, Department of Chemical Engineering and Geoscience, Division of Minerals and Metallurgical Engineering, University of Technology, Luleå, Sweden, 2010.

37. Stark, J.; Bollmann, K. Delayed Ettringite Formation in Concrete Nordic, Concrete Research (NCR), Bauhaus-University Weimar Publication No.23, Germany. Available online: https://www.imxtechnologies. com/storage/app/media/uploaded-files/ettringite.pdf (accessed on 2 July 2020).

38. Altun, A.; Yilmaz, I. Study on steel furnace slags with high MgO as additive in Portland cement. *Cem. Concr. Res.* **2002**, *32*, 1247–1249. [CrossRef]

39. Xuequan, W.; Hong, Z.; Xinkai, H.; Husen, L. Study on steel slag and fly ash composite Portland cement. *Cem. Concr. Res.* **1999**, *29*, 1103–1106. [CrossRef]

40. Chatterjee, A.K. High belite cements-present status and future technological options: Part I. *Cem. Concr. Res.* **1996**, *26*, 1213–1225. [CrossRef]

41. Petrounias, P.; Giannakopoulou, P.P.; Rogkala, A.; Kalpogiannaki, M.; Koutsovitis, P.; Damoulianou, M.-E.; Koukouzas, N. Petrographic Characteristics of Sandstones as a Basis to Evaluate Their Suitability in Construction and Energy Storage Applications. A Case Study from Klepa Nafpaktias (Central Western Greece). *Energies* **2020**, *13*, 1119. [CrossRef]

42. Arvanitis, A.; Koutsovitis, P.; Koukouzas, N.; Tyrologou, P.; Karapanos, D.; Karkalis, C.; Pomonis, P. Potential Sites for Underground Energy and CO_2 Storage in Greece: A Geological and Petrological Approach. *Energies* **2020**, *13*, 2707. [CrossRef]

43. Parron-Rubio, M.E.; Perez-Garcia, F.; Gonzalez-Herrera, A.; Oliveira, M.J.; Rubio-Cintas, M.D. Slag Substitution as a Cementing Material in Concrete: Mechanical, Physical and Environmental Properties. *Materials* **2019**, *12*, 2845. [CrossRef] [PubMed]

 *applied sciences*

Article

Role of Internal Stress in the Early-Stage Nucleation of Amorphous Calcium Carbonate Gels

Qi Zhou [1], Tao Du [2], Lijie Guo [3,*], Gaurav Sant [4] and Mathieu Bauchy [5,*]

[1] Physics of AmoRphous and Inorganic Solids Laboratory (PARISlab), University of California, Los Angeles, CA 90095-1593, USA; qi1197@ucla.edu

[2] Key Lab of Structures Dynamic Behavior and Control (Harbin Institute of Technology), Ministry of Education, Harbin 150090, China; dutaohit@gmail.com

[3] BGRIMM Technology Group, Beijing 100160, China

[4] Laboratory for the Chemistry of Construction Materials (LC2), California Nanosystems Institute (CNSI), Institute for Carbon Management (ICM), Department of Civil and Environmental Engineering, University of California, Los Angeles, CA 90095, USA; gsant@ucla.edu

[5] Physics of AmoRphous and Inorganic Solids Laboratory (PARISlab), California Nanosystems Institute (CNSI), Institute for Carbon Management (ICM), University of California, Los Angeles, CA 90095-1593, USA

* Correspondence: guolijie@bgrimm.com (L.G.); bauchy@ucla.edu (M.B.)

Received: 8 May 2020; Accepted: 24 June 2020; Published: 25 June 2020

Featured Application: This work offers new insights into the atomic-scale mechanism governing the early-stage nucleation of amorphous calcium carbonate gels, which is the key to the acceleration of the development of new methods, enabling low-cost CO_2 utilization by mineralization.

Abstract: Although calcium carbonate ($CaCO_3$) precipitation plays an important role in nature, its mechanism remains only partially understood. Further understanding the atomic driving force behind the $CaCO_3$ precipitation could be key to facilitate the capture, immobilization, and utilization of CO_2 by mineralization. Here, based on molecular dynamics simulations, we investigate the mechanism of the early-stage nucleation of an amorphous calcium carbonate gel. We show that the gelation reaction manifests itself by the formation of some calcium carbonate clusters that grow over time. Interestingly, we demonstrate that the gelation reaction is driven by the existence of some competing local molecular stresses within the Ca and C precursors, which progressively get released upon gelation. This internal molecular stress is found to originate from the significantly different local coordination environments exhibited by Ca and C atoms. These results highlight the key role played by the local stress acting within the atomic network in governing gelation reactions.

Keywords: calcium carbonate; molecular dynamics; carbon utilization; gelation

---

1. Introduction

Calcium carbonate ($CaCO_3$) is ubiquitous in nature and plays an important role in biomineralization [1]. For instance, $CaCO_3$ binding phases are commonly formed by organisms that produce an exoskeleton, e.g., snails, clams, and mollusks, and in cemented granular soils formed by bacteria [2,3]. On the technological side, calcium carbonate is key for CO_2 utilization approaches relying on mineralization [4], which offer a promising route to turn CO_2 (usually considered as a waste) into a resource, e.g., concrete binders [5,6].

The development of novel carbonation routes enabling low-cost and low-energy CO_2 capture has thus far been largely limited by a lack of understanding of the physical and chemical features that govern the carbonation process [4]. Indeed, despite its ubiquitous nature, the mechanism of the precipitation of calcium carbonate remains poorly understood [7]. The carbonation process usually occurs via a dissolution-precipitation reaction, wherein a solid phase (e.g., $Ca(OH)_2$) dissolves in an aqueous environment, reacts with dissolved carbonate species, and reprecipitates as an amorphous, hydrated precursor calcium carbonate gel phase [8]. Crystalline $CaCO_3$ then forms upon the drying and crystallization of the amorphous precursor [9]. Notably, the nucleation of amorphous calcium carbonate is barrierless in highly supersaturated calcium carbonate solutions and does not follow conventional nucleation pathways [10]. Although the existence of some stable prenucleation clusters has been suggested, the atomic scale mechanism of calcium carbonate nucleation in an aqueous condition remains largely unknown [11,12]. Since amorphous calcium carbonate often acts as a precursor before the formation of subsequent crystalline carbonate phases, it is important to understand the formation mechanism of the amorphous calcium carbonate gel [13,14].

As an alternative route to experiments, molecular dynamics (MD) simulations can offer some valuable insights into the atomic mechanism governing the precipitation of gels [15,16]. Indeed, although MD simulations are limited to small systems and timescales, they can offer a direct access to the time-dependent structure of calcium carbonate during its early-stage nucleation [17]. In particular, Raiteri et al. recently developed a thermodynamically-consistent forcefield, allowing the full modeling of the nucleation of carbonates, ranging from aqueous complexes to solid carbonates [18,19]. This forcefield was found to accurately reproduce the dynamics and thermodynamics of alkaline-earth carbonate solutions and to yield an excellent agreement with experimental data. These recent developments now make it possible to investigate the atomic mechanism behind the carbonation reaction.

Here, based on MD simulations, we investigate the early-stage nucleation of an amorphous calcium carbonate gel in supersaturated conditions. We show that the gelation reaction manifests itself by the formation of some calcium carbonate clusters that grow over time. As a key novelty resulting from this study, we demonstrate the existence of some local, competing atomic stress acting in calcium carbonate gels, which arises from the significantly different coordination environments exhibited by Ca and C atoms. Such internal stress is found to be progressively released upon gelation and acts as a driving force for early-stage nucleation.

2. Methods

2.1. Simulation Methodology

The calcium carbonate gelation simulations are conducted based on the method presented by Cormack et al. and described in the following [20]. A $CaCO_3$ hydrated gel is simulated with the large-scale atomic/molecular massively parallel simulation (LAMMPS) package [21]. The system comprises 11,840 atoms with a $CaCO_3/H_2O$ molar ratio of 1/60, which is highly supersaturated. Although this concentration might not mimic realistic solution conditions, such a high concentration allows us to accelerate the gelation reaction. First, Ca^{2+} ions and CO_3^{2-} and H_2O molecules are randomly placed in a cubic box of 55 Å in length, with periodic boundary conditions, while ensuring the absence of any unrealistic overlap between atoms using the Packmol package [22]. In the following, Ow and Oc refer to the O atoms that belong to H_2O and CO_3^{2-} molecules, respectively. A snapshot of initial configuration is shown in Figure 1a. The system is first relaxed at zero pressure and 300 K for 25 ps in the isothermal-isobaric (*NPT*) ensemble, using a Nosé-Hoover thermostat and the barostat developed by Melchionma et al. [23,24] This duration is found to be long enough to ensure a convergence of pressure and volume. The early-stage gelation dynamics of the system are then studied by conducting a 10 ns MD run in the *NVT* ensemble. The velocity-Verlet integration algorithm is employed for the description of the atomic motion, with a time step of 0.25 fs. For all the simulations, we adopt the forcefield parameterized by Raiteri et al., which models the atoms as rigid ions with fixed charges [18]. Although more complex forcefields are available (e.g., ReaxFF [25,26] or fluctuating charge model [27]), they come with a significantly increased computational cost, which would not allow us to simulate such a large system over such an extended timescale. In addition, the forcefield developed by Raiteri et al. presents the advantage of offering an excellent description of the dynamics and thermodynamics of carbonates—both in solution and as solids [18]. The full parameterization of this forcefield can be found in Ref. [18]. For illustration purposes, Figure 1b,c show some snapshots of the simulated configuration after 0, 5, and 10 ns of dynamics.

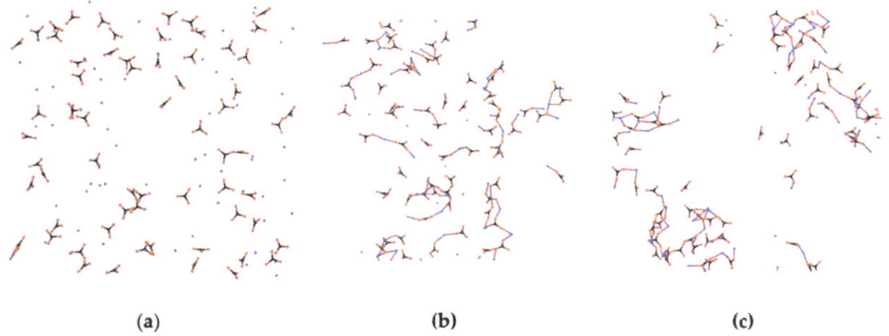

(a)   (b)   (c)

Figure 1. Snapshots of the simulated amorphous calcium carbonate system after (a) 0, (b) 5, and (c) 10 ns of gelation. Carbon, oxygen, and calcium atoms are indicated in black, red, and blue, respectively. Water molecules are here omitted for clarity.

2.2. Structural Analysis

To describe the time evolution of the atomic topology of the amorphous calcium carbonate gel, we compute the coordination number of each atom by enumerating the number of neighbors present in its first coordination shell. The distance cutoffs are determined for each atom as the first minimum distance of the partial pair distribution functions (i.e., 1.5 and 3.0 Å for C–O and Ca–O bonds, respectively). Based on the identification of each C–O and Ca–O bond, we then compute the size and number of carbonate clusters, wherein a cluster is defined as a group of interconnected C, Ca, and O atoms. The size of each cluster is then given by the number of C atoms belonging to the cluster. The clustering analysis is conducted by using the OVITO software [28].

2.3. Computation of the Local Atomic Stress

To assess the existence of some local instabilities within the atomic network, we adopt the concept of "stress per atom" introduced by Egami [29,30]. Note that, strictly speaking, stress is intrinsically a macroscopic property and, hence, is ill-defined for individual atoms [31]. Nevertheless, an atomic-level stress tensor $\sigma_i^{\alpha\beta}$ can be defined for each atom i by expressing the contribution of each atom to the virial of the system [32]:

$$\sigma_i^{\alpha\beta} = \frac{1}{V_i}\Sigma_j r_{ij}^\alpha \cdot F_{ij}^\beta, \tag{1}$$

where V_i is the volume of atom i, r_{ij} the distance between atoms i and j, F_{ij} the interatomic force applied by atom j on atom i, and the indexes α and β refer to the projections of these vectors along the Cartesian directions x, y, or z. The volume V_i of each atom is defined as the Voronoi volume. By convention, a positive stress represents here a local state of tension, whereas a negative one represents a local state of compression. This approach was recently used to quantify the internal stress exhibited by stressed–rigid atomic networks [33] and mixed alkali glasses [32,34–36]. It should be noted that, in the thermodynamic sense, stress is only properly defined for a large ensemble of atoms, so that the physical meaning of the "stress per atom" is unclear. Nevertheless, this quantity can conveniently capture the existence of local instabilities within the gel, due to competitive interatomic forces [37].

3. Results

We first focus on the connectivity of the Ca atoms upon gelation. Figure 2 shows the evolution of the average number of Ca–O bonds per Ca atom over time. We find that, overall, the average coordination number of the Ca atoms remains fairly constant over time throughout the gelation process, namely, around 6.5. However, we observe that the type of O atoms belonging to the first coordination shell of Ca atoms changes over time. As expected, Ca atoms are initially entirely surrounded by Ow atoms, that is, O atoms belonging to water molecules. At this point, Ca^{2+} ions act as isolated hydrated cations. As gelation proceeds, we observe that Ca–Ow bonds are gradually replaced by Ca–Oc, that is, Ca atoms gradually start to form some bonds, with some O atoms belonging to CO_3^{2-} molecules. This signals a gradual increase in the degree of polymerization of the calcium carbonate gel (see Figure 1).

The formation of Ca–Oc–C bonds (see Figure 2) results in the appearance of carbonate clusters, that is, some groups of C atoms that are interconnected via Ca atoms (i.e., C–Oc–Ca–Oc–C bonds). Figure 3a shows the number of carbonate clusters as a function of time. As expected, the system initially comprises a large number of small isolated clusters. As gelation proceeds, the increase in the number of bonds between CO_3^{2-} groups results in a decrease in the number of carbonate clusters, which is concurrent with an increase in their average size (see Figure 1). As shown in Figure 3b,c, both the size of the largest carbonate cluster (as described by the number of C atoms it comprises) and the average size of the carbonate clusters quickly increase over time. This suggests that the carbonate gelation process is percolative in nature.

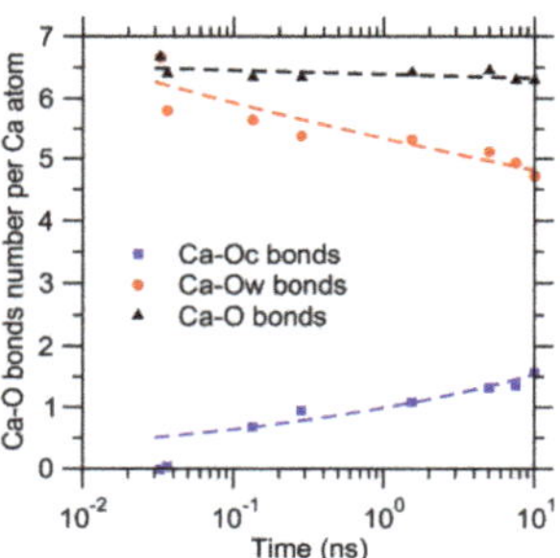

Figure 2. Evolution of the average number of Ca–O bonds per Ca atom as a function of time. The data include the contributions of Oc (i.e., O connected to a C atom) and Ow atoms (i.e., O belonging to a water molecule). The lines are to guide the eye.

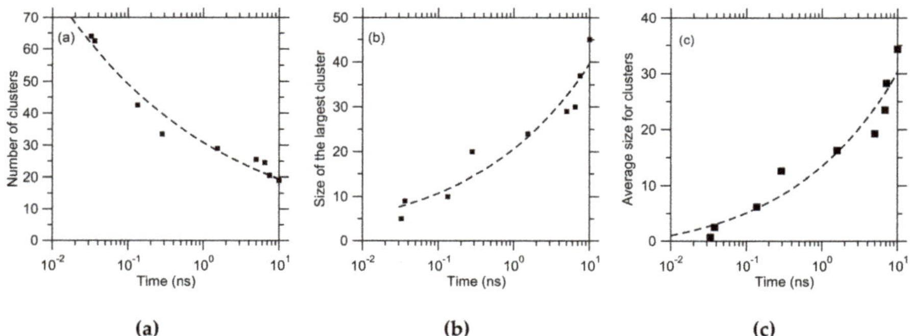

Figure 3. (a) Number of carbonate clusters, (b) size of the largest cluster (expressed in terms of the number of C atoms), and (c) average size of the carbonate clusters (expressed in terms of the number of C atoms), as a function of time during the early-stage nucleation of the amorphous calcium carbonate gel. The lines are to guide the eye.

4. Discussion

We now discuss the local stability of the structure by computing the magnitude of the local stress undergone by each atom in the structure. Figure 4 shows the evolution of the local stress undergone by Ca, Ow, C, and Oc atoms upon gelation. We first focus on the state of stress of the Ca atoms. We observe that, initially, Ca atoms are experiencing a local state of tension (i.e., positive stress, see Figure 4a). On the other hand, Ow atoms are under compression (i.e., negative stress, see Figure 4b). The origin of this state of stress can be understood in terms of the number of neighboring water molecules around Ca atoms. In detail, we find that the average O–O distance around Ca cations is 2.72 Å, which is smaller than the average O–O distance in bulk water (2.80 Å) [38]. Such O–O distances match with experimental data [11]. This indicates that, by being attracted by the central Ca cation, neighboring water molecules tend to get closer to each other than they are in bulk water. As a result, the neighboring water molecules enter a state of local compression (since they repulse each other)—as evidenced in Figure 4b. In turn, the mutual repulsion between water molecules tends to stretch the central Ca atom, so that Ca atoms experience a state of tension—as evidenced in Figure 4a. These local tensile and compressive forces mutually compensate each other, so that the CaO_6 polytope is overall at a (metastable) mechanical equilibrium. This behavior is schematically illustrated in Figure 5a. The state of tension of Ca atoms echoes the case of silicate glasses, wherein Si atoms also undergo a local tensile stress [33]. Nevertheless, we note that it remains whether this feature (e.g., the lowering of the O–O distance around Ca atoms) is generic or specific to this system [11,39,40]. It should also be noted that the coordination number of Ca atoms may depend on the solution pH and Ca concentration [41], which, in turn, could affect their state of stress. Clearly, the effect of the solution chemistry on the state of stress of the gel precursors would deserve follow-up investigations.

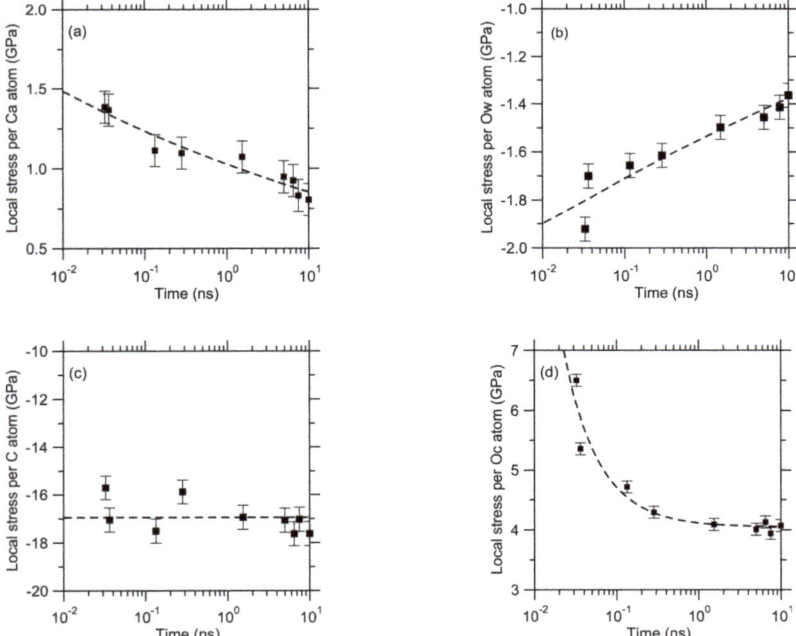

Figure 4. Evolution of the local stress experienced by (**a**) Ca, (**b**) Ow, (**c**) C, and (**d**) Oc atoms as a function of time. Positive and negative stress values indicate a state of local tension and compression, respectively.

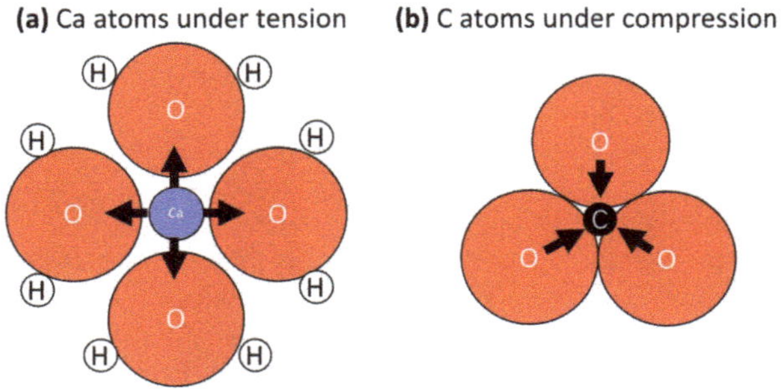

Figure 5. Schematic description of the origin of the (a) tensile and (b) compressive stress experienced by Ca and C atoms, respectively. Note that 6-fold coordinated Ca atoms also exhibit 2 additional O atoms above and below this plane. The circles are intended to represent the radii of the atoms (which are not at scale to enhance readability).

In contrast, C and Oc present a state of compression and tension, respectively (see Figure 4c,d), which contrasts with the case of Ca and Ow atoms. This can be understood from the fact that, in contrast to Ca, C atoms have a low coordination number (around 3), so that their first coordination shell exhibits a low density of Oc atoms. As such, in contrast with the case of the water molecules around Ca cations, no significant mutual coulombic repulsion among O atoms is observed around C atoms. In fact, we find that the average C–O distance around C atoms is 1.30 Å (in agreement with experimental data [11]), which is notably shorter than the equilibrium C–O distance of 1.56 Å (based on the sum of their ionic radii) [42]. Hence, the fact that C atoms are locally under compression (as evidenced by Figure 4c) arises from the fact that they are surrounded by very close O neighbors, which forces their effective radius to be smaller than their equilibrium ionic radius. In turn, the Oc neighbors around the central C atom undergo a state of tension (as evidenced by Figure 4d) for the CO_3 polytope to be at an overall (metastable) mechanical equilibrium. This behavior is schematically illustrated in Figure 5b.

These results highlight the fact that, although the entire gel is at zero pressure, the network locally exhibits some localized compressive and tensile stress at the atomic level, which arises from the significantly different coordination numbers of C and Ca atoms. However, as shown in Figure 4, such internal stress gradually gets released as the gelation proceeds. This suggests that the local stress that is initially present in the system acts as a driving force that drives the gelation transformation. This can be understood as follows. The existence of local stress around C and Ca atoms comes with an elastic energy penalty. As the gelation reaction proceeds, the formation of linkages among Ca and C atoms—which initially experience opposite states of stress—reduces the magnitude of such stress. This arises from the gradual replacement of Ow by Oc atoms in the first coordination shell of the Ca atoms, so that O atoms initially experiencing some states of compression and tension are mutually able to release each other's stress. Eventually, we find that such a combination of atomic species initially experiencing opposite states of stress mostly benefits Ca atoms—which experience a significant drop in their internal stress—whereas the stress experienced by C atoms remains largely unaffected. This likely arises from the difference in stiffness between Ca–O and C–O bonds.

We now investigate how the release of the local internal stress upon gelation manifests itself in the structure of the gel. To this end, we first compute the evolution of the distribution of the O–O neighbor-neighbor distances around Ca and C atoms (see Figure 6). We first note that, in both cases, the distributions gradually become sharper, which echoes the fact that the gel becomes more and more stable. Importantly, we find that the average O–O distance around Ca cations tends to slightly increase over time (see Figure 6a). This is consistent with the fact that the compressive stress experienced by the O atoms around Ca cations tends to decrease over time—since water molecules gradually get further away from each other and overlap less with each other. In contrast, we note that the average O–O distance around C atoms tends to slightly decrease over time (see Figure 6b). This is consistent with the fact that the tensile stress experienced by the O atoms around C atoms tends to decrease over time.

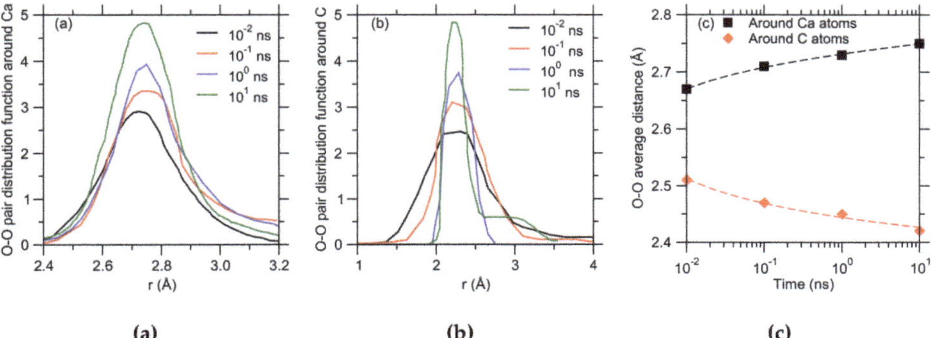

Figure 6. Distribution of the O–O distances around (**a**) Ca and (**b**) C atoms, computed at select times during the polymerization of the calcium carbonate gel. (**c**) Average O–O distance around Ca and C atoms as a function of time. The lines are to guide the eye.

Finally, we focus on the structural manifestation of the release of the local tensile stress experienced by Ca atoms. To this end, we compute the evolution of the Ca–Oc and Ca–Ow partial pair distribution functions (see Figure 7). The average Ca–O distance obtained herein agrees with previous experimental data [11,39,40]. We observe that the average Ca–Ow bond length tends to decrease over time, while the average Ca–Oc bond length tends to increase (see Figure 7c). Eventually, the average Ca–Oc bond length (2.256 Å) becomes notably larger than the average Ca–Ow bond length (2.242 Å), which indicates that the CaO_6 polytope becomes less spherical and symmetric (i.e., since different Ca–O bonds exhibit different lengths). The decoupling between Ca–Ow and Ca–Oc bond lengths is at the origin of the increase in the average O–O distance around Ca atoms (see Figure 6c). Overall, due to the predominance of Ca–Ow bonds, the average Ca–O bond length tends to decrease over time (see Figure 7c). The decrease in the Ca–O distance indicates that the central Ca atoms become less and less stretched over time. This is consistent with the fact that the magnitude of the local tensile stress tends to decrease over time. Overall, these results highlight the fact that the formation of calcium carbonation clusters (i.e., the formation of Ca–O–C linkages through the replacement of Ca–Ow by Ca–Oc bonds) directly results in the lowering of the local stress exhibited by Ca atoms. This establishes gelation as an efficient mechanism to release the internal stress that is initially present within the precursors. A similar mechanism was also suggested to occur in hydrated calcium aluminosilicate gels [43], wherein the formation of Si–O–Al linkages was also noted to result in a decrease in the magnitude of the internal stress exhibited by Si (experiencing local tension) and Al (experiencing local compression) precursors. This suggests that the relationship between local internal stress and gelation might apply to a broad array of hydrated systems [15,43].

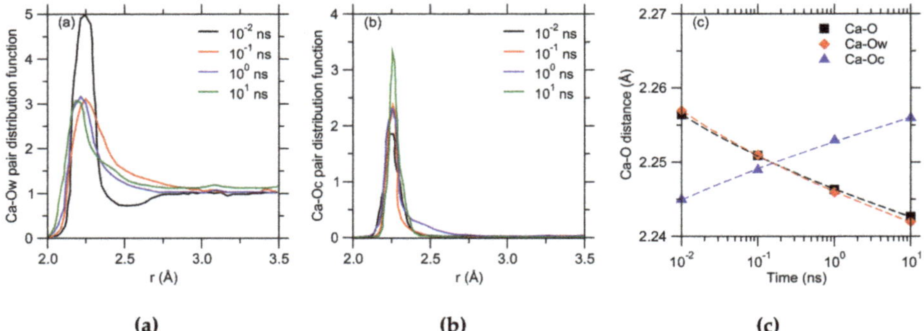

Figure 7. (**a**) Ca–Ow and (**b**) Ca–Oc partial pair distribution functions computed at select times during the polymerization of the calcium carbonate gel. (**c**) Average Ca–O, Ca–Ow, and Ca–Oc bond length as a function of time. The lines are to guide the eye.

5. Conclusions

Overall, these results highlight the role played by the internal stress in governing the early-stage nucleation of amorphous calcium carbonate. We find that, although the system is, overall, at zero pressure, the CaO_6 and CO_3 polytope precursors initially experience some competing internal stress, which arises from their significantly different coordination states. Such stress acts as an elastic energy penalty that thermodynamically promotes the polymerization of the gel—so that the combination of polytopes undergoing some opposite states of stress results in an overall decrease in the magnitude of the local stress acting in the atomic network. These results highlight the key role played by local atomic instabilities in the precursor species in driving sol–gel reactions at the atomic level.

Author Contributions: Conceptualization, G.L., G.S. and M.B.; Formal analysis, Q.Z.; Funding acquisition, M.B.; Investigation, Q.Z.; Methodology, T.D.; Project administration, M.B.; Resources, G.L.; Software, Q.Z. and T.D.; Supervision, M.B.; Visualization, Q.Z.; Writing—original draft, Q.Z.; Writing—review and editing, Q.Z., T.D., G.L., G.S. and M.B. All authors have read and agreed to the published version of the manuscript.

Funding: This research was supported by the Beijing General Research Institute of Mining & Metallurgy (BGRIMM) Technology Group and the U.S. National Science Foundation (DMREF: 1922167).

Conflicts of Interest: The authors declare no conflict of interest. The funders had no role in the design of the study; in the collection, analyses, or interpretation of data; in the writing of the manuscript, or in the decision to publish the results.

References

1. Demichelis, R.; Raiteri, P.; Gale, J.D.; Quigley, D.; Gebauer, D. Stable prenucleation mineral clusters are liquid-like ionic polymers. *Nat. Commun.* **2011**, *2*, 590. [CrossRef] [PubMed]

2. Mujah, D.; Shahin, M.A.; Cheng, L. State-of-the-Art Review of Biocementation by Microbially Induced Calcite Precipitation (MICP) for Soil Stabilization. *Geomicrobiol. J.* **2017**, *34*, 524–537. [CrossRef]

3. Addadi, L.; Joester, D.; Nudelman, F.; Weiner, S. Mollusk Shell Formation: A Source of New Concepts for Understanding Biomineralization Processes. *Chem. Eur. J.* **2006**, *12*, 980–987. [CrossRef]

4. Committee on Developing a Research Agenda for Utilization of Gaseous Carbon Waste Streams; Board on Chemical Sciences and Technology; Division on Earth and Life Studies; National Academies of Sciences, Engineering, and Medicine. *Gaseous Carbon Waste Streams Utilization: Status and Research Needs*; National Academies Press: Washington, DC, USA, 2019; ISBN 978-0-309-48336-0.

5. Vance, K.; Falzone, G.; Pignatelli, I.; Bauchy, M.; Balonis, M.; Sant, G. Direct Carbonation of $Ca(OH)_2$ Using Liquid and Supercritical CO_2: Implications for Carbon-Neutral Cementation. *Ind. Eng. Chem. Res.* **2015**, *54*, 8908–8918. [CrossRef]

6. Wei, Z.; Wang, B.; Falzone, G.; La Plante, E.C.; Okoronkwo, M.U.; She, Z.; Oey, T.; Balonis, M.; Neithalath, N.; Pilon, L.; et al. Clinkering-free cementation by fly ash carbonation. *J. CO2 Util.* **2018**, *23*, 117–127. [CrossRef]

7. Wang, Y.-W.; Kim, Y.-Y.; Stephens, C.J.; Meldrum, F.C.; Christenson, H.K. In Situ Study of the Precipitation and Crystallization of Amorphous Calcium Carbonate (ACC). *Cryst. Growth Des.* **2012**, *12*, 1212–1217. [CrossRef]

8. An X-ray absorption spectroscopy study of the structure and transformation of amorphous calcium carbonate from plant cystoliths. *Proc. R. Soc. Lond. B Biol. Sci.* **1993**, *252*, 75–80. [CrossRef]

9. Ogino, T.; Suzuki, T.; Sawada, K. The formation and transformation mechanism of calcium carbonate in water. *Geochim. Cosmochim. Acta* **1987**, *51*, 2757–2767. [CrossRef]

10. Tribello, G.A.; Bruneval, F.; Liew, C.; Parrinello, M. A Molecular Dynamics Study of the Early Stages of Calcium Carbonate Growth. *J. Phys. Chem. B* **2009**, *113*, 11680–11687. [CrossRef]

11. Michel, F.M.; MacDonald, J.; Feng, J.; Phillips, B.L.; Ehm, L.; Tarabrella, C.; Parise, J.B.; Reeder, R.J. Structural Characteristics of Synthetic Amorphous Calcium Carbonate. *Chem. Mater.* **2008**, *20*, 4720–4728. [CrossRef]

12. Henzler, K.; Fetisov, E.O.; Galib, M.; Baer, M.D.; Legg, B.A.; Borca, C.; Xto, J.M.; Pin, S.; Fulton, J.L.; Schenter, G.K.; et al. Supersaturated calcium carbonate solutions are classical. *Sci. Adv.* **2018**, *4*, eaao6283. [CrossRef] [PubMed]

13. Gebauer, D.; Völkel, A.; Cölfen, H. Stable Prenucleation Calcium Carbonate Clusters. *Science* **2008**, *322*, 1819–1822. [CrossRef] [PubMed]

14. Wallace, A.F.; Hedges, L.O.; Fernandez-Martinez, A.; Raiteri, P.; Gale, J.D.; Waychunas, G.A.; Whitelam, S.; Banfield, J.F.; Yoreo, J.J.D. Microscopic Evidence for Liquid-Liquid Separation in Supersaturated CaCO$_3$ Solutions. *Science* **2013**, *341*, 885–889. [CrossRef]

15. Du, T.; Li, H.; Sant, G.; Bauchy, M. New insights into the sol–gel condensation of silica by reactive molecular dynamics simulations. *J. Chem. Phys.* **2018**, *148*, 234504. [CrossRef]

16. Wang, M.; Smedskjaer, M.M.; Mauro, J.C.; Bauchy, M. Modifier clustering and avoidance principle in borosilicate glasses: A molecular dynamics study. *J. Chem. Phys.* **2019**, *150*, 044502. [CrossRef] [PubMed]

17. Du, T.; Li, H.; Zhou, Q.; Wang, Z.; Sant, G.; Ryan, J.V.; Bauchy, M. Atomistic origin of the passivation effect in hydrated silicate glasses. *NPJ Mater. Degrad.* **2019**, *3*, s41529–s019. [CrossRef]

18. Raiteri, P.; Demichelis, R.; Gale, J.D. Thermodynamically Consistent Force Field for Molecular Dynamics Simulations of Alkaline-Earth Carbonates and Their Aqueous Speciation. *J. Phys. Chem. C* **2015**, *119*, 24447–24458. [CrossRef]

19. Raiteri, P.; Gale, J.D. Water Is the Key to Nonclassical Nucleation of Amorphous Calcium Carbonate. *J. Am. Chem. Soc.* **2010**, *132*, 17623–17634. [CrossRef]

20. Adkins, L.; Cormack, A. Large-scale simulations of sodium silicate glasses. *J. Non-Cryst. Solids* **2011**, *357*, 2538–2541. [CrossRef]

21. Plimpton, S. Fast Parallel Algorithms for Short–Range Molecular Dynamics. *J. Comput. Phys.* **1995**, *117*, 1–19. [CrossRef]

22. Martínez, L.; Andrade, R.; Birgin, E.G.; Martínez, J.M. PACKMOL: A package for building initial configurations for molecular dynamics simulations. *J. Comput. Chem.* **2009**, *30*, 2157–2164. [CrossRef] [PubMed]

23. Shinoda, W.; Shiga, M.; Mikami, M. Rapid estimation of elastic constants by molecular dynamics simulation under constant stress. *Phys. Rev. B* **2004**, *69*. [CrossRef]

24. Tuckerman, M.E.; Alejandre, J.; López-Rendón, R.; Jochim, A.L.; Martyna, G.J. A Liouville-operator derived measure-preserving integrator for molecular dynamics simulations in the isothermal–isobaric ensemble. *J. Phys. Math. Gen.* **2006**, *39*, 5629–5651. [CrossRef]

25. Gale, J.; Raiteri, P.; Van Duin, A.C.T. A reactive force field for aqueous-calcium carbonate systems. *Phys. Chem. Chem. Phys.* **2011**, *13*, 16666–16679. [CrossRef]

26. Wang, N.; Feng, Y.; Guo, X.; Van Duin, A.C.T. Insights into the Role of H$_2$O in the Carbonation of CaO Nanoparticle with CO$_2$. *J. Phys. Chem. C* **2018**, *122*, 21401–21410. [CrossRef]

27. Costa, M.F. Molecular dynamics of molten Li$_2$CO$_3$–K$_2$CO$_3$. *J. Mol. Liq.* **2008**, *138*, 61–68. [CrossRef]

28. Stukowski, A. Visualization and analysis of atomistic simulation data with OVITO—the Open Visualization Tool. *Model. Simul. Mater. Sci. Eng.* **2010**, *18*, 015012. [CrossRef]

29. Vitek, V.; Egami, T. Atomic Level Stresses in Solids and Liquids. *Phys. Status Solidi B* **1987**, *144*, 145–156. [CrossRef]

30. Egami, T. Atomic level stresses. *Prog. Mater. Sci.* **2011**, *56*, 637–653. [CrossRef]

31. Thompson, A.P.; Plimpton, S.J.; Mattson, W. General formulation of pressure and stress tensor for arbitrary many-body interaction potentials under periodic boundary conditions. *J. Chem. Phys.* **2009**, *131*, 154107. [CrossRef]

32. Yu, Y.; Wang, M.; Anoop Krishnan, N.M.; Smedskjaer, M.M.; Deenamma Vargheese, K.; Mauro, J.C.; Balonis, M.; Bauchy, M. Hardness of silicate glasses: Atomic-scale origin of the mixed modifier effect. *J. Non-Cryst. Solids* **2018**, *489*, 16–21. [CrossRef]

33. Li, X.; Song, W.; Smedskjaer, M.M.; Mauro, J.C.; Bauchy, M. Quantifying the internal stress in over-constrained glasses by molecular dynamics simulations. *J. Non-Cryst. Solids X* **2019**, *1*, 100013. [CrossRef]

34. Yu, Y.; Wang, M.; Smedskjaer, M.M.; Mauro, J.C.; Sant, G.; Bauchy, M. Thermometer Effect: Origin of the Mixed Alkali Effect in Glass Relaxation. *Phys. Rev. Lett.* **2017**, *119*, 095501. [CrossRef]

35. Yu, Y.; Mauro, J.C.; Bauchy, M. Stretched exponential relaxation of glasses: Origin of the mixed-alkali effect. *Am. Ceram. Soc. Bull.* **2017**, *96*, 34–36.

36. Yu, Y.; Wang, M.; Zhang, D.; Wang, B.; Sant, G.; Bauchy, M. Stretched Exponential Relaxation of Glasses at Low Temperature. *Phys. Rev. Lett.* **2015**, *115*, 165901. [CrossRef] [PubMed]

37. Ioannidou, K.; Del, G.E.; Ulm, F.-J.; Pellenq Roland, J.-M. Inhomogeneity in Cement Hydrates: Linking Local Packing to Local Pressure. *J. Nanomech Micromech* **2017**, *7*, 04017003. [CrossRef]

38. Mähler, J.; Persson, I. A Study of the Hydration of the Alkali Metal Ions in Aqueous Solution. *Inorg. Chem.* **2012**, *51*, 425–438. [CrossRef] [PubMed]

39. Becker, A.; Bismayer, U.; Epple, M.; Fabritius, H.; Hasse, B.; Shi, J.; Ziegler, A. Structural characterisation of X-ray amorphous calcium carbonate (ACC) in sternal deposits of the crustacea Porcellio scaber. *Dalton Trans.* **2003**, 551–555. [CrossRef]

40. Levi-Kalisman, Y.; Raz, S.; Weiner, S.; Addadi, L.; Sagi, I. Structural Differences Between Biogenic Amorphous Calcium Carbonate Phases Using X-ray Absorption Spectroscopy. *Adv. Funct. Mater.* **2002**, *12*, 43–48. [CrossRef]

41. Avaro, J.; Moon, E.M.; Rose, J.; Rose, A.L. Calcium coordination environment in precursor species to calcium carbonate mineral formation. *Geochim. Cosmochim. Acta* **2019**, *259*, 344–357. [CrossRef]

42. Shannon, R.D. Revised effective ionic radii and systematic studies of interatomic distances in halides and chalcogenides. *Acta Crystallogr. Sect. A* **1976**, *32*, 751–767. [CrossRef]

43. Zhao, C.; Zhou, W.; Zhou, Q.; Zhang, Y.; Liu, H.; Sant, G.; Liu, X.; Guo, L.; Bauchy, M. Precipitation of Calcium–Alumino–Silicate–Hydrate Gels: The Role of the Internal Stress. *J. Chem. Phys.* **2020**, *124*, 105806.

Article

Numerical Prediction of the Behavior of CO_2 Bubbles Leaked from Seafloor and Their Convection and Diffusion near Southeastern Coast of Korea

Se-Min Jeong [1], Seokwon Ko [2] and Wu-Yang Sean [3,*]

[1] Department of Naval Architecture and Ocean Engineering, Chosun University, Gwangju 61452, Korea; smjeong@chosun.ac.kr
[2] Department of Naval Architecture and Ocean Engineering, Graduate School of Chosun University, Gwanju 61452, Korea; seokwon2486@gmail.com
[3] Department of Environmental Engineering, Chung Yuan Christian University, Taoyuan 32023, Taiwan
* Correspondence: wysean@cycu.org.tw

Received: 31 May 2020; Accepted: 18 June 2020; Published: 20 June 2020

Abstract: Among various carbon capture and storage technologies to mitigate global warming and ocean acidification due to greenhouse gases, ocean geological storage is considered the most feasible for Korea due to insufficient inland space to store CO_2. However, the risk of CO_2 leakage and the behavior and environmental effects of the leaked CO_2 need to be assessed for its successful implementation. Therefore, the behavior of CO_2 bubbles/droplets dissolving into the surrounding seawater and the diffusion of dissolved CO_2 by ocean flows should be accurately predicted. However, finding corresponding research has been difficult in Korea. Herein, the behavior and convection-diffusion of CO_2 that was assumed to have leaked from the seafloor near the southeastern coast of Korea were numerically predicted using a multi-scale ocean model for the first time. In the simulation region, one of the pilot projects of CO_2 ocean geological storage had started but has been temporarily halted. In the ocean model, hydrostatic approximation and the Eulerian–Lagrangian two-phase model were applied for meso- and small-scale regions, respectively. Parameters for the simulations were the leakage rate and the initial diameter of CO_2. Results revealed that all leaked and rising CO_2 bubbles were dissolved into the seawater before reaching the free surface; further, the change in the partial pressure of CO_2 did not exceed 500 ppm during 30 days of leakage for all cases.

Keywords: carbon capture and storage; CO_2 ocean geological storage; multi-scale ocean model; hydrostatic approximation; Eulerian–Lagrangian two-phase model; environmental impact

---

1. Introduction

Carbon capture and storage (CCS) is one of the technologies used to mitigate global warming and ocean acidification. Various CCS methods are available and can be categorized into inland or ocean geological storage, direct injection and so on. Among these, only inland and ocean geological storage are considered viable following the prohibition of the direct injection method due to its uncertain environmental impacts. CO_2 geological storage is a method for capturing CO_2 from power plants or industrial processes without releasing it into the atmosphere, transporting it to sites suitable for geological storage, and storing it stably underground for the long term. Furthermore, it has the advantage of large-scale CO_2 reduction. Geological storage has been mainly carried out in countries with sufficient inland storage space. However, for countries with limited inland space such as Korea, Japan, Norway, and the United Kingdom, ocean geological storage is considered the most feasible option.

The Sleipner CO_2 injection project in Norway was the world's first industrial offshore CO_2 ocean geological storage project, through which more than 16 Mt of CO_2 was injected from 1996 to 2014.

The injection rate was approximately 0.9 Mt/y during the early years and reduced slightly during later years due to reduced gas flow from the Sleipner Vest [1,2]. From 2008 to 2012, 1.6 Mt of CO_2 was injected into the Snøhvit gas field. Although the injection was occasionally halted due to operational challenges at the LNG plant during that period, approximately 0.5 Mt of CO_2 was stored in the well, and injection has continued since 2011 [3]. A large-scale CCS demonstration project in Japan's Tomakomai area, which can store 0.1 Mt/y of CO_2 in two reservoirs that lie 1100 m and 2400 m below the seabed, is being undertaken by the Japanese government [4]. In the case of Korea, a CO_2 storage project that can store 1 Mt/y was being constructed near the southeastern coast of Korea, but was temporarily stopped in 2018.

To utilize any kind of CCS technology, public acceptance must be ensured through the risk assessment of CO_2 leakage, monitoring the behavior of leaked and dissolved CO_2 (DCO_2), and environmental assessment of the leaked CO_2. The Quantifying and Monitoring Potential Ecosystem Impacts of Geological Carbon Storage (QICS) project in the UK was created to assess the environmental impact of CO_2 release experiments in Ardmucknish Bay [5]. Sellami et al. [6] investigated the dynamics of leaked CO_2 bubbles in a plume in the bay through observational data obtained from the QICS project. Because of the limitations of field experiments, studies utilizing computational fluid dynamics (CFD) have been carried out to estimate the behaviors of leaked CO_2 and DCO_2. The highly complex phenomena that occur when liquid CO_2 is injected into the deep ocean was numerically studied by Eulerian-Eulerian two-phase CFD simulations by Alendal and Drange [7] and Eulerian–Lagrangian simulations by Sato and Sato [8]. Jeong et al. [9] developed a multi-scale ocean model, which builds a bridge between smaller near-field scale and larger regional-scale models to predict the fate of DCO_2 in the deep ocean. Kano et al. [10] conducted numerical simulations on the behavior of CO_2 bubbles and droplets leaked from the seafloor into water columns in uniform flows perpendicular to the leakage band. They showed that CO_2 dissolved in seawater before returning to the air under their simulation conditions, which indicates that the ocean can play the role of a buffer that does not allow CO_2 to return to the atmosphere. Kano et al. [11] developed a multi-scale numerical method to predict the behavior of dissolvable CO_2 bubbles leaked from the seafloor and dissolved mass in the ocean. A simulation using this model was conducted with real topography and tidal currents near the Japanese coastline. Mori et al. [12] used the model developed by Kano et al. [11] and conducted case studies in which the ratio of CO_2 seepage in the dissolved phase and the proportion remaining in the sediment were changed to predict CO_2 concentration distributions in Ardmucknish Bay. However, in Korea, finding corresponding research has been difficult, and the few studies, such as that by Kang et al. [13], have only focused on the behavior inside CO_2 reservoirs and interactions between the reservoir and the surface of seabed through the fault. Such studies cannot be applied to consider ocean flows nor assess environmental impacts in the ocean.

In this study, the behavior and convection-diffusion of CO_2, which was assumed to have leaked from the seafloor near the southeastern coast of Korea, where a pilot project of CO_2 ocean geological storage started but was temporarily stopped, were numerically predicted using a multi-scale ocean model for the first time. In the ocean model, hydrostatic approximation and Eulerian–Lagrangian two-phase models were applied for the meso- and small-scale regions, respectively. Numerical simulations involving changes in the main parameters, such as the initial diameter of CO_2 and leakage rate, were carried out to investigate its effects on the simulation results, especially regarding environmental impacts.

2. Methodology

The multi-scale ocean model used in this study is an improved version of the original Maritime Environment Committee (MEC) ocean model developed by the Japan Society of Naval Architecture and Ocean Engineers (JASNAOE). In the MEC model, hydrostatic approximation and non-hydrostatic (i.e., full-3D) models are applied for the meso- and small-scale domains, respectively. In the mesoscale domain, tidal flow is generated under hydrostatic (pressure) approximation to reduce the computational time. The applicability and accuracy of the hydrostatic model of the MEC model can be found in the research of Kano et al. [11], Lee et al. [14], and so on. A full-3D small-scale model is necessary when the vertical flow component cannot be ignored. The spatial connection between the two models is such that the full-3D model domain matches one grid-column of the hydrostatic model. Kano et al. [11] modified the MEC model by adopting a two-phase model for the small-scale domain, where the continuous liquid phase (seawater) and dispersed gas/liquid phase (individual CO_2 bubble/droplet) are solved using Eulerian and Lagrangian methods, respectively. This model, hereafter termed the "MEC-CO_2 model", was used in the present study.

The governing equations of the mesoscale model adopting the hydrostatic approximation are as follows.

$$\frac{\partial u}{\partial x} + \frac{\partial v}{\partial y} + \frac{\partial w}{\partial z} = 0 \tag{1}$$

$$\frac{\partial u}{\partial t} + u\frac{\partial u}{\partial x} + v\frac{\partial u}{\partial y} + w\frac{\partial u}{\partial z} = -\frac{1}{\rho}\frac{\partial p}{\partial x} + fv + A_M\left(\frac{\partial^2 u}{\partial x^2} + \frac{\partial^2 u}{\partial y^2}\right) + \frac{\partial}{\partial z}\left(K_M\frac{\partial u}{\partial z}\right) \tag{2}$$

$$\frac{\partial v}{\partial t} + u\frac{\partial v}{\partial x} + v\frac{\partial v}{\partial y} + w\frac{\partial v}{\partial z} = -\frac{1}{\rho}\frac{\partial p}{\partial y} - fu + A_M\left(\frac{\partial^2 v}{\partial x^2} + \frac{\partial^2 v}{\partial y^2}\right) + \frac{\partial}{\partial z}\left(K_M\frac{\partial v}{\partial z}\right) \tag{3}$$

$$0 = -\frac{\partial p}{\partial z} - \rho g \tag{4}$$

where u, v, and w are the velocity components in the x, y, and z directions, respectively. p, ρ, g, and f are the pressure, density of seawater, gravity acceleration, and Coriolis force coefficient, respectively. A_M and K_M are the horizontal and vertical eddy viscosities, respectively.

The transport equations of scalar properties ϕ (i.e., temperature (T), salinity (S), and DCO$_2$ (C)) are as follows:

$$\frac{\partial \phi}{\partial t} = A_D\left(\frac{\partial^2 \phi}{\partial x^2} + \frac{\partial^2 \phi}{\partial y^2}\right) + \frac{\partial}{\partial z}\left(K_D\frac{\partial \phi}{\partial z}\right) \tag{5}$$

where A_D and K_D are the horizontal and vertical eddy diffusivities, respectively.

As mentioned earlier, the Eulerian–Lagrangian two-phase model is applied for small-scale full-3D regions. The continuity and Navier–Stokes equations for the continuous phase are as follows:

$$\frac{\partial}{\partial t}(\alpha_c\rho_c) + \nabla\cdot(\alpha_c\rho_c\boldsymbol{u_c}) = \frac{1}{V_{cell}}\sum_{n=1}^{n_d}\Gamma \tag{6}$$

$$\frac{\partial}{\partial t}(\alpha_c\rho_c\boldsymbol{u_c}) + \nabla\cdot(\alpha_c\rho_c\boldsymbol{u_c}\otimes\boldsymbol{u_c}) + \frac{1}{V_{cell}}\sum_{n=1}^{n_d}\frac{D_d}{Dt}(\rho_d V_d\boldsymbol{u_d}) = -\nabla P + \nabla\cdot[\alpha_c\rho_c(v_c + v_t)\boldsymbol{d_c}] \tag{7}$$

where,

$$\alpha_d + \alpha_c = 1 \tag{8}$$

$$\alpha_d = \frac{1}{V_{cell}}\sum_{n=1}^{n_d}V_d \tag{9}$$

$$\boldsymbol{d_c} = \nabla\boldsymbol{u_c} + (\nabla\boldsymbol{u_c})^T \tag{10}$$

where α is the volume fraction and $\boldsymbol{u}$ is the velocity vector. n_d, V, and V_{cell} are the number of bubbles in a computational cell, volume, and the volume of a computational cell, respectively. Γ is the mass transfer from a bubble at the interface and P is the pressure. κ and v are the thermal diffusivity and kinematic viscosity, respectively. The subscripts c and d denote the continuous water phase and dispersed CO_2 bubble phase, respectively. ρ_d and ρ_c are given by Pitzer and Sterner [15] and Alendal and Drange [7], respectively, and v_t is determined by Smagorinsky's model [16].

The transport equations of T, S, and C are also solved as given below:

$$\frac{\partial}{\partial t}(\rho_c c_c T_c) + \nabla \cdot (\rho_c c_c T_c \boldsymbol{u_c}) = \nabla \cdot \left[\rho_c c_c \left(\kappa + \frac{\nu_t}{Pr_t} \right) \nabla T_c \right] + \Phi \tag{11}$$

$$\frac{\partial}{\partial t}(\alpha_c S) + \nabla \cdot (\alpha_c S \boldsymbol{u_c}) = \nabla \cdot \left[\alpha_c \left(D_s + \frac{\nu_t}{Sc_t} \right) \nabla S \right] \tag{12}$$

$$\frac{\partial}{\partial t}(\alpha_c C) + \nabla \cdot (\alpha_c C \boldsymbol{u_c}) = \nabla \cdot \left[\alpha_c \left(D_C + \frac{\nu_t}{Sc_t} \right) \nabla C \right] + \frac{1}{V_{cell}} \sum_{n=1}^{n_d} \Gamma \tag{13}$$

$$\boldsymbol{d_c} = \nabla \boldsymbol{u_c} + (\nabla \boldsymbol{u_c})^T \tag{14}$$

$$\Gamma = \pi d_e^2 k (C_I - C_{cell}) \tag{15}$$

where C_I and C_{cell} are C at the bubble surface and in a computational cell, respectively. D is the diffusion coefficient. Pr_t and Sc_t are the turbulent Prandtl and Schmidt numbers, respectively. Hirai et al. [17] found that C_I matches the solubility of CO_2, which was measured by Weiss [18]. The mass transfer rate k is modeled as in Chen et al. [19] and d_e is the equivalent diameter of a bubble.

For the dispersed phase, the mass conservation and motion equation are solved for each bubble in the Lagrangian frame as:

$$\frac{D_d}{Dt}(\rho_d V_d) = -\Gamma \tag{16}$$

$$\frac{D_d}{Dt}[(\rho_d + \beta \rho_c) V_d \boldsymbol{u_d}] = V_d [-\nabla P + (\rho_c - \rho_d) \boldsymbol{g} - \boldsymbol{f_D} - \boldsymbol{f_L}] \tag{17}$$

where,

$$\boldsymbol{f_D} = \frac{1}{2} C_D \frac{3}{2 d_e} \rho_c |\boldsymbol{u_r}| \boldsymbol{u_r} \tag{18}$$

$$\boldsymbol{f_L} = C_L \rho_c \boldsymbol{u_r} \times \boldsymbol{\omega_c} \tag{19}$$

$$\boldsymbol{u_r} = \boldsymbol{u_d} - \boldsymbol{u_c} \tag{20}$$

$$\boldsymbol{\omega_c} = \nabla \times \boldsymbol{u_c} \tag{21}$$

where β is the coefficient of the added mass of a bubble and $\boldsymbol{g}$ is the gravity vector. $\boldsymbol{f_D}, \boldsymbol{f_L}$, and C_D, C_L are the drag and lift forces and the coefficients of a bubble, respectively. $\boldsymbol{u_r}$ is the relative velocity of the dispersed phase to the continuous phase, and $\boldsymbol{\omega_c}$ is the vorticity of a bubble. β and C_L are both set to 0.5. The details on the modeling C_D of a bubble can be found in [19].

The governing equations were discretized by the finite volume method using orthogonal and staggered grids: velocity components were defined at the cell faces, and the other variables were defined at the cell centers. For spatial discretization, third-order up-winding and second-order central differencing schemes were adopted for the advection and diffusion terms, respectively. The second-order Adams–Bashforth method was used for the time integration in the full-3D model.

3. Simulation Conditions

Figure 1 shows the target area of the present simulations near the southeastern coast of Korea, where one of the candidate sites for CO_2 ocean geological storage is located. The data for inland and seafloor topographies were obtained from the Global 30 Arc-Second Elevation Dataset (GTOPO30) of the United States Geological Survey's Earth Resources Observation and Science Center and the JODC-Expert Grid Data for Geography-500m (J-EGG500) of the Japan Oceanographic Data Center, respectively. Grid systems for the mesoscale domain were generated using the preprocessor of the MEC model with these data. The dimensions of the computational domain were approximately 100 km × 50 km × 1 km, and the number of grid cells were 50 × 50 × 63 in the x, y, and z directions, respectively, as shown in Figure 2a. The vertical cell column contains the full-3D small-scale domain, of which the size is 2 km × 1 km × 0.18 km with 100 × 100 × 28 grid cells and is shown by a closed cell in Figure 2b. The vertical grid levels are listed in Table 1, where layers 1 and 28 match the free surface and the depth of CO_2 leakage, that is, 190 m, respectively.

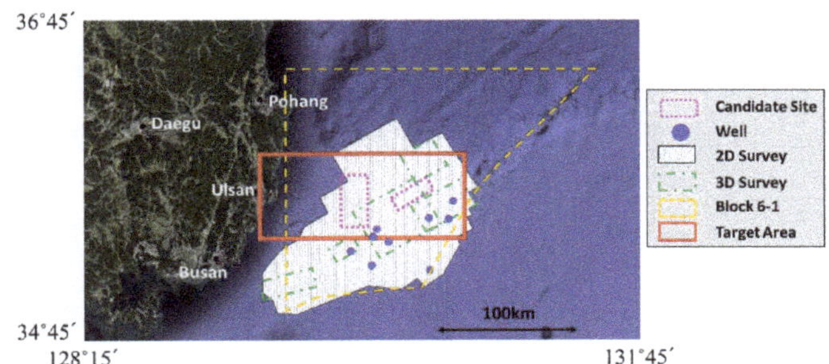

Figure 1. Target area for numerical simulation (source: https://www.google.com/maps).

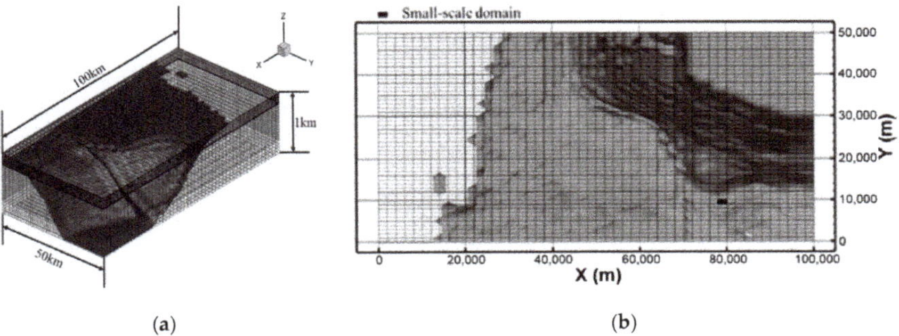

(a) (b)

Figure 2. (**a**) Perspective and (**b**) horizontal view of grid systems (one cell filled with black color in the mesoscale domain is the computational domain for the full-3D small-scale domain).

Table 1. Layer numbers, thicknesses, and depths of vertical grids.

Layer Number	Thickness (m)	Depth (m)
1(surface)–2	5	0–10
3–18	10	20–170
19–28	2	172–190
29	10	200
30–59	20	220–800
60–63	50	900–1000

To reproduce proper ocean flows, the major tidal components of M2, O1, K1, and S2, which were obtained from the NAO99b model [20] and listed in Table 2, were interpolated and imposed on the open boundaries of the mesoscale model domain with nonreflecting boundary conditions by Hino and Nakaze [21].

Table 2. Tidal components used for open boundary conditions.

Tidal Components	Period (h)	Amplitude (m)	Phase (°)
M2	12.42	0.0689	356.07
O1	25.82	0.0522	182.96
K1	23.93	0.0476	220.93
S2	12.00	0.0386	102.91

Figure 3 shows the initial conditions of T, S, and DCO_2 obtained from the data of the Array for Real-time Geostrophic Oceanography project by the National Institute of Meteorological Sciences of Korea (http://argo.nims.go.kr).

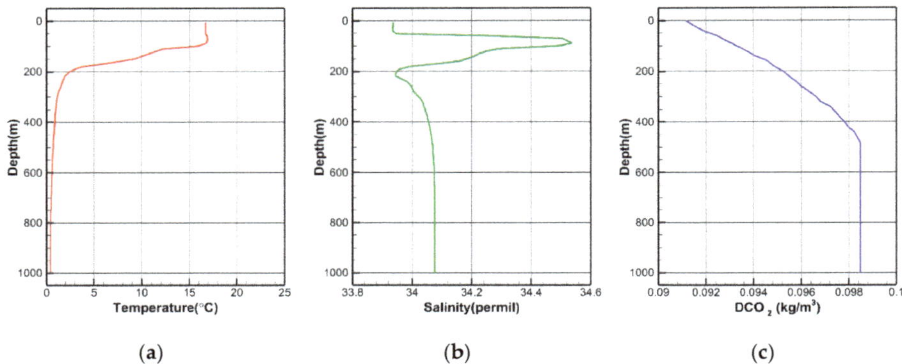

(a)   (b)   (c)

Figure 3. Initial conditions of (**a**) temperature, (**b**) salinity, and (**c**) DCO_2.

To calculate the fluxes of temperature and salinity at the free surface, climate data, which were the temporal average of observed data by the National Aeronautics and Space Administration (NASA, https://power.larc.nasa.gov/data-access-viewer/), were used, as depicted in Table 3.

Table 3. Climate conditions.

Climate Conditions	Temporal Average
Albedo (-)	0.06
Injection rate (-)	0.97
Cloud amount coefficient (-)	0.65
Global solar radiation (W/m^2)	148.1
Cloud amount (0–10)	6.07
Precipitation (mm/h)	0.1647
Water vapor pressure (hPa)	12.8
Wind speed (m/s)	3.10
Air temperature (°C)	16.72

The leakage rate and initial diameter of the CO_2 bubble were selected as the main parameters. The assumed leakage rates for the present study were 3800, 50,000, and 100,000 t/y, based on the study of Kano et al. [11], where two cases for the leakage rate were studied: an extreme case, 94,600 t/y, which assumed that a large fault accidentally connected the CO_2 reservoir and the seafloor; and a reasonable case, 3800 t/y, based on the seepage rate of an existing enhanced oil recovery (EOR) site [22]. The initial diameters of the CO_2 bubble were set to 5, 10, and 20 mm. The simulation cases and leakage are listed and illustrated in Table 4 and Figure 4, respectively.

Table 4. Simulation cases.

Case	Leakage Rate (t/y)	Diameter of CO$_2$ Bubble (mm)	Leakage Area (m^2)
1	3800	20	20,000
2	50,000	20	20,000
3	100,000	20	20,000
4	100,000	10	20,000
5	100,000	5	20,000

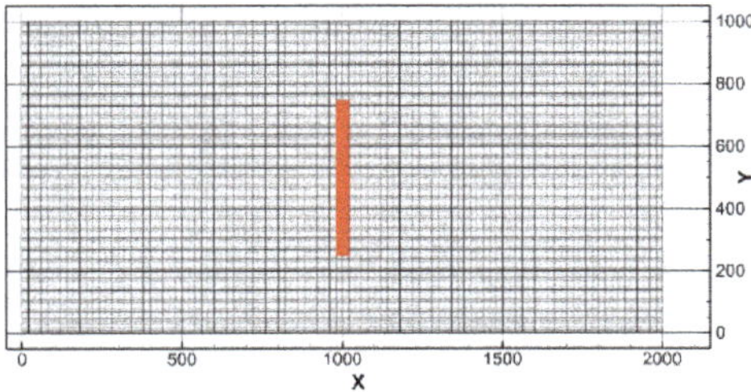

Figure 4. Distribution of leakage area at the bottom surface of the small-scale domain, of which the horizontal grid sizes are 20 m × 10 m.

Quantitative criteria are needed to assess the environmental impact when CO_2 leakage occurs. Kikkawa et al. [23] elucidated that the biological impacts of CO_2 in the ocean should not be related solely to pH, but also to the partial pressure of CO_2 (pCO_2). Kita and Watanabe [24] collected LC50 and LT50 data against pCO_2 for various marine species and proposed that the change in pCO_2 (ΔpCO_2) of 5000 ppm is the no-observed effect concentration (NOEC) and that 500 ppm is the predicted no-effect concentration (PNEC). Although these values have not been authorized, we refer to them as tentative standards in this study.

4. Results

Figure 5 shows the temporal change of the estimated tidal level in the full-3D domain, which is one cell of the mesoscale domain. The upper and lower envelopes are similar, and periodic patterns are observed. Inside the computational domain, complicated flows are generated owing to the interaction between tidal flows from three open boundaries and the bottom topography, as shown in Figure 6.

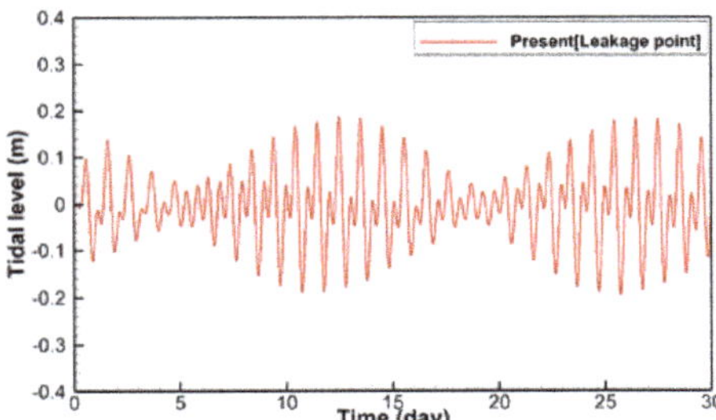

Figure 5. Time history of calculated tidal level in a small-scale domain.

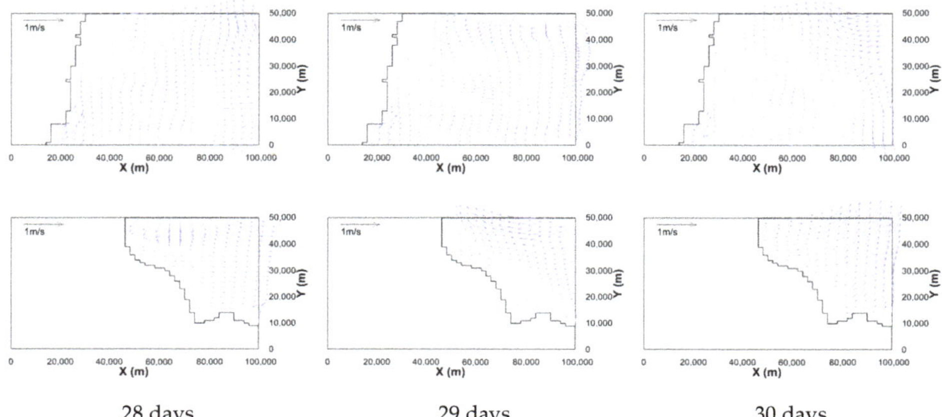

28 days   29 days   30 days

Figure 6. Velocity vector fields on the horizontal planes (upper rows: near free surface, lower rows: z = −187 m) at 28, 29, and 30 days after the start of simulation (data are skipped by 2).

4.1. Effect of Leakage Rate (Cases 1, 2, and 3/Bubble Size = 20 mm, Leakage Area = 20,000 m)

Figure 7 shows contour maps of some variables on the center plane in the x direction (x = 1000 m) of the small-scale domain, 30 days after the start of leakage. From the distribution of void rate and cell-averaged number densities of CO_2 bubbles, it can be seen that all CO_2 bubbles leaking from the depth of 190 m would be dissolved at a depth of 70 m before reaching the free surface. Because the initial diameter of the CO_2 bubbles are the same, but the leakage rates are different for these cases, the void rates of the bubbles near the leakage depth are higher than other depths and become lower as the depth becomes shallower due to the rising and dissolution of the bubbles to the surrounding seawater. The reason for the higher values of the number densities near the depth of 100 m is that the density of the bubbles is almost the same as that of seawater, and the volume of the bubbles is very small, which makes the buoyant force zero. Although the velocity vector fields or contour maps of the velocities are not presented, one may know that complex flows exist owing to the rising of the CO_2 bubbles with surrounding seawater and tidal flows from the distribution of the DCO_2 shown in Figure 7c. Since more bubbles exist and dissolution lasts for a longer time as the leakage rate increases, relatively higher values and wider areas of dispersed DCO_2 are observed, as shown in the figure. Comparing the contour maps of DCO_2 with those of the changes in DCO_2 (ΔDCO_2) due to the dissolution of the bubbles, no large discrepancy is observed because the background DCO_2 is much smaller.

The contour maps of ΔpCO_2 on the center planes of the x, y, and z directions of the small-scale domain are depicted in Figure 8, where not only the diffusion but also the advection of ΔpCO_2 by the tide are clearly shown. A high ΔpCO_2 is observed near the leakage depth, and the maximum ΔpCO_2 is lower than 200 ppm and does not exceed the PNEC. It can also be seen that the distributions of ΔpCO_2 shown in Figure 8a are similar to those of ΔDCO_2 on the same plane (Figure 7c).

The contour maps of ΔpCO_2 on the constant z planes of the mesoscale domain with respect to depths and elapsed time after leakage for the most extreme case are illustrated in Figures 9 and 10, respectively, where the black rectangular box indicates the full-3D small-scale domain including the leakage area. Both figures show that high ΔpCO_2 regions are placed at depths between 150 and 170 m, not in the leakage depth. This might be due to the convection of DCO_2 by the geostrophic flows. It can also be found that diffusion is more dominant than convection as water depth becomes deeper. The time change of maximum ΔpCO_2 in the mesoscale domain is not significant, as shown in Figure 10. The maximum ΔpCO_2 is lower than 50 ppm and does not exceed the PNEC.

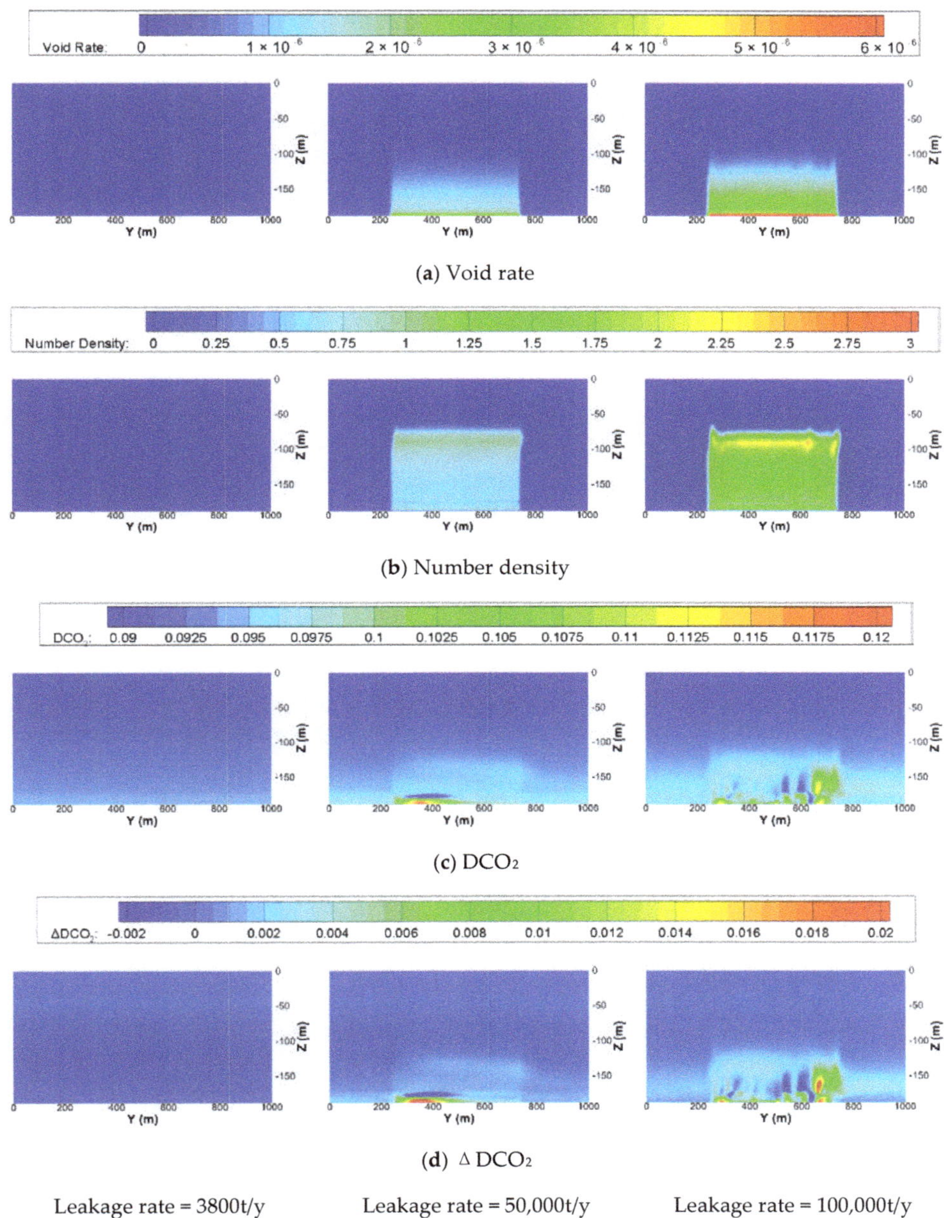

Figure 7. Contour maps of (**a**) void rate, (**b**) number density of undissolved CO_2 bubbles, (**c**) DCO_2, and (**d**) ΔDCO_2 on the center plane in the x direction (x = 1000 m) of the small-scale domain at 30 days after the start of leakage with the same initial diameter of bubble of 20 mm and leakage area of 20,000 m^2 but different leakage rates.

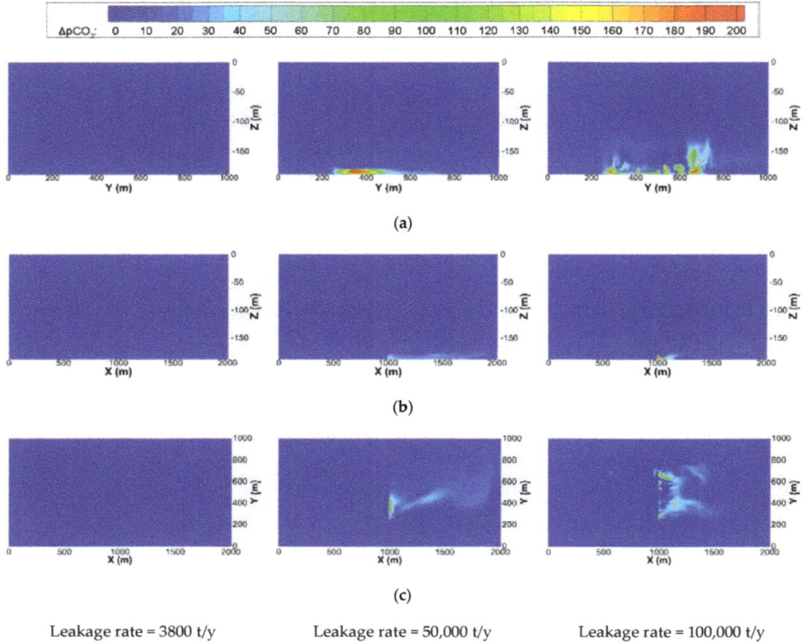

Figure 8. Contour maps of ΔpCO_2 on the center planes of the x, y, and z directions of the small-scale domain at 30 days after the start of leakage with the same initial diameter of bubble of 20 mm and leakage area of 20,000 m^2 but with different leakage rates. (**a**) Center plane in x direction (x = 1000 m); (**b**) Center plane in y direction (y = 500 m); (**c**) xy plane near leakage area (z = −180 m)

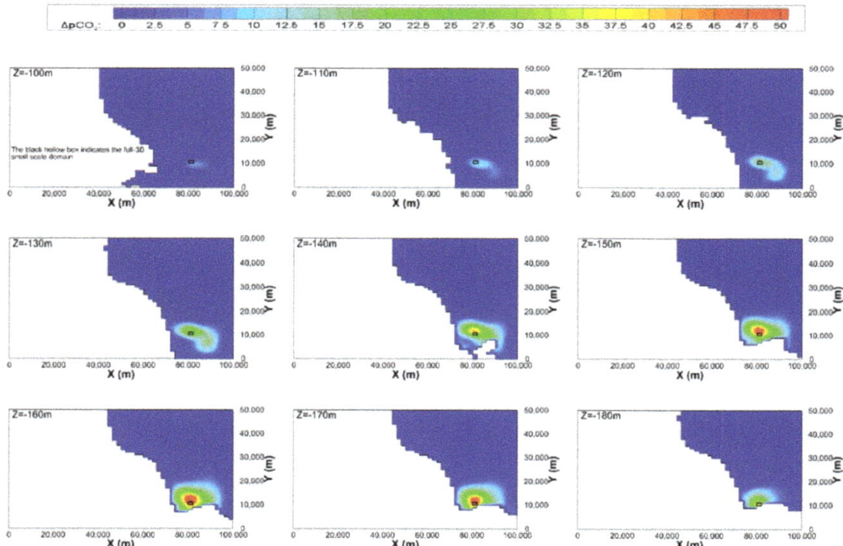

Figure 9. Contour maps of ΔpCO_2 in the constant z planes of the mesoscale domain at 30 days after the start of leakage for case 3 (leakage rate = 100,000 t/y).

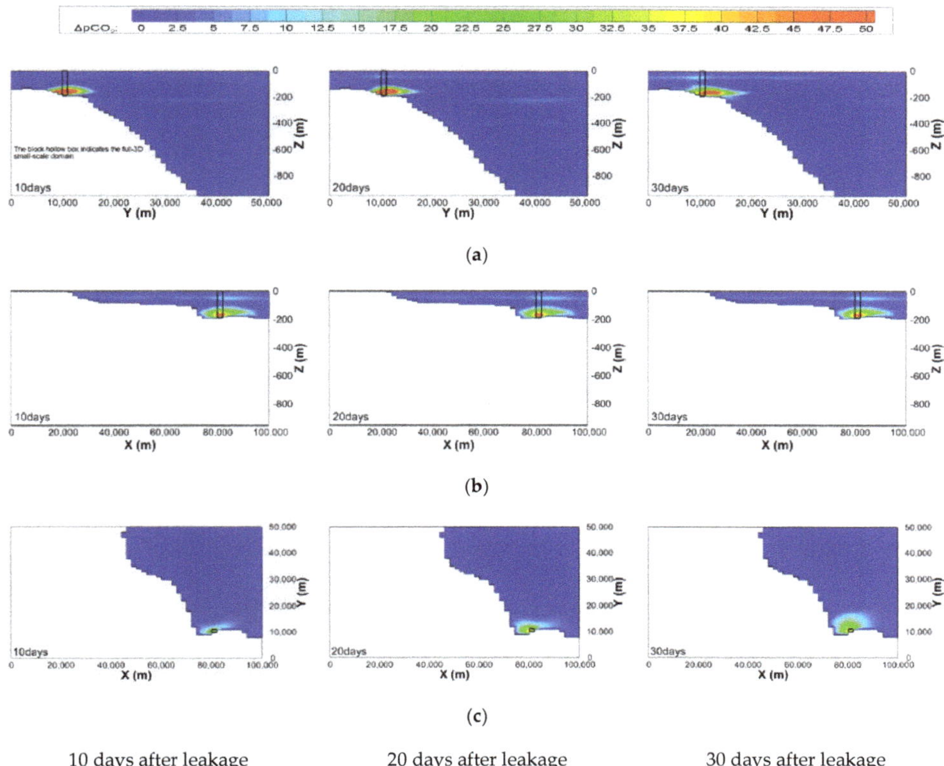

10 days after leakage   20 days after leakage   30 days after leakage

Figure 10. Time change of contour maps of ΔpCO_2 on the planes, where the full-3D small-scale domain is located, of the mesoscale domain for case 3 (leakage rate = 100,000 t/y). (**a**) yz plane (x = 81,000 m); (**b**) zx plane (y = 10,500 m); (**c**) xy plane (z = 180 m)

4.2. Effect of Initial Diameter of CO_2 Bubble (Cases 5, 4, and 3/Leakage Rate = 100,000 t/y, Leakage Area = 20,000 m)

Contour maps of some variables on the center plane in the x direction (x = 1000 m) of the small-scale domain at 30 days after the start of leakage are shown in Figure 11. As seen in the distribution of void rate and cell-averaged number densities of CO_2 bubbles, as the initial diameter of the bubbles increases, the rising distance of the bubbles increases, which results from the larger buoyant force of the relatively large volume of the bubble. Because the leakage rates are the same for these cases, the number of bubbles is high within the short band when the initial diameter is 5 mm. Comparing the distributions of DCO_2 or ΔDCO_2 among the cases, one can find that the distribution when the diameter is 10 mm is quite different from those of 5 and 20 mm. The reason for this phenomenon is that the mass transfer rate of a CO_2 bubble is largely affected by the shape of the bubble. The total mass transfer per unit volume decreases as the diameter of a bubble decreases. However, if the diameter exceeds a certain value and changes to the spherical cap, there is a sudden jump, and it decreases as the diameter increases. The threshold is approximately 18 mm. Therefore, the mass transfer rates when the diameters are 5 and 20 mm are similar, but the rate becomes smaller when the diameter is 10 mm, as discussed by Kano et al. (2009). The lower dissolution of the bubbles results in a low DCO_2 and ΔDCO_2 when the diameter is 10 mm.

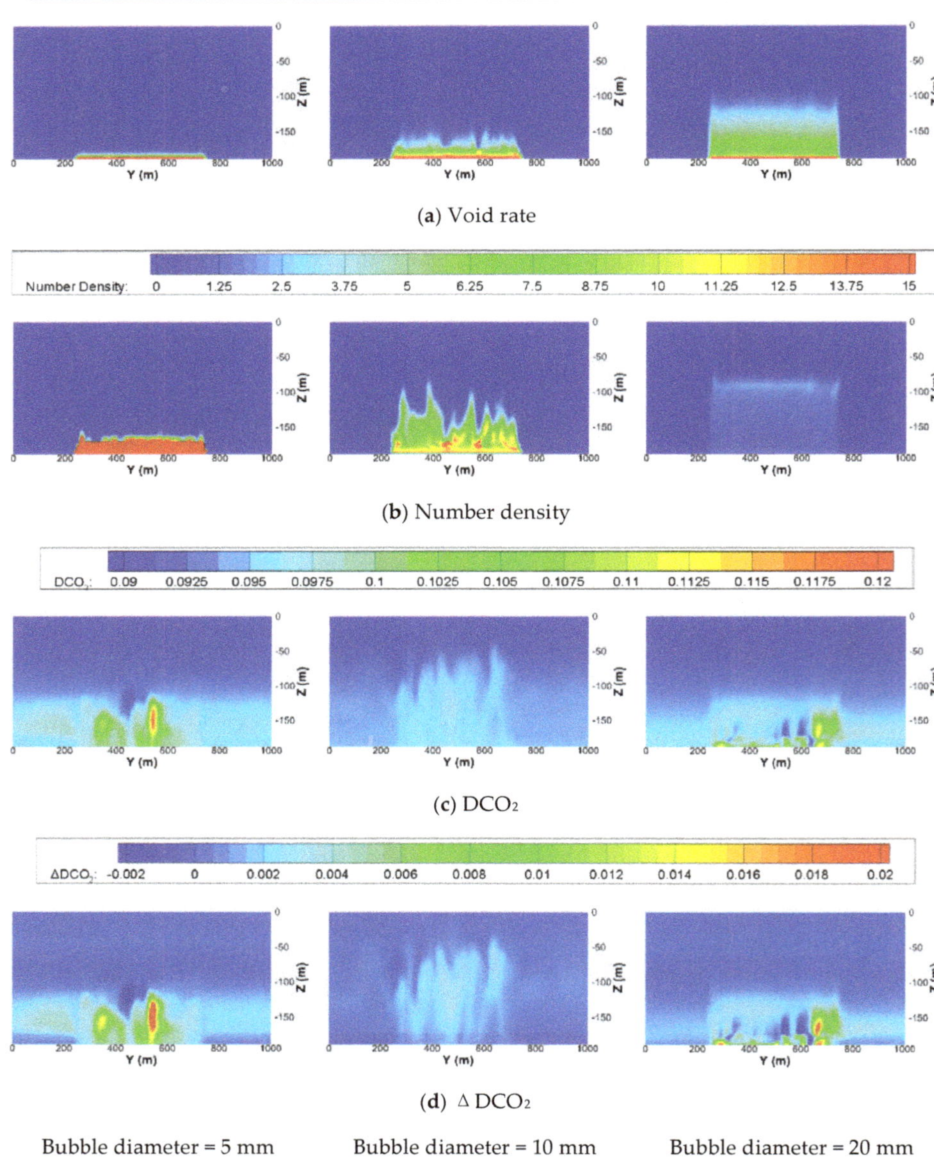

(**a**) Void rate

(**b**) Number density

(**c**) DCO₂

(**d**) Δ DCO₂

Bubble diameter = 5 mm   Bubble diameter = 10 mm   Bubble diameter = 20 mm

Figure 11. Contour maps of (**a**) void rate, (**b**) number density of undissolved CO_2 bubbles, (**c**) DCO_2, and (**d**) ΔDCO_2 on the center plane in the x direction (x = 1000 m) of the small-scale domain at 30 days after the start of leakage with the same leakage rates of 100,000 t/y and leakage area of 20,000 m^2 but different initial diameter of bubble.

Figure 12 shows the contour maps of ΔpCO_2 on the center planes of the x, y, and z directions of the small-scale domain. The maximum ΔpCO_2 for cases 5 and 3 are similar but do not exceed 500 ppm, which is also clearly seen in Figure 13, where those on the constant x, y, and z planes of the mesoscale domain are illustrated.

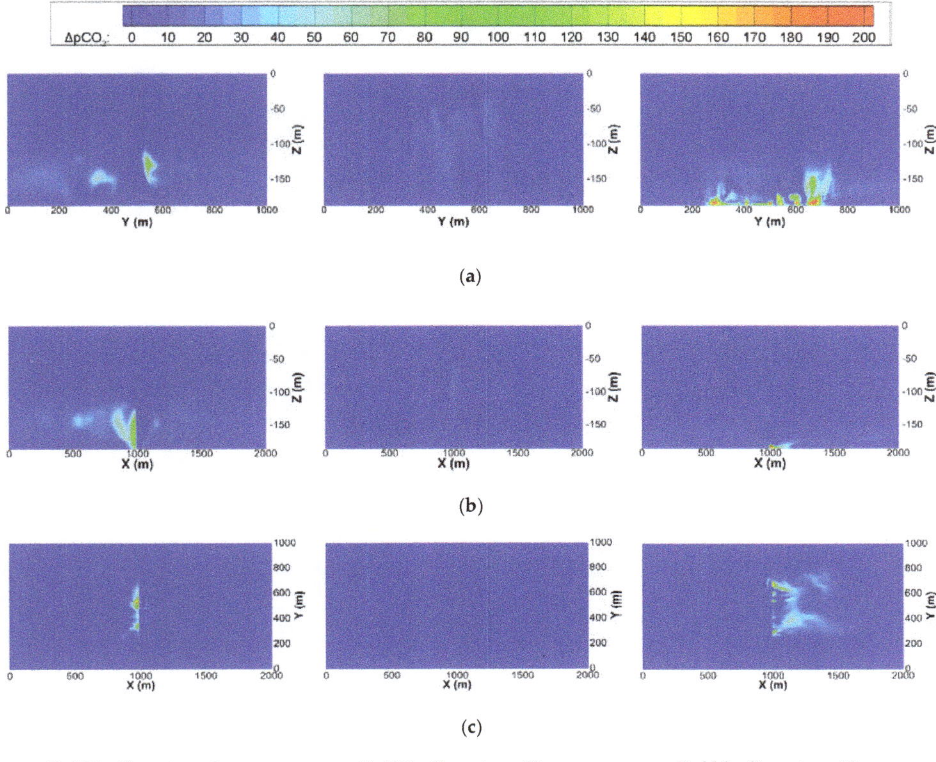

Bubble diameter = 5 mm   Bubble diameter = 10 mm   Bubble diameter = 20 mm

Figure 12. Contour maps of ΔpCO_2 on the center planes of the x, y, and z directions of the small-scale domain at 30 days after the start of leakage with the same leakage rates of 100,000 t/y and leakage area of 20,000 m^2 but different initial diameters of the bubble. (**a**) Center plane in x direction (x = 1000 m); (**b**) Center plane in y direction (y = 500 m); (**c**) xy plane near leakage area (z = −180 m).

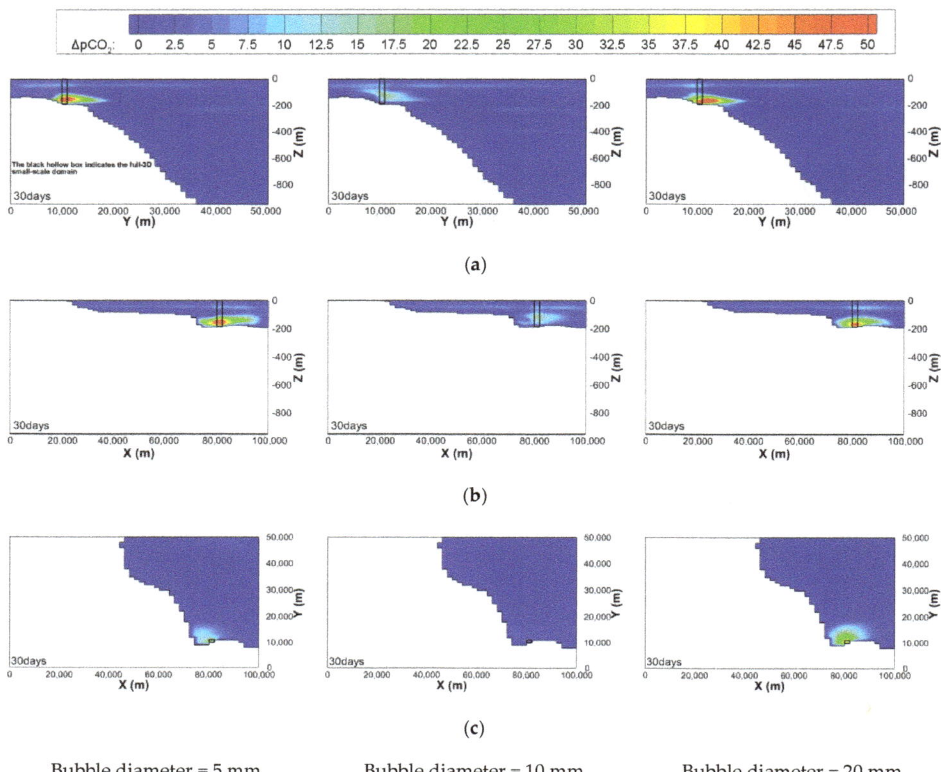

(a)

(b)

(c)

Figure 13. Contour maps of ΔpCO_2 on the planes, where the full-3D small-scale domain is located, of the mesoscale domain at 30 days after the start of leakage with the same leakage rates of 100,000 t/y and leakage area of 20,000 m^2 but different initial diameters of the bubble. (**a**) yz plane (x = 81,000 m); (**b**) zx plane (y = 10,500 m); (**c**) xy plane near leakage area (z = 180 m)

5. Conclusions

In this study, the behavior and diffusion of CO_2, assumed to have leaked from the seafloor, were numerically predicted for the first time near the one of the candidate sites for CO_2 ocean geological storage in Korea. The behavior of leaked CO_2 bubbles were analyzed numerically using a multi-scale ocean model to assess its environmental impacts, which is essential for obtaining public acceptance. The main parameters chosen for the simulation were leakage rates and initial diameters of CO_2 bubbles. The former was assumed to be 3800, 50,000, and 100,000 t/y and the latter to be 5, 10, and 20 mm, respectively. A total of five simulations were carried out by combining these parameters.

From the simulation results, it was found that all CO_2 bubbles were dissolved into the seawater before reaching the free surface in all cases. As the leakage rate increased, relatively higher values were concentrated near the leakage depth, and wider areas of ΔpCO_2 were observed because more bubbles existed and dissolution lasted for a longer time. As the initial diameter of the bubbles increased, the rising distance of the bubbles also increased due to the larger buoyant force of the relatively large volume of the bubble. However, small ΔpCO_2 values were estimated when the diameter was 10 mm. For the mass transfer rates, though they were similar for the 5 and 20 mm diameters, they were reduced when the diameter was 10 mm.

The estimated maximum ΔpCO_2 for the extreme case was approximately 200 ppm during 30 days of leakage under the present simulation conditions. Therefore, it can be said that the environmental impact caused by the leakage of CO_2 was not significant.

The present numerical model is expected to be useful and applicable not only for estimation of the behaviors and environmental impact of leaked CO_2 but for various purposes, such as determination of the proper location for leakage monitoring devices during ocean geological storage of CO_2.

Scenario-based simulations and parametric studies will be performed in the near future. The main parameters, such as the leakage depth closely related to the phase of the leaked CO_2, the vertical distributions of the scalar properties according to the seasonal change, and the shape of the leakage area, will be considered. Furthermore, validation and improvement of adopted sub-models will be continued as uncertainties and assumptions remain, although well-known or validated ones were implemented in the present model.

Author Contributions: Conceptualization, S.-M.J. and W.-Y.S.; Methodology, S.-M.J. and W.-Y.S.; Validation, S.K.; Formal Analysis, S.K.; Investigation, W.-Y.S. and S.-M.J.; Resources, S.-M.J. and S.K.; Writing—Original Draft Preparation, S.-M.J.; Writing—Review and Editing, S.-M.J. and W.-Y.S.; Visualization, S.K.; Supervision, S.-M.J. and W.-Y.S.; Funding Acquisition, S.-M.J. All authors have read and agreed to the published version of the manuscript.

Funding: This research was supported by the Basic Science Research Program through the National Research Foundation of Korea (NRF) funded by the Ministry of Education (No. 2017R1D1A3B03030031).

Conflicts of Interest: The authors declare no conflict of interest.

References

1. Furre, A.-K.; Eiken, O.; Alnes, H.; Vevatne, J.N.; Kiær, A.F. 20 Years of Monitoring CO_2-injection at Sleipner. *Energy Procedia* **2017**, *114*, 3916–3926. [CrossRef]
2. Bentham, M.; Kirby, M. CO_2 Storage in Saline Aquifers. *Oil Gas. Sci. Technol.* **2006**, *60*, 559–567. [CrossRef]
3. Hansen, O.; Gilding, D.; Nazarian, B.; Osdal, B.; Ringrose, P.; Kristoffersen, J.-B.; Eiken, O.; Hansen, H. Snøhvit: The History of Injecting and Storing 1 Mt CO_2 in the Fluvial Tubåen Fm. *Energy Procedia* **2013**, *37*, 3565–3573. [CrossRef]
4. Tanase, D.; Sasaki, T.; Yoshii, T.; Motohashi, S.; Sawada, Y.; Aramaki, S.; Yamanouchi, Y.; Tanaka, T.; Ohkawa, S.; Inowaki, R. Tomakomai CCS Demonstration Project in Japan. *Energy Procedia* **2013**, *37*, 6571–6578. [CrossRef]
5. Blackford, J.C.; Stahl, H.; Bull, J.M.; Berges, B.J.P.; Cevatoglu, M.; Lichtschlag, A.; Connelly, D.; James, R.H.; Kita, J.; Long, D.; et al. Detection and impacts of leakage from sub-seafloor carbon dioxide storage. *Nat. Clim. Chang.* **2014**, *4*, 1011–1016. [CrossRef]
6. Sellami, N.; Dewar, M.; Stahl, H.; Chen, B. Dynamics of rising CO_2 bubble plumes in the QICS field experiment. *Int. J. Greenh. Gas. Control* **2015**, *38*, 44–51. [CrossRef]
7. Alendal, G.; Drange, H. Two-phase, near-field modeling of purposefully released CO_2 in the ocean. *J. Geophys. Res. Oceans* **2001**, *106*, 1085–1096. [CrossRef]
8. Sato, T.; Sato, K. Numerical prediction of the dilution process and its biological impacts in CO_2 ocean sequestration. *J. Mar. Sci. Tech.-Jpn.* **2002**, *6*, 169–180. [CrossRef]
9. Jeong, S.M.; Sato, T.; Chen, B.; Tabeta, S. Numerical simulation on multi-scale diffusion of CO_2 injected in the deep ocean in a practical scenario. *Int. J. Greenh. Gas. Control* **2010**, *4*, 64–72. [CrossRef]
10. Kano, Y.; Sato, T.; Kita, J.; Hirabayashi, S.; Tabeta, S. Model prediction on the rise of pCO_2 in uniform flows by leakage of CO_2 purposefully stored under the seabed. *Int. J. Greenh. Gas. Control* **2009**, *3*, 617–625. [CrossRef]
11. Kano, Y.; Sato, T.; Kita, J.; Hirabayashi, S.; Tabeta, S. Multi-scale modeling of CO_2 dispersion leaked from seafloor off the Japanese coast. *Mar. Pollut. Bull.* **2010**, *60*, 215–224. [CrossRef] [PubMed]
12. Mori, C.; Sato, T.; Kano, Y.; Oyama, H.; Aleynik, D.; Tsumune, D.; Maeda, Y. Numerical study of the fate of CO_2 purposefully injected into the sediment and seeping from seafloor in Ardmucknish Bay. *Int. J. Greenh. Gas Control* **2015**, *38*, 153–161. [CrossRef]
13. Kang, K.; Huh, C.; Kang, S.-G. A numerical study on the CO_2 leakage through the fault during offshore carbon sequestration. *J. Korean Soc. Mar. Environ. Energy* **2015**, *18*, 94–101. [CrossRef]

14. Lee, H.S.; Lee, B.H.; Kim, K.; Kim, S.Y.; Park, J.C. Tidal current simulation around the straits of korea and its application to a speed trial. *Int. J. Nav.* **2019**, *11*, 474–481. [CrossRef]

15. Pitzer, K.S.; Sterner, S.M. Equations of state valid continuously from zero to extreme pressures for H_2O and CO_2. *J. Chem. Phys.* **1994**, *101*, 3111–3116. [CrossRef]

16. Smagorinsky, J. Some historical remarks on the use of nonlinear viscosities. Large Eddy Simulation of Complex. Engineering and Geophysical Flows. In *Large Eddy Simulation of Complex Engineering and Geophysical Flows*; Galperin, B., Orszag, S.A., Eds.; Cambridge University Press: Cambridge, UK; New York, NY, USA; Melbourne, Australia; Madrid, Spain; Cape Town, South Africa; Singapore; São Paulo, Brazil; Delhi, India; Dubai, UAE; Tokyo, Japan, 1993; pp. 3–36.

17. Hirai, S.; Okazaki, K.; Tabe, Y.; Hijikata, K. Mass transport phenomena of liquid CO_2 with hydrate. *J. Waste Manag.* **1997**, *17*, 353–360. [CrossRef]

18. Weiss, R.F. Carbon dioxide in water and seawater: The solubility of a non-ideal gas. *Mar. Chem.* **1974**, *2*, 203–215. [CrossRef]

19. Chen, B.; Song, Y.; Nishio, M.; Someya, S.; Akai, M. Modeling near-field dispersion from direct injection of carbon dioxide into the ocean. *J. Geophys. Res.* **2005**, *110*, C09S15. [CrossRef]

20. Matsumoto, K.; Takanezawa, T.; Ooe, M. Ocean tide models developed by assimilating TOPEX/POSEIDON altimeter data into hydrodynamical model: A global model and a regional model around Japan. *J. Oceanogr.* **2000**, *56*, 567–581. [CrossRef]

21. Hino, M.; Nakaza, E. Tests of a new numerical scheme on a "non-reflection and free-transmission" open-sea boundary for longwaves. *Fluid Dyn. Res.* **1989**, *4*, 305. [CrossRef]

22. Klusman, R.W. Rate measurements and detection of gas microseepage to the atmosphere from an enhanced oil recovery/sequestration project, Rangely, CO, USA. *Appl. Geochem.* **2003**, *18*, 1825–1838. [CrossRef]

23. Kikkawa, T.; Kita, J.; Ishimatsu, A.J.M.P.B. Comparison of the lethal effect of CO_2 and acidification on red sea bream (Pagrus major) during the early developmental stages. *Mar. Pollut. Bull.* **2004**, *48*, 108–110. [CrossRef]

24. Kita, J.; Watanabe, Y. Impact assessment of high-CO_2 environment on marine organisms. In Proceedings of the 8th International Conference on Greenhouse Gas Control Technologies, (GHGT-8), Trondheim, Norway, 19–22 June 2006. CD-ROM.

 *applied sciences*

Article

Durability and Mechanical Characteristics of Blast-Furnace Slag Based Activated Carbon-Capturing Concrete with Respect to Cement Content

Seungwon Kim [1,2,*] and Cheolwoo Park [1,2,*]

[1] KIIT (Kangwon Institute of Inclusive Technology), Kangwon National University, 1 Gangwondaegil, Chuncheon 24341, Korea

[2] Department of Civil Engineering, Kangwon National University, 346 Jungang-ro, Samcheok 25913, Korea

* Correspondence: inncoms@kangwon.ac.kr (S.K.); tigerpark@kangwon.ac.kr (C.P.); Tel.: +82-33-570-6518 (S.K.); +82-33-570-6515 (C.P.)

Received: 18 February 2020; Accepted: 16 March 2020; Published: 19 March 2020

Abstract: The recent abnormal temperature phenomena such as the rise of global mean temperature and sea level due to global climate change are clear threats that can no longer be overlooked to the human beings who have pursued indiscriminate development and rapid growth. Climate change has emerged as a serious risk that threatens the survival of the entire human race from the environmental and ecological aspects, despite international efforts for several decades. The CO_2 concentration in the atmosphere has increased by approximately 39% since the industrial revolution. Even if carbon emissions are stopped right now, it is expected to take at least 50–200 years to return to the CO_2 level before the industrial revolution. Therefore, we conducted an experimental study to develop a carbon-capturing concrete that has active as well as passive carbon reduction functions using blast-furnace slag, an industrial byproduct, instead of cement. For active carbon reduction, we used calcium hydroxide and sodium silicate as carbon capture activators, and conducted tests on mechanical properties and durability characteristics.

Keywords: climate change; carbon emission; carbon-capturing concrete; carbon capture activator; carbon reduction

---

1. Introduction

Climate change has emerged as a serious risk factor that threatens the survival of the entire human race from the environmental and ecological aspects, despite exerting the international efforts for several decades [1]. The CO_2 concentration in the atmosphere has increased by approximately 39% since the industrial revolution. Even if carbon emissions are stopped right now, it is expected to take at least 50–200 years to return to the CO_2 level before the industrial revolution [1]. To cope with this problem, global greenhouse gas (GHG) reduction measures have been made, such as the Kyoto Protocol to the United Nations Framework Convention on Climate Change to tackle global warming [2]. According to the fourth report of the Intergovernmental Panel on Climate Change (IPCC), the global mean temperature has risen by 0.74 °C for the last 100 years and is expected to rise by up to 6.4 °C by the end of the 21st century [1,3].

As such, climate change is a critical global issue, and as one of the world's top 10 countries in GHG emissions, South Korea has set up "carbon reduction technology development" as a new pillar of national development vision, developed related technologies and established laws since 2008 [4]. In addition, South Korea showed the highest annual increasing rate of 6.2% in the transportation sector of GHG emissions from 1990 to 2000 among the Organization for Economic Co-operation and Development (OECD) member countries. To take part in the development of carbon reduction technologies in the road sector, which accounts for 16% of the total GHG emissions, the need for the development of design and development technologies to absorb and reduce CO_2 to minimize CO_2 emissions is emerging [4]. Recently, the secondary environmental pollution problem is appearing due to the increase of industrial byproducts such as mining byproducts and coal incineration ashes [4]. For efficient processing of industrial byproducts, some of the industrial byproducts are mixed with existing cement or asphalt materials and these mixtures are used as construction materials [4]. Comprehensive Assessment System for Building Environmental Efficiency (CASBEE) in Japan has evaluated the GHG emissions of building materials using the CO_2 emission database of major materials since 2010 [5–7]. Building performance is now a major concern of professionals in the building industry and environmental building performance assessment has emerged as one of the major issues in sustainable construction [8–11]. Efforts to reduce CO_2 emission are being made through the use of procurement systems for low CO_2 emission materials and the application of high-strength concrete at construction sites. As a part of such research, there is a demand for the establishment of a quantitative assessment method of CO_2 emission from concrete production to site procurement and a proposal of a CO_2 reduction plan at construction sites [12]. However, the utilization of materials for CO_2 capture is still insufficient. The development of concrete applying CO_2 reduction technology is considered to be a leading technology for the society as a whole, as well as for the construction sector. The MIT Technology Review selected "Green Concrete", which is a CO_2 reducing concrete, one of top 10 new technologies in 2010 [4,13].

Therefore, in this study, we aim to develop carbon-capturing concrete using industrial byproducts that can satisfy not only passive carbon reduction function but also active carbon reduction function through the utilization of industrial byproducts. To develop a carbon-capturing concrete using blast-furnace slag powder, which is one of the representative industrial byproducts, we conducted an experimental study on the mechanical and durability characteristics of carbon-capturing concrete composed of blast-furnace slag using carbon capture activator.

2. Research Trends of Carbon-Capturing Concrete

The generation amount of blast-furnace slags, a representative industrial byproduct, has been continuously increasing with the accelerated development of the steel industry. If blast-furnace slag powder is used as a raw material of concrete by mixing with cement, approximately 40 million tons of cement per year can be reduced as well as CO_2 emissions [14,15].

To reduce CO_2 emissions, the main cause of GHGs, the construction industry is striving to develop environment-friendly concretes. Consequently, cementless binders that use industrial byproducts such as blast-furnace slag, fly ash, and metakaolin instead of ordinary Portland cement (OPC) are being actively researched in and outside the country. Technologies related to cementless binders include geopolymer technology using alkaline reaction with clay minerals and alkali-activated binder technology that develops hardened body by stimulating the latent hydraulic activity of slag byproducts [14,15].

The economic efficiency of 50 MPa or lower alkali-activated concrete is moderately higher than that of the equivalent OPC, but the cost of alkali-activated concrete for high-strength concrete is known to be lower by approximately 10–40% compared to that of the OPC concrete with replaced silica fume [14,15].

Furthermore, the alkali-activated concrete can have advantage over the OPC concrete in terms of economic aspect because it uses industrial byproducts, which do not require a firing process for cement production. Moreover, using an alkali stimulant that is easy to handle can further improve economic efficiency. Therefore, the development of a carbon-capturing concrete through the research of geopolymer concrete and alkali-activated concrete of a non-plastic inorganic binder is expected to effectively reduce CO_2 emissions [14,15].

3. Experiment Overview

3.1. Used Materials

This study used type 1 OPC and Table 1 outlines the physical properties and chemical compositions of this cement. Furthermore, blast-furnace slag powder, which is a representative industrial byproduct, was used to reduce the content of OPC. Table 2 outlines the physical properties and chemical composition of the used blast-furnace slag powder.

Table 1. Physical properties and chemical composition of ordinary Portland cement.

Physical Properties					
Specific Gravity	Fineness (cm^2/g)	Stability (%)	Setting Time (min)		Ig-Loss (%)
			Initial	Final	
3.15	3400	0.10	230	410	2.58
Chemical Composition					
SiO_2 (%)	CaO (%)	MgO (%)	SO_3 (%)	Al_2O_3 (%)	
21.95	60.12	3.32	2.11	6.59	

Table 2. Physical properties and chemical composition of blast-furnace slag.

Physical Properties			
Specific Gravity	Fineness (cm^2/g)	Flow Ratio (%)	Ig-loss (%)
2.90	4314	104	0.22
Chemical Composition			
MgO (%)	SO_3 (%)	Chloride Ion (%)	Basicity
3.82	1.58	0.003	1.76

The maximum size of 25 mm was used for the coarse aggregates in the carbon-capturing concrete. Tables 3 and 4 outline the physical properties of the used coarse and fine aggregates, respectively.

Table 3. Physical properties of the used coarse aggregates.

Gmax (mm)	Specific Gravity	Water Absorption (%)	Fineness Modulus
25	2.76	0.45	6.72

Table 4. Physical properties of the used fine aggregates.

Specific Gravity	Water Absorption (%)	Fineness Modulus
2.52	1.45	2.62

To improve the carbon absorption performance of the carbon-capturing concrete composed of blast-furnace slag, calcium hydroxide ($Ca(OH)_2$) and powder-type sodium silicate (Na_2SiO_3) with 95% or higher purity were used as carbon capture activators.

For chemical admixture, a polycarboxylate high range water reducer (HRWR) that has excellent dispersion effect and can achieve fluidity even at a low water-cement ratio was used. Table 5 outlines the properties of the used polycarboxylate HRWR.

Table 5. Properties of polycarboxylate high range water reducer (HRWR).

Principal Component	Specific Gravity	pH	Alkali Content (%)	Chloride Content (%)
Polycarboxylate	1.05 ± 0.05	5.0 ± 1.5	less than 0.01	less than 0.01

3.2. Mixing

A literature review for the development of the carbon-capturing concrete composed of blast-furnace slag revealed that replacing part of the blast-furnace slag powder with cement had positive effects in terms of the durability and carbon absorption performance [3,4,15]. Furthermore, the difference in strength of the carbon-capturing concretes by curing temperature was insignificant. Hence, the experimental specimens were cured in a constant temperature and humidity room (23 °C, relative humidity 50%).

Table 6 shows the mix table used in this study. For the carbon-capturing concrete satisfying the mechanical and durability characteristics, the cement replacement ratio relative to the weight of the blast-furnace slag was increased in 10% intervals from 10% to 40%. The water–binder ratio was fixed at 0.325, the fine aggregate ratio at 0.45, and the replacement ratio of carbon capture activator at 40% (calcium hydroxide 20% and sodium silicate 20%) for this experiment.

Table 6. Mix variables of carbon-capturing concrete.

| Variables | W/B (%) | S/a (%) | Unit Weight (kg/m^3) | | | | | | | Admixture (HRWR) |
			W	BFS	Cement	Activator Ca(OH)$_2$	Na$_2$SiO$_3$	F.A.	C.A.	
S90Ca20Na20				495	55			472.44	630.12	
S80Ca20Na20	0.325	0.45	250.3	440	110	110	110	474.19	632.40	10.01
S70Ca20Na20				385	165			475.85	634.68	
S60Ca20Na20				330	220			477.54	636.90	

W/B: Water/Binder (BFS+C+Activator), W: Water, BFS: Blast-Furnace Slag, C: OPC, Ca(OH)$_2$: Calcium Hydroxide, Na$_2$SiO$_3$: Sodium Silicate. F.A.: Fine Aggregate, C.A.: Coarse Aggregate. Admixture: Polycarboxylate High Range Water Reducer. S00: BFS/C, Ca: Ca(OH)$_2$/BFS+C, Na: Na$_2$SiO$_3$/BFS+C.

3.3. Experiment Method

An experiment was conducted to analyze the properties of carbon-capturing concrete using blast-furnace slag powder, which is a representative industrial byproduct, and calcium hydroxide and sodium silicate as carbon capture activators before and after curing.

3.3.1. Properties of Concrete before Curing

Slump and air content tests were performed to examine the properties of concrete before curing.

3.3.2. Properties of Concrete after Curing

Exposure Conditions

A literature review revealed that in the high concentration exposure condition of 20% (200,000 ppm) CO_2, the carbon capture depth of alkali-activated concrete was approximately 5 mm or lower in 1 month and approximately 20 mm or lower in 12 months [16]. As shown in this literature review, the change of carbon capture depth was insignificant even though the concrete was exposed to a high concentration of CO_2 for a long time. In the present study, a CO_2 concentration of 10% (100,000 ppm) was set as the high concentration CO_2 exposure condition for the accelerated carbonation tester. The relative humidity and temperature inside the accelerated carbonation tester during exposure were 50% and 23 ± 2 °C, respectively.

Furthermore, in the case of the high-purity air exposure conditions for comparison with high concentration CO_2 exposure conditions, an exposure experiment was performed in a constant temperature and humidity room at the relative humidity and temperature of 50% and 23 ± 2 °C using a plastic chamber that can be fully sealed by applying a urethane foam solution. The used high-purity air was composed of only nitrogen and oxygen and contained less than 0.01% of other components ($N_2+O_2 > 99.99\%$), which was considered as a control for comparison with the high concentration CO_2 exposure conditions.

Compressive Strength

Cylindrical specimens of ϕ100 mm × 200 mm were fabricated for the compressive strength experiment. To measure the compressive strength for each variable and each exposure condition, the compressive strength characteristics of carbon-capturing concrete were analyzed using the specimens at 1, 7, 28, 56, and 90 days of exposure.

Flexural Strength

Prismatic specimens of 100 mm × 100 mm × 400 mm were fabricated for the flexural strength experiment. The flexural strength characteristics of the specimens were analyzed at 7 and 28 days of exposure.

Carbon Capture Depth

To analyze the carbon capture depth of carbon-capturing concrete, the carbon capture depth was measured with respect to the progress of exposure period for each exposure conditions and variables at 1, 7, 28, 56, and 90 days.

Freeze–Thaw Resistance

A freeze–thaw resistance experiment was conducted as an indoor experiment to measure the freeze–thaw resistance of concrete specimens in rapid iterative cycles. The same prismatic specimens of 100 mm × 100 mm × 400 mm as those used in the flexural strength experiment were fabricated and cured for 14 days. Then their freeze–thaw resistance was evaluated in accordance with ASTM C 666, Standard Test Method for Resistance of Concrete to Rapid Freezing and Thawing [17].

Chloride Ion Penetration Resistance

For the specimens, the penetration resistance of chloride ion was evaluated in accordance with ASTM C 1202 'Standard Test Method for Electrical Indication of Concrete's Ability to Resist Chloride Ion Penetration' [18], which is a standard test method that determines the electric conductivity to measure the resistance of concrete to chloride ion penetration in a short time. The same cylindrical specimens of ϕ100 mm × 200 mm as the compressive strength specimens were fabricated and the experiment was performed by cutting the specimens to a thickness of 50±3 mm after curing for 28 days. To prevent the water evaporation from the inside of the specimens, the edges of the specimens were coated with a concrete protective coating, and the resistance voltage was measured every 30 min for 6 h. In addition, the passed charge was calculated using Equation (1). Table 7 outlines the evaluation criteria for chloride ion penetration resistance presented in ASTM C 1202 [18].

$$Q = 900 \times (I_0 + 2I_{30} + 2I_{60} + \cdots + 2I_{300} + 2I_{330} + 2I_{360}) \tag{1}$$

Table 7. Evaluation criteria for chloride ion penetration resistance [18].

Passed Charge (C)	Chloride Ion Penetrability
>4000	High
2000–4000	Moderate
1000–2000	Low
100–1000	Very Low
<100	Negligible

Here, Q is the passed charge (C), I_0 is the current (A) immediately after applying the voltage, and I_t is the current (A) at t min after applying the voltage [18].

Image Analysis

The air-void structure of the carbon-capturing concrete for each variable with respect to the cement replacement ratio was analyzed using the linear traverse method, which is the A experiment method in ASTM C 457, 'Standard Test Method for Microscopical Determination of Parameters of the Air-Void System in Hardened Concrete' [19]. Furthermore, the air-void structure analysis of the concrete after curing is a method that can be used for indirect evaluation of freeze–thaw resistance.

The image analysis experiment method is often used for rapid measurement and error reduction in comparison with the conventional analysis method by naked eye. This experiment can obtain the total air volume, specific surface area, spacing factor, air volume per air void size, and the number of voids by measuring the sizes and positions of air voids on the images captured by the analysis system. Figure 1 shows the image analysis process.

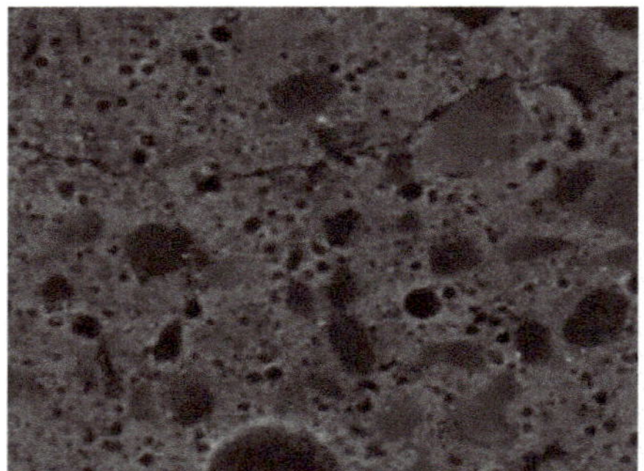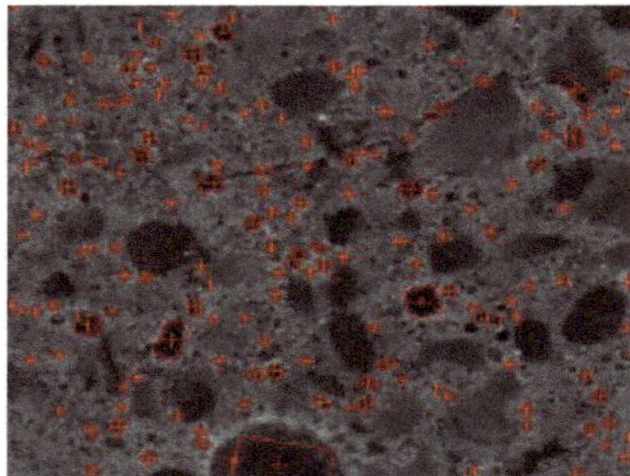

Figure 1. Image analysis process.

4. Experiment Results and Analysis

4.1. Properties of Concrete before Curing

In the slump experiment results of each variable with respect to the cement replacement ratio of carbon-capturing concrete composed of blast-furnace slag, the S90Ca20Na20 variable showed the highest slump value of approximately 45 mm, where the cement replacement ratio was 10%. In the variables where the cement replacement ratio increased in 10% unites from 20% to 40%, the slump values were similar although they increased with the rising cement replacement ratio. Thus, the slump differences by the variable of the carbon-capturing concrete with respect to the cement replacement ratio were insignificant in general.

The S60Ca20Na20 variable, which is the highest cement replacement ratio of the carbon-capturing concrete composed of blast-furnace slag relative to the weight of blast-furnace slag, showed the highest air volume of 3.6%. The changes in the air volume of the carbon-capturing concrete with rising cement replacement ratio were around 3.0% in general. As with the slump experiment results, the differences in air volume by the cement replacement ratio were insignificant. The reason for this phenomenon is that the polycarboxylate HRWR used as an admixture in this study did not have a positive effect on the acquisition of air volume when compared with admixtures exhibiting air entraining performance like air-entraining admixtures. Figure 2 shows the slump and air content test results for each variable.

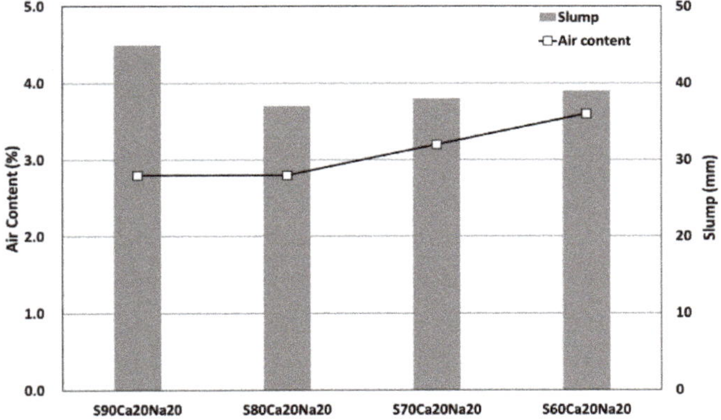

Figure 2. Slump and air content test results.

4.2. Properties of Hardened Concrete

4.2.1. Compressive Strength

Figure 3 shows the results of compressive strength experiment with respect to the cement replacement ratio of the carbon-capturing concrete composed of blast-furnace slag exposed to high concentration CO_2. The S90Ca20Na20 variable with the lowest cement replacement ratio relative to the weight of blast-furnace slag showed the lowest compressive strength of the range from 1 to 90 days of exposure among all the variables. This seems to be due to the differences in the strength expression by the low cement replacement ratio.

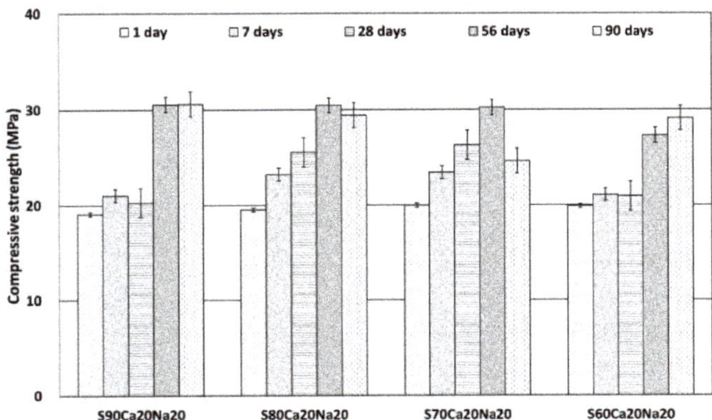

Figure 3. Compressive strength experiment results of high concentration CO_2 exposure conditions.

The variables with 20%, 30%, and 40% cement replacement ratios showed compressive strengths of around 20 MPa until 90 days of exposure. The increase in the strength with respect to the progress of the exposure period was generally insignificant. The S60Ca20Na20 variable with 40% cement replacement ratio showed 25 MPa or higher compressive strengths at 90 days of exposure. This is due to the differences in strength depending on the higher cement replacement ratio compared to other variables.

Figure 4 shows the results of compressive strength experiment with respect to the cement replacement ratio of the carbon-capturing concrete composed of blast-furnace slag exposed to the high-purity air. In the experiment results at 1 day of exposure, every variable showed a compressive strength of 20 MPa. Subsequently, until 28 days of exposure, the compressive strength was approximately 25 MPa in the S80Ca20Na20 and S70Ca20Na20 variables with cement replacement ratios of 20% and 30%, respectively. However, the S60Ca20Na20 and S90Ca20Na20 variables showed an insignificant increase in strength.

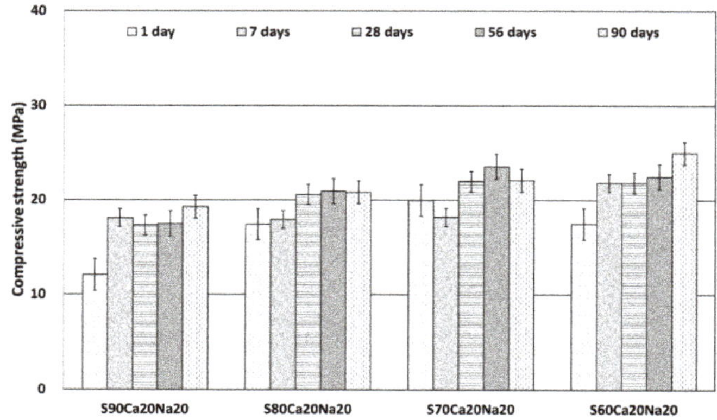

Figure 4. Compressive strength experiment results of high-purity air exposure conditions.

At 56 days of exposure, the strength improved by approximately 15% compared to 28 days of exposure. At 90 days of exposure, the strength increase over the exposure period compared to 56 days of exposure was generally insignificant. The S70Ca20Na20 variable with 30% cement replacement ratio relative to the weight of blast-furnace slag showed a decrease of strength at 90 days of exposure. This seems to be due to poor compaction, which is a factor of the specimen fabrication process.

The increase of the compressive strength of the carbon-capturing concrete with respect to the cement replacement ratio and exposure conditions over the exposure period was generally insignificant. The variable with the high-purity air condition showed a greater increase of the compressive strength than the variable of exposure in high concentration CO_2 condition. This is considered to be due to the difference in the relative strength expression speed with the variable of high-purity air exposure generated by contraction due to the surface carbon absorption of the specimen exposed to high concentration CO_2.

4.2.2. Flexural Strength

Figure 5 shows the results of flexural strength experiment for each exposure condition and variable of carbon-capturing concrete composed of blast-furnace slag at 7 days of age. In the case of 7 days of exposure, the flexural strength in the high concentration CO_2 exposure condition was approximately 3.0 MPa. The difference in flexural strength with the rising cement replacement ratio was insignificant. Furthermore, the variable of high-purity air exposure condition showed a flexural strength of approximately 4.0 MPa, which is higher than that of the CO_2 exposure condition.

Figure 6 shows the flexural strength experiment results for each exposure condition and variable of carbon-capturing concrete composed of blast-furnace slag at 28 days of age. In the case of 28 days of exposure, the flexural strength in the high concentration CO_2 exposure condition was approximately 3.0 MPa. The S60Ca20Na20 variable of 40% cement replacement ratio showed the highest flexural strength of 3.2 MPa. Furthermore, the strength increase compared to the result of 7 days of exposure for the high concentration CO_2 condition was insignificant.

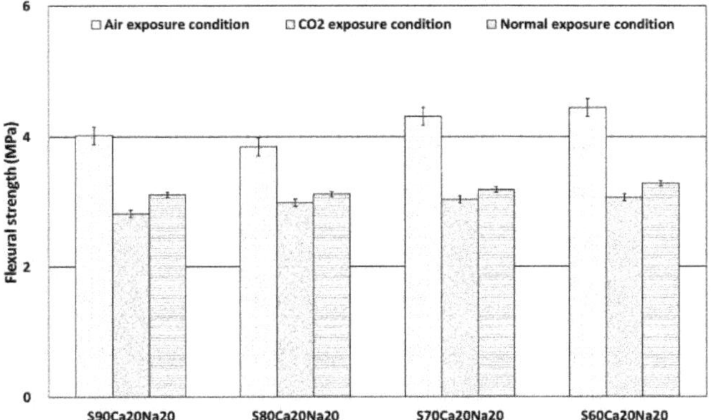

Figure 5. Flexural strength experiment results (7 days of age).

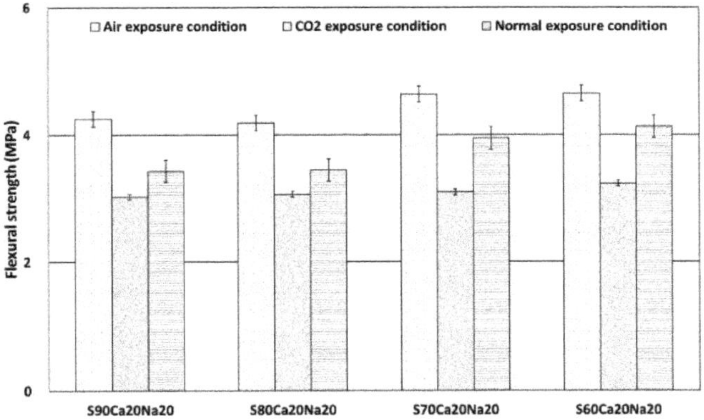

Figure 6. Flexural strength experiment results (28 days of age).

In the high-purity air exposure condition, the flexural strength was measured above approximately 4.0 MPa in general. The variable of a higher cement replacement ratio also showed a higher flexural strength. When compared to 7 days of exposure, the strength improved by approximately 5%. The increase of flexural strength for each variable with respect to the exposure condition and cement replacement ratio of the carbon-capturing concrete over the exposure period was moderately insignificant. Furthermore, compared to the high concentration CO_2 exposure condition, the increase of the variable for measurement in high-purity air exposure condition was higher compared to the high concentration CO_2 exposure condition. This seems to be due to the difference in the strength expression speed generated because the effect of carbon absorption is smaller than the high-concentration CO_2 exposure condition. Furthermore, the difference in flexural strength due to the increasing cement replacement ratio of the carbon-capturing concrete was moderately insignificant.

4.2.3. Carbon Capture Depth

Figure 7 shows the carbon capture depth measurement result for each exposure period of the exposed variable that was expressed in high concentration CO_2. No change in depth according to carbon capture did not appear at one day, but at 7 days, a carbon capture depth of more than approximately 5 mm was measured. At 28 days of exposure, a carbon capture depth of approximately

10 mm appeared. Until 90 days of exposure, approximately 20 mm carbon capture depth appeared in the S60Ca20Na20 variable with the highest 40% cement replacement ratio.

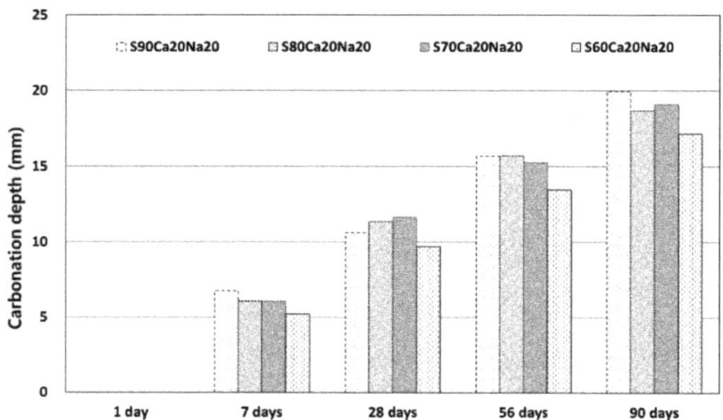

Figure 7. Carbon capture depth measurement result (exposure to high concentration CO_2).

Furthermore, the replacement ratio of blast-furnace slag powder increased with the exposure period, and the carbon capture depth increased with declining cement replacement ratio. This is considered to be due to the increased carbonation resistance through the generation of calcium hydroxide by calcium silicate by the rising cement replacement ratio.

The carbon capture depth measurement result of the high-purity air exposure conditions variable showed that no change in the carbon capture depth occurred in every variable from 1 to 90 days of exposure. This is considered to be due to no interference from external CO_2 resulting from the injection of air consisting of N_2 and O_2 only in the sealed chamber in the case of high-purity air exposure condition.

The carbon capture depth did not change at 1 day of exposure in the case of the high concentration CO_2 exposure condition. However, the maximum carbon capture depth was approximately 20 mm from the 7 to 90 days of exposure during which carbon capture from the surface occurred continuously. In the case of the high-purity air exposure condition, the carbon capture did not change due to the progress of exposure until 90 days of exposure. This result is considered to be due to the carbonation resistance resulting from not receiving direct effect of CO_2 exposure compared to the high-concentration CO_2 exposure condition.

4.2.4. Freeze–Thaw Resistance

Figure 8 shows the results of freeze–thaw resistance experiment with respect to the cement replacement ratio of the carbon-capturing concrete. The relative dynamic modulus of elasticity increased to much higher than 80% for 300 freeze–thaw cycles, but no difference in the freeze–thaw resistance was caused by the change of the cement replacement ratio.

This is considered to be due to the positive freeze–thaw resistance resulting from the satisfaction of the spacing factor (200 μm or less) and specific surface area (25 mm^2/mm^3 or higher) conditions as in the result of the image analysis (air-void property analysis).

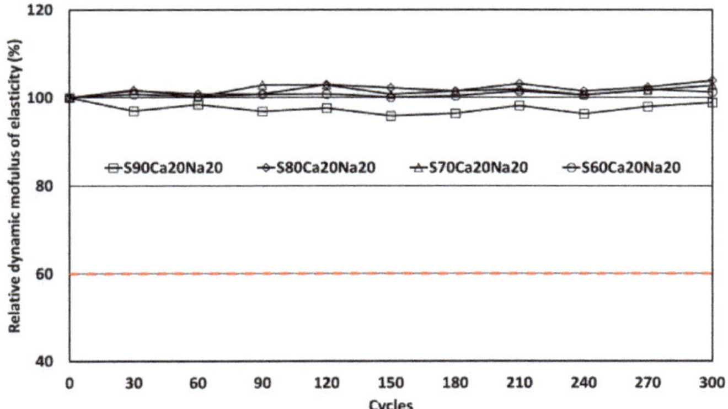

Figure 8. Freeze–thaw resistance experiment results.

4.2.5. Chloride Ion Penetration Resistance

Figure 9 shows the experiment results for penetration resistance of chloride ion in the carbon-capturing concrete with respect to the cement replacement ratio. The S90Ca20Na20 variable with the lowest cement replacement ratio of 10% relative to the weight of blast-furnace slag showed the lowest total passed charge of 1264 (coulombs). Subsequently, the total passed charge tended to increase with a declining cement replacement ratio. In the case of the S60Ca20Na20 variable with the highest cement replacement ratio of 40%, the total passed charge was measured at 1544 (coulombs), which is higher by approximately 22% than that of the S90Ca20Na20 variable.

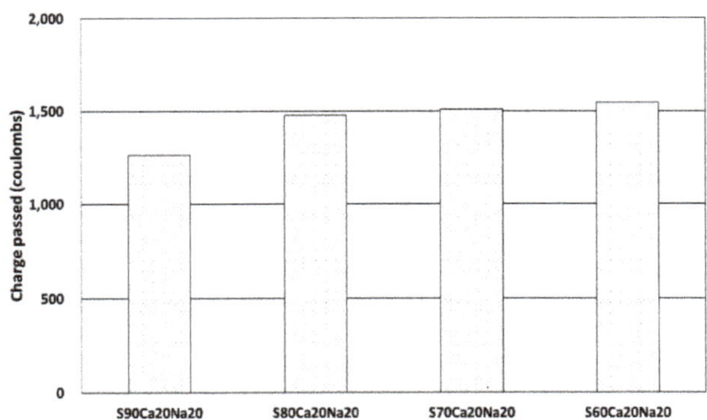

Figure 9. Chloride ion penetration resistance experiment results.

The penetration resistance of chloride ion in the carbon-capturing concrete was evaluated as 'low' as the passed charge ranged from 1000 to 2000. It was analyzed that the increased water-tightness of the voids in the concrete due to the mixing of blast-furnace slag powder increased the chloride ion penetration resistance. Furthermore, the differences in chloride ion penetration due to the increase of cement replacement ratio resulting from the use of blast-furnace slag above a certain level were generally insignificant.

4.2.6. Image Analysis

Figure 10 shows the spacing factor and specific surface area measurement results obtained through the image analysis of carbon-capturing concrete with respect to the cement replacement ratio.

The spacing factor was generally lower than 185 μm, and the change of the spacing factor with respect to the cement replacement ratio was moderately insignificant. Furthermore, every variable showed a specific surface area above 30 mm^2/mm^3, and similar levels of specific surface areas were measured in general.

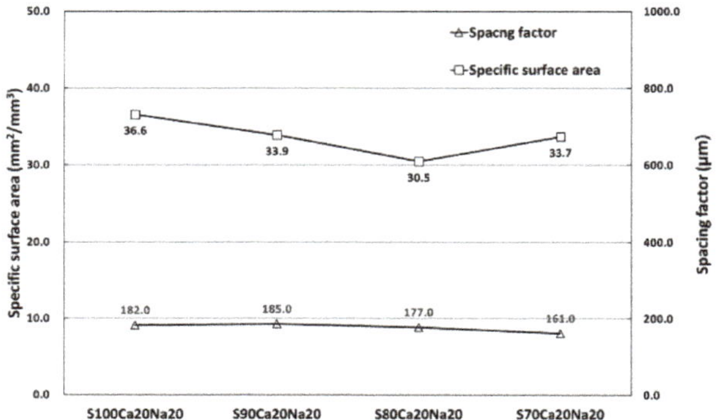

Figure 10. Image analysis experiment results.

The spacing factor and specific surface area of the voids of the concrete after curing are factors for forming good air-void structure of concrete. Sidney Mindess [20] reported that the spacing factor of concrete after curing must be 200 μm (0.2 mm) or lower and the specific surface area 25 mm^2/mm^3 or higher to secure sufficient freeze–thaw resistance [20]. Therefore, the freeze–thaw resistance of the carbon-capturing concrete with respect to the cement replacement ratio is considered to be excellent based on the measured spacing factor and specific surface area.

5. Conclusions

An experimental study was conducted to develop a carbon-capturing concrete composed of blast-furnace slag that can perform passive as well as active carbon reductions by replacing cement with industrial byproducts. To examine the applicability of this carbon-capturing concrete, we conducted experiments about the mechanical and durability characteristics with respect to the cement replacement ratio. The conclusions of this study are as follows.

(1) The compressive strength experiment results with respect to the exposure conditions and cement replacement ratio showed that the compressive strength of the specimen exposed to high concentration CO_2 decreased very little over the exposure period in general. In the high-purity air exposure condition, the compressive strength increased greater than that of the high concentration CO_2 exposure condition. However, the increase of compressive strength with respect to the cement replacement ratio was insignificant.

(2) The flexural strength experiment results showed that the increase of the flexural strength with respect to the cement replacement ratio was insignificant as in the compressive strength experiment results. The high-purity air exposure condition showed a higher flexural strength than that of the high concentration CO_2 exposure condition. This is considered to be due to the difference in the strength expression speed caused by the relative lack of the carbon absorption effect.

(3) The carbon capture depth experiment result of the high concentration CO_2 exposure condition showed that the carbon capture depth changed very little with the rising cement replacement ratio, and almost no carbon capture occurred in the high-purity air exposure condition.

(4) The freeze–thaw resistance experiment results showed that the air volume of the concrete before curing was approximately 3%, but the relative dynamic modulus of elasticity increased to much

higher than 80% until 300 freeze–thaw cycles. No change was caused by the cement replacement ratio. The image analysis for the air-void structure showed that every variable satisfied the condition of 200 μm (0.2 mm) or lower spacing factor and 25 mm^2/mm^3 or higher specific surface area, which are the criteria for securing freeze–thaw resistance. Despite somewhat low air volume experiment result, the freeze–thaw resistance was excellent owing to the spacing factor of air bubbles and the entrained air.

(5) The experiment results of chloride ion penetration resistance showed that the passed charge increased with rising cement replacement ratio, but the difference was not large. Furthermore, every variable showed a 'low' value in the total passed charge range of 1000–2000 coulomb, indicating excellent chloride ion penetration resistance. This is considered to be due to the increased water-tightness of concrete by the mixing of a large quantity of blast-furnace slag despite the cement replacement.

(6) To summarize the experiment results on the mechanical and durability characteristics of the carbon-capturing concrete composed of blast-furnace slag, the carbon-capturing concrete had mechanical and durability characteristics above the appropriate levels for concrete. Furthermore, this study found that the active as well as passive carbon reduction functions can be achieved through the use of blast-furnace slag, an industrial byproduct, and the development of the carbon-capturing concrete can effectively reduce CO_2 emissions.

Author Contributions: Conceptualization, S.K. and C.P.; methodology, S.K. and C.P.; writing—original draft preparation, S.K. and C.P.; writing—review and editing, S.K. and C.P. All authors have read and agreed to the published version of the manuscript.

Funding: This study was conducted under research project 'Development of High-Performance Concrete Pavement Maintenance Technology to Extend Roadway Life (20POQW-B146693-03)' funded by the Ministry of Land, Infrastructure and Transport (MOLIT) and the Korea Agency for Infrastructure Technology Advancement (KAIA).

Conflicts of Interest: The authors declare no conflicts of interest.

References

1. Working Group III of Intergovernmental Panel on Climate Change. Climate Change 2007: Mitigation of Climate Change. In *Fourth Assessment Report*; Intergovernmental Panel on Climate Change (IPCC): Geneva, Switzerland, 2007.

2. United Nations. *Kyoto Protoco to the United Nations Framework Convention on Climate Change*; United Nations: New York, NY, USA, 1998.

3. Cho, H.M.; Kim, S.W.; Song, J.H.; Park, H.M.; Park, C.W. Experimental Study on Mechanical Properties of Carbon-Capturing Concrete Composed of Blast Furnace Slag with Changes in Cement Content and Exposure. *Int. J. Highw. Eng.* **2015**, *17*, 41–51. [CrossRef]

4. Jang, B.J.; Kim, S.W.; Song, J.H.; Park, H.M.; Ju, M.K.; Park, C. Fundamental Characteristics of Carbon-Capturing and Sequestering Activated Blast-Furnace Slag Mortar. *Int. J. Highw. Eng.* **2013**, *15*, 95–103. [CrossRef]

5. Ding, G.K.C. Sustainable construction—The role of environmental assessment tools. *J. Environ. Manag.* **2008**, *86*, 451–464. [CrossRef] [PubMed]

6. Lee, W.L. A comprehensive review of metrics of building environmental assessment schemes. *Energy Build.* **2013**, *62*, 403–413. [CrossRef]

7. Lee, W.L. Benchmarking energy use of building environmental assessment schemes. *Energy Build.* **2012**, *45*, 326–334. [CrossRef]

8. Crawley, D.; Aho, I. Building environmental assessment methods: Application and development trends. *Build. Res. Inf.* **1999**, *27*, 300–308. [CrossRef]

9. Cole, R.J. Emerging trends in building environmental assessment methods. *Build. Res. Inf.* **1998**, *26*, 3–16. [CrossRef]

10. Cooper, I. Which focus for building assessment methods—Environmental performance or sustainability? *Build. Res. Inf.* **1999**, *27*, 321–331. [CrossRef]

11. Holmes, J.; Hudson, G. An evaluation of the objectives of the BREEAM scheme for offices: A Local Case Study. In Proceedings of the RICS Cutting Edge 2000 Conference, London, UK, 6–8 September 2000.

12. Department of Climate Policy. *Climate Change Handbook, Korea Meteorological Administration*; Korea Meteorological Administration: Seoul, Korea, 2009.

13. MIT Technology Review. Green Concrete. 10 Breakthrough Technologies. 2010. Available online: http://www2.technologyreview.com/news/418542/tr10-green-concrete/ (accessed on 20 April 2019).

14. Ministry of Land. Infrastructure and Transport. The planning about Construction Material with CO_2 Reduction. In *Hanyang University R&D Report*; Ministry of Land: Dhaka, Bangladesh, 2010.

15. Ministry of Land. Infrastructure and Transport. Development of Concrete Materials for the Effective Carbon Capture and Sequestration Using Recycled Industrial Wastes. In *Infrastructure R&D Report*; Ministry of Land: Dhaka, Bangladesh, 2015.

16. Caijun, S.; Della, R.; Pavel, K. *Alkali-Activated Cements and Concretes*; Taylor & Francis Group: Boca Raton, FL, USA, 2006.

17. American Society for Testing and Materials (ASTM) International. *Standard Test Method for Resistance of Concrete to Rapid Freezing and Thawing*; ASTM C 666: West Conshohocken, PA, USA, 2015.

18. American Society for Testing and Materials (ASTM) International. *Standard Test Method for Electrical Indication of Concrete's Ability to Resist Chloride Ion Penetration*; ASTM C 1202: West Conshohocken, PA, USA, 2019.

19. American Society for Testing and Materials (ASTM) International. *Standard Test Method for Microscopical Determination of Parameters of the Air-Void System in Hardened Concrete*; ASTM: West Conshohocken, PA, USA, 2016.

20. Mindess, S.; Young, F.J.; Darwin, D. *Concrete*, 2nd ed.; Prentice Hall: Upper Saddle River, NJ, USA, 2003.

Article

Large-Scale Screening and Machine Learning to Predict the Computation-Ready, Experimental Metal-Organic Frameworks for CO$_2$ Capture from Air

Xiaomei Deng, Wenyuan Yang, Shuhua Li, Hong Liang *, Zenan Shi * and Zhiwei Qiao *

Guangzhou Key Laboratory for New Energy and Green Catalysis, School of Chemistry and Chemical Engineering, Guangzhou University, Guangzhou 510006, China; 1705200045@e.gzhu.edu.cn (X.D.); 2111705055@e.gzhu.edu.cn (W.Y.); lish@gzhu.edu.cn (S.L.)
* Correspondence: lhong@gzhu.edu.cn (H.L.); zenanshi@126.com (Z.S.); zqiao@gzhu.edu.cn (Z.Q.)

Received: 19 December 2019; Accepted: 10 January 2020; Published: 13 January 2020

Abstract: The rising level of CO$_2$ in the atmosphere has attracted attention in recent years. The technique of capturing CO$_2$ from higher CO$_2$ concentrations, such as power plants, has been widely studied, but capturing lower concentrations of CO$_2$ directly from the air remains a challenge. This study uses high-throughput computer (Monte Carlo and molecular dynamics simulation) and machine learning (ML) to study 6013 computation-ready, experimental metal-organic frameworks (CoRE-MOFs) for CO$_2$ adsorption and diffusion properties in the air with very low concentrations of CO$_2$. First, the law influencing CO$_2$ adsorption and diffusion in air is obtained as a structure-performance relationship, and then the law influencing the performance of CO$_2$ adsorption and diffusion in air is further explored by four ML algorithms. Random forest (RF) was considered the optimal algorithm for prediction of CO$_2$ selectivity, with an R value of 0.981, and this algorithm was further applied to analyze the relative importance of each metal-organic framework (MOF) descriptor quantitatively. Finally, 14 MOFs with the best properties were successfully screened out, and it was found that a key to capturing a low concentration CO$_2$ from the air was the diffusion performance of CO$_2$ in MOFs. When the pore-limiting diameter (PLD) of a MOF was closer to the CO$_2$ dynamic diameter, this MOF could possess higher CO$_2$ diffusion separation selectivity. This study could provide valuable guidance for the synthesis of new MOFs in experiments that capture directly low concentration CO$_2$ from the air.

Keywords: CO$_2$ capture; Monte Carlo; machine learning; metal–organic framework; adsorption; diffusion

1. Introduction

It is well known that the amount of CO$_2$ discharged into the atmosphere increases with the rapid development of industry and population growth. In addition, deforestation, the large amount of CO$_2$ and other gases generated by the burning of fossil fuels such as coal, oil, and natural gas directly discharged into the atmosphere, and the emission of limestone roasting to produce cement have resulted in global carbon dioxide emissions increasing by 3.8% [1]. All of the above factors have aggravated carbon dioxide emissions, thereby increasing the urgency of counteracting the greenhouse effect and its associated global warming. The Kyoto Protocol and the Paris Agreement aim to control greenhouse gas emissions under the United Nations Framework Convention on Climate Change (UNFCCC), in which CO$_2$ is listed as a major greenhouse gas that needs to be mitigated or recycled [2]. The greenhouse gases include more than CO$_2$, however; in fact, the global warming potentials of CH$_4$ and N$_2$O are 25 times and 298 times that of CO$_2$, respectively. Nevertheless, due to its relatively large emission levels, CO$_2$ accounts for approximately 55% of the total greenhouse gas contribution [3,4].

Thus, it is obvious that the adsorption and separation of carbon dioxide from the air is particularly important. In addition, the successful capture of CO_2 could have multifaceted practical values: first, oil recovery could be improved through appropriate reservoir engineering; second, the captured CO_2 could be used to produce industrial chemicals, including concrete, paint, and fertilizer; third, the CO_2 in the atmosphere could be captured and combined with hydrogen for direct synthesis into liquid hydrocarbons, which could then be utilized in fuel synthesis and supply, including gasoline and diesel. The use of raw materials can reduce the proportion of fossil energy to further control CO_2 emissions, ultimately achieving carbon neutrality or even net negative carbon emissions [5].

Recently, carbon engineering has developed a series of capture technologies that remove carbon dioxide directly from the air. Carbon dioxide can be removed from the atmosphere using biological, chemical, or physical processes [6]. These methods have certain limitations, however. For example, biological processes are very economical, but they are usually very slow and ineffective. As for chemical processes, the waste of carbon resources and volatilization of organic solvents during these actions lead to further environmental pollution, equipment corrosion, and complex post-treatment issues. The traditional technique for separating carbon dioxide is solvent washing, such as the use of an alcohol amine solution [7–10]. Although this conventional method can reduce the concentration of carbon dioxide in the air, it is extremely expensive, the solvent is difficult to regenerate, the operation is complicated, and it consumes a great deal of energy [11]. In fact, the energy consumption of solvent washing is 3 to 4 times that of CO_2 captured from exhaust gas [12]. Given these drawbacks, there is an urgent need to find a more efficient, convenient, and energy-saving technique to replace the traditional carbon dioxide capture method. Adsorption separation is a potential technique. It is not only inexpensive, but also simple in terms of operation and equipment, and relatively low in energy consumption when the adsorbent is regenerated (the regeneration process of adsorbents is to desorb the adsorbed substances). Conventional adsorbents, however, including activated carbon, zeolite, silica gel, and metal oxides have poor scavenging effects on carbon dioxide in the air due to inferior separation selectivity and regeneration difficulty. For example, silica gel, which has amorphous properties, does not have a continuous uniform porous structure and exhibits unfavorable diffusion properties [11]. Therefore, the development of a new type of adsorbent is imperative. In recent years, studies have shown that the use of metal-organic frameworks (MOFs) to adsorb and separate carbon dioxide can not only make up for the shortcomings of the above adsorbents, but also feature the advantages of high selectivity and being non-polluting. The MOF is an organic–inorganic hybrid material with intramolecular pores formed by the self-assembly of organic ligands and inorganic metal ions or clusters by coordination bonds [13]. Compared with common adsorbents, MOFs exhibit many advantages such as various structures and properties, large specific surface area, high porosity, and structural control. Therefore, they are widely used in gas adsorption [11] and separation [14–19], as well as general materials in processes including storage [20], optics [21], catalysis [22–25], and drug delivery [26,27]. To date, thousands of MOFs have been synthesized, some of which have been utilized in the attempt to capture CO_2 from the air. Peng et al. [28] designed and synthesized 2 incorporated MOFs to study their stability and ability to capture CO_2 from the air. Liu et al. [11] used an amine-functionalized MOF and an ultra-microporous MOF to capture CO_2 directly from the air, and further investigated the performance of CO_2 capture and the reproducibility of MOFs under humid conditions. Osama et al. [29] synthesized an isomorphic MOF SIFSIX-3-Cu with uniform adsorption sites for capturing CO_2 from the air. Since CO_2 capture from the air has a very high selectivity of MOF, when the traditional approach is used to screen MOFs for the best-performing candidates, it not only consumes a great deal of manpower and material resources, but also has an extended study period and causes pollution to a certain extent. With the continuous advancement and development of computers, molecular simulation is playing an increasingly important role in the field of materials science [30]. Some studies have used high-throughput molecular simulation calculation methods to screen large numbers of MOFs in a database, thereby successfully screening MOFs with high selectivity and high working capacity based on different target performance requirements. For example, Wilmer et al. adsorbed pure carbon dioxide, nitrogen, and methane using more than 130,000 hypothetical MOFs, and proposed a relationship between structural properties (pore size, volume, and surface area) and chemical functions, as well as evaluation criteria for the separation of carbon dioxide from adsorbents [31]. In the presence of nickel dilution, Watanabe et al. combined pore size analysis with classical simulation to screen 1163 MOFs as membrane materials for CO_2/N_2 separation [32]. Lin et al. screened hundreds of thousands of theoretically predicted zeolites and zeolite MOFs and identified a number of potential materials for capturing carbon dioxide [33]. Based on 105 MOFs, Wu et al. proposed the relationship of CO_2/N_2 adsorption selectivity with porosity and the isosteric heat of adsorption [34]. Fernandez et al. [35] used advanced machine-learning (ML) algorithms to quickly identify 292,050 hypothetical high-performance MOFs for pure CO_2 adsorption (0.15 bar and 1 bar). These screening studies, however, were aimed at capturing high concentrations of CO_2. Given that the concentration of CO_2 in the atmosphere is comparatively low relative to the concentrations of natural gas and other components, it is undoubtedly a challenge to discover efficient MOF materials that can directly capture CO_2 from the air.

To date, given that there have been 6013 MOFs reported, finding the appropriate MOFs for a specific system in such a large database is undoubtedly a daunting task. This study focused on the aforementioned MOF simulation of the adsorption and diffusion performances of CO_2, N_2, and O_2 in infinite dilutions in order to identify materials with excellent performance in terms of both static adsorption and kinetic adsorption. The influencing factors affecting the adsorption and diffusion of CO_2 were obtained by univariate analysis. Next, multivariate analyses, namely 4 ML algorithms (back propagation neural network (BPNN), decision tree (DT), random forest (RF), and support vector machine (SVM)), were explored in depth. Finally, we adopted the optimal algorithm model. The parameters affecting CO_2 selectivity were predicted, and 14 types of MOFs with the same diffusion selectivity and adsorption selectivity were selected.

2. Materials and Methods

2.1. Molecular Model

In this work, we used molecular simulation to screen the capability of 6013 computation-ready, experimental MOFs (CORE-MOFs version 2) [36] to capture CO_2 from the air. Their crystal structures were derived from the Cambridge Crystallographic Data Centre (CCDC), and their parameters were compiled and verified by Chung et al. [37]. We removed all solvent and ligand molecules prior to running the simulation. Each MOF used 5 structural parameters, namely, volumetric surface area (VSA), largest cavity diameter (LCD), pore-limiting diameter (PLD), porosity ϕ, density ρ, and an energy parameter: heat of adsorption. Both LCD and PLD were calculated using the Zeo++ software package [38]. The VSA and ϕ were calculated using the N_2 of 3.64 Å and He of 2.58 Å as probes in the RASPA software package [39]. If VSA is close to or equal to 0, this indicates that the MOF cannot accommodate N_2 molecules [40]. We used NVT-Monte Carlo (NVT-MC) simulation, where N is the number of particles, V is the volume of the system, and T is the temperature of the system. The Q_{st} of each gas was calculated in an infinite dilution state.

The force field parameters for the 3 gas components $CO_2/N_2/O_2$ were from the transferable potentials for phase equilibria (TraPPE) force field [41] and are listed in Table S2 The CO_2 molecule has a C-O bond length of 1.16 Å and a bond angle $\angle OCO$ of 180°. N_2 is considered as a 3-point model, and the bond length of N-N is 1.10 Å. Oxygen is also a 3-point atom, and the O-O bond length is 1.21 Å. The models of 3 gases are shown in Figure S1, The atomic charge of MOF was estimated using the MOF electrostatic-potential-optimized charge scheme (MEPO-Qeq) method [42], which accurately evaluated electrostatic interactions. Due to the advantages of the MEPO-Qeq method with fast and accurate, it is widely used in various systems of adsorption-MOF [43–45]. The Lennard–Jones (LJ) electrostatic parameters were obtained from the universal force field (UFF) [46] and are listed in Table S1 Data from previous studies had shown that the UFF–TraPPE force field combination could accurately predict the adsorption and diffusion behaviors of these 3 gases in MOFs [40,47,48]. The Lorentz–Berthelot combination rule was used to calculate the cross-LJ parameters.

2.2. Screening Methods

In MOFs, the values of Henry's constant K and the diffusion coefficient D of CO_2, N_2, and O_2 were estimated using Monte Carlo (MC) and molecular dynamics (MD) simulations with the same set, respectively. In principle, a single gas molecule should be added to an MOF to simulate infinite dilution, while in reality, we added 30 gas molecular models to each MOF, ignoring the force between the gas models, thus being equivalent to the independent simulation of each gas molecule. Ultimately, the simulation results of the 30 independent molecules were statistically averaged. Throughout the simulation, the MOF frame was assumed to be rigid and the simulation elements were extended to at least 24 Å along the three-dimensional periodic boundary conditions. A 12 Å spherical cutoff with long-range correction was used to calculate the LJ interaction, while the Ewald sum was used to calculate the electrostatic interaction. In each MOF, the MC simulation ran 100,000 cycles, with the first 50,000 used for balancing and the last 50,000 used for overall averaging. Each cycle consisted of n trials (n: number of adsorbed molecules), including translation, rotation, regeneration, and exchange (exchange movement, including insertion and deletion). In the MD simulation, the 30 gas molecules had an MD duration of 10 ns at each MOF, and 5 ns was ultimately selected for statistical averaging. After the sampling analysis of dozens of MOFs, it was found that further increases of cycle time and MD duration had little effect on the simulation results. All MCs and MDs were simulated using the RASPA software package [39].

3. Results and Discussion

3.1. Univariate Analysis

In order to investigate the relationship of CO_2 adsorption and diffusion properties in N_2+O_2 with the MOF structure during static adsorption and kinetic adsorption, we first analyzed the relationship among adsorption selectivity S_{ads}, diffusion selective S_{diff}, and the LCD of CO_2/N_2+O_2, as shown in Figure 1. Obviously, most MOFs with large adsorption selectivity and diffusion selectivity have relatively small LCDs. Figure 1a indicates that when the LCD is 2.8–6.5 Å, the adsorption selectivity of CO_2/N_2+O_2 decreases, and when the LCD is >15 Å, the adsorption selectivity gradually becomes stable, tending to 5, as depicted by the red line in Figure 1 This is because CO_2 has a strong quadrupole moment, and even in infinitely large pores it is preferentially adsorbed compared to N_2 and O_2. The trend of gas separation is consistent with the trends of previous reports [49,50]. Figure 1b presents the relationship between S_{diff} and LCD. Similar to S_{ads}, the larger diffusion selectivity (S_{diff} >1) only occurs in the LCDs ranging from 2.4–5 Å, since the kinetic diameter of CO_2 is less than the kinetic diameters of O_2 and N_2 (the kinetic diameters of CO_2, O_2, and N_2 are 3.3, 3.46, and 3.64 Å, respectively). When the LCD of an MOF is small, the CO_2 molecules with smaller diameters diffuse faster, so the diffusion selectivity $S_{diff\,(CO2/N2+O2)}$ is larger. As the LCD increases, the diffusion selectivity gradually decreases. When the LCD is >15 Å, the diffusion selectivity tends to be stable and fluctuates at around 0.36. Comparison of Figure 1a,b reveals that the adsorption selectivity is generally >1, while it is rare for the diffusion selectivity to be >1. Because CO_2 has a strong quadrupole distance, it has a strong interaction force with MOF molecules, thus hindering the diffusion of CO_2 and resulting in a slower diffusion rate, which may be even smaller than the diffusion rates of N_2 and O_2.

Figure 1c,d show the relationships of S_{ads} and S_{diff} to the PLD, respectively. Comparing the panels in Figure 1 reveals that the PLD and LCD display the same trend in their relationships to the S_{ads} and S_{diff} of CO_2/N_2+O_2. Larger S_{ads} and S_{diff} values appear when the LCD and PLD are small, and S_{ads} and S_{diff} both decrease with increasing PLD or LCD, eventually tending toward stability. Therefore, there is a greater possibility of finding MOFs with simultaneously high S_{ads} and S_{diff} among MOFs with small PLDs and LCDs.

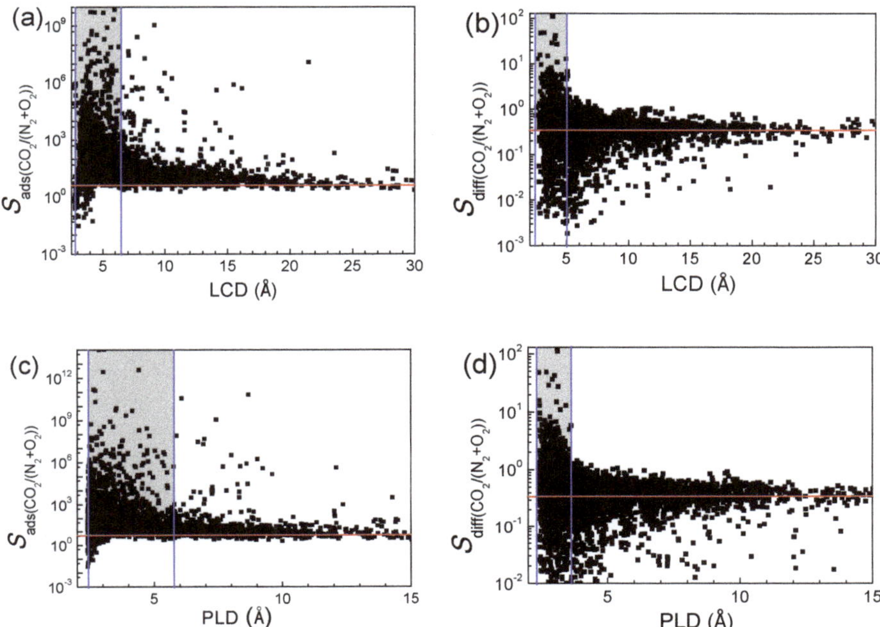

Figure 1. The relationships of (**a**) $S_{ads\,(CO2/N2+O2)}$ and (**b**) $S_{diff\,(CO2/N2+O2)}$ with (**c**) largest cavity diameter (LCD) and (**d**) pore-limiting diameter (PLD).

Figure 2a shows that $S_{ads\,(CO2/N2+O2)}$ increases monotonically with increasing Q_{st}, indicating that Q_{st} may be the main parameter during the adsorption process. Since the concentration of CO_2 in the atmosphere is low, it is close to the infinite dilution state. Hence, the selectivity is strongly dependent on the isosteric heat of adsorption of CO_2 in the infinite dilution state. The larger $S_{diff\,(CO2/N2+O2)}$ in Figure 2b occurs when the VSA is close to zero. As the VSA continues to increase, $S_{diff\,(CO2/N2+O2)}$ gradually decreases, and eventually stabilizes. This is because when the VSA is close to zero, the MOF molecule either cannot pass any or only passes a small amount of CO_2 molecules. When the VSA is large, all the gas molecules can pass through the MOF molecule. Therefore, the separation of CO_2 cannot be achieved, i.e., the diffusion selectivity is substantially unchanged. Figure S11b,c indicate the relationship of adsorption selectivity with porosity and VSA, respectively. It can be observed that both of these parameters exert weak influences on adsorption selectivity.

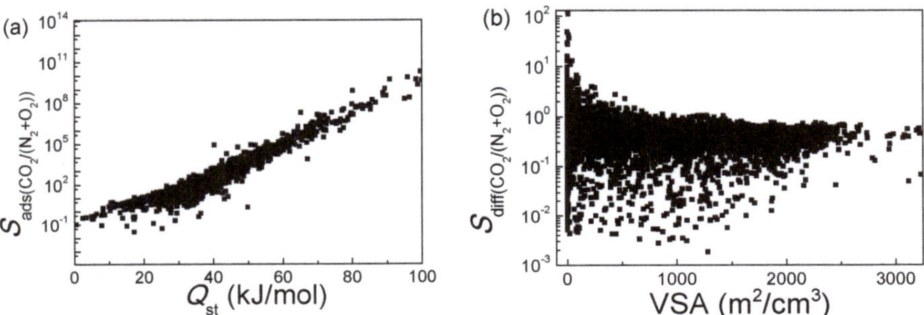

Figure 2. The relationships between (**a**) $S_{ads\,(CO2/N2+O2)}$ and Q_{st}, (**b**) $S_{diff\,(CO2/N2+O2)}$ and volumetric surface area (VSA).

In addition to adsorption and diffusion selectivity, the Henry coefficient of CO_2 reflects the adsorption performance of CO_2 in the infinite dilution state, helping to explain the capture performance of MOFs for air with very low CO_2 concentration. Figure 3a clearly shows the tendency of K_{N2} to change with enthalpy. When the porosity ϕ is small, the MOF has no space due to the limited pore volume, and only a small amount of N_2 can be adsorbed; therefore, K_{N2} is small. When ϕ is in the range of 0–0.29, K_{N2} increases significantly with increasing ϕ. When $\phi > 0.29$, K_{N2} slows down and gradually stabilizes with increasing ϕ. Figure 3b compares the Henry coefficients of the 3 gases. It can be seen that the trends of the Henry coefficients of N_2 and O_2 are almost the same; however, CO_2 is different. First, in most MOFs, the Henry coefficient values of CO_2 are basically larger than the Henry coefficient values of N_2 and O_2. Second, when LCD >20 Å, the K_{CO2} value tends to be level, and eventually stabilizes. The Henry coefficient of CO_2 is still higher than the coefficients of N_2 and O_2, which also leads to MOF selectivity >1 when the LCD is infinite, as seen in Figure 1a Finally, it can be observed that only a few MOFs can be identified for which the CO_2 Henry coefficient can be $>10^{-1}$ mmol/g/Pa. Observing these MOF structures reveals that most have smaller or open metal sites. The above univariate analysis can only determine the relationship between individual parameters and performance. Q_{st}, PLD and LCD are considered to have dramatic impacts on adsorption selectivity and diffusion selectivity, but their variable influences cannot be analyzed quantitatively. We will further utilize 4 types of ML algorithms to obtain additional information about structure-performance.

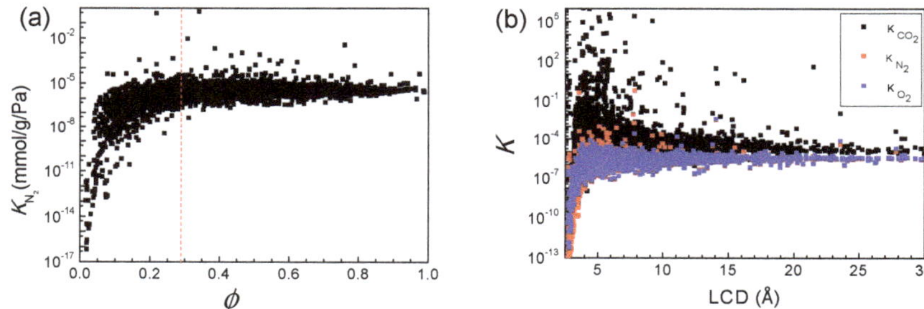

Figure 3. Henry coefficient K_{N2} *versus* (**a**) ϕ and (**b**) LCD.

3.2. Machine Learning

Currently, machine learning has been used to predict the performance of materials and to filter high-performance materials from large databases [51]. Aiming to discover a better machine prediction method suitable for this system, we individually compared the simulations of the 4 ML algorithms commonly used in big data analysis, i.e., the BPNN, DT, RF, and SVM. Among them, BPNN is a kind of forward signal propagation with error back propagation in which the gradient descent algorithm continuously adjusts the weight and threshold until the error is less than a set threshold. Some parameters of BPNN were set: the training function is Levenberg–Marquardt, the transfer function is a hyperbolic tangent sigmoid transfer function, and the performance evaluation function is the mean square error (MSE). The number of hidden layer neurons was 18, the maximum number of training was 1000, the training required an accuracy of 0.001, and the learning rate was 0.01. DT is a traditional method for data classification and screening. Under the condition that the probability of occurrence takes place in various situations, probability analysis is employed to analyze data with the dendritic model to obtain the expected values. The random forest algorithm is composed of multiple decision trees. The setting parameters of DT were: standard CART (classification and regression tree) used to select the best split predictor at each node. The criteria of splitting and pruning are the MSE function. After optimizing and pruning, the minimum number of branch node observations was 10, the minimum number of leaf node observations was 4, and the maximal number of decision splits was 1. RF uses the method of randomly selecting split attribute sets to construct a decision tree. The parameters for RF were set as: number of trees 200, minimum leaf size 10. The number of variables randomly selected in the variable subset of the node split in each tree was 2. SVM is an algorithm for binary classification of data through supervised learning, and employs mathematical transformation methods to divide data with a certain centralized structure into rules. We chose the support vector machine regression model of Statistics and Machine Learning Toolbox in MATLAB 2016b to predict, where the kernel function is radial basis function (Gaussian), the kernel scale parameter is set as "auto", and the loss function is epsilon-insensitive. The box constraint (also called the penalty coefficient, C) was 0.0567, and the half width of epsilon-insensitive band (ε) was set as 0.0057. In the radial basis kernel function, Gamma = $1/(2\sigma^2)$, where σ is the parameter of the kernel function, which can affect the complexity of the SVM regression algorithm. In our study, the value of gamma was 7.125. The solver of convex quadratic programming is sequential minimal optimization (SMO). Before training and testing, we first processed the data set out-of-order, and then randomly divide it into training and testing sets based on a ratio of 7:3. More detailed descriptions of ML algorithms are listed in the supporting information, and the corresponding diagrams of each algorithm are shown in Figures S2–S5.

We used BPNN, RF, DT, and SVM to predict the adsorption selectivity, and took the logarithm of the adsorption selectivity in order to reduce the differences associated with the varying magnitudes of the data. The 4 ML for predicting the correlation coefficient R value of the adsorption selectivity are listed in Table 1 The results of the testing and training are shown in Figure 4 and Figure S12. The distribution trends of the points in Figure 4a–d are all straight lines inclined upward. The different colors from top to bottom in the figure represent an increase in the number of points, and most of the points are concentrated on the diagonal, indicating that the prediction results are good. Figure 4 reveals that RF has the highest correlation coefficient value (0.982), while the support vector machine algorithm has the lowest (0.886). Thus, the prediction accuracy obtained by the RF algorithm is the highest. Therefore, among the 4 ML algorithms, the structure-performance relationship of RF on adsorption selectivity obtains more information, and the prediction results are the best. RF has good generalization ability and strong model learning ability, and this type of ML is suitable for the system. To verify the accuracy of four ML algorithms, we performed 5 repeated predictions, listed in Table S3. The repeated prediction results do not vary significantly, confirming that the RF algorithm is a suitable model. Because RF introduces two kinds of randomness (sampling randomness and feature randomness), it has strong generalization ability. In previous studies, random forest algorithms also exhibited the best prediction results [52,53]. Whether the overfitting of the model is an important issue. We used combinations of different descriptors and 5 times 5-fold cross-validation to verify the RF model. The results showed that the selected model was not overfitting. During the material screening process, the relative importance of parameters may affect the ultimate screening results. We selected the best RF algorithm to predict the relative importance of each descriptor. The relative importance percentages are shown in Figure 5 and Table S7. We used mean squared error (MSE) to evaluate the relative importance of the 6 descriptors; the greater the percentage of relative importance of the resulting descriptors, the higher the relative contribution of the specific descriptor. According to the results presented in Figure 5, the percentage of Q_{st} is the largest, thus indicating that Q_{st} exerts the greatest impact on the adsorption selectivity. The relative importance of the MOF descriptors to the adsorption selectivity is $Q_{st} > \rho > LCD > VSA \geq \phi > PLD$. The importances of VSA and ϕ are very close, indicating that the effects of two descriptors on adsorption selectivity are roughly equal. From a material science point of view, the larger the ϕ, the larger the VSA. This may be the reason why the effects of these parameters are essentially the same.

Table 1. The 4 ML algorithms for predicting the correlation coefficient R value of the adsorption selectivity.

ML Algorithms	R Value	
	Train	Test
BPNN	0.982	0.979
RF	0.994	0.981
DT	0.985	0.969
SVM	0.915	0.886

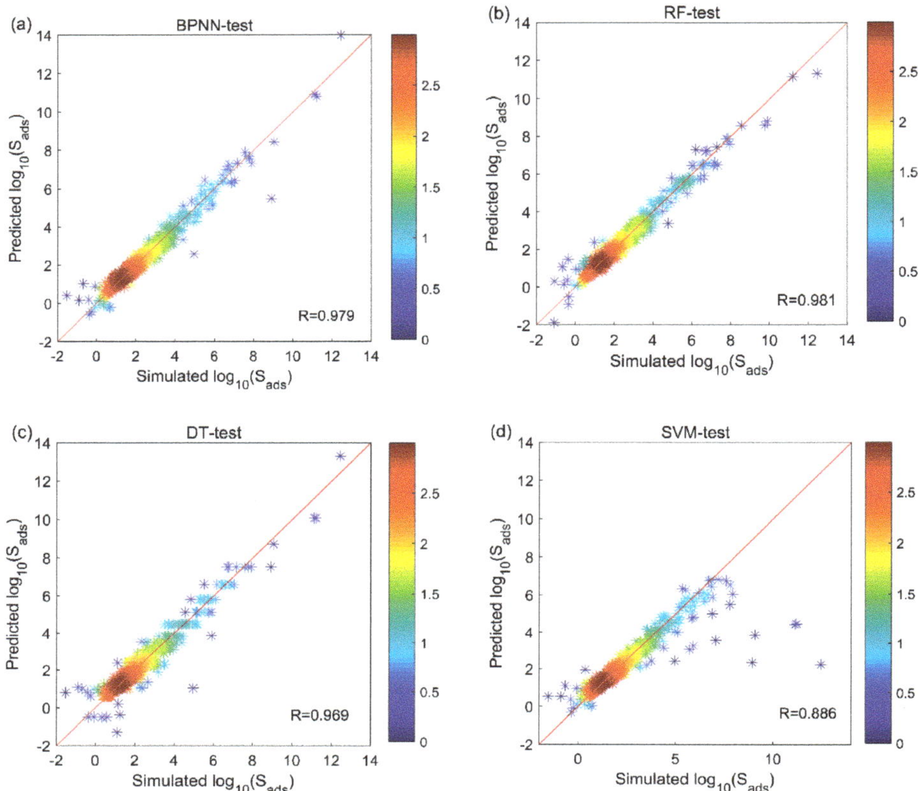

Figure 4. The test results of four machine-learning (ML) algorithms predicted versus simulated $S_{ads(CO2/N2+O2)}$ using ρ, ϕ, VSA, LCD, PLD and heat of adsorption. (**a**) back propagation neural network (BPNN), (**b**) random forest (RF), (**c**) decision tree (DT), (**d**) support vector machine (SVM). The color of point represents the number of metal-organic frameworks (MOFs), and the unit of number is a base-10 logarithm of MOF numbers.

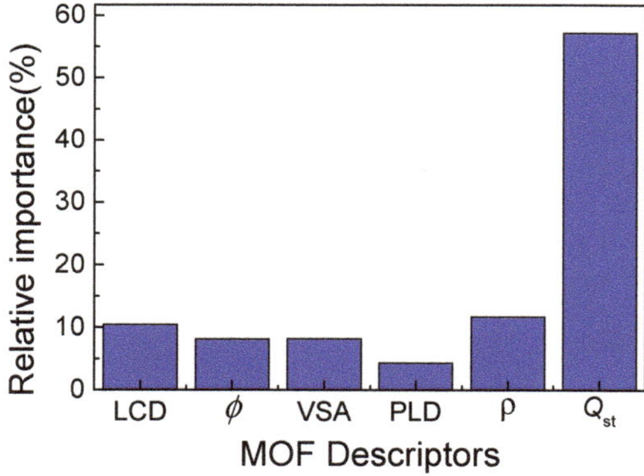

Figure 5. Predicted by the Random Forest, the relative importance of the six descriptors for adsorption selectivity.

4. Best Metal-Organic Frameworks (MOFs)

We selected 5 limiting conditions for $S_{ads\ (CO2/N2+O2)}$ and $S_{diff\ (CO2/N2+O2)}$, and chose 14 optimal MOFs from the 6013 MOFs, as listed in Table 2 of the 14 materials, HIQPEE exhibited the largest S_{diff}, which was as much as 62.27 Å. NORGOS displayed the largest S_{ads}, which also corresponded to its maximum heat of adsorption. In comparison, the optimal MOF selected by this study is also more selective at higher Q_{st} (4712.33) under the same conditions than that predicted by Wu et al. (433) [30] at Q_{st} = 47.8 kJ/mol and Ravichandar Babarao et al. (500) [54]. It was discovered that diffusion selectivity is generally lower than adsorption selectivity. The diffusion of CO_2 is the key property in determining the performance of MOFs for low concentrations of CO_2 during the kinetic adsorption process. For these 14 MOFs, the LCD, ϕ, and PLD ranges of the six descriptors also corresponded to those in the previous univariate analysis. Especially for PLDs, the optimal range of 2.66–3.64 Å only spans 1 Å, which is also very close to the kinetic diameter of the 3 gases. In such strictly restricted channels, only CO_2 molecules can enter and be adsorbed, greatly increasing the probability of CO_2 being captured at low concentrations. Therefore, the analysis of the optimal MOF revealed that a PLD with a kinetic diameter close to CO_2 is a key condition for good CO_2 diffusion performance, further resulting in the excellent performance of the MOF in capturing CO_2 from the air, and thus providing effective guidance for the design and synthesis of new MOFs.

Table 2. Best computation-ready, experimental metal-organic frameworks (CoRE-MOFs).

No	CSD Code [a]	LCD [b]	ϕ	VSA [c] (m²/cm³)	PLD [d] (Å)	P (kg/m³)	Q_{st_CO2} (kJ/mol)	$S_{diffCO2/(N2+O2)}$	$S_{adsCO2/(N2+O2)}$
1	REYCEF	3.75	0.08	0	2.83	1646.40	26.02	8.21	6.25
2	JAHNEM	3.50	0.05	0	2.66	1714.62	27.01	6.18	6.77
3	HOJLEY	3.63	0.14	0	2.83	1410.99	32.65	5.00	7.45
4	KASPOL	4.04	0.19	16.26	2.93	1665.95	34.05	36.55	7.68
5	XUNJOG	3.25	0.07	0	2.67	1737.75	26.41	5.31	11.12
6	HIQPII	3.87	0.15	9.46	3.12	1472.40	35.20	55.67	11.24
7	YUBFUX	4.58	0.16	98.56	3.64	1786.84	30.52	5.79	11.51
8	HIQPEE	3.84	0.15	7.68	3.12	1440.14	35.31	62.27	12.39
9	FEJKEM	3.46	0.09	0.33	3.09	2132.02	30.00	15.01	15.72
10	FALQIU	5.08	0.12	82.58	3.14	1977.64	34.40	5.38	22.97
11	FALQOA	6.06	0.12	83.34	3.07	2004.41	35.03	6.76	24.10
12	TOXNAX	3.80	0.28	4.42	2.83	1346.52	42.65	5.02	25.15
13	OFIWIK	4.20	0.05	5.99	3.14	1866.86	39.64	27.42	25.66
14	NORGOS	4.95	0.13	45.40	3.43	1728.90	51.67	12.92	4712.33

[a] CSD Code is the code of MOFs in the Cambridge Structure Database; [b] LCD: largest cavity diameter; [c] VSA: volumetric surface area; [d] PLD: pore-limiting diameter.

5. Conclusions

Firstly, we simulated the adsorption and diffusion properties of CO_2, N_2, and O_2 in 6013 CoRE-MOFs using high-throughput MC + MD. Then, we investigated the correlation among adsorption selectivity and diffusion selectivity for CO_2 and MOF descriptors by the univariate analysis. The Q_{st} and PLD were considered to be the most important for $S_{ads\ (CO2/N2+O2)}$ and $S_{diff\ (CO2/N2+O2)}$, respectively. In conjunction with multivariate analysis, a comparison of 4 ML algorithms revealed that the RF had the best prediction results for adsorption selectivity, with an R value of 0.982. This indicated that the RF method was the most suitable for the predictions of the capture of low CO_2 concentrations in MOF. The relative importance analysis of the RF algorithm quantitatively indicated that the relative importance of the MOF descriptors on adsorption selectivity is $Q_{st} > \rho > LCD > VSA \geq \phi > PLD$. It was also confirmed that Q_{st} is the most important parameter, while the VSA and ϕ are relatively less important. Through this high-throughput screening, 14 types of MOFs with optimal adsorption selectivity and diffusion selectivity were obtained. After comparison, it was found that their adsorption selectivity was generally higher than their diffusion selectivity. The diffusion separation performance of CO_2 is the key property in determining the performance of MOFs on low concentrations of CO_2 during the kinetic adsorption process. This study provides experimental guidance for the determination of MOFs that effectively capture CO_2 from the air, and indicates that advanced ML algorithms can accelerate the research and development of new materials.

Supplementary Materials: The following are available online at http://www.mdpi.com/2076-3417/10/2/569/s1: Figure S1: Models of N_2, O_2 and CO_2, Figure S2: Back propagation neural network, Figure S3: Decision tree, Figure S4: Random forest, Figure S5: Support vector machine, Figure S6: Diffusion coefficient D_i *versus* (a) LCD, (b–d) Q_{st} (D_i: the diffusion coefficient of different gases, (a,b) CO_2, (c) N_2, (d) O_2), Figure S7: Henry coefficient K_{CO2} *versus* (a) ϕ, (b) PLD, (c) Q_{st} and (d) VSA, Figure S8: Henry coefficient K_{N2} *versus* (a) ϕ, (b) PLD, (c) Q_{st} and (d) VSA, Figure S9: Henry coefficient K_{O2} *versus* (a) ϕ, (b) PLD, (c) Q_{st} and (d) VSA, Figure S10: Diffusion selectivity $S_{diff\,(CO2/(N2+O2))}$ *versus* (a) LCD, (b) ϕ, (c) PLD and (d) VSA, Figure S11: Adsorption selectivity $S_{ads\,(CO2/N2+O2)}$ *versus* (a) LCD, (b) ϕ, (c) PLD and (d) VSA, Figure S11: The training results of four machine learning predicted *versus* MC simulated $\log_{10}(S_{ads\,(CO2/N2+O2)})$ using density, void fraction, volumetric surface area, density, LCD, PLD and heat of adsorption. (a) BPNN, (b) RF, (c) DT and (d) SVM. The color of point represents the number of MOF, and the unit of mumber is a base-10 logarithm of MOF numbers, Table S1: Lennard-Jones parameters of MOFs, Table S2: Lennard-Jones parameters and charges of adsorbates, Table S3: The training and testing R values of adsorption selectivity using repeat 5 time-four ML. Table S4 Prediction using RF models with different descriptor combinations. Table S5 Prediction using repeat 5 times-RF models with different descriptor combinations. Table S6 The results of predicted RF with k times k-fold cross validation. Table S7: Predicted by the RF the relative importance of the six descriptors for adsorption selectivity. Formula S1: $S_{ads\,(CO2/N2+O2)}$ indicates the adsorption selectivity of CO_2/N_2+O_2; K_i represents the Henry coefficient of component i (CO_2, N_2 and O_2), Formula S2: $S_{diff\,(CO2/(N2+O2))}$ represents the diffusion selectivity of CO_2/N_2+O_2; D_i represents the diffusion coefficient of component i.

Author Contributions: X.D., Z.S., H.L. and Z.Q. conceived the idea. Z.Q. calculated all the materials' structural parameters and obtained valid data about the structure descriptors and performance. X.D., W.Y. and Z.S. analyzed the relationship between structure descriptors and performance. X.D. and S.L. used univariate analysis to obtain the influence law of affecting CO_2 adsorption and diffusion in air and Z.S. used ML algorithms to predict the MOF performance. X.D. and Z.S. wrote the original draft. H.L. and Z.Q. wrote the manuscript with contributions from all authors. All authors have read and agreed to the published version of the manuscript.

Funding: This research was funded by the National Natural Science Foundation of China (Nos. 21978058, 21676094, and 21576058).

Acknowledgments: We gratefully thank the National Natural Science Foundation of China (Nos. 21978058, 21676094, and 21576058) for financial support.

Conflicts of Interest: The authors declare no conflict of interest.

References

1. Canadell, J.G.; Le Quere, C.; Raupach, M.R.; Field, C.B.; Buitenhuis, E.T.; Ciais, P.; Conway, T.J.; Gillett, N.P.; Houghton, R.A.; Marland, G. Contributions to accelerating atmospheric CO_2 growth from economic activity, carbon intensity, and efficiency of natural sinks. *Proc. Natl. Acad. Sci. USA* **2007**, *104*, 18866–18870. [CrossRef] [PubMed]

2. Zhang, C.; Zeng, G.; Huang, D.; Lai, C.; Chen, M.; Cheng, M.; Tang, W.; Tang, L.; Dong, H.; Huang, B.; et al. Biochar for environmental management: Mitigating greenhouse gas emissions, contaminant treatment, and potential negative impacts. *Chem. Eng. J.* **2019**, *373*, 902–922. [CrossRef]

3. Fan, R.; Chen, C.-L.; Lin, J.-Y.; Tzeng, J.-H.; Huang, C.-P.; Dong, C.; Huang, C.P. Adsorption characteristics of ammonium ion onto hydrous biochars in dilute aqueous solutions. *Bioresour. Technol.* **2019**, *272*, 465–472. [CrossRef] [PubMed]

4. Fang, G.; Liu, C.; Wang, Y.; Dionysiou, D.D.; Zhou, D. Photogeneration of reactive oxygen species from biochar suspension for diethyl phthalate degradation. *Appl. Catal. B Environ.* **2017**, *214*, 34–45. [CrossRef]

5. Available online: https://carbonengineering.com/ (accessed on 10 July 2019).

6. Nibleus, K.; Lundin, R. Climate Change and Mitigation. *Ambio* **2010**, *39*, 11–17. [CrossRef]

7. Boyd, P.G.; Chidambaram, A.; García-Díez, E.; Ireland, C.P.; Daff, T.D.; Bounds, R.; Gładysiak, A.; Schouwink, P.; Moosavi, S.M.; Maroto-Valer, M.M.; et al. Data-driven design of metal–organic frameworks for wet flue gas CO_2 capture. *Nature* **2019**, *576*, 253–256. [CrossRef]

8. Faig, R.W.; Popp, T.M.O.; Fracaroli, A.M.; Kapustin, E.A.; Kalmutzki, M.J.; Altamimi, R.M.; Fathieh, F.; Reimer, J.A.; Yaghi, O.M. The Chemistry of CO_2 Capture in an Amine-Functionalized Metal-Organic Framework under Dry and Humid Conditions. *J. Am. Chem. Soc.* **2017**, *139*, 12125–12128. [CrossRef]

9. Haszeldine, R.S. Carbon Capture and Storage: How Green Can Black Be? *Science* **2009**, *325*, 1647–1652. [CrossRef]

10. McDonald, T.M.; Mason, J.A.; Kong, X.; Bloch, E.D.; Gygi, D.; Dani, A.; Crocellà, V.; Giordanino, F.; Odoh, S.O.; Drisdell, W.S.; et al. Cooperative insertion of CO_2 in diamine-appended metal-organic frameworks. *Nature* **2015**, *519*, 303–308. [CrossRef]

11. Liu, J.; Wei, Y.; Zhao, Y. Trace Carbon Dioxide Capture by Metal-Organic Frameworks. *ACS Sustain. Chem. Eng.* **2019**, *7*, 82–93. [CrossRef]

12. Zhao, R.; Liu, L.; Zhao, L.; Deng, S.; Li, S.; Zhang, Y.; Li, H. Thermodynamic exploration of temperature vacuum swing adsorption for direct air capture of carbon dioxide in buildings. *Energy Convers. Manag.* **2019**, *183*, 418–426. [CrossRef]

13. Batten, S.R.; Champness, N.R.; Chen, X.-M.; Garcia-Martinez, J.; Kitagawa, S.; Ohrstrom, L.; O'Keeffe, M.; Suh, M.P.; Reedijk, J. Terminology of metal-organic frameworks and coordination polymers (IUPAC Recommendations 2013). *Pure Appl. Chem.* **2013**, *85*, 1715–1724. [CrossRef]

14. Murray, L.J.; Dinca, M.; Long, J.R. Hydrogen storage in metal-organic frameworks. *Chem. Soc. Rev.* **2009**, *38*, 1294–1314. [CrossRef] [PubMed]

15. Sculley, J.; Yuan, D.; Zhou, H.-C. The current status of hydrogen storage in metal-organic frameworks-updated. *Energy Environ. Sci.* **2011**, *4*, 2721–2735. [CrossRef]

16. Li, J.-R.; Kuppler, R.J.; Zhou, H.-C. Selective gas adsorption and separation in metal-organic frameworks. *Chem. Soc. Rev.* **2009**, *38*, 1477–1504. [CrossRef] [PubMed]

17. Verma, S.; Mishra, A.K.; Kumar, J. The Many Facets of Adenine: Coordination, Crystal Patterns, and Catalysis. *Acc. Chem. Res.* **2010**, *43*, 79–91. [CrossRef]

18. Li, J.-R.; Sculley, J.; Zhou, H.-C. Metal-Organic Frameworks for Separations. *Chem. Rev.* **2012**, *112*, 869–932. [CrossRef]

19. Bae, Y.-S.; Snurr, R.Q. Development and Evaluation of Porous Materials for Carbon Dioxide Separation and Capture. *Angew. Chem. Int. Ed.* **2011**, *50*, 11586–11596. [CrossRef]

20. Wu, X.-J.; Zhao, P.; Fang, J.-M.; Wang, J.; Liu, B.-S.; Cai, W.-Q. Simulation on the Hydrogen Storage Properties of New Doping Porous Aromatic Frameworksl. *Acta Phys. Chim. Sin.* **2014**, *30*, 2043–2054. [CrossRef]

21. Wu, P.; He, C.; Wang, J.; Peng, X.; Li, X.; An, Y.; Duan, C. Photoactive Chiral Metal-Organic Frameworks for Light-Driven Asymmetric alpha-Alkylation of Aldehydes. *J. Am. Chem. Soc.* **2012**, *134*, 14991–14999. [CrossRef]

22. Farrusseng, D.; Aguado, S.; Pinel, C. Metal-Organic Frameworks: Opportunities for Catalysis. *Angew. Chem. Int. Ed.* **2009**, *48*, 7502–7513. [CrossRef] [PubMed]

23. Ma, L.; Abney, C.; Lin, W. Enantioselective catalysis with homochiral metal-organic frameworks. *Chem. Soc. Rev.* **2009**, *38*, 1248–1256. [CrossRef] [PubMed]

24. Lee, J.; Farha, O.K.; Roberts, J.; Scheidt, K.A.; Nguyen, S.T.; Hupp, J.T. Metal-organic framework materials as catalysts. *Chem. Soc. Rev.* **2009**, *38*, 1450–1459. [CrossRef] [PubMed]

25. Farha, O.K.; Shultz, A.M.; Sarjeant, A.A.; Nguyen, S.T.; Hupp, J.T. Active-Site-Accessible, Porphyrinic Metal-Organic Framework Materials. *J. Am. Chem. Soc.* **2011**, *133*, 5652–5655. [CrossRef] [PubMed]

26. Della Rocca, J.; Liu, D.; Lin, W. Nanoscale Metal-Organic Frameworks for Biomedical Imaging and Drug Delivery. *Acc. Chem. Res.* **2011**, *44*, 957–968. [CrossRef]

27. Bernini, M.C.; Fairen-Jimenez, D.; Pasinetti, M.; Ramirez-Pastor, A.J.; Snurr, R.Q. Screening of bio-compatible metal-organic frameworks as potential drug carriers using Monte Carlo simulations. *J. Mater. Chem. B* **2014**, *2*, 766–774. [CrossRef]

28. Peng, Y.-W.; Wu, R.-J.; Liu, M.; Yao, S.; Geng, A.-F.; Zhang, Z.-M. Nitrogen Coordination to Dramatically Enhance the Stability of In-MOF for Selectively Capturing CO_2 from a CO_2/N_2 Mixture. *Cryst. Growth Des.* **2019**, *19*, 1322–1328. [CrossRef]

29. Shekhah, O.; Belmabkhout, Y.; Chen, Z.; Guillerm, V.; Cairns, A.; Adil, K.; Eddaoudi, M. Made-to-order metal-organic frameworks for trace carbon dioxide removal and air capture. *Nat. Commun.* **2014**, *5*. [CrossRef]

30. Jain, A.; Shyue Ping, O.; Hautier, G.; Chen, W.; Richards, W.D.; Dacek, S.; Cholia, S.; Gunter, D.; Skinner, D.; Ceder, G.; et al. Commentary: The Materials Project: A materials genome approach to accelerating materials innovation. *APL Mater.* **2013**, *1*. [CrossRef]

31. Furukawa, H.; Cordova, K.E.; O'Keeffe, M.; Yaghi, O.M. The Chemistry and Applications of Metal-Organic Frameworks. *Science* **2013**, *341*, 1230444. [CrossRef]

32. Watanabe, T.; Sholl, D.S. Accelerating Applications of Metal-Organic Frameworks for Gas Adsorption and Separation by Computational Screening of Materials. *Langmuir* **2012**, *28*, 14114–14128. [CrossRef] [PubMed]

33. Lin, L.-C.; Berger, A.H.; Martin, R.L.; Kim, J.; Swisher, J.A.; Jariwala, K.; Rycroft, C.H.; Bhown, A.S.; Deem, M.W.; Haranczyk, M.; et al. In silico screening of carbon-capture materials. *Nat. Mater.* **2012**, *11*, 633–641. [CrossRef] [PubMed]

34. Wu, D.; Yang, Q.; Zhong, C.; Liu, D.; Huang, H.; Zhang, W.; Maurin, G. Revealing the Structure-Property Relationships of Metal-Organic Frameworks for CO_2 Capture from Flue Gas. *Langmuir* **2012**, *28*, 12094–12099. [CrossRef] [PubMed]

35. Fernandez, M.; Boyd, P.G.; Daff, T.D.; Aghaji, M.Z.; Woo, T.K. Rapid and Accurate Machine Learning Recognition of High Performing Metal Organic Frameworks for CO_2 Capture. *J. Phys. Chem. Lett.* **2014**, *5*, 3056–3060. [CrossRef] [PubMed]

36. Available online: https://github.com/gregchung/gregchung.github.io/blob/master/CoRE-MOFs/structure-doi-CoRE-MOFsV2.0.csv (accessed on 15 January 2019).

37. Chung, Y.G.; Camp, J.; Haranczyk, M.; Sikora, B.J.; Bury, W.; Krungleviciute, V.; Yildirim, T.; Farha, O.K.; Sholl, D.S.; Snurr, R.Q. Computation-Ready, Experimental Metal-Organic Frameworks: A Tool to Enable High-Throughput Screening of Nanoporous Crystals. *Chem. Mater.* **2014**, *26*, 6185–6192. [CrossRef]

38. Willems, T.F.; Rycroft, C.; Kazi, M.; Meza, J.C.; Haranczyk, M. Algorithms and tools for high-throughput geometry-based analysis of crystalline porous materials. *Microporous Mesoporous Mater.* **2012**, *149*, 134–141. [CrossRef]

39. Dubbeldam, D.; Calero, S.; Ellis, D.E.; Snurr, R.Q. RASPA: Molecular simulation software for adsorption and diffusion in flexible nanoporous materials. *Mol. Simul.* **2016**, *42*, 81–101. [CrossRef]

40. Yang, W.; Liang, H.; Peng, F.; Liu, Z.; Liu, J.; Qiao, Z. Computational Screening of Metal-Organic Framework Membranes for the Separation of 15 Gas Mixtures. *Nanomaterials* **2019**, *9*, 467. [CrossRef]

41. Potoff, J.J.; Siepmann, J.I. Vapor-liquid equilibria of mixtures containing alkanes, carbon dioxide, and nitrogen. *AIChE J.* **2001**, *47*, 1676–1682. [CrossRef]

42. Kadantsev, E.S.; Boyd, P.G.; Daff, T.D.; Woo, T.K. Fast and Accurate Electrostatics in Metal Organic Frameworks with a Robust Charge Equilibration Parameterization for High-Throughput Virtual Screening of Gas Adsorption. *J. Phys. Chem. Lett.* **2013**, *4*, 3056–3061. [CrossRef]

43. Shi, Z.; Liang, H.; Yang, W.; Liu, J.; Liu, Z.; Qiao, Z. Machine learning and in silico discovery of metal-organic frameworks: Methanol as a working fluid in adsorption-driven heat pumps and chillers. *Chem. Eng. Sci.* **2020**, *214*, 115430. [CrossRef]

44. Qiao, Z.; Xu, Q.; Jiang, J. Computational screening of hydrophobic metal-organic frameworks for the separation of H2S and CO2 from natural gas. *J. Mater. Chem. A* **2018**, *6*, 18898–18905. [CrossRef]

45. Bian, L.; Li, W.; Wei, Z.; Liu, X.; Li, S. Formaldehyde Adsorption Performance of Selected Metal-Organic Frameworks from High-throughput Computational Screening. *Acta Chim. Sin.* **2018**, *76*, 303–310. [CrossRef]

46. Rappe, A.K.; Casewit, C.J.; Colwell, K.S.; Goddard, W.A.; Skiff, W.M. UFF, a full periodic table force field for molecular mechanics and molecular dynamics simulations. *J. Am. Chem. Soc.* **1992**, *114*, 10024–10035. [CrossRef]

47. Qiao, Z.; Xu, Q.; Jiang, J. High-throughput computational screening of metal-organic framework membranes for upgrading of natural gas. *J. Membr. Sci.* **2018**, *551*, 47–54. [CrossRef]

48. Babarao, R.; Jiang, J. Diffusion and separation of CO_2 and CH_4 in silicalite, C_{168} schwarzite, and IRMOF-1: A comparative study from molecular dynamics simulation. *Langmuir* **2008**, *24*, 5474–5484. [CrossRef]

49. Qiao, Z.; Zhang, K.; Jiang, J. In silico screening of 4764 computation-ready, experimental metal-organic frameworks for CO_2 separation. *J. Mater. Chem. A* **2016**, *4*, 2105–2114. [CrossRef]

50. Wilmer, C.E.; Farha, O.K.; Bae, Y.-S.; Hupp, J.T.; Snurr, R.Q. Structure-property relationships of porous materials for carbon dioxide separation and capture. *Energy Environ. Sci.* **2012**, *5*, 9849–9856. [CrossRef]

51. Takahashi, K.; Tanaka, Y. Materials informatics: A journey towards material design and synthesis. *Dalton Trans.* **2016**, *45*, 10497–10499. [CrossRef]

52. Pardakhti, M.; Moharreri, E.; Wanik, D.; Suib, S.L.; Srivastava, R. Machine Learning Using Combined Structural and Chemical Descriptors for Prediction of Methane Adsorption Performance of Metal Organic Frameworks (MOFs). *ACS Comb. Sci.* **2017**, *19*, 640–645. [CrossRef]

53. Wu, X.; Xiang, S.; Su, J.; Cai, W. Understanding Quantitative Relationship between Methane Storage Capacities and Characteristic Properties of Metal-Organic Frameworks Based on Machine Learning. *J. Phys. Chem. C* **2019**, *123*, 8550–8559. [CrossRef]
54. Babarao, R.; Jiang, J. Unprecedentedly High Selective Adsorption of Gas Mixtures in rho Zeolite-like Metal-Organic Framework: A Molecular Simulation Study. *J. Am. Chem. Soc.* **2009**, *131*, 11417–11425. [CrossRef]

Article

Comparative Kinetic Analysis of CaCO$_3$/CaO Reaction System for Energy Storage and Carbon Capture

Larissa Fedunik-Hofman [1], Alicia Bayon [2,*] and Scott W. Donne [1]

[1]   Discipline of Chemistry, University of Newcastle, Callaghan, NSW 2308, Australia;
      Larissa.Fedunik-Hofman@uon.edu.au (L.F.-H.); scott.donne@newcastle.edu.au (S.W.D.)
[2]   CSIRO Energy, P.O. Box 330, Newcastle, NSW 2300, Australia
*   Correspondence: alicia.bayonsandoval@csiro.au

Received: 20 September 2019; Accepted: 23 October 2019; Published: 29 October 2019

Featured Application: Kinetic parameters for the development of CaCO$_3$/CaO reactor systems for carbon capture and storage and thermochemical energy storage.

Abstract: The calcium carbonate looping cycle is an important reaction system for processes such as thermochemical energy storage and carbon capture technologies, which can be used to lower greenhouse gas emissions associated with the energy industry. Kinetic analysis of the reactions involved (calcination and carbonation) can be used to determine kinetic parameters (activation energy, pre-exponential factor, and the reaction model), which is useful to translate laboratory-scale studies to large-scale reactor conditions. A variety of methods are available and there is a lack of consensus on the kinetic parameters in published literature. In this paper, the calcination of synthesized CaCO$_3$ is modeled using model-fitting methods under two different experimental atmospheres, including 100% CO$_2$, which realistically reflects reactor conditions and is relatively unstudied kinetically. Results are compared with similar studies and model-free methods using a detailed, comparative methodology that has not been carried out previously. Under N$_2$, an activation energy of 204 kJ mol^{-1} is obtained with the R2 (contracting area) geometric model, which is consistent with various model-fitting and isoconversional analyses. For experiments under CO$_2$, much higher activation energies (up to 1220 kJ mol^{-1} with a first-order reaction model) are obtained, which has also been observed previously. The carbonation of synthesized CaO is modeled using an intrinsic chemical reaction rate model and an apparent model. Activation energies of 17.45 kJ mol^{-1} and 59.95 kJ mol^{-1} are obtained for the kinetic and diffusion control regions, respectively, which are on the lower bounds of literature results. The experimental conditions, material properties, and the kinetic method are found to strongly influence the kinetic parameters, and recommendations are provided for the analysis of both reactions.

Keywords: kinetics; solid–gas reactions; carbonate looping; calcium looping; thermochemical energy storage; carbon capture and storage

---

1. Introduction

The calcium looping cycle is an important process which can reduce greenhouse gases emissions into the atmosphere through a variety of applications. Research into the use of calcium looping (CaL) has found the system to be highly suitable for processes such as thermochemical energy storage (TCES) [1–4] and carbon capture and storage (CCS) [1,5–9].

The CaL cycle consists of two reactions, which are theoretically reversible. In the endothermic decomposition (calcination) reaction, calcium carbonate (CaCO$_3$) absorbs energy to produce a metal oxide (CaO or lime) and CO$_2$. The exothermic carbonation reaction occurs at a lower temperature and/or higher CO_2 partial pressure and releases thermal energy, which can be used to drive a power cycle [1,10]. The reactions are described by the following expression:

$$CaCO_3 \text{ (s)} \rightleftharpoons CaO \text{ (s)} + CO_2 \text{ (g)} \tag{1}$$

where the reaction enthalpy at standard conditions is −178.4 kJ mol^{-1}.

The study of reaction kinetics is highly applicable to the practical implementation of the CaL cycle for use in energy storage or carbon capture. Reaction kinetics are needed for the design of reactors for CaL, for which several system configurations exist. A typical configuration consists of two interconnected circulating fluidized bed reactors, a calciner and carbonator [11–14]. Kinetic reaction models suitable for the conditions of interest in CaL have been specifically studied [12,15–19] and have helped in the design of reactors [11,13,14,20–22].

There are two key challenges associated with performing a kinetic analysis of CaL systems, although the reactions have been extensively studied (i.e., [23–27]). The first is the disparity in the activation energies suggested for both calcination and carbonation [28], and the second is the lack of consensus on the reaction mechanisms [29,30].

The calcination of $CaCO_3$ has been studied using a variety of kinetic analysis methods, including the Coats–Redfern method, the Agarwal and Sivasubramanium method, the Friedman (isoconversional) method, and generalized methods such as pore models and grain models [30]. Calcination activation energies varying between 164 and 225 kJ mol^{-1} have been obtained for various forms of limestone and synthetic $CaCO_3$ under inert atmospheres [30], while studies carried out under CO_2 have produced values of activation energy (E_a, kJ mol^{-1}) which are as high as 2105 kJ mol^{-1} [31]. A consensus on the reaction mechanism has not been established. The intrinsic chemical reaction is considered to be the rate-limiting step by most authors [30]. However, some consider the initial diffusion of CO_2 as a rate-limiting step [32] and some studies indicate that mass transport is significant [33–35].

Numerous studies suggest that the carbonation reaction takes place in two stages: An initial rapid conversion (kinetic control region) followed by a slower plateau (diffusion control region) [29,30]. The most commonly applied reaction models which are used to determine kinetic parameters include the so-call generalized methods, which include shrinking core models, pore models, grain models, and apparent (semi-empirical) models [28,30,36,37]. Modeling of the carbonation reaction considers several mechanisms, such as nucleation and growth, impeded CO_2 diffusion, or geometrical constraints related to the shape of the particles and pore size distribution of the powder [26].

For the initial kinetic control region of carbonation, typical values of E_a are 19–29 kJ mol^{-1} in synthetic CaO or natural lime [30], although slightly larger values (i.e., 39–46 [38,39] and 72 kJ mol^{-1} [37]) have been reported. In the kinetic control region, E_a has been reported to be independent of material properties and morphology (although variation in kinetic parameters for synthetic CaO and natural lime may suggest otherwise [30]), while morphological effects have a greater influence in the diffusion control region, leading to a greater disparity in diffusion activation energies [28]. The diffusion region control region takes place after a compact layer of the product $CaCO_3$ develops on the outer region of the CaO particle at the product–reactant interface [40]. There is a lack of consensus on the diffusion mechanism (gas or solid state diffusion), as well as the diffusing species (CO_2 gas molecules, CO_3^{2-} ions, or O^{2-} ions) [29]. It is suggested that the diffusion mechanism may change depending on whether the sample is porous or non-porous [29] and values of E_a vary between 100 and 270 kJ mol^{-1} in synthetic CaO or natural lime [30].

The effect of experimental conditions on kinetic parameters is also an important consideration to address. Thermogravimetric experiments in the literature are typically performed using an inert atmosphere for calcination and a mixed inert/CO_2 atmosphere for carbonation [30]. However, this may not replicate reactor conditions for CCS and TCES applications. The coupling of concentrating solar power with CaL using a closed CO_2 cycle has been proposed for its high thermoelectric efficiencies [1,23]. This system conducts both calcination and carbonation under pure CO_2 and has been tested in a small number of studies [23,24,41].

This paper presents a kinetic analysis of synthesized $CaCO_3$ and CaO using several methods: Coats–Redfern method, master plots, and generalized methods. Synthetic materials have been used to eliminate the effects of impurities. Experiments were carried out using two types of experiments: calcination under inert atmosphere and carbonation under mixed/inert atmosphere; and calcination and carbonation under 100% CO_2. The kinetic analysis of calcination and carbonation under 100% CO_2 is particularly relevant in some reactor configurations and is rarely carried out [24]. In order to complete the calcination reaction within a short residence time, pilot plants for CO_2 capture currently employ high CO_2 partial pressures in the calciner (70–90%) and high temperatures (>900 °C) [42,43], which are chosen to ensure a practical calcination rate [44]. However, most laboratory-scale tests are performed under inert gas atmospheres or with low concentrations of CO_2, as opposed to under high CO_2 volume concentrations, which accelerates material sintering [23]. This study is unique in that it uses realistic reactor conditions using a sintering-resistant material.

As a means of validation, the kinetic parameters obtained with different methods were compared against each other and with published literature. There are no prior studies which directly compare model-fitting methods such as Coats–Redfern and master plot methods to isoconversional and generalized methods (to the best of the authors' knowledge). The objectives were therefore to: First, compare kinetic parameters and mechanisms obtained by different methods of kinetic analysis, and second, to compare kinetic parameters and mechanisms obtained by different experimental conditions (reaction atmospheres) in order to find the best description of the reaction mechanisms.

2. Materials and Methods

2.1. Material Synthesis

Pechini-synthesized CaO (denoted as P-CaO or P-$CaCO_3$ in its carbonate form) was prepared following the steps described by Jana, de la Peña O'Shea [45], which is also detailed in the experimental procedure of Fedunik-Hofman et al. [24]. First, 48 g of citric acid (CA; $C_6H_8O_7$; Sigma-Aldrich, 99.5% purity; 0.25 mol) was added to 100 mL of distilled water (Milli-Q >18.2 MΩ cm). The mixture was stirred at 70 °C until totally dissolved. Then, 11.8 g of the metal precursor $Ca(NO_3)_2$ (Sigma-Aldrich; 99.0% purity; 0.05 mol) was then added to the solution with a precursor to CA molar ratio of 1:5. The solution was stirred for 3 hours before the temperature was increased to 90 °C. Next, 9.23 mL of ethylene glycol (EG; $HOCH_2CH_2OH$; Sigma-Aldrich; 99.8% purity; 0.033 mol) was added with a molar ratio of CA to EG of 3:2. The resulting solution was further stirred at the same temperature to remove the excess solvent until the solution became a viscous resin. This resin was subsequently dried in an oven at 180 °C for 5 hours. The sample was then ground with an agate mortar and pestle to achieve fine particles and then calcined in a tube furnace. The furnace was programmed to heat from ambient temperature to 400 °C (10 °C min^{-1} heating rate), where it was held for 2 hours, before heating to 900 °C and held for 4 hours. After calcination, the sample was ground again to ensure very fine particles were obtained. The experimental procedure is described in the schematic in Figure 1.

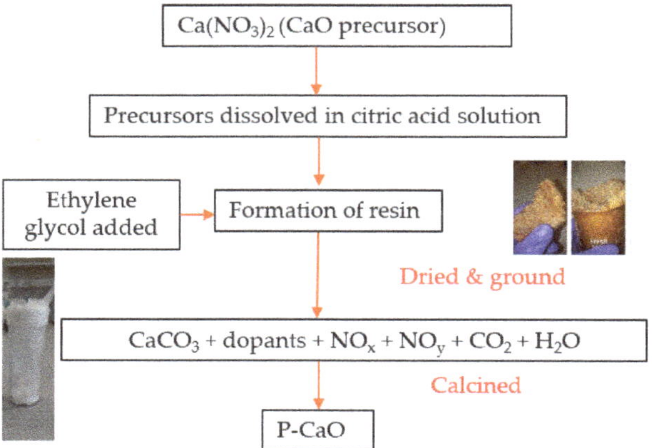

Figure 1. Materials synthesis procedure using Pechini synthesis.

2.2. Cycling Analysis

The extent of conversion with cycling was monitored by thermogravimetric analysis (TGA) using a SETSYS Evolution 1750 TGA-DSC from Setaram. Two different experimental conditions were performed in order to carry out the different kinetic analysis methods.

To carry out a kinetic analysis using the Coats–Redfern method, non-isothermal calcination and carbonation was performed under two atmospheric conditions: Calcination under inert atmosphere (100% N$_2$; maximum temperature 850 °C) followed by carbonation under a mixed atmosphere (25% v/v CO$_2$); and calcination and carbonation under 100% CO$_2$ (maximum temperature 1000 °C). A total gas flow of 20 mL min^{-1} was used for both conditions. Experimental conditions are detailed in Table 1. For each experiment, ~18 mg of P-CaO (powder sample) was placed into a 100 µL alumina crucible and subjected to a temperature ramp from ambient to the maximum temperature, followed by subsequent cooling to ambient temperature. The experiment was repeated for four different heating rates (2.5, 5, 10, and 15 °C min^{-1}).

To implement generalized methods of kinetic analysis (i.e., intrinsic chemical reaction rate model and apparent model), it was necessary to carry out carbonation under isothermal conditions. For this case, P-CaO was heated to the set temperature (see Table 1) under 100% N$_2$ and held isothermally for 20 minutes. After this time, the gaseous atmosphere was adjusted to introduce 25% v/v CO$_2$ and held isothermally for 40 minutes to allow for carbonation. The sample was then cooled to ambient temperature. The heating and cooling rates were 10 °C min^{-1} and a constant gas flow of 20 mL min^{-1} was maintained.

For all experiments, a blank was performed under the same conditions and subtracted from each experiment to correct weight signal drifts. For the decomposition reaction of P-CaCO$_3$ to P-CaO, the extent of reaction, α, was determined using the fractional weight loss:

$$\alpha = \frac{m_i - m(t)}{m_i - m_f} \tag{2}$$

where m_i is the initial mass of the sample in grams, $m(t)$ is the mass of the sample in grams after t minutes and m_f is the final mass of the sample in grams (after calcination). The carbonation conversion of P-CaO to P-CaCO$_3$ was calculated as follows:

$$X(t) = \left(\frac{m(t) - m_i}{m_{X=1}}\right) \times \left(\frac{M_{CaCO_3}}{M_{CaCO_3} - M_{CaO}}\right) \tag{3}$$

where $m(t)$ is the mass of the sample in grams after t minutes under the carbonation atmosphere, m_i is the initial mass of the sample in grams (before carbonation), $m_{X=1}$ is the theoretical mass of the sample in grams after 100% carbonation conversion (43.97% mass gain), and M is the molar mass in g mol^{-1}.

This expression differs slightly from the means of calculating carbonation conversion usually utilized in literature [46]. The rationale is that prior to calcination in the thermogravimetric analyzer, P-CaO may still include $CaCO_3$ phases due to incomplete calcination in the furnace and adsorption of atmospheric CO_2. As a result, $m(t)$ must be measured with reference to the theoretical mass after 100% carbonation conversion (which would occur during the first temperature ramp), as opposed to the initial mass of P-CaO.

Table 1. Experimental conditions for kinetic analyses.

Experiment	Calcination			Carbonation		
	Temp. (°C)	Atmosphere	Isotherm (min)	Temp. (°C)	Atmosphere	Isotherm (min)
Non-isothermal (mixed N_2/CO_2)	850	N_2	0	–	75 v/v% N_2 25 v/v% CO_2	0
Non-isothermal (100% CO_2)	1000	CO_2	0	–	CO_2	0
Isothermal	650/700/750	N_2	20	650/700/750	75 v/v% N_2 25 v/v% CO_2	40

3. Kinetic Analysis Methodology

Methods for solid–gas reactions are either differential, based on the differential form of the reaction model, $f(\alpha)$, or integral, based on the integral form $g(\alpha)$ [47]. The expression for $g(\alpha)$ is the general expression of the integral form:

$$g(\alpha) = \int_0^\alpha \frac{d\alpha}{f(\alpha)} = A \int_0^t exp\left(\frac{-E_a}{RT}\right)h(P)dt \tag{4}$$

where A is the pre-exponential factor (min^{-1}), R is the universal gas constant (kJ mol^{-1} K^{-1}), T is the temperature (K), and $h(P)$ is the pressure dependence term (dimensionless). Non-isothermal experiments employ a heating rate (β in K min^{-1}) which is constant with time, leading to Equations (5) and (6) [48].

$$\frac{d\alpha}{dT} = \frac{A}{\beta} exp\left(\frac{-E_a}{RT}\right)h(P)f(\alpha) \tag{5}$$

$$g(\alpha) = \frac{A}{\beta} \int_0^T exp\left(\frac{-E_a}{RT}\right)h(P)dT \tag{6}$$

Equation (5) is the starting point for differential methods of kinetic analysis, such as the Friedman (isoconversional) method. This method calculates E_a independently of a reaction model and requires multiple data sets [49]. It is referred to as a model-free technique as it avoids assumptions about a reaction mechanism and can identify multi-step reactions. This method is explained in detailed elsewhere [24,30].

Equation (6) is the starting equation for many integral methods of evaluating non-isothermal kinetic parameters with constant β. One example is the Coats–Redfern integral approximation method, referred to as a model-fitting method of kinetic analysis.

Model-fitting methods aim to determine the kinetic parameters of the reaction model integral by fitting obtained data to various known solid-state kinetic models and generally omit the pressure dependence term $h(P)$ [47]. Kinetic models for reaction mechanisms can be categorized using different algebraic functions for $f(\alpha)$ and $g(\alpha)$, which can be found in Fedunik-Hofman et al. [30]. The general principle of model-fitting methods is to minimize the difference between the experimentally measured and calculated data for the given reaction rate expression [47].

3.1. Coats–Redfern Integral Approximation Method

Integral approximation methods replace the integral in Equation (6) with an integral approximation function, $Q(x)$, where x is equal to E_a/RT. The integral can therefore be expressed by the following expression (complete derivations can be obtained in previously published works [30]):

$$g(\alpha) = \frac{AR}{E_a\beta}T^2 exp\left(\frac{-E_a}{RT}\right)Q(x) \tag{7}$$

Many approximations for $Q(x)$ exist and the Coats–Redfern approach is a commonly-used example:

$$Q(x) = \frac{(x-2)}{x} \tag{8}$$

$Q(x)$ changes slowly with values of x and is close to unity [47]. Plots of $\ln\left[g(\alpha)/T^2\right]$ versus $1/T$ (Arrhenius plots) will result in a straight line for which the slope and intercept are E_a and A [50]:

$$ln\left[\frac{g(\alpha)}{T^2}\right] = -\left(\frac{E_a}{R}\right)\left(\frac{1}{T}\right) + ln\left[\frac{AR}{E_a\beta}\right]\left(1 - \frac{2RT_{av}}{E_a}\right) \tag{9}$$

where T_{av} is the average temperature over the course of the reaction [51].

In order to find the kinetic parameters, Arrhenius plots for each $g(\alpha)$ mechanism can be produced. One pair of E_a and A for each mechanism can be obtained by linear data-fitting and the mechanism is chosen from the data fit with the best linear correlation coefficient (R^2) [47].

Coats–Redfern cannot be applied to simultaneous multi-step reactions, although it can adequately represent a multi-step process with a single rate-limiting step [47]. For reactions with consecutive steps, the reaction can be split into multiple steps and model-fitting can be carried out for each step [47].

3.2. Master Plots

Master plots are characteristic curves independent of the condition of measurement which are obtained from experimental data [52]. In this paper, master plots were used to validate the reaction mechanism suggested by the Coats–Redfern method. One type of master plot is the $Z(\alpha)$ method, which is derived from a combination of the differential and integral forms of the reaction mechanism. $Z(\alpha)$ is defined as follows:

$$Z(\alpha) = f(\alpha)g(\alpha) \tag{10}$$

An alternative integral approximation for $g(\alpha)$ to the Coats–Redfern approximation in Equation (9) is $\frac{\pi(x)}{x}$, where $\pi(x)$ is a polynomial function of x, for which several approximations exist [53,54]; i.e.:

$$\pi(x) = \frac{x^3 + 18x^2 + 88x + 96}{x^4 + 20x^3 + 120x^2 + 240x + 120} \tag{11}$$

Experimental values of $Z(\alpha)$ can be obtained with the following equation, which is derived in previously published works [30]:

$$Z(\alpha) = \frac{d\alpha}{dt}T^2\left[\frac{\pi(x)}{\beta T}\right] \tag{12}$$

For each of the reaction mechanisms, theoretical master plot curves can be plotted with Equation (10) (using the approximate algebraic functions $f(\alpha)$ described elsewhere [30]) and experimental values with Equation (12). The experimental values will provide a good fit for the theoretical curve with the same mechanism. As the experimental points have not been transformed into functions of the kinetic models, no prior assumptions are made for the kinetic mechanism [30].

3.3. Compensation Effect

Isoconversional methods, such as the Friedman method, do not explicitly suggest a reaction model. However, if the reaction model can be reasonably approximated with single-step reaction kinetics (if E_a does not vary significantly with α), the reaction mechanism can be determined using the compensation effect, which is described in detail elsewhere [24,30]. In this method, the isoconversional analysis of $CaCO_3$ calcination in Fedunik-Hofman et al. [24] is used to determine the average kinetic parameters (E_0 and A_0), which are used to reconstruct the reaction model; i.e.:

$$g(\alpha) = \frac{A_0}{\beta} \int_0^{T_a} exp\left(\frac{-E_0}{RT}\right) dT \tag{13}$$

Equation (13) is then plotted against the theoretical reaction models (see Section 3.1). In this way, experimental and theoretical curves can be compared to suggest a reaction mechanism.

3.4. Generalized Models for Kinetic Analysis of CaO Carbonation

In CaL, the calcination reaction commonly follows a thermal decomposition process that can be approximated by model-fitting and model-free methods. However, the carbonation reaction is more complex and is believed to be controlled by several mechanisms such as nucleation and growth, impeded CO_2 diffusion, or geometrical constraints related to the shape of the particles and pore size distribution of the powder [26]. Therefore, instead of using a purely kinetics-based approach for analysis of carbonation, functional forms of $f(\alpha)$ have been proposed to reflect these diverse mechanisms. This approach is referred to by Pijolat et al. as a generalized approach [55] and leads to a rate equation with both thermodynamic variables (e.g., temperature and partial pressures) and morphological variables [55].

Numerous studies suggest that the carbonation reaction takes place in two stages: An initial rapid conversion (kinetic control region) followed by a slower plateau (diffusion control region) [29]. These stages are generally analyzed separately using different models, and this approach is taken to perform a kinetic analysis for the carbonation reaction. Interested readers are directed to a previously published review paper for an extended discussion of generalized methods [30].

3.4.1. Chemical Reaction Control Region: Intrinsic Reaction Rate Model

The kinetic control region of the carbonation reaction is generally considered to be limited by heterogeneous surface chemical reaction kinetics, with the driving force for the reaction being the difference between bulk CO_2 pressure and equilibrium CO_2 pressure [56]. This region is typically described by the kinetic reaction rate constant k_s (m^4 mol^{-1} s^{-1}), which is considered to be an intrinsic property of the material [7].

In this paper, the carbonation of P-CaO is modeled using the grain model used by Sun et al. to determine the intrinsic rate constants of the $CaO-CO_2$ reaction [56]. Under kinetic control, the reaction rate, R (min^{-1}), is described by:

$$R = \frac{dX}{dt(1 - X(t))^{2/3}} = 3r \tag{14}$$

where r is the grain model reaction rate (min^{-1}), which is assumed to be constant over the kinetic control region. In integral form the reaction rate can be expressed as:

$$\left[1 - (1 - X(t))^{1/3}\right] = r \times t \tag{15}$$

Plotting $1-(1-X(t))^{\frac{1}{3}}$ versus t will produce a straight line plot due to the constant reaction rate. Values of r can be determined for each isothermal experiment and are taken to represent the true reaction rate at the zero conversion point [56], i.e.:

$$r_0 = r \tag{16}$$

The specific reaction rate can also be expressed in power law form as:

$$R = 3r(1 - X(t))^{-1/3} = 56k_s\left(P_{CO_2} - P_{CO_{2,eq}}\right)^n S \tag{17}$$

where k_s is the intrinsic chemical reaction rate constant (mol m^{-2} s^{-1} kPa^{-n}), $P_{CO_2} - P_{CO_{2,eq}}$ is the difference between the equilibrium and partial pressure of CO_2, n is the reaction order, and S is the specific surface area (m^2 g^{-1}). Sun et al. determined that at CO_2 partial pressures greater than 10 kPa, the reaction order n is zero-order [56]. At time 0, the following relationships are therefore established:

$$k_s = k_0 exp\left(\frac{-E_a}{RT}\right) \tag{18}$$

where k_0 is the pre-exponential factor (mol. m^2 s^{-1}). An Arrhenius plot of slope lnr_0 vs $1/T$ can therefore be produced to determine the kinetic parameters E_a and k_0:

$$ln_{r_0} = ln\left(\frac{56k_0 S_0}{3}\right) - \frac{E_a}{RT} \tag{19}$$

The intrinsic chemical reaction rate constant k_s (calculated using Equation (18)) can be converted to the rate constant k (min^{-1}) by means of:

$$k = k_s S M_{CaO} \tag{20}$$

3.4.2. Diffusion Control Region: Apparent Model

Following the kinetic control region, the reaction slows and becomes diffusion-limited [40]. A simplified generalized model (referred to as an apparent model) has been developed for the carbonation of CaO by Lee [37]. This model does not involve the use of morphological parameters and is described by the following equations:

$$X(t) = X_u\left[1 - exp\left(-\frac{k}{X_u}t\right)\right] \tag{21}$$

$$X(t) = \frac{X_u t}{(X_u/k) + t} \tag{22}$$

where X_u is the ultimate conversion of CaO to $CaCO_3$ and k is the reaction rate (min^{-1}). A constant b is introduced to represent the time taken to attain half the ultimate conversion: $X = X_u/2$ at $t = b$. X_u can then be expressed as:

$$X_u = kb \tag{23}$$

Substituting Equation (23) into Equation (22) leads to:

$$X(t) = \frac{kbt}{b+t} \tag{24}$$

To determine the constants k and b using the least squares method, the equation is written in linear form:

$$\frac{1}{X(t)} = \frac{1}{k}\left(\frac{1}{t}\right) + \frac{1}{kb} \tag{25}$$

An Arrhenius plot can then be produced using the following relationship [57]:

$$ln(k) = -\frac{E_a}{R}\left(\frac{1}{T}\right) + lnA \tag{26}$$

The kinetic parameters E_a (kJ mol^{-1}) and A (min^{-1}) can therefore be obtained, as well as the diffusion rate constant k (min^{-1}).

4. Results of Kinetic Modeling

4.1. Calcination Kinetics Comparative Analysis

Calcination kinetics were modeled using Coats–Redfern, master plots, and isoconversional methods with compensation effects for two different atmospheres, as explained previously (see Section 3.1). The Coats–Redfern analysis under 100% N$_2$ will be discussed first. Figure 2a displays the Coats–Redfern Arrhenius plots for all of the theoretical reaction models (see Section 3.1), while Figure 2b shows selected reaction models with the highest correlation coefficients, which are also presented in Table 2.

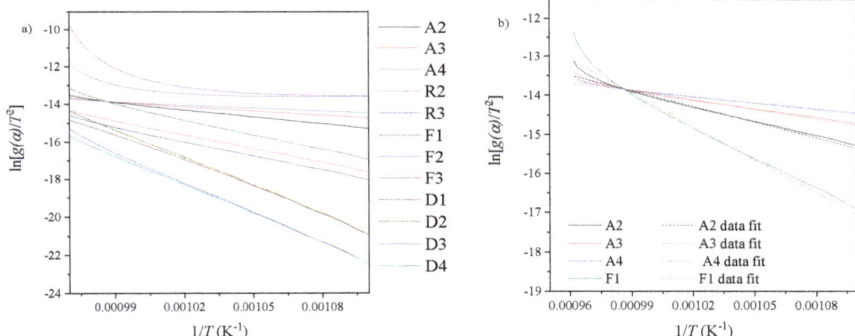

Figure 2. Coats–Redfern Arrhenius plots for P-CaCO$_3$ (10 K min^{-1}) under 100% N$_2$: (**a**) All models; (**b**) selected models with highest R^2 values.

Table 2. Kinetic parameters for calcination obtained from P-CaCO$_3$ (100% N$_2$ atmosphere) (average over four heating rates) compared with literature results (CaCO$_3$ decomposition under inert atmosphere).

Kinetic Model	E_a (kJ mol^{-1})	A (min^{-1})	Correlation Coefficient (R^2)	Material and Reference
A2	108.73	3.5×10^4	0.9924	P-CaCO$_3$ This work
A3	67.11	1.7×10^2	0.9913	This work
A4	46.30	1.1×10^1	0.9900	This work
R2	203.57	2.1×10^9	0.9995	This work
R3	213.12	5.0×10^9	0.9982	This work
F1	233.59	2.3×10^{11}	0.9933	This work
D1	371.14	1.0×10^{18}	0.9985	This work
R2	180.12	1.2×10^7	-	CaCO$_3$ [50]
F1	190.46	3.4×10^7	-	CaCO$_3$ [58]
R3	187.1	-	-	CaCO$_3$ [54]
D1	224.21	3.0×10^4	-	CaCO$_3$ [59]

The kinetic models are classified as follows. A2–4 are the Avrami–Erofeyev nucleation models, where 2–4 refer to the exponential in the algebraic function. R2–3 are the geometric contraction models, where R2 uses a contracting area model and R3 uses a contracting volume model. F1 refers to a first-order reaction model and D1 to a first-order diffusion model.

Correlation coefficients are seen to be very high for all reaction mechanisms, while calculated activation energies vary between 47 and 370 kJ mol^{-1}. By a small margin, the highest correlation coefficient is produced by the R2 geometric model, which determined an E_a of 204 kJ mol^{-1}. These kinetic parameters are typical for the calcination of $CaCO_3$ under an inert atmosphere. Literature results generally produce values of E_a between 180 and 224 kJ mol^{-1} [30], showing a lack of consensus on the calcination reaction model (see Table 2). Physical descriptions of the reaction models referenced in Table 2 are provided in a review of kinetics applied to CaL [30].

As for the experiment under N_2, Coats–Redfern Arrhenius plots (Figure 3) and kinetic parameters for selected reaction models (Table 3) are presented for an experiment under 100% CO_2. Under this atmosphere, calculated activation energies are much higher, ranging from 300 to 600 kJ mol^{-1} when modeled by the nucleation models to 1200 kJ mol^{-1} when fitted with a first-order model. This is due to the high gradients in the Arrhenius plots under 100% CO_2, which mathematically leads to high activation energies. High values of E_a are also obtained in the few literature studies carried out under 100% CO_2 [31,60]. Caldwell et al. attribute this to the fact that the temperature range of decomposition becomes higher and narrower as the percentage of CO_2 increases, resulting in a higher apparent activation energy [31].

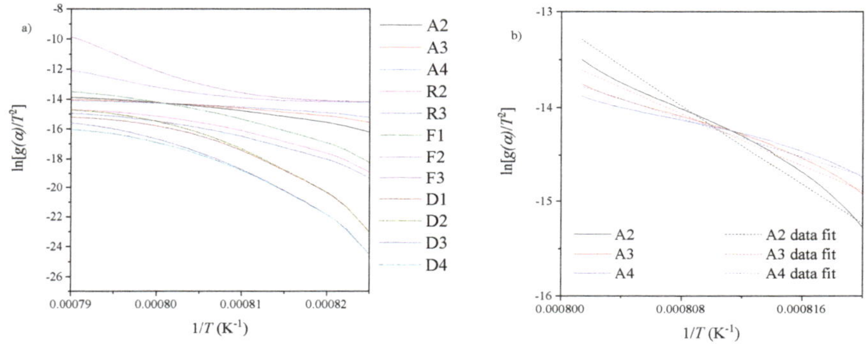

Figure 3. Coats–Redfern Arrhenius plots for P-CaCO$_3$ (10 K min^{-1}) under 100% CO_2: (**a**) All models; (**b**) selected models with highest R^2 values.

Table 3. Kinetic parameters for calcination obtained from P-CaO material (100% CO_2 atmosphere; average over four heating rates) compared with literature results.

Kinetic Model	E_a (kJ mol^{-1})	A (min^{-1})	Correlation Coefficient (R^2)	Material and Reference
A2	599.6	3.4×10^{27}	0.9788	P-CaCO$_3$ This work
A3	419.9	4.5×10^{19}	0.9781	This work
A4	309.9	6.2×10^{14}	0.9774	This work
F1	1219.2	2.9×10^{54}	0.9795	This work
R3	1037.6	3.9×10^{40}	-	CaCO$_3$ [60]
2nd order	2104.6	10^{90}	-	CaCO$_3$ [31]

Examining the correlation coefficients suggests that the reaction mechanism of P-CaCO$_3$ decomposition under CO_2 can be modeled by a first-order reaction or Avrami nucleation A2 and A3 (see Table 3). Correlation coefficients are seen to be lower than the models under N_2 (see Table 2). As for the inert atmosphere, there is no consensus on the reaction model in published literature.

Intraparticle and transport resistances may affect reaction rates and reaction mechanisms and have been suggested as the cause of large values of E_a for CaCO$_3$ calcination [60]. In a decomposition mechanism where the reaction advances inwards from the outside of the particle, smaller particles will decompose more quickly [30,61]. If particle sizes are too large, it has been suggested that the reaction may become limited by mass transport, as opposed to the chemical reaction. However, the influence of particle sizes has been found to be small at high CO_2 partial pressures [33], so intraparticle mass transport effects on the kinetic parameters can be ruled out as the cause of the overestimated kinetic parameters for the experiments under 100% CO_2. [7].

From the comparison of experiments under different atmospheres, it is evident that calculated activation energies can be highly variable. Although there is believed to be a single, intrinsic activation energy for the calcination reaction, values determined experimentally with model-fitting methods have been described as "effective" activation energies [48]. For the experiment under 100% CO_2, in particular, the large disparity in activation energies determined by different reaction models (310–1220 kJ mol^{-1}) and the lower correlation coefficients suggest that model-fitting methods may be unsuitable for calcination under 100% CO_2. An alternative is the use of model-free methods to determine kinetic parameters, as in the previous study of Fedunik-Hofman et al., which used the isoconversional Friedman method to obtain kinetic parameters which vary over the course of the reaction [24].

For both experimental atmospheres, rate constants could not be evaluated using model-fitting methods, as the Coats–Redfern method produced unreasonably large values of $k(T)$. This is because both $k(T)$ and $f(\alpha)$ vary simultaneously under non-isothermal conditions. This causes the Arrhenius parameters for non-isothermal experiments to be highly variable and exhibit a strong dependence on the reaction model (see Table 3) [62]. It has been established that the rate constant cannot be viably predicted by model-fitting methods due to the ambiguity of the kinetic triplet [62]. As suggested previously, isoconversional methods are an alternative means of determining rate constants [24].

As a means of comparison and to aid in a selection of a reaction mechanism, a master plot was produced using the experimental data (non-isothermal experiments carried out under both experimental atmospheres).

Examining the $Z(\alpha)$ master plot in Figure 4a for calcination under N_2, there is good correlation between the experimental data points and the R3 geometric model, which is most pronounced for the experiments performed at 10 and 15 K min^{-1}. The lower heating rates experiments show a poorer fit, which could be the result of heat transfer effects. This result correlates with the mechanism suggested by the Coats–Redfern method, which produces high correlation coefficients for both geometric models R2 and R3, as well as literature results (see Table 2).

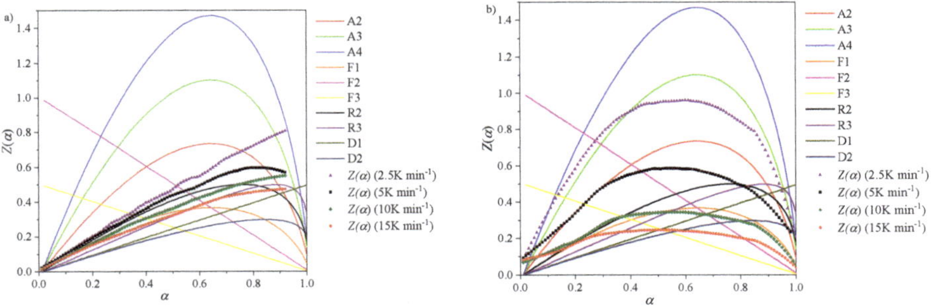

Figure 4. Master plot for P-CaCO3 calcination: (**a**) Under 100% N_2; (**b**) under 100% CO_2.

For calcination under CO_2 (Figure 4b), a good correlation exists for the experimental data points and the Avrami curves A2 and A3, as well as the first-order chemical reaction model F1. Different experimental heating rates are seen to suggest different mechanisms. For example, the 2.5 K min^{-1} heating rate suggests the mechanism is A3, while 5 K min^{-1} suggests A2 and 10 K min^{-1} suggests F1. It is important to note that F1 is a special case of Avrami nucleation [63], as can be seen by the identical shape of the curves, which only differ in magnitude due to the different coefficient n in $g(\alpha)$. Differences in the magnitude of the curves for different heating rates could be the result of heat transfer effects. Therefore, it can be concluded that the calcination reaction under CO_2 is a nucleation process. The reaction mechanism suggested by the master plot generally reflects those in the literature carried out using generalized (non-model-fitting) methods, which are summarized elsewhere [30]. Studies typically suggest that the decomposition reaction can be modeled by either Avrami nucleation [26] or first-order chemical reaction F1 [32,58], which is logical when taking into account that F1 is a special case of nucleation.

As a further means of comparison, the model-fitting results presented previously were compared with published results which employed the isoconversional Friedman method, carried out by the authors using the same experimental data [24]. The results of the analysis are summarized in Table 4, which also presents average values of E_a over the course of the reaction (E_o). A detailed explanation of the assumption of single-step reaction kinetics is presented elsewhere [30]. The equilibrium pressure, P_0 (kPa), of the gaseous product is seen to have a significant influence on $d\alpha/dt$ and, hence, the activation energies under the 100% CO_2 atmosphere [24] (see Table 4).

Table 4. Kinetic parameters for calcination obtained from P-CaO material using the Friedman method [24].

Experimental Atmosphere	CO_2	N_2
E_a (kJ mol^{-1})	430–171	171–147
E_o (kJ mol^{-1})	307	164
A_α (min^{-1})	3.8×10^{34}–2.9×10^5	1.5×10^7–3.7×10^4
A_o (min^{-1})	–	$3.13 \cdot 10^6$
Max k (min^{-1})	0.023	0.012

The compensation effect was used as an alternative means of determining a reaction mechanism for calcination. This method was only applied for calcination under N_2, due to the large variation in kinetic parameters under CO_2 [24] (see Section 3.3). Figure 5 shows a comparison of the experimental and theoretical reaction models, which are compared to suggest a reaction mechanism. The plot shows that the best fitting reaction mechanism is the R2 (contracting area) geometric reaction model, which produces a correlation coefficient of 0.9919. This model also produced a high correlation coefficient in the Coats–Redfern analysis ($R^2 > 0.999$; see Table 2). The activation energy determined with Coats–Redfern is 30% higher than the average Friedman value (213 compared to 164 kJ mol^{-1}), although the values are still within the range obtained in the literature for $CaCO_3$ calcination [30]. In conclusion, the R2 model (contracting area) is the mechanism which most accurately models the calcination of $CaCO_3$ under pure N_2, in agreement with the literature [50], which has been verified using both Coats–Redfern and compensation effect methods.

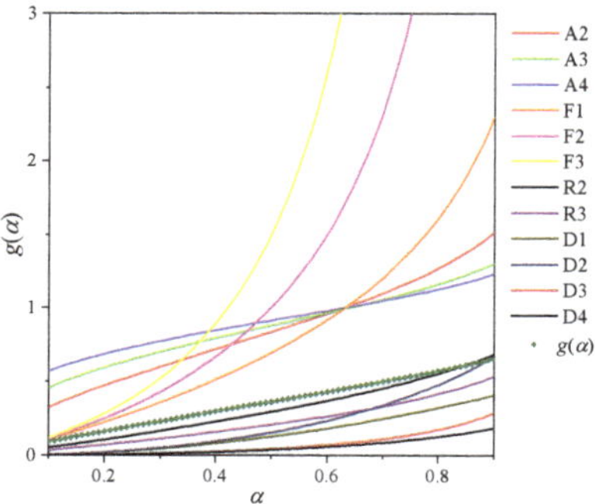

Figure 5. Comparison of theoretical reaction models and experimental model-fitting curve (compensation effect) for calcination under N_2.

It is important to note that the nucleation models and the R2 reaction model are proposed as the mechanisms most consistent with kinetic data for calcination of $CaCO_3$ under 100% CO_2 and N_2, respectively. As set out by Vyazovkin and Wight in their discussion of the solid state reactions and the extraction of Arrhenius parameters from thermal analysis, one of the fundamental tenets of chemical kinetics is that no reaction mechanism can ever be proved on the basis of kinetic data alone [64]. Despite not allowing for complete isolation of the experimental reaction without influence from physical processes (i.e., diffusion, adsorption, desorption), the suggested reaction models are useful for drawing reasonable mechanistic conclusions.

4.2. Carbonation Kinetics Comparative Analysis

In order to implement generalized kinetic models, isothermal experiments were carried out, as detailed in Section 2.2. The chemical reaction control region was modeled using an intrinsic reaction rate model, while the diffusion control region was modeled using an apparent model.

4.2.1. Chemical Reaction Control Region (Intrinsic Reaction Rate Model)

Using the methodology in Section 3.4.1, plots of the carbonation conversion were produced for three isothermal experiments (see Figures 6 and 7). Specific reaction rates for each isothermal experiment are presented in Table 5. The kinetic parameters were obtained by performing data fitting on the Arrhenius plot (Figure 7b). The correlation coefficients produced by linear data fitting are high for all experimental temperatures ($R^2 > 0.99$; see Table 5), which confirms that the chemical reaction region is well-modeled by the intrinsic reaction rate model. An activation energy of 17.45 kJ mol^{-1} was obtained for this region.

Table 5. Carbonation reaction: Intrinsic reaction rate model parameters.

T ($^\circ$C)	r_o (s^{-1} × 10^{-4})	R^2	k_0 (mol m^{-2} s^{-1} × 10^{-5})	k (min^{-1})
650	6.41	0.9950	2.88	0.137
700	7.33	0.9927	3.23	0.173
750	8.00	0.9999	3.59	0.213

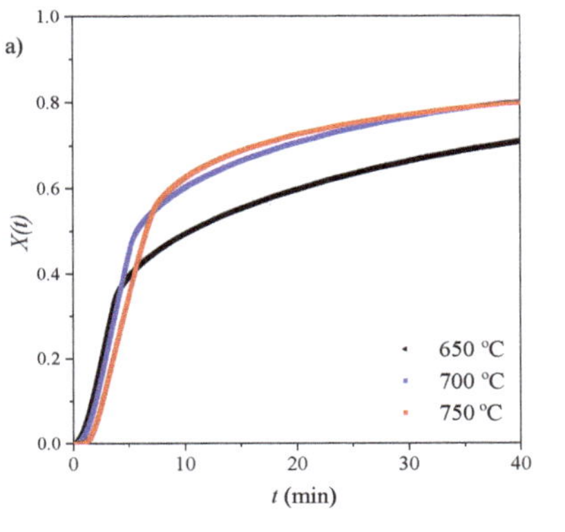

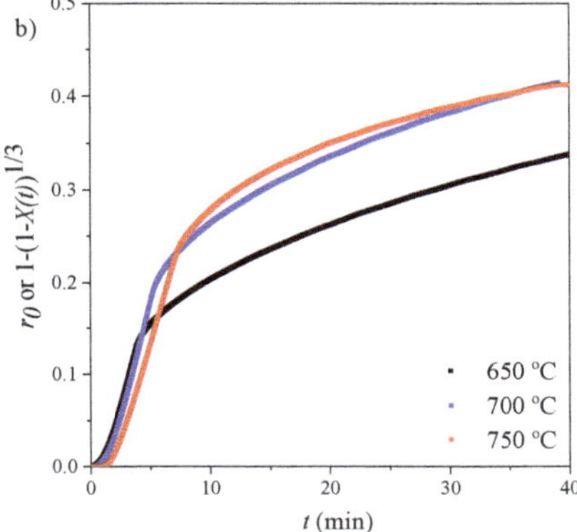

Figure 6. Intrinsic reaction rate model: (**a**) $X(t)$ versus time; (**b**) application of model to carbonation: r_0 versus time.

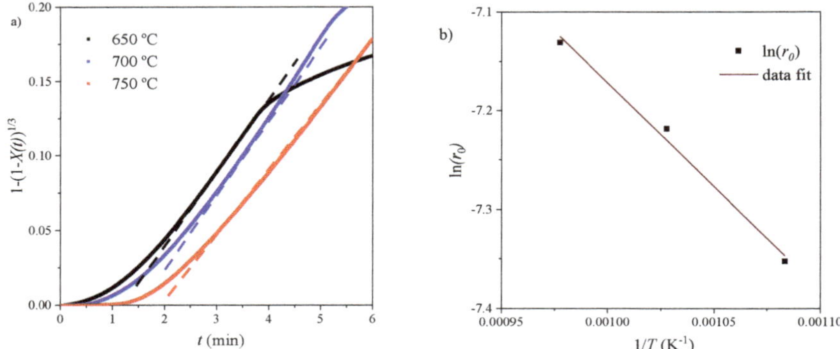

Figure 7. Intrinsic reaction rate model: (**a**) Data fitting with model; (**b**) Arrhenius plot.

4.2.2. Diffusion Control Region (Apparent Model)

Using the methodology described in Section 3.4.2, an apparent model was used to determine the kinetic parameters of the diffusion control region. Data fitting plots are shown in Figures 8 and 9 and apparent model parameters are recorded in Table 6.

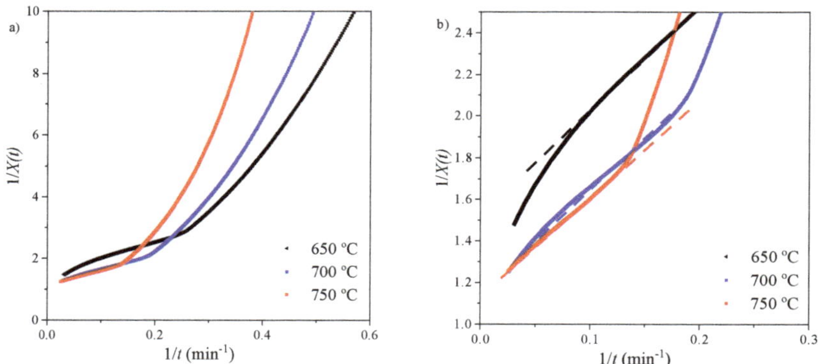

Figure 8. Apparent model: (**a**) $1/X(t)$ versus time; (**b**) data fitting for diffusion control region.

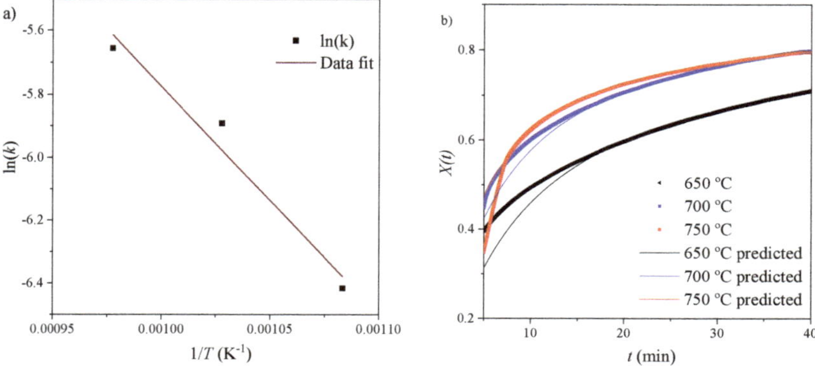

Figure 9. Apparent model: (**a**) Arrhenius plot for diffusion region; (**b**) conversion predicted using apparent model compared to original data.

Table 6. Carbonation reaction: Apparent model parameters.

T (°C)	k (min^{-1})	b (min)	X_u
650	0.098	8.80	0.863
700	0.158	5.54	0.875
750	0.210	4.16	0.872

In order to validate the model, the obtained values of X_u and k were used to calculate $X(t)$ (Equation (23)) and plotted against the isothermal data for the diffusion region. Figure 9b shows a good agreement between experimental data and the model. The kinetic parameters were obtained by performing data fitting on the Arrhenius plot (Figure 9a). An activation energy of 59.95 kJ mol^{-1} was calculated for this region.

Kinetic parameters for both the kinetic and diffusion control regions are presented in Table 7 and compared with selected literature results. Lee states that k can be regarded as the intrinsic chemical reaction rate constant k_s [37], as well as the diffusion constant D, although the different units mean that it cannot be directly compared with reaction rate constants for studies such as those using pore models [17].

Table 7. Carbonation reaction: Kinetic parameters. k_s and D are averages over the temperature range 650, 700, and 750 °C. S is provided in as-synthesized state or after one cycle.

Kinetic Control Region					
Reference	**This Work**	**Sun et al. [56]**	**Lee [37] Data: Bhatia and Perlmutter [25]**	**Lee [37] Data: Gupta and Fan [39]**	**Grasa et al. [17]**
Material	P-CaO	Strassburg limestone	Calcined limestone	Calcined precipitated CaCO$_3$	Katowice limestone
Method	Intrinsic reaction rate model	Intrinsic reaction rate model	Apparent model	Apparent model	Pore model
T (°C)	k (min^{-1})				
650	0.24	0.97	0.925 *	0.858	–
700	0.27	0.81	–	–	–
750	0.30	0.67	–	–	–
S (m^2 g^{-1})	13.8	29	15.6	12.8	12.9 **
E_a (kJ mol^{-1})	17.45	24.0	72.2	72.7	19.2
k_s (m^4 kmol^{-1} s^{-1})	3.34×10^{-6}	4.68×10^{-6}	6.0×10^{-7}	–	5.29×10^{-6}
Diffusion control region					
Method	Apparent model	–	Apparent model	Apparent model	Pore model
T (°C)	k (min^{-1})				
650	0.098	–	0.344 *	0.357	–
700	0.158	–	–	–	–
750	0.210	–	–	–	–
E_a (kJ mol^{-1})	59.95	–	189.3	102.5	163
D (m^2 s^{-1})	–	–	–	4.32×10^{-6}	–

* Calculated at 655 °C [37]; ** calculated from specific surface area (m^2 m^{-3}) assuming ρ (non-porous CaCO$_3$) of 2710 kg m^{-3}.

For both the chemical reaction and diffusion control regions, the reaction rates and values of X_u (see Table 6) tend to increase with increasing temperature, which is consistent with reaction kinetics increasing with higher temperatures (as described by Arrhenius relationships; i.e., Equation (4)). The reaction rates for P-CaO are seen to be slightly slower than those in the literature. The activation energies of P-CaO are also slightly lower than expected for both the kinetic control region (literature values typically fall between 20 and 32 kJ mol^{-1}) and the diffusion region (literature values are typically between 80 and 200 kJ mol^{-1}) [30].

The variation in the activation energies has been attributed to the differences in morphology and texture of the CO$_2$ sorbent [46,56]. While it has been suggested that morphology of CaO does not greatly influence kinetics in the kinetic control region [28], outlying values have been attributed to diffusion limitations in the material [37]. Lee postulated that lower activation energies are associated with microporous materials that are susceptible to pore plugging, which would limit diffusion of CO_2, and is also associated with low carbonation conversion [37]. A study by Grasa et al. determined a much higher diffusion E_a of 163 kJ mol^{-1} for mesoporous limestone [17], which has been associated with resistance to pore plugging [39]. In the case of P-CaO, however, BET analysis showed that the material was largely nonporous, with the majority of pore volume in the macroporous region [24]. Nevertheless, the lower diffusion activation energy could be the result of incomplete carbonation conversion (see Table 6).

Cycling experiments with P-CaO performed elsewhere [24] show good performance with cycling, which suggests that the reaction rates, although initially slower than other materials, will be retained with subsequent cycles. Future analysis could involve calculation of kinetic parameters after multiple isothermal cycles. Carrying out a kinetic analysis with another generalized method is also recommended, as the apparent method uses only a small number of data points to produce an Arrhenius equation.

5. Conclusions

This paper has presented a comparative kinetic analysis of the reactions for CaL, using both inert atmospheres/low CO_2 concentrations, as is typically performed in the literature, and the more rarely-used 100% CO_2 atmosphere. Additionally, it has discussed the validity of model-fitting methods in direct comparison with isoconversional and generalized methods, which has not been carried out previously. The experimental conditions, material properties, and the kinetic method have been found to strongly influence the kinetic parameters for both calcination and carbonation reactions. The main conclusions of the study are summarized as follows:

For the calcination of synthesized $CaCO_3$ under N_2, an E_a of 204 and 233.6 kJ mol^{-1} was obtained by model-fitting using the R2 geometric model and F1 model, respectively, which is consistent with the literature results. Calcination of the same material determined an average E_a of 164 kJ mol^{-1} using an isoconversional method. The compensation effect also suggested that the best reaction model was R2. For calcination under CO_2, much higher values of E_a were obtained, i.e., 310 kJ mol^{-1} with A4 (Avrami nucleation) and 1220 kJ mol^{-1} with the F1 model, while isoconversional results ranged between 170 and 530 kJ mol^{-1}, which was attributed to the major morphological changes in the material under CO_2 [24].

From this study, we can conclude that a multiple heating-rate isoconversional method, such as Friedman, should first be carried out to arrive at non-mechanistic parameters. It will also show whether the kinetic parameters vary over the course of the reaction. Then, if the activation energy does not vary significantly under the experimental atmosphere (as is the case for calcination of $CaCO_3$ under inert gas), the compensation effect can be used to suggest a reaction model. As a means of comparison, a model-fitting method such as Coats–Redfern can be applied over the whole reaction to determine kinetic parameters based on several reaction models. This can be corroborated with master plots. Where applicable, the reaction model with the highest correlation coefficient can be compared with the results of the isoconversional analysis. If the same kinetic parameters are obtained, greater confidence can be placed in the results of the analysis.

If the activation energy does vary significantly under the experimental atmosphere (as is the case for calcination of $CaCO_3$ under CO_2), the Coats–Redfern method is not recommended, as the discrete values of kinetic parameters will not capture the variation in reaction kinetics. Additionally, the compensation effect cannot be applied, as it requires single values of kinetic parameters. In this case, an isoconversional analysis is the best option for obtaining kinetic parameters.

The carbonation of synthesized CaO was modeled using an intrinsic chemical reaction rate model (for the kinetic control region) and apparent model (for the diffusion region), and subsequently compared with the literature. An activation energy of 17.45 kJ mol^{-1} was obtained for the kinetic control region, and 59.95 kJ mol^{-1} for the diffusion region. Lower values of E_a compared to the literature could be attributed to the incomplete carbonation conversion. For this synthesized material, the generalized models proved to model the two carbonation regions with high accuracy. For further studies, material characterization (i.e., scanning/transmission electron microscopy, physisorption characterization, and/or porosimetry) is recommended prior to selecting a generalized kinetic analysis method.

Author Contributions: Conceptualization, L.F.-H. and A.B.; methodology, L.F.-H. and A.B.; formal analysis, L.F.-H.; investigation, L.F.-H.; writing—original draft preparation, L.F.-H.; writing—review and editing, L.F.-H. and A.B.; supervision, A.B. and S.W.D.

Funding: This research received funding from the Australian Solar Thermal Research Institute (ASTRI), a program supported by the Australian Government through the Australian Renewable Energy Agency (ARENA).

Conflicts of Interest: The authors declare no conflicts of interest. The funders had no role in the design of the study; in the collection, analyses, or interpretation of data; in the writing of the manuscript; or in the decision to publish the results.

References

1. Bayon, A.; Bader, R.; Jafarian, M.; Fedunik-Hofman, L.; Sun, Y.; Hinkley, J.; Miller, S.; Lipiński, W. Techno-economic assessment of solid–gas thermochemical energy storage systems for solar thermal power applications. *Energy* **2018**, *149*, 473–484. [CrossRef]

2. Edwards, S.; Materić, V. Calcium looping in solar power generation plants. *Sol. Energy* **2012**, *86*, 2494–2503. [CrossRef]

3. Angerer, M.; Becker, M.; Härzschel, S.; Kröper, K.; Gleis, S.; Vandersickel, A.; Spliethoff, H. Design of a MW-scale thermo-chemical energy storage reactor. *Energy Rep.* **2018**, *4*, 507–519. [CrossRef]

4. Ströhle, J.; Junk, M.; Kremer, J.; Galloy, A.; Epple, B. Carbonate looping experiments in a 1 MWth pilot plant and model validation. *Fuel* **2014**, *127*, 13–22. [CrossRef]

5. André, L.; Abanades, S.; Flamant, G. Screening of thermochemical systems based on solid-gas reversible reactions for high temperature solar thermal energy storage. *Renew. Sustain. Energy Rev.* **2016**, *64*, 703–715. [CrossRef]

6. Stanmore, B.R.; Gilot, P. Review: Calcination and carbonation of limestone during thermal cycling for CO_2 sequestration. *Fuel Process. Technol.* **2005**, *86*, 1707–1743. [CrossRef]

7. Grasa, G.; Abanades, J.C.; Alonso, M.; González, B. Reactivity of highly cycled particles of CaO in a carbonation/calcination loop. *Chem. Eng. J.* **2008**, *137*, 5615–5667. [CrossRef]

8. King, P.L.; Wheeler, V.M.; Renggli, C.J.; Palm, A.B.; Wilson, S.A.; Harrison, A.L.; Morgan, B.; Nekvasil, H.; Troitzsch, U.; Mernagh, T.; et al. Gas–Solid Reactions: Theory, Experiments and Case Studies Relevant to Earth and Planetary Processes. *Rev. Mineral. Geochem.* **2018**, *84*, 1–56. [CrossRef]

9. Boot-Handford, M.E.; Abanades, J.C.; Anthony, E.J.; Blunt, M.J.; Brandani, S.; Mac Dowell, N.; Fernández, J.R.; Ferrari, M.-C.; Gross, R.; Hallett, J.P.; et al. Carbon capture and storage update. *Energy Environ. Sci.* **2014**, *7*, 130–189. [CrossRef]

10. Hanak, D.P.; Manovic, V. Calcium looping with supercritical CO_2 cycle for decarbonisation of coal-fired power plant. *Energy* **2016**, *102*, 343–353. [CrossRef]

11. Shimizu, T.; Hirama, T.; Hosoda, H.; Kitano, K.; Inagaki, M.; Tejima, K. A Twin Fluid-Bed Reactor for Removal of CO_2 from Combustion Processes. *Chem. Eng. Res. Des.* **1999**, *77*, 62–68. [CrossRef]

12. Martínez, I.; Grasa, G.; Murillo, R.; Arias, B.; Abanades, J.C. Kinetics of Calcination of Partially Carbonated Particles in a Ca-Looping System for CO_2 Capture. *Energy Fuels* **2012**, *26*, 1432–1440. [CrossRef]

13. Diego, M.; Martinez, I.; Alonso, M.; Arias, B.; Abanades, J. Calcium looping reactor design for fluidized-bed systems. In *Calcium and Chemical Looping Technology for Power Generation and Carbon Dioxide (CO_2) Capture*; Fennell, P., Anthony, B., Eds.; Woodhead Publishing: Cambridge, UK, 2015; pp. 107–138.

14. Alonso, M.; Rodriguez, N.; Gonzalez, B.; Arias, B.; Abanades, J.C. Capture of CO_2 during low temperature biomass combustion in a fluidized bed using CaO. Process description, experimental results and economics. *Energy Procedia* **2011**, *4*, 795–802. [CrossRef]

15. Bouquet, E.; Leyssens, G.; Schönnenbeck, C.; Gilot, P. The decrease of carbonation efficiency of CaO along calcination–carbonation cycles: Experiments and modelling. *Chem. Eng. Sci.* **2009**, *64*, 2136–2146. [CrossRef]

16. Sun, P.; Grace, J.R.; Lim, C.J.; Anthony, E.J. A discrete-pore-size-distribution-based gas-solid model and its application to the CaO + CO_2 reaction. *Chem. Eng. Sci.* **2008**, *63*, 57–70. [CrossRef]

17. Grasa, G.; Murillo, R.; Alonso, M.; Abanades, J.C. Application of the random pore model to the carbonation cyclic reaction. *AIChE J.* **2009**, *55*, 1246–1255. [CrossRef]

18. Yue, L.; Lipiński, W. Thermal transport model of a sorbent particle undergoing calcination–carbonation cycling. *AIChE J.* **2015**, *61*, 2647–2656. [CrossRef]

19. Yue, L.; Lipiński, W. A numerical model of transient thermal transport phenomena in a high-temperature solid–gas reacting system for CO_2 capture applications. *Int. J. Heat Mass Transf.* **2015**, *85*, 1058–1068. [CrossRef]

20. Lasheras, A.; Ströhle, J.; Galloy, A.; Epple, B. Carbonate looping process simulation using a 1D fluidized bed model for the carbonator. *Int. J. Greenh. Gas Control* **2011**, *5*, 686–693. [CrossRef]

21. Romano, M. Coal-fired power plant with calcium oxide carbonation for postcombustion CO_2 capture. *Energy Procedia* **2009**, *1*, 1099–1106. [CrossRef]

22. Martínez, I.; Murillo, R.; Grasa, G.; Abanades, J.C. Integration of a Ca-looping system for CO_2 capture in an existing power plant. *Energy Procedia* **2011**, *4*, 1699–1706. [CrossRef]

23. Sarrión, B.; Perejón, A.; Sánchez-Jiménez, P.E.; Pérez-Maqueda, L.A.; Valverde, J.M. Role of calcium looping conditions on the performance of natural and synthetic Ca-based materials for energy storage. *J. CO2 Util.* **2018**, *28*, 374–384. [CrossRef]

24. Fedunik-Hofman, L.; Bayon, A.; Hinkley, J.; Lipiński, W.; Donne, S.W. Friedman method kinetic analysis of CaO-based sorbent for high-temperature thermochemical energy storage. *Chem. Eng. Sci.* **2019**, *200*, 236–247. [CrossRef]

25. Bhatia, S.K.; Perlmutter, D.D. Effect of the product layer on the kinetics of the CO_2-lime reaction. *AIChE J.* **1983**, *29*, 79–86. [CrossRef]

26. Valverde, J.M.; Sanchez-Jimenez, P.E.; Perez-Maqueda, L.A. Limestone calcination nearby equilibrium: Kinetics, CaO crystal structure, sintering and reactivity. *J. Phys. Chem. C* **2015**, *119*, 1623–1641. [CrossRef]

27. Ingraham, T.R.; Marier, P. Kinetic studies on the thermal decomposition of calcium carbonate. *Can. J. Chem. Eng.* **1963**, *41*, 170–173. [CrossRef]

28. Salaudeen, S.A.; Acharya, B.; Dutta, A. CaO-based CO_2 sorbents: A review on screening, enhancement, cyclic stability, regeneration and kinetics modelling. *J. CO2 Util.* **2018**, *23*, 179–199. [CrossRef]

29. Kierzkowska, A.M.; Pacciani, R.; Müller, C.R. CaO-based CO_2 sorbents: From fundamentals to the development of new, highly effective materials. *ChemSusChem* **2013**, *6*, 1130–1148. [CrossRef]

30. Fedunik-Hofman, L.; Bayon, A.; Donne, S.W. Kinetics of Solid-Gas Reactions and Their Application to Carbonate Looping Systems. *Energies* **2019**, *12*, 2981. [CrossRef]

31. Caldwell, K.M.; Gallagher, P.K.; Johnson, D.W. Effect of thermal transport mechanisms on the thermal decomposition of $CaCO_3$. *Thermochim. Acta* **1977**, *18*, 15–19. [CrossRef]

32. Rodriguez-Navarro, C.; Ruiz-Agudo, E.; Luque, A.; Rodriguez-Navarro, A.B.; Ortega-Huertas, M. Thermal decomposition of calcite: Mechanisms of formation and textural evolution of CaO nanocrystals. *Am. Mineral.* **2009**, *94*, 578–593. [CrossRef]

33. Khinast, J.; Krammer, G.F.; Brunner, C.; Staudinger, G. Decomposition of limestone: The influence of CO_2 and particle size on the reaction rate. *Chem. Eng. Sci.* **1996**, *51*, 623–634. [CrossRef]

34. Dai, P.; González, B.; Dennis, J.S. Using an experimentally-determined model of the evolution of pore structure for the calcination of cycled limestones. *Chem. Eng. J.* **2016**, *304*, 175–185. [CrossRef]

35. García-Labiano, F.; Abad, A.; de Diego, L.F.; Gayán, P.; Adánez, J. Calcination of calcium-based sorbents at pressure in a broad range of CO_2 concentrations. *Chem. Eng. Sci.* **2002**, *57*, 2381–2393. [CrossRef]

36. Li, Z.; Sun, H.; Cai, N. Rate equation theory for the carbonation reaction of CaO with CO_2. *Energy Fuels* **2012**, *26*, 4607–4616. [CrossRef]

37. Lee, D.K. An apparent kinetic model for the carbonation of calcium oxide by carbon dioxide. *Chem. Eng. J.* **2004**, *100*, 71–77. [CrossRef]

38. Nouri, S.M.M.; Ebrahim, H.A. Kinetic study of CO_2 reaction with CaO by a modified random pore model. *Pol. J. Chem. Technol.* **2016**, *18*, 93–98. [CrossRef]

39. Gupta, H.; Fan, L.-S. Carbonation–Calcination Cycle Using High Reactivity Calcium Oxide for Carbon Dioxide Separation from Flue Gas. *Ind. Eng. Chem. Res.* **2002**, *41*, 4035–4042. [CrossRef]

40. Mess, D.; Sarofim, A.F.; Longwell, J.P. Product layer diffusion during the reaction of calcium oxide with carbon dioxide. *Energy Fuels* **1999**, *13*, 999–1005. [CrossRef]

41. Bouineau, V.; Pijolat, M.; Soustelle, M. Characterisation of the chemical reactivity of a $CaCO_3$ powder for its decomposition. *J. Eur. Ceram. Soc.* **1998**, *18*, 1319–1324. [CrossRef]

42. Manovic, V.; Charland, J.-P.; Blamey, J.; Fennell, P.S.; Lu, D.Y.; Anthony, E.J. Influence of calcination conditions on carrying capacity of CaO-based sorbent in CO_2 looping cycles. *Fuel* **2009**, *88*, 1893–1900. [CrossRef]
43. Valverde, J.M.; Sanchez-Jimenez, P.E.; Perez-Maqueda, L.A. Calcium-looping for post-combustion CO_2 capture. On the adverse effect of sorbent regeneration under CO_2. *Appl. Energy* **2014**, *126*, 161–171. [CrossRef]
44. Symonds, R.T.; Lu, D.Y.; Manovic, V.; Anthony, E.J. Pilot-Scale Study of CO_2 Capture by CaO-Based Sorbents in the Presence of Steam and SO_2. *Ind. Eng. Chem. Res.* **2012**, *51*, 7177–7184. [CrossRef]
45. Jana, P.; de la Peña O'Shea, V.A.; Coronado, J.M.; Serrano, D.P. Cobalt based catalysts prepared by Pechini method for CO_2-free hydrogen production by methane decomposition. *Int. J. Hydrogen Energy* **2010**, *35*, 10285–10294. [CrossRef]
46. Zhou, Z.; Xu, P.; Xie, M.; Cheng, Z.; Yuan, W. Modeling of the carbonation kinetics of a synthetic CaO-based sorbent. *Chem. Eng. Sci.* **2013**, *95*, 283–290. [CrossRef]
47. Vyazovkin, S.; Burnham, A.K.; Criado, J.M.; Pérez-Maqueda, L.A.; Popescu, C.; Sbirrazzuoli, N. ICTAC Kinetics Committee recommendations for performing kinetic computations on thermal analysis data. *Thermochim. Acta* **2011**, *520*, 1–19. [CrossRef]
48. Urbanovici, E.; Popescu, C.; Segal, E. Improved iterative version of the Coats-Redfern method to evaluate non-isothermal kinetic parameters. *J. Therm. Anal. Calorim.* **1999**, *58*, 683–700. [CrossRef]
49. Sanders, J.P.; Gallagher, P.K. Kinetic analyses using simultaneous TG/DSC measurements Part I: Decomposition of calcium carbonate in argon. *Thermochim. Acta* **2002**, *388*, 115–128. [CrossRef]
50. Ninan, K.N.; Krishnan, K.; Krishnamurthy, V.N. Kinetics and mechanism of thermal decomposition of insitu generated calcium carbonate. *J. Therm. Anal.* **1991**, *37*, 1533–1543. [CrossRef]
51. Vyazovkin, S.; Wight, C. Model-Free and Model-Fitting Approaches to Kinetic Analysis of Isothermal and Nonisothermal Data. *Thermochim. Acta* **1999**, *340*, 53–68. [CrossRef]
52. Criado, J.M.; Málek, J.; Ortega, A. Applicability of the master plots in kinetic analysis of non-isothermal data. *Thermochim. Acta* **1989**, *147*, 377–385. [CrossRef]
53. Vyazovkin, S. Estimating Reaction Models and Preexponential Factors. In *Isoconversional Kinetics of Thermally Stimulated Processes*; Springer International Publishing: Berlin/Heidelberg, Germany, 2015; pp. 41–50.
54. Yue, L.; Shui, M.; Xu, Z. The decomposition kinetics of nanocrystalline calcite. *Thermochim. Acta* **1999**, *335*, 121–126. [CrossRef]
55. Pijolat, M.; Favergeon, L. Kinetics and mechanisms of solid-gas reactions. In *Handbook of Thermal Analysis and Calorimetry*, 2nd ed.; Sergey, V., Nobuyoshi, K., Christoph, S., Eds.; Elsevier: Amsterdam, The Netherlands, 2018; Volume 6, pp. 173–212.
56. Sun, P.; Grace, J.R.; Lim, C.J.; Anthony, E.J. Determination of intrinsic rate constants of the CaO-CO_2 reaction. *Chem. Eng. Sci.* **2008**, *63*, 47–56. [CrossRef]
57. Lee, D.K.; Min, D.Y.; Seo, H.; Kang, N.Y.; Choi, W.C.; Park, Y.K. Kinetic Expression for the Carbonation Reaction of K_2CO_3/ZrO_2 Sorbent for CO_2 Capture. *Ind. Eng. Chem. Res.* **2013**, *52*, 9323–9329. [CrossRef]
58. Maitra, S.; Chakrabarty, N.; Pramanik, J. Decomposition kinetics of alkaline earth carbonates by integral approximation method. *Ceramica* **2008**, *54*, 268–272. [CrossRef]
59. Maitra, S.; Bandyopadhyay, N.; Das, S.; Pal, A.J.; Pramanik, M.J. Non-isothermal decomposition kinetics of alkaline earth metal carbonates. *J. Am. Ceram. Soc.* **2007**, *90*, 1299–1303. [CrossRef]
60. Gallagher, P.K.; Johnson, D.W. Kinetics of the thermal decomposition of $CaCO_3$ in CO_2 and some observations on the kinetic compensation effect. *Thermochim. Acta* **1976**, *14*, 255–261. [CrossRef]
61. Lee, J.-T.; Keener, T.; Knoderera, M.; Khang, S.-J. Thermal decomposition of limestone in a large scale thermogravimetric analyzer. *Thermochim. Acta* **1993**, *213*, 223–240. [CrossRef]
62. Vyazovkin, S.; Wight, C.A. Isothermal and non-isothermal kinetics of thermally stimulated reactions of solids. *Int. Rev. Phys. Chem.* **1998**, *17*, 407–433. [CrossRef]
63. Khawam, A.; Flanagan, D.R. Solid-state kinetic models: Basics and mathematical fundamentals. *J. Phys. Chem. B* **2006**, *110*, 17315–17328. [CrossRef]
64. Vyazovkin, S.; Wight, C.A. Kinetics in Solids. *Annu. Rev. Phys. Chem.* **1997**, *48*, 125–149. [CrossRef] [PubMed]

Article

An Experimental and Numerical Study of CO_2–Brine-Synthetic Sandstone Interactions under High-Pressure (P)–Temperature (T) Reservoir Conditions

Zhichao Yu [1,2], **Siyu Yang** [1,3,*], **Keyu Liu** [4], **Qingong Zhuo** [1,2] and **Leilei Yang** [5]

[1]   PetroChina Exploration and Development Research Institute, Beijing 100083, China
[2]   Key Laboratory of Basin Structure and Hydrocarbon Accumulation, CNPC, Beijing 100083, China
[3]   Department of Middle East E & P, CNPC, Beijing 100083, China
[4]   School of Geosciences, China University of Petroleum, Qingdao 266580, China
[5]   Enhanced Oil Recovery Research Institute, China University of Petroleum, Beijing 100083, China
*   Correspondence: yangsiy@petrochina.com.cn; Tel.: +86-10-8359-2410; Fax: +86-10-8359-2410

Received: 28 June 2019; Accepted: 12 August 2019; Published: 15 August 2019

Abstract: The interaction between CO_2 and rock during the process of CO_2 capture and storage was investigated via reactions of CO_2, formation water, and synthetic sandstone cores in a stainless-steel reactor under high pressure and temperature. Numerical modelling was also undertaken, with results consistent with experimental outcomes. Both methods indicate that carbonates such as calcite and dolomite readily dissolve, whereas silicates such as quartz, K-feldspar, and albite do not. Core porosity did not change significantly after CO_2 injection. No new minerals associated with CO_2 injection were observed experimentally, although some quartz and kaolinite precipitated in the numerical modelling. Mineral dissolution is the dominant reaction at the beginning of CO_2 injection. Results of experiments have verified the numerical outcomes, with experimentally derived kinetic parameters making the numerical modelling more reliable. The combination of experimental simulations and numerical modelling provides new insights into CO_2 dissolution mechanisms in high-pressure/temperature reservoirs and improves understanding of geochemical reactions in CO_2-brine-rock systems, with particular relevance to CO_2 entry of the reservoir.

Keywords: CO_2 sequestration; physical simulation; Numerical modelling; dissolution; precipitation

---

1. Introduction

Carbon dioxide emissions from fossil-fuel combustion are projected to increase from 13 Gt yr^{-1} in 2010 to 20–24 Gt yr^{-1} in 2050 [1]. CO_2 capture and storage (CCS) technologies beneficially affect the lifecycle of greenhouse gases emitted from fossil-fuel power plants [2,3], with CCS expected to account for up to 19% of global CO_2 emission reductions by 2050, making it the most significant technology worldwide in this area [4]. Suitable geological formations for CCS include depleted oil and gas reservoirs, un-mineable coal seams, salt caverns, and deep saline aquifers [5,6]. After CO_2 injection, the initial physico-chemical equilibrium between saline formation fluid and reservoir rocks can be disturbed by the triggering of reactions between CO_2, fluid (brine), and reservoir rock [2]. Such interactions could lead to the dissolution of carbonates, feldspars, and clay cement in the aquifers [7,8]. In the absence of dynamic forces, such mineral dissolution could increase porosity and permeability by etching new pore spaces or widening narrow pore channels, temporarily increasing injectivity [9,10]. However, while sequestration of CO_2 in carbonate minerals can contribute to long-term storage security [11], rapid mineral dissolution, especially of carbonates, could corrode caprocks, wellbores, and fault seals, potentially leading to migration of CO_2 into overlying formations. Study of CO_2-fluid-rock interactions is thus crucial for us to understand the physico-chemical processes involved.

Laboratory experiments can reveal the mineralogical and chemical changes resulting from CO_2-brine-rock interactions, how they impact the lithological porosity and permeability of the geological sequence, and the effects on CCS potential [12–15]. However, experiments are limited to short-term effects of CO_2 injection, whereas CCS is a long-term geochemical issue. Numerical modelling or simulation is useful for longer-tern studies. Several reactive geochemical transport models have been developed to simulate CCS, including NUFT [16], PFLOTRAN [17], CMG-GEM [18], STOMP [19], and TOUGHREACT [20,21]. The TOUGHREACT program has been widely used in studying geological CO_2 sequestration [22–26]. However, simulations are less reliable without the availability of parameters derived from laboratory studies, so a combination of physical experiments and numerical simulation is the optimal choice for investigating the geochemical effects following CO_2 injection.

In this study, both laboratory experiments (physical simulation) and numerical modelling were used to study geochemical interactions between CO_2-induced fluids and reservoir rock during CCS. In the physical simulation, synthetic cores with composition consistent with geological samples were used to avoid interference from other geological factors such as sedimentary processes and diagenesis. The numerical simulation involved the same conditions of sample compositions, temperature, pressure, and fluid composition, with the two simulation types being mutually authenticating. Both numerical and physical simulations were used to document the process of short-term geochemical interactions after CO_2 injection. A consistency of results would indicate the reliability of the simulations, with outcomes expected to be similar to those pertaining to actual geological conditions.

2. Samples and Methods

2.1. Sample Descriptions

Six synthetic sandstones were prepared for the physical simulation, with mineralogical compositions similar to sandstones of the Cretaceous Bashijiqike Formation (K_1bs) of the Kuqa Depression, Tarim Basin, and western China. In order to identify mineralogical compositions of K_1bs sandstones, the sandstone samples were prepared in thin sections and examined petrographically by point counting 300 to 400 points per section. In addition, these sandstones were also measured using quantitative X-ray diffraction analysis (D/max2500, Rigaku, Tokyo, Japan), which can provide quantitative mineralogical results within ±0.1 weight percentage (wt. %). The detail analysis processes can be found in Yu et al. (2012) [15]. The analytical results indicated that K_1bs sandstones are fine- to medium-grained lithic sandstones with particle sizes of 0.25~0.5 mm, comprising mainly quartz (average ~37.5 wt. %), plagioclase (~20.8 wt. %), K-feldspar (~23.3 wt. %), calcite (~9.5 wt. %), dolomite (~7.4 wt. %), and kaolinite (~1.5 wt. %) (Table 1). According to Yu et al. (2015) [27], the K_1bs reservoir sandstones were at the stage of mesogenetic diagenetic phase. Then we used the fine- to medium-grained mineral powders (particle size of 0.25~0.5 mm), having the above-mentioned mineralogical compositions, to reconstruct the six synthetic cores under the condition of mesogenetic diagenesis.

Table 1. The mineral composition of synthetic core samples.

Mineral Types	Quartz	K-Feldspar	Albite	Calcite	Kaolinite	Dolomite
Content (wt. %)	37.5	23.3	20.8	9.5	1.5	7.4

2.2. Physical Experimental Conditions

The experimental condition is outlined as the following: (1) 48.45 MPa back-pressure (pore fluid pressure), (2) 60 MPa confining pressure, (3) 150 °C reaction autoclave temperature (formation temperature). The injection solutions were prepared by dissolving NaCl in deionized water saturated with CO_2 at 150 °C and 48.45 MPa, similar to actual K_1bs conditions. The injection solutions had a salinity of 14,182 mg L^{-1}, approximating K_1bs formation water. Here we only used the NaCl solution as the injection fluids and did not employ the imitate reservoir brines, because an amount of divalent cations, such as Ca^{2+} and Fe^{2+}, were present in the reservoir bines. After CO_2 induced fluid injection into the autoclaves, some carbonates will precipitate and affect experimental results. Thus, pure NaCl solution, having a similar salinity with K_1bs formation water, would be the most appropriate.

Under the experimental condition (P = 48.45 MPa and T = 150 °C), the injection solution was saturated with CO_2. For the solution with a salinity of 14,000 mg L^{-1}, the solubility of CO_{-2} was 1.5451 mol/Kg, according to the CO_{-2} solubility in bine of Duan and Sun (2003) [28]. During the experiment, the injected V_{brine} (brine volume), and the volume of CO_2 injected into the cylinder was V_{CO_2}. Based on the equation of sate (EOS) for gas, PV = ZnRT, where Z is the compressibility, n is the mole number of CO_2 (n_{CO_2}) in the injection solution, R is gas constant, and T is temperature, we can obtain the volume of CO_2 ($V_{CO_2 soluble}$) dissolved in the injection solutions under the experimental condition (P = 48.45 MPa and T = 150 °C). Thus we can calculate the volume of the CO_2 gas cap ($V_{CO_2 gc}$) in the intermediate container. The derivation is as follows:

$$V_{CO_2 gc} = V_{CO_2} - V_{CO_2 soluble} \tag{1}$$

$$V_{CO_2} = 1030 - V_{kerosene} - V_{brine} \tag{2}$$

$$V_{CO_2 soluble} = Zn_{CO_2} RT/P \tag{3}$$

Based on the above, it is possible to calculate the volume of CO_2 in the gas cap of the intermediate container, which was ca.190.56 mL. Therefore the brine was CO_2 saturated throughout the experiments.

2.3. Experimental Apparatus

The physical simulation experiment was conducted at the Key Laboratory of Basin Structure and Hydrocarbon Accumulation of the China National Petroleum Corporation, Beijing, China. A reservoir diagenesis modelling system with six identical reaction autoclaves was employed (Figure 1). The system includes six modules: heating furnace, pressure system, fluid-injection system, sampling system, control panel, and auxiliary system. In addition, a corrosion-resistant HP/HT CFR-50-100 cylinder (1030 mL) from TEMC, USA was used as an intermediate container for storing the CO_2-bearing experimental solution. The six reaction autoclaves (Huaxing Company, Nan tong, Jiangxi Province, China) have a working pressure of 165 MPa and temperature of 300 °C. The pressure and fluid injection systems are controlled by the injection syringe pump and a back-pressure regulator. The 100DX syringe pump (Teledyne ISCO, Lincoln NE, USA) was used to control the fluid injection system, which consists of two separate systems (A and B), each of which has a capacity of 103 mL (Figure 1). It is capable of injecting at rates of 0.001~60 mL/min, with a precision of 0.5% of set point. The pump can handle pressure from 0.1 to 68.97 MPa. The advantage of this pump is its capability of continuous injection of any fluids including supercritical CO_2. The pore-fluid pressure was controlled by the back-pressure regulator (DBRP-005, Honeywell, USA), which has a high precision and operating pressure range up to 51.72 MPa. All experimental parameters including the injection pressure, pore fluid pressure, and temperature were monitored.

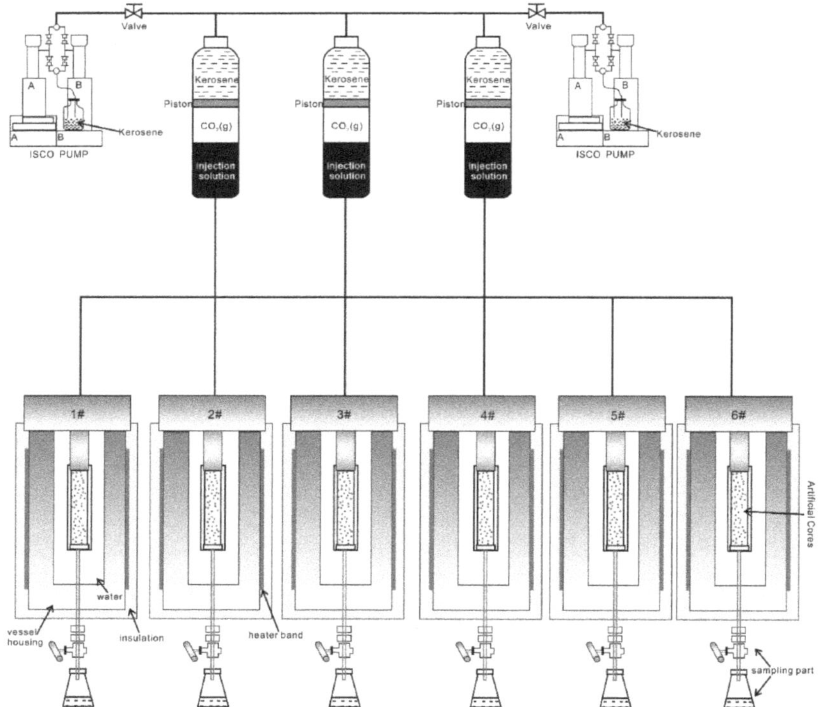

Figure 1. Schematic diagram of CO_2-formation water-rock physical experiment.

2.4. Physical Simulation Workflow and Analysis

The experiment was undertaken in two steps: preparation of the synthetic core, and geochemical reaction between the core and CO_2 fluids. During the first step, the selected mineral powder (particle size 0.25~0.5 mm) was blended with distilled water and placed in six columnar autoclaves (diameter 3.0 cm, length 11 cm, volume 77.7 cm^3). The six core samples (# 1 to # 6) were used in the experiment over 5 days under P/T conditions equivalent to mesogenetic diagenesis (Figure 2). The syringe-pump injection system injected synthetic formation water saturated with CO_2 into the six synthetic core pores at 150 °C and 48.45 MPa, after which temperature and pressure were kept constant for 4 d (# 1), 7 d (# 3), 10 d (# 4), 13 d (# 5), and 16 d (# 6), while # 2 was used as a blank.

During the experiment, the temperature and pressure of each autoclave were monitored automatically by the control system. After reaction, core and fluid samples were analyzed for ion contents, mineralogical changes, and porosity. The producing fluid was measured for its pH values using an Orion4 STAR acidity meter from Thermo within 6 h of each sampling. The ionic compositions of the water were analyzed after being spiked with 1 mol/L HCl in order to avoid carbonates precipitation, and measured using an OPTIMA 7300DX ICP-OES (Inductively Coupled Plasma–Optical Emission Spectrometry) with an analytical precision of 10^{-3}~10^{-9}. Mineralogical changes were examined using a JSM6700F scanning electron microscope from JEOL with EDS (Energy Dispersive Spectrometer) from INCA software (Oxford Company, Oxford, England). The porosity changes were analyzed using visual porosity estimation, which is an image analysis technique. Firstly, core samples were impregnated with blue epoxy and then polished and made into casting thin sections. Then, combined high-resolution images of these thin sections were taken under the optical petrographic microscope; the image analysis software can delineate different types of porosity and calculate the percentages of these porosities in the thin sections with an accuracy of up to 0.01%.

Figure 2. Synthetic core samples made by the physical experiment.

2.5. Numerical Simulation

The program TOUGHREACT was used in the numerical simulations. This program is a non-isothermal, multiphase reactive transport simulation code that was used here to simulate fluid-rock interactions [21]. The kinetic data used during the simulation are shown in Table 2.

Table 2. List of minerals considered and parameters for calculating the kinetic rate constants.

Mineral	A/(cm²/g)	Geochemical Kinetic Rate Constants		
		K_{25}/(mol/(m²·s))	E_a/(kJ/mol)	^{n} H$^+$
Quartz	9.8			
Kaolinite	151.6	4.9×10^{-12}	65.9	0.8
Illite	151.6	1.0×10^{-11}	23.6	0.3
K-feldspar	9.8	8.7×10^{-11}	51.7	0.5
albite	9.8	6.9×10^{-11}	65.0	0.5
Chlorite	9.8	7.8×10^{-12}	88.0	0.5
Calcite	9.8	5.0×10^{-1}	14.4	1.0
Dolomite	9.8	6.5×10^{-4}	36.1	0.5
Siderite	9.8	6.5×10^{-4}	36.1	0.5
Ankerite	9.8	1.6×10^{-4}	36.1	0.5
Dawsonite	9.8	1.6×10^{-4}	36.1	0.5
Magnesite	9.8	4.2×10^{-7}	14.4	1.0
Pyrite	12.9	3.0×10^{-8}	56.9	−0.5

Note that: (1) All rate constants are listed for dissolution; (2) A is specific surface area, k_{25} is kinetic constant at 25 °C, E_a is activation energy, and n is the power term (Equation (A1) in Appendix A); (3) The power terms n for acid mechanisms are with respect to H$^+$. Data from Palandri and Kharaka (2004) [29].

According to the columnar autoclaves employed by the physical simulation, three identical cubic grids with volumes of 77.7 cm³ were used to construct the model (Figure 3). The upper and lower grids were used as boundary cells, while the middle grid was the objective model grid for simulating the processes of injection and sampling. The numerical model simulated six autoclave reactions, corresponding to the laboratory experiment, with the same mineralogical cores, temperature, pressure, and pore fluids. We used the simulation duration to mimic the six numbered autoclaves. The entire simulation ran for 16 days with intermittent sampling on day 0, 4, 7, 10, 13, and 16, corresponding to the physical simulation. At the start of simulation (Day zero), the numerical model had an initial mineralogical composition and visual porosity, which corresponded to Autoclave # 2. In the same way, Day 4 corresponded to Autoclave # 1, Day 7 corresponded to Autoclave # 3, Day 10 corresponded to Autoclave # 4, Day 13 corresponded to Autoclave # 5, and Day 16 corresponded to Autoclave # 6. Accordingly, these results of different simulation duration from the numerical models can be used for comparison with the results from the physical simulations. The boundary cell here is an "inactive" element, whose thermodynamic conditions do not change at all from fluid or heat exchange with finite-size blocks (numerical model cell) in the flow domain. The boundary cell can confine geochemical interactions that only occur in the numerical model, which makes the results more reasonable.

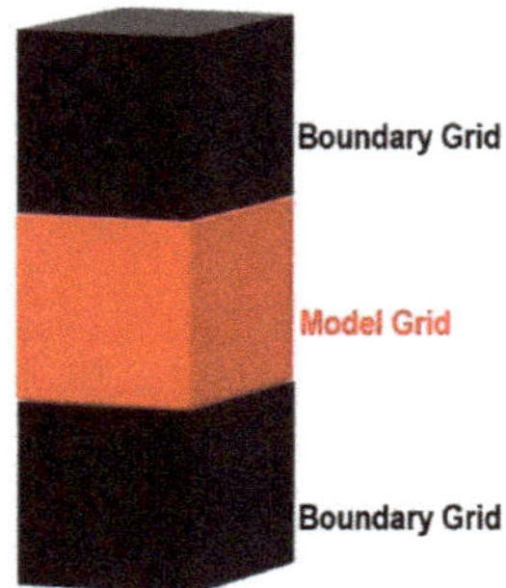

Figure 3. Schematic diagram of CO_2-formation water-rock numerical simulation.

3. Results

3.1. Changes in Fluid Chemistry

Results of physical and numerical analyses of reaction products are summarized in Table 3. Significant changes in solution chemistry were observed in both sets of experiments (Figure 4). In the physical simulation, the pH continued to increase during the 16 d of the experiment, from 5.86 to 6.44 (Figure 4). In the numerical simulation, the pH first decreased to ~2.8 within 12 d, then increased to 4.6 over the next 4 d (Figure 4).

Fluid Si and K contents show similar changes in both simulations (Figure 4), with concentrations continuing to increase with reaction time (Figure 4). Fluid Ca and Mg concentrations increased with reaction time in the physical simulation, but were more constant in the numerical simulation (Figure 4). The Al content exhibited a distinct trend (Figure 4), reaching maximum values after 7 d and 10 d for the physical and numerical simulations, respectively, and then decreasing during further reaction. Absolute value of ion concentration differed between the simulations, with the numerical simulation set generally being higher, not including pH and Al (Figure 4).

Table 3. Chemical composition of outlet solutions.

	Reaction Time (d)	pH	K	Si	Ca	Mg	Al
			mol/L	mol/L	mol/L	mol/L	mol/L
Physical Simulation	0	5.86	0.000000	0.000000	0.000000	0.000000	0.000000
	4	5.97	0.000046	0.001000	0.001360	0.000554	0.000148
	7	5.94	0.000810	0.001004	0.002500	0.000879	0.000667
	10	6.01	0.000854	0.001832	0.003125	0.001079	0.000852
	13	6.27	0.001987	0.002943	0.006825	0.002396	0.000500
	16	6.44	0.002000	0.004500	0.009575	0.004583	0.000200

Table 3. *Cont.*

	Reaction Time (d)	pH	K	Si	Ca	Mg	Al
			mol/L	mol/L	mol/L	mol/L	mol/L
Numerical simulation	0	4.01	0.000000	0.000000	0.000000	0.000000	0.000000
	3	3.93	0.000252	0.000625	0.000631	0.000298	0.000182
	4	3.09	0.000616	0.001471	0.001344	0.000648	0.000450
	5	2.94	0.000964	0.002277	0.002119	0.000985	0.000664
	6	2.87	0.001226	0.002881	0.002733	0.001238	0.000795
	7	2.84	0.001417	0.003323	0.003178	0.001423	0.000825
	9	2.85	0.001678	0.003922	0.003846	0.001676	0.000675
	10	2.85	0.001749	0.004083	0.004035	0.001744	0.000656
	11	2.88	0.001818	0.004243	0.004192	0.001812	0.000518
	12	3.41	0.002036	0.004758	0.004568	0.002029	0.000435
	13	4.08	0.002144	0.005024	0.004732	0.002137	0.000321
	14	4.40	0.002201	0.005174	0.004819	0.002196	0.000211
	15	4.57	0.002245	0.005295	0.004891	0.002242	0.000194
	16	4.68	0.002282	0.005395	0.004953	0.002281	0.000100

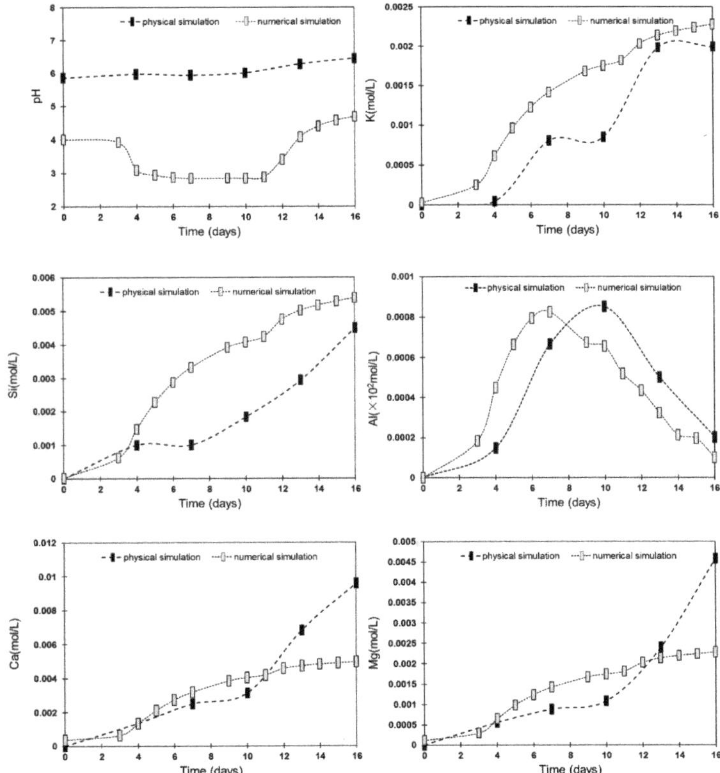

Figure 4. Changes of pH and typical ion concentrations over the physical and numerical simulations.

3.2. Changes in Mineral Morphology during the Physical Simulation

Scanning electron microscope (SEM) analyses of core samples before and after physical simulations showed that minerals such as quartz, K-feldspar, albite, and dolomite dissolved after CO_2 injection, with feldspar and dolomite showing pronounced dissolution and quartz weak dissolution. Before the experiment, mineral surfaces of quartz grains were generally smooth with terraced growth patterns (Figure 5A), with dissolution effects and corrosion pits being evident afterwards (Figure 5B). Initially, the albite surface was relatively flat and exhibited no obvious dissolution, but dissolution pits and fissures along cleavage surfaces were evident after the experiment (Figure 5C,D). K-feldspar was partially dissolved after the experiment, with the formation of corrosion pits (Figure 5E,F). The dissolution of K-feldspar was stronger than that of quartz and weaker than that of albite. Carbonates exhibited stronger dissolution than silicates, with entire dolomite particles being dissolved into a cloud-like phase and showing a paste-like flow structure (Figure 5G,H). Calcite was not observed after the experiment, indicating that it was completely dissolved.

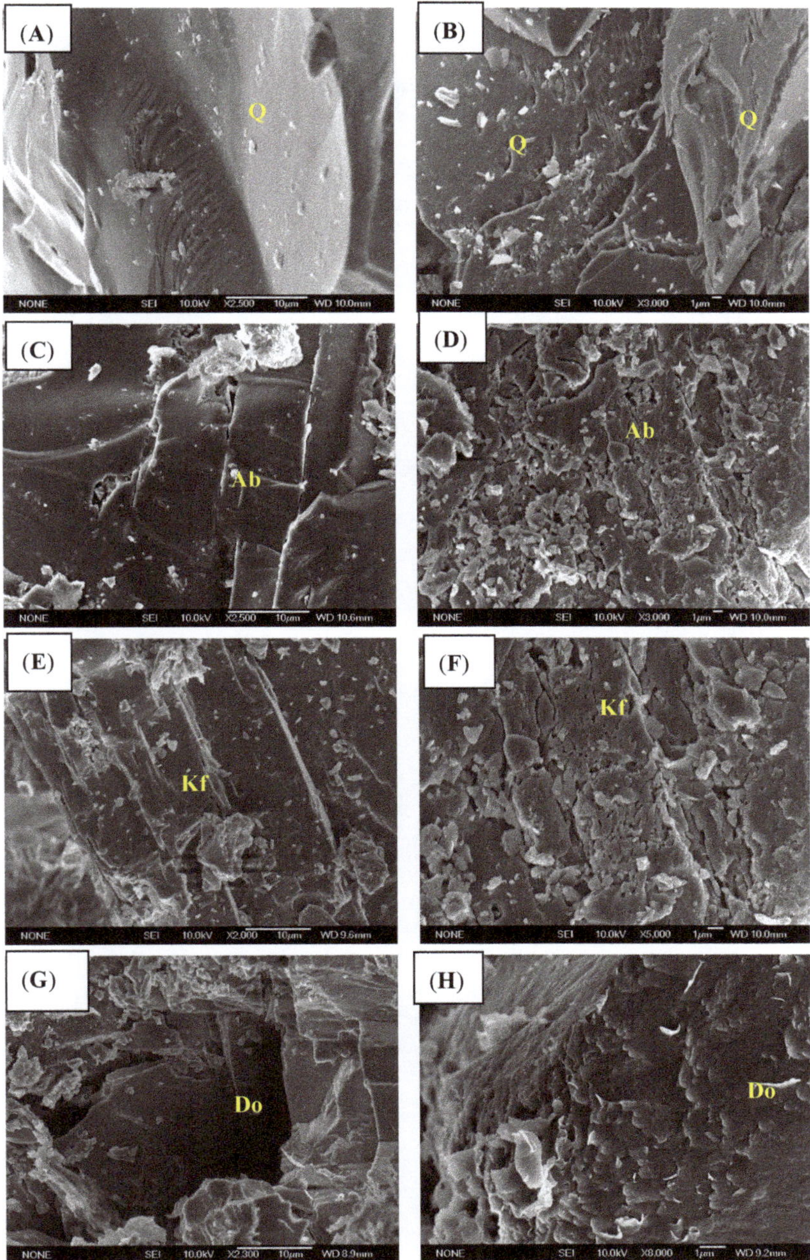

Figure 5. Scanning electron photomicrographs of pre-and post-experimental cores. (**A**) Quartz before the experiment; (**B**) Quartz after the experiment; (**C**) detrital albite before the experiment; (**D**). detrital albite after the experiment; (**E**) K-feldspar before the experiment; (**F**) K-feldspar after the experiment; (**G**) dolomite before the experiment; (**H**) dolomite after the experiment. Q—quartz; Ab—detrital albite; Kf—K-feldspar; Do—dolomite.

3.3. Changes in Porosity

Surface porosity of the synthetic cores remained relatively constant at 12.64% during the physical simulation, with no variation being observed (perhaps limited by the analytical method). Similarly, porosity changes were not evident in the numerical simulation (Figure 6), with porosity being constant up to 8 d of reaction, then increasing with carbonate dissolution to only 12.646% over the next 8 d (Figure 6).

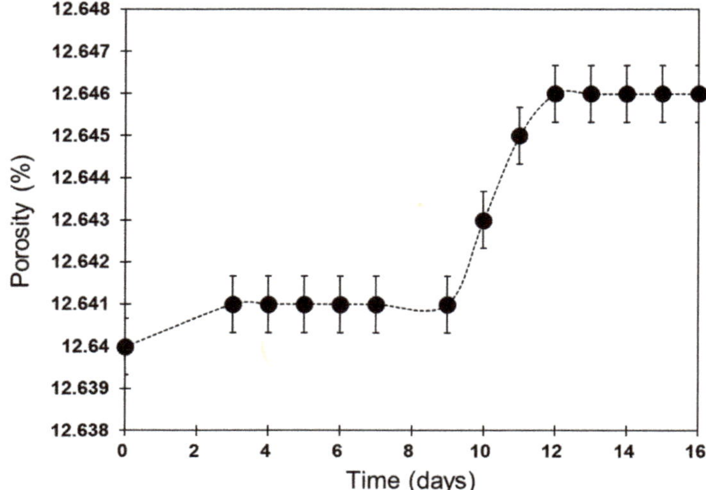

Figure 6. Porosity changes over time in the numerical simulation.

4. Discussion

4.1. Mineral Dissolution and Precipitation

Feldspars and carbonates are known to be easily corroded by acidic fluids during CO_2 injection [2,30,31], as confirmed by numerical simulations [32–34], in situ, real-time field monitoring [35,36], and natural analogies [37,38]. Changes in fluid ion contents and SEM core observations in the physical simulation confirm that feldspar and carbonate were altered by CO_2 injection. This is consistent with the numerical simulation, which also indicated dissolution of feldspars and carbonates (Figure 7). Both simulations indicate that Si and K, and Ca and Mg exhibit similar trends with ongoing reaction (Figure 4). Statistical analysis of Si, K, Ca, and Mg data using SPSS (Statistical Program for Social Sciences) software indicates correlation coefficients >0.5 (Table 4). Ion contents are thus likely controlled by a common reaction mechanism, as follows.

Table 4. Correlation coefficient matrix of the outlet solution ions.

Correlation Matrix	K	Ca	Mg	Si	Fe	Al
K	1.000					
Ca	0.943	1.000				
Mg	0.989	0.979	1.000			
Si	0.921	0.932	0.955	1.000		
Fe	0.877	0.978	0.938	0.931	1.000	
Al	−0.035	−0.143	−0.126	−0.301	−0.330	1.000

The main mechanism controlling these reactions involves the formation of H_2CO_3 from dissolved CO_2, causing the formation water to become acidic (Equation (4)), with reducing pH. Reactions between the acidic fluid and core minerals, especially carbonates and feldspars (Equations (5)–(8), below), buffer formation-water pH, causing an increase in pH of fluid produced during the experiments [39]. This process is described by the following equations:

$$CO_2 + H_2O \rightarrow H^+ + HCO_3^- \tag{4}$$

$$CaCO_3 \text{ (calcite)} + H^+ \rightarrow Ca^{2+} + HCO_3^- \tag{5}$$

$$CaMg (CO_3)_2 \text{ (dolomite)} + 2H^+ \rightarrow Ca^{2+} + Mg^{2+} + 2HCO_3^- \tag{6}$$

$$2KAlSi_3O_8 \text{ (K-feldspar)} + 2H^+ + 9H_2O \rightarrow Al_2Si_2O_5(OH)_4 \text{ (kaolinite)} + 2K^+ + 4H_4SiO_4(aq) \tag{7}$$

$$\Delta G^0 = 18 \text{ KJ mol}^{-1}, \Delta S^0 = 73 \text{J mol}^{-1}$$

$$NaAlSi_3O_8 \text{ (albite)} + CO_2 + H_2O \rightarrow NaAlCO_3 (OH)_2 \text{ (dasownite)} + 3SiO_2 \text{ (chalcedony)} \tag{8}$$

$$\Delta G^0 = -132 \text{ KJ mol}^{-1}, \Delta S^0 = -101 \text{J mol}^{-1}$$

where ΔG^0 is the Gibbs free-energy change and ΔS^0 is the entropy change.

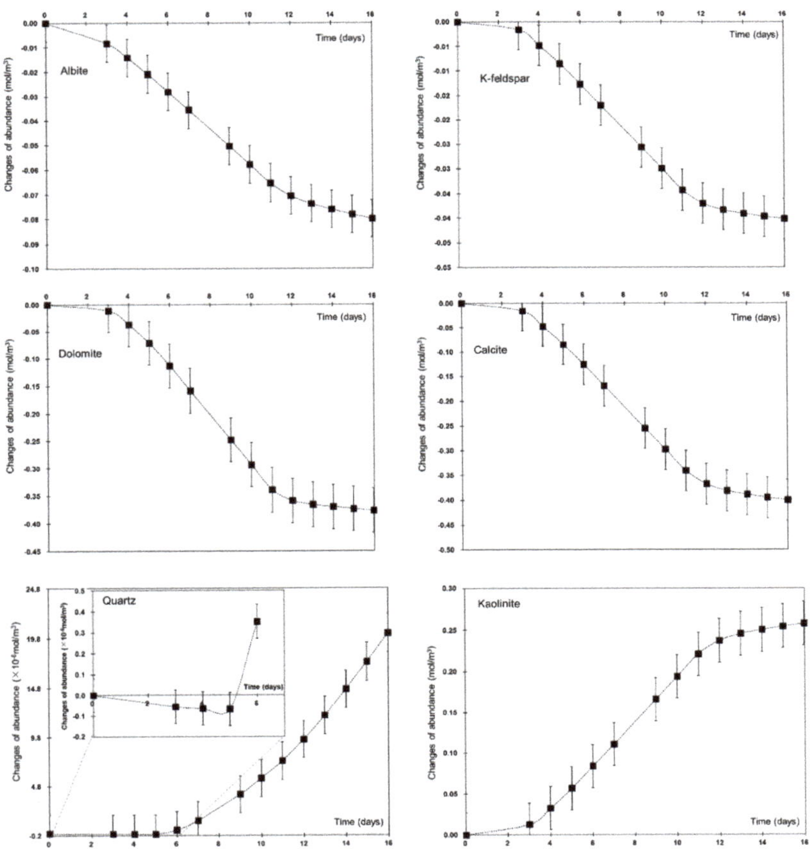

Figure 7. Mineral changes over time in the numerical simulation.

The degree of dissolution of albite is significantly greater (by a factor of ~2) than that of K-feldspar (Figure 7), possibly due to differences in their ΔG^0 and ΔS^0 values. Albite ΔG^0 and ΔS^0 values are both negative, with albite therefore needing little energy to dissolve, whereas K-feldspar values are positive, with more energy input needed for dissolution. The numerical simulation indicated that some kaolinite (up to 0.25 mol m^{-3}) and quartz (up to 19.8 × 10^{-6} mol m^{-3}) precipitated after 4 d of reaction (Figure 7), although quartz precipitation can obviously be ignored. Equations (7) and (8) indicate that kaolinite precipitation restrains the reactions, leading to reduced K-feldspar dissolution. This is consistent with the results of other studies [15,40].

The precipitation of carbonate minerals is common during CO_2-induced reactions [41], and our physical and numerical results indicate that the concentrations of carbonate minerals, calcite, and dolomite all decreased significantly during reaction. In particular, dolomite was almost completely dissolved, with no carbonate minerals remaining after the experiments. This is consistent with previous experimental findings [36,37,42]. However, the numerical simulations indicate that calcite and dolomite have similar dissolution tendencies (Figure 8), whereas calcite was completely dissolved in the physical simulations. We infer that under actual geological conditions, CO_2 fluids react first with the most reactive minerals until they are exhausted before reacting with other minerals. In contrast, in the numerical simulations the reactions followed normal geochemical dynamic processes associated with the different minerals. Carbonate minerals did not precipitate during reaction (but produced minor amounts of kaolinite and quartz) because under the experimental conditions the reaction liquid was unsaturated with carbonates (Figure 8). Similarly, results akin to the above-mentioned calculations have also been presented by Ketzer et al. (2009) [43] and Tutolo et al. (2015) [44]. Quartz dissolution began after 5 d, and it precipitated later (Figure 8), but this reaction was very weak and is ignored here. Kaolinite was the predominant precipitated mineral (Figures 7 and 8), consistent with the results of Yu et al. (2012) [15]. However, carbonate precipitation is usually observed in CO_2-formation-water–rock autoclave experiments conducted in closed systems over extended periods. For example, in an experiment using Triassic Sherwood Sandstone and sea water, Pearce et al. (1996) [37] observed calcite precipitation on the sample surface in an autoclave reaction under reservoir P/T conditions after almost eight months.

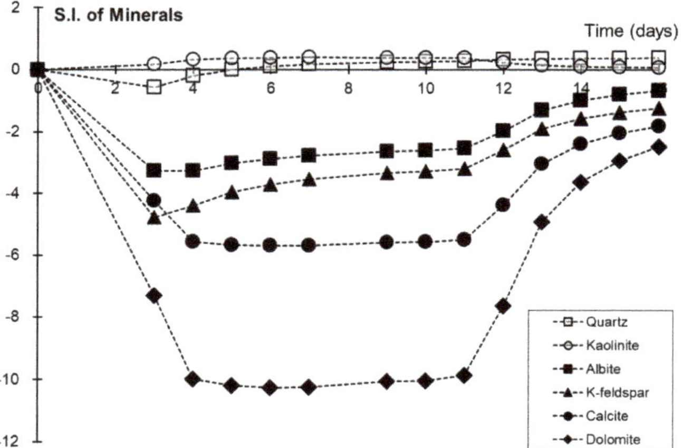

Figure 8. Saturation indices of carbonate minerals vs. reaction time in the numerical simulation.

Equations (4)–(6) indicate that calcite is the main reaction product, with some dolomite dissolving rapidly in the CO_2-saturated formation water at the beginning of the experiments. However, the silicate minerals (mainly detrital albite and K-feldspar) also gradually become unstable and start dissolving. Precipitation of clay minerals such as kaolinite occurs under acidic conditions during the reaction process. Details of the reaction process are indicated by Equations (7) and (8). These reactions lead to a rapid increase in pH of liquid produced during the initial stage, but with pH gradually reaching a stable equilibrium value, as also observed by Bowker and Shuler (1991) [35].

4.2. Porosity Changes

No obvious porosity changes were observed in the synthetic core after the physical simulation, with plane porosity being constant (within measurement uncertainty) at 12.64%. This was also observed in the numerical simulation (Figure 6) where porosity only increased from 12.64% to 12.646% (Figure 6). Minor changes in mineral contents after CO_2 injection lead to minor changes in core porosity, as confirmed by changes in ion content in the physical simulation and other mineral changes in the numerical simulation (Figures 4 and 7). However, there were variations in porosity in the numerical simulations, where after six days of reaction the dissolution of minerals was very weak and porosity did not change noticeably, but over the following four days mineral dissolution increased with marked changes in porosity (Figures 4 and 7). Especially, a notable changes happened in porosity (Figure 6). This is due to the remarkable changes in the mineral dissolution (Figure 7). After nine days, the dissolution of feldspars and carbonates reached their peaks, indicating that the dissolution volume induced by the CO_2-fluid injection increased to its maximum. A large number of newly added pore spaces lead to the porosity increase. By Day 10, minerals such as kaolinite and quartz began to precipitate, with porosity becoming less variable (Figures 4, 6 and 7). Overall, porosity varied little, indicating limited dissolution and precipitation during short-term CO_2 injection.

The lack of reduction in porosity is common in CO_2-induced reactions in sandstone [7,14,40,45], with a reduction of permeability being the dominant result of short-term CO_2 injection. The precipitation of kaolinite, solid-phase materials, and clay particles released by the dissolution of carbonate cement may account for the non-reduction of porosity and the reduction of permeability. Shiraki and Dunn (2000) [40] considered that the precipitation of kaolinite crystals in pores is the main reason for the reduction of permeability after CO_2 displacement reactions, while Luquot et al. (2012) [14] considered that newly formed minerals of amorphous carbon cause the reduction in permeability. Our results also indicate that precipitation of new minerals is related to the non-reduction of porosity. In both the physical and numerical simulations, the concentration of Al increased over the first six days before decreasing over the following 10 days. In the numerical simulation, the precipitation of kaolinite occurred after six days of reaction, with this requiring large amounts of Al (Equation (7)). While minor kaolinite was precipitated during the reaction, core porosity remained almost unchanged, for two possible reasons: (1) the dissolution of minerals was very weak in short-term CO_2-induced reactions, with few changes occurring in feldspars and carbonates after CO_2 injection (<1% mol m^{-3} variation); and (2) the precipitation of minerals was limited. Kaolinite content varied by a few percent, while changes in quartz content were negligible, with porosity being unchanged during such weak reactions.

The physical simulation was an autoclave experiment with the inlet connected to an injection pump (an open system), and with the outlet being a closed system opened only during sampling at the end of the experiment. The reaction system was therefore a semi-closed system. Under conditions of deep burial in semi-closed space, dissolution of carbonates rarely occurs or is very weak [46]. Regarding the volumes of water required to increase porosity through calcite or dolomite dissolution, the problem is essentially the inverse of the effect on porosity loss in limestones of calcite cementation caused by dissolved calcium carbonate from external sources [47–50]. For example, to increase the porosity of a 100 m thick limestone bed by 1%, 1 m^3 of calcite must be dissolved for each m^2 of bedding surface. For pore water that is undersaturated by 100 ppm, ~27,000 volumes of water are required to dissolve one volume of calcite. Increasing the porosity by 1% in 100 m thick limestone thus requires 27,000 m^3 of water per square meter of surface. Even if the limestone was underlain by 5 km of sediments in which an average porosity loss of 10% of total rock volume occurred, the pore water released from the underlying sediments would not exceed 500 m^3 m^{-2} [46], which, in an actual geological reservoir, would not be sufficient to dissolve the carbonates. In our experiment, the autoclave volume was 77.7 cm^3, and it was impossible to provide sufficient water for carbonate dissolution. However, it is certain that dissolution and precipitation are very weak at the beginning of CO_2 fluid-rock interactions, with our physical and numerical simulations confirming that only limited geochemical reactions, including dissolution and precipitation, occur during short-term CO_2 injections, with no sharp variations in core porosity or permeability. Similar results were also found by Tutolo et al. (2015), which confirmed that only very weak geochemical reactions could happen during the reaction of CO_2 and feldspar-rich sandstone [51]. For long-term CO_2 injections, however, dissolution and precipitation are the dominant geochemical processes occurring between CO_2-induced fluids and sandstones [51–53]. Our study of short-term geochemical interactions in a semi-closed system therefore showed no remarkable changes in the porosity of cores.

5. Conclusions

(1)   No significant short-term CO_2-rock-formation-water geochemical reactions are induced by CO_2 injection.
(2)   Neither physical nor numerical simulation found significant core porosity variations after CO_2 injection.
(3)   Minor amounts of kaolinite and quartz were precipitated during the numerical modelling but were not observed in the physical simulation.
(4)   Physical and numerical simulations conducted in tandem can be used to verify each other and improve their reliability.

Author Contributions: Z.Y. collected and analyzed the data and wrote the original draft. S.Y. reviewed and edited the paper. K.L. conducted the English editing work of this paper. Q.Z. contributed to the constructive discussions. L.Y. contributed to the numerical simulations.

Funding: This research and APC were both funded by the National Hydrocarbon Accumulation, Distribution and Favorable Areas Evaluation in Foreland Thrust Belts and Complex Tectonic Zones (No. 2016ZX05003-002).

Acknowledgments: We also thank the Key Laboratory of Basin Structure and Hydrocarbon Accumulation for allowing us to carry out laboratory experiments and access its rock characterization facilities.

Conflicts of Interest: The authors declare no conflict of interest.

Appendix A. Kinetic Rate Law for Mineral Dissolution and Precipitation

The general rate expression used in TOUGHREACT is taken from Lasaga et al. (1994) [54]:

$$r_n = \pm k_n A_n \left|1 - \left(\frac{Q_n}{K_n}\right)^{\theta}\right|^{\eta} \tag{A1}$$

where n denotes the kinetic mineral index, positive values of r_n indicate dissolution, while negative values indicate precipitation; k_n is the rate constant (moles per unit mineral surface area and unit time) and is temperature dependent; A_n is the specific reactive surface area per kg H_2O; K_n is the equilibrium constant for the mineral-water reaction for the destruction of one mole of mineral n; and Q_n is the reaction quotient. The parameters θ and η must be determined from experiments. However, they are usually, but not always, set to 1.

For many minerals, the kinetic rate constant k can be summed from three mechanisms (Palandri and Kharaka, 2004) [29]:

$$k = k_{25}^{nu} \exp\left[-\frac{E_a^{nu}}{R}\left(\frac{1}{T} - \frac{1}{298.15}\right)\right] + k_{25}^{H} \exp\left[-\frac{E_a^{H}}{R}\left(\frac{1}{T} - \frac{1}{298.15}\right)\right] a_{H}^{n_H} + k_{25}^{OH} \exp\left[-\frac{E_a^{OH}}{R}\left(\frac{1}{T} - \frac{1}{298.15}\right)\right] a_{OH}^{n_{OH}} \tag{A2}$$

where superscripts or subscripts nu, H, and OH indicate neutral, acidic, and alkaline mechanisms, respectively; E_a is the activation energy; k_{25} is the rate constant at 25 °C; R is gas constant; T is the absolute temperature; a is the activity of the species; and n is an exponent (constant).

References

1. Intergov Panel on Climate Change. *Climate change 2014: Mitigation of Climate Change Working Group III Contribution to the Fifth Assessment Report of the Intergovernmental Panel on Climate Change*; Cambridge University Press: Cambridge, UK; New York, NY, USA, 2014; pp. 10013–12473.
2. Fischer, S.; Liebscher, A.; Wandrey, M.; the CO_2 SINK Group. CO_2–brine–rock interaction—First results of long-term exposure experiments at in situ P–T conditions of the Ketzin CO_2 reservoir. *Chem. Erde* **2010**, *70*, 155–164. [CrossRef]
3. Hutcheon, I.; Shevalier, M.; Durocher, K.; Bloch, J.; Johnson, G.; Nightingale, M.; Mayer, B. Interactions of CO_2 with formation waters, oil and minerals and CO_2 storage at the Weyburn IEA EOR site, Saskatchewan, Canada. *Int. J. Greenh. Gas Control* **2016**, *53*, 354–370. [CrossRef]
4. IEA. *CO_2 Emissions from Fuel Combustion*, 2012th ed.; OECD/IEA: Paris, France, 2012; p. 136. Available online: http://www.iea.org/publications/freepublications/publication/CO2emissionfromfuelcombustionH IGHLIGHTS.pdf (accessed on 25 November 2012).
5. Bachu, S.; Gunter, W.D.; Perkins, E.H. Aquifer disposal of CO_2: Hydrodynamic and mineral trapping. *Energy Convers. Manag.* **1994**, *35*, 269–279. [CrossRef]
6. Hitchon, B.; Gunter, W.D.; Gentzis, T.; Baileyc, R.T. Sedimentary basins and greenhouse gases: A serendipitous association. *Energy Convers. Manag.* **1999**, *40*, 825–843. [CrossRef]
7. Sayegh, S.G.; Krause, F.F.; Girard, M.; DeBree, C. Rock/fluid interactions of carbonated brines in a sandstone reservoir: Pembina Cardium, Alberta, Canada. *SPE Form. Eval.* **1990**, *5*, 399–405. [CrossRef]
8. Tobergte, D.R.; Curtis, S. Experimental perspectives of mineral dissolution and precipitation due to carbon dioxide-water-rock interactions. *J. Chem. Inf. Model.* **2013**, *53*, 1689–1699.
9. Lombard, J.M.; Azaroual, M.; Pironon, J.; Broseta, D.; Egermann, P.; Munier, G.; Mouronval, G. CO_2 injectivity in geological storages: An overview of program and results of the GeoCarbone-Injectivity project. *Oil Gas Sci. Technol.* **2010**, *65*, 533–539. [CrossRef]
10. Miri, R.; van Noort, R.; Aagaard, P.; Hellevang, H. New insights on the physics of salt precipitation during injection of CO_2 into saline aquifers. *Int. J. Greenh. Gas Control* **2015**, *43*, 10–21. [CrossRef]
11. Kampman, N.; Bickle, M.; Wigley, M.; Dubacq, B. Fluid flow and CO_2–fluid–mineral interactions during CO_2-storage in sedimentary basins. *Chem. Geol.* **2014**, *369*, 22–50. [CrossRef]
12. Pokrovsky, O.S.; Golubev, S.V.; Schott, J.; Castillo, A. Calcite, dolomite and magnesite dissolution kinetics in aqueous solutions at acid to circumneutral pH, 25 to 150 C and 1 to 55 atm pCO_2: New constraints on CO_2 sequestration in sedimentary basins. *Chem. Geol.* **2009**, *265*, 20–32. [CrossRef]
13. Alemu, B.L.; Aker, E.; Soldal, M.; Johnsen, Ø.; Aagaard, P. Effect of sub-core scale heterogeneities on acoustic and electrical properties of a reservoir rock: A CO_2 flooding experiment of brine saturated sandstone in a computed tomography scanner. *Geophys. Prospect* **2013**, *61*, 235–250. [CrossRef]
14. Luquot, L.; Andreani, M.; Gouze, P.; Camps, P. CO_2 percolation experiment through chlorite/zeolite-rich sandstone (Pretty Hill Formation–Otway Basin–Australia). *Chem. Geol.* **2012**, *294–295*, 75–88. [CrossRef]
15. Yu, Z.; Liu, L.; Yang, S.; Li, S.; Yang, Y. An experimental study of CO_2–brine–rock interaction at in situ pressure–temperature reservoir conditions. *Chem. Geol.* **2012**, *326–327*, 88–101. [CrossRef]
16. Nitao, J.J. *User's Manual for the USNT Module of NUFT Code*; Version 2.0. LLNL.; Lawrence Livermore National Laboratory Report UCRL-MA-130653; Lawrence Livermore National Laboratory: Livermore, CA, USA, 1998.
17. Lichtner, P.C. *FLOWTRAN User Manual*; Tech. Report, LA-Ur-01-2349; Las Alamos National Laboratory: Las Alamos, NM, USA, 2001.
18. Nghiem, L.; Sammon, P.; Grabenstetter, J.; Ohkuma, H. Modeling CO_2 storage in aquifers with fully-coupled geochemical EOS compositional simulator. *Soc. Pet. Eng. J.* **2004**, 89474. [CrossRef]
19. White, M.D.; Oostrom, M. *STOMP: Subsurface Transport over Multiple Phases*; Version 4.0; User's Guide; Pacific Northwest National Laboratory: Richland, WA, USA, 2006; p. 120.

20. Xu, T.; Sonnenthal, E.L.; Spycher, N.; Pruess, K. TOUGHREACT: A simulation program for non-isothermal multiphase reactive geochemical transport in variably saturated geologic media. *Comput. Geosci.* **2006**, *32*, 145–165. [CrossRef]

21. Xu, T.; Spycher, N.; Sonnenthal, E.L.; Zhang, G.; Zheng, L.; Pruess, K. TOUGHREACT Version 2.0: A simulator for subsurface reactive transport under non-isothermal multiphase flow conditions. *Comput. Geosci.* **2011**, *37*, 763–774. [CrossRef]

22. Audigane, P.; Gaus, I.; Czernichowski-Lauriol, I.; Pruess, K.; Xu, T. Two dimensional reactive transport modeling of CO_2 injection in a saline aquifer at the Sleipner Site. *Am. J. Sci.* **2007**, *307*, 974–1008. [CrossRef]

23. Gherardi, F.; Xu, T.; Pruess, K. Numerical modeling of self-limiting and self-enhancing cap rock alteration induced by CO_2 storage in a depleted gas reservoir. *Chem. Geol.* **2007**, *244*, 103–129. [CrossRef]

24. Okuyama, Y.; Sasaki, M.; Nakanishi, S.; Norifumi, T.; Ajima, S. Geochemical CO_2 trapping in open aquifer storage—The Tokyo Bay model. *Energy Procedia* **2009**, *1*, 3253–3258. [CrossRef]

25. Zhang, L.; Apps, J.A.; Zhang, Y.Q.; Xu, T.; Birkholzer, J.T. Reactive Transport Simulations to Study Groundwater Quality Changes in Response to CO_2 Leakage from Deep Geological Storage. *Phys. Procedia* **2009**, *1*, 1887–1894. [CrossRef]

26. Xu, T.; Kharaka, Y.; Doughty, C.; Freifeld, B.M.; Daley, T.M. Reactive transport modeling to study changes in water chemistry induced by CO_2 injection at the Frio-I brine pilot. *Chem. Geol.* **2010**, *271*, 153–164. [CrossRef]

27. Yu, Z.; Liu, K.; Zhao, M.; Liu, S.; Zhuo, Q.; Lu, X. Characterization of Diagenesis and the Petroleum Charge in Kela 2 Gas Field, Kuqa Depression, Tarim Basin. *Earth Sci.* **2015**, *41*, 533–545, (In Chinese with English Abstract).

28. Duan, Z.H.; Sun, R. An improved model calculating CO_2 solubility in pure water and aqueous NaCl solutions from 273 to 533 K and from 0 to 2000 bar. *Chem. Geol.* **2003**, *193*, 257–271. [CrossRef]

29. Palandri, J.; Kharaka, Y.K. *A Compilation of Rate Parameters of Water–Mineralinteraction Kinetics for Application to Geochemical Modeling*; US Geological Survey Open File Report (of 2004–1068); USGS: Reston, VA, USA, 2004; p. 64.

30. Bertier, P.; Swennen, R.; Laenen, B.; Lagrou, D.; Dreesen, R. Experimental identification of CO_2–water–rock interactions caused by sequestration of CO_2 in Westphalia and Buntsandstein sandstones of the Campine Basin (NE-Belgium). *J. Geochem. Explor.* **2006**, *89*, 10–14. [CrossRef]

31. Wandrey, M.; Fischera, S.; Zemkea, K.; Liebschera, A.; Scherfb, A.K.; Hillebrandb, A.V.; Zettlitzerc, M.; Würdemann, H. Monitoring petrophysical, mineralogical, geochemical and microbiological effects of CO_2 exposure–Results of long-term experiments under in situ conditions. *Energy Procedia* **2011**, *4*, 3644–3650. [CrossRef]

32. Xu, T.; Apps, J.A.; Pruess, K. Numerical simulation of CO_2 disposal by mineral trapping in deep aquifers. *Appl. Geochem.* **2004**, *19*, 917–936. [CrossRef]

33. Xu, T.; Apps, J.A.; Pruess, K. Mineral sequestration of carbon dioxide in a sandstone–shale system. *Chem. Geol.* **2005**, *217*, 295–318. [CrossRef]

34. Zerai, B.; Saylor, B.Z.; Matisoff, G. Computer simulation of CO_2 trapped through mineral precipitation in the Rose Run Sandstone. *Ohio Appl. Geochem.* **2006**, *21*, 223–240. [CrossRef]

35. Bowker, K.A.; Shuler, P.J. Carbon dioxide injection and resultant alteration of the Weber sandstone, Rangely Field, Colorado. *Am. Assoc. Pet. Geol. Bull.* **1991**, *75*, 1489–1499.

36. Assayag, N.; Matter, J.; Ader, M.; Goldberg Agrinier, P. Water–rock interactions during a CO_2 injection field-test: Implications on host rock dissolution and alteration effects. *Chem. Geol.* **2009**, *265*, 227–235. [CrossRef]

37. Pearce, J.M.; Holloway, S.; Wacker, H.; Nelis, M.K.; Rochelle, C.; Bateman, K. Natural occurrences as analogues for the geological disposal of carbon dioxide. *Energy Convers.* **1996**, *37*, 1123–1128. [CrossRef]

38. Wilkinson, M.; Haszeldine, R.S.; Fallick, A.E.; Odling, N.; Stoker, S.J.; Gatliff, R.W. CO_2-mineral reaction in a natural analogue for CO_2 storage–implications for modeling. *J. Sediment. Res.* **2009**, *79*, 486–494. [CrossRef]

39. Gaus, I. Role and impact of CO_2–rock interactions during CO_2 storage in sedimentary rocks. *Int. J. Greenh. Gas Control* **2010**, *4*, 73–89. [CrossRef]

40. Shiraki, R.; Dunn, T.L. Experimental study on water–rock interactions during CO_2 flooding in the Tensleep Formation, Wyoming, USA. *Appl. Geochem.* **2000**, *15*, 265–279. [CrossRef]

41. Weibel, R.; Kjoller, C.; Bateman, K.; Nielsen, L.H.; Frykman, P.; Springer, N.; Laier, T. Mineral changes in CO_2 experiments–Examples from Danish onshore saline aquifers. *Energy Procedia* **2011**, *4*, 4495–4502. [CrossRef]

42. Flukiger, F.; Bernard, D. A new numerical model for pore scale dissolution of calcite due to CO_2 saturated water flow in 3D realistic geometry: Principles and first results. *Chem. Geol.* **2009**, *265*, 171–180. [CrossRef]

43. Ketzer, J.M.; Iglesias, R.; Einloft, S.; Dullius, J.; Ligabue, R.; Lima, V. Water–rock–CO_2 interactions in saline aquifers aimed for carbon dioxide storage: Experimental and numerical modeling studies of the Rio Bonito Formation (Permian), southern Brazil. *Appl. Geochem.* **2009**, *24*, 760–767. [CrossRef]

44. Tutolo, B.M.; Luhmann, A.J.; Kong, X.Z.; Saar, M.O.; Seyfried, W.E. CO_2 sequestration in feldspar-rich sandstone: Coupled evolution of fluid chemistry, mineral reaction rates, and hydrogeochemical properties. *Geochim. Cosmochim. Acta* **2015**, *160*, 132–154. [CrossRef]

45. Ross, G.D.; Todd, A.C.; Tweedie, J.A.; Will, A.G.S. *The Dissolution Effects of CO_2-Brine Systems on the Permeability of U.K. and North Sea Calcareous Sandstones*; Proc.3rd Joint SPE/DOE Symposium Enhanced Oil Recovery; (SPE/DOE 10685); Society of Petroleum Engineers Enhanced Oil Recovery Symposium: Tulsa, OK, USA, 1982; pp. 149–162.

46. Ehrenberg, S.; Walderhaug, O.; Bjørlykke, K. Carbonate porosity creation by mesogenetic dissolution: Reality or illusion? *AAPG* **2012**, *96*, 217–233. [CrossRef]

47. Dunham, R.J. Early vadose silt in Townsend mound (reef), New Mexico. In *Depositional Environments in Carbonate Rocks: A Symposium*; Friedman, G.M., Ed.; SEPM Special Publication: Los Angeles, CA, USA, 1969; Volume 14, pp. 139–181.

48. Bathurst, R.G.C. *Carbonate Sediments and their Diagenesis*; Elsevier: Amsterdam, The Netherlands, 1975; p. 658.

49. Scholle, P.A.; Halley, R.B. Burial diagenesis: Out of sight, out of mind! In *Carbonate Cements*; Mann, N.S., Harris, P.M., Eds.; SEPM Special Publication: Malden, MA, USA, 1985; Volume 36, pp. 309–334.

50. Morse, J.W.; Mackenzie, F.T. Geochemical constraints on $CaCO_3$ transport in subsurface sedimentary environments. *Chem. Geol.* **1993**, *105*, 181–196. [CrossRef]

51. Tutolo, B.M.; Kong, X.Z.; Seyfried, W.E.; Saar, M.O. High performance reactive transport simulations examining the effects of thermal, hydraulic, and chemical (THC) gradients on fluid injectivity at carbonate CCUS reservoir scales. *Int. J. Greenh. Gas Control* **2015**, *39*, 285–301. [CrossRef]

52. Zhang, W.; Li, Y.; Xu, T.; Cheng, H.; Zheng, Y.; Xiong, P. Long-term variations of CO_2 trapped in different mechanisms in deep saline formations: A case study of the Songliao Basin, China. *Int. J. Greenh. Gas Control* **2009**, *3*, 161–180. [CrossRef]

53. Yang, L.; Xu, T.; Liu, K.; Peng, B.; Yu, Z.; Xu, X. Fluid–rock interactions during continuous diagenesis of sandstone reservoirs and their effects on reservoir porosity. *Sedimentology* **2017**, *64*, 1303–1321. [CrossRef]

54. Lasaga, A.C.; Steefel, C.I. A coupled model for transport of multiple chemical species and kinetic precipitation/dissolution reactions with applications to reactive flow in single phase hydrothermal system. *Am. J. Sci.* **1994**, *294*, 529–592.

 *applied sciences*

Article

Carbon Spheres as CO$_2$ Sorbents

P. Staciwa [1], U. Narkiewicz [1,*], D. Sibera [1], D. Moszyński [1], R. J. Wróbel [1] and R. D. Cormia [2]

[1] Institute of Chemical and Environment Engineering, Faculty of Chemical Technology and Engineering, West Pomeranian University of Technology in Szczecin, Pulaskiego 10, 70-322 Szczecin, Poland

[2] Engineering Faculty, Chemistry Department, Foothill College, 12345 El Monte Road, Los Altos Hills, CA 94022, USA

* Correspondence: urszula.narkiewicz@gmail.com

Received: 27 June 2019; Accepted: 12 August 2019; Published: 15 August 2019

Abstract: Microporous nanocarbon spheres were prepared by using a microwave assisted solvothermal method. To improve the carbon dioxide adsorption properties, potassium oxalate monohydrate and ethylene diamine (EDA) were employed, and the influence of carbonization temperature on adsorption properties was investigated. For nanocarbon spheres containing not only activator, but also EDA, an increase in the carbonization temperature from 600 °C to 800 °C resulted in an increase of the specific surface area of nearly 300% (from 439 to 1614 m^2/g) and an increase of the CO$_2$ adsorption at 0 °C and 1 bar (from 3.51 to 6.21 mmol/g).

Keywords: carbon nanospheres; nanocarbon spheres; carbon dioxide uptake; EDA

---

1. Introduction

The global economy requires a great amount of energy, which is produced primarily by the combustion of fossil fuels. Carbon dioxide emissions are a significant negative side-effect of this activity. Transportation also requires a great amount of petroleum and is responsible for significant emissions of greenhouse gases [1,2]. The cumulative emission of CO$_2$ strongly contributes to climate change and is the greatest single contributor to the greenhouse effect [3,4]. The average concentration of CO$_2$ in the earth's atmosphere in 2018 was 407 ppm, which is about 40% higher than in the preindustrial age [5]. The effort to develop technologies that will reduce CO$_2$ emissions is very important for both the global economy and the environment.

Recently, methods of CO$_2$ capture from flue gas have been based on absorption into liquids (e.g., amines [6] or methanol [7]). These technologies are energy intensive and not environmentally sound. Solid sorbents offer an alternative solution. There are a number of criteria that must be met for a successful sorbent material, namely: high selectivity and adsorption capacity for CO$_2$, fast adsorption/desorption kinetics, efficient regeneration of sorbents, and low cost [8].

In recent years, a number of materials have been investigated as solid state adsorbents for CO$_2$, such as: zeolites [9], silica [10], porous polymer materials [11], metal organic frameworks [12], and carbon materials [13–16]. The most efficient for CO$_2$ adsorption are carbon materials, which exhibit a high surface area, large porous volume, chemical stability, affinity for carbon dioxide, low cost, and the possibility of modification with heteroatoms [17]. The weak side of carbon sorbents is their poor selectivity.

The application of carbon materials for CO$_2$ uptake has been widely investigated. There are many sources of carbon that can be used for the production of activated carbon: polymers, biomass, or resins. Some examples are shown in Table 1. Potassium compounds, namely potassium hydroxide or potassium oxalate, are most often used as chemical activators. Special attention should be paid to resins. Gradual growth of the polymer chain allows incorporating modificators into the carbon matrix. Homogeneously distributed activator can improve not only the surface area of the material (impregnation), but its whole volume. A resorcinol–formaldehyde resin mixture could be a suitable carbon source for CO_2 adsorption.

Table 1. Comparison of CO_2 uptakes on various carbon adsorbents.

Carbon Source	Activator	CO_2 Adsorption 0 °C [mmol/g] 1 bar	CO_2 Adsorption 25 °C [mmol/g] 1 bar	Reference
Resorcinol–formaldehyde resin	$K_2C_2O_4$	7.67	4.95	[18]
Resorcinol–formaldehyde resin	$K_2C_2O_4$	6.6	-	[19]
Resorcinol–formaldehyde resin	KOH	7.34	4.83	[20]
Resorcinol–formaldehyde resin/EDA	EDA; CO_2	6.2	4.1	[21]
Furfuryl alcohol	KOH	5.8	3.3	[22]
Polyacrylonitryle	KOH	-	2.74	[23]
Waste wool	KOH	3.73	2.81	[24]
Starch	KOH	6.6	4.3	[25]

According to the results presented in Table 1, resins are very promising as a carbon source to produce solid sorbents for carbon dioxide capture.

Among nanocarbon materials, spherical structures have been widely studied. The most popular method to obtain porous nanocarbon spheres is the method of Stöber, using resins as a carbon source. The application of this method to produce carbon spheres was described in the work of Liu et al. [26], where the source of carbon was a resorcinol–formaldehyde resin. Thanks to the development of the Stöber method, researchers discovered a simple method to produce polymer beads. The product was in the form of spherical regular particles. Since then, the phenol derivatives were widely used as a carbon source [27,28]. In work by Zhao and co-workers [28], using 3-aminophenol as a precursor, highly monodisperse material were obtained. They also proved that changing different parameters allowed for tuning spherical size in a very broad range.

To enhance the surface area and porous volume, various processes of activation are employed, with two primary methods to activate carbon materials. First, physical activation carried out through carbonization in the presence of proper gases [29]. Second, chemical activation is induced by the addition of a strong base, i.e., potassium oxalate [19], potassium hydroxide [20,30,31], and potassium carbonate [32]. In the work of Choma et al. [18], chemical activation with potassium oxalate resulted in a large increase in the surface area of carbon materials (from 680 m^2/g to 1490 m^2/g) and an increase of CO_2 uptake from 3.03 mmol/g to 7.67 mmol/g in 0 °C at 1 atm. This example showed how modification with potassium oxalate can significantly enhance specific surface area and CO_2 adsorption of carbon spheres.

In order to improve synthesis conditions of carbon nanospheres, microwave assisted solvothermal reactor has been used [33,34]. Performing the reaction in a common autoclave takes a significant amount of time, often several hours, while the reaction in microwave assisted solvothermal reactor is very fast, about 15 min. The temperature gradient using a microwave in the reactor volume is very low and can be negligible. The microwave's influence on the behavior of polar solvents in the reaction is significant, and volume nucleation points are created rapidly.

In this work, the influence of the concentration of activator, potassium oxalate, carbonization temperature, and influence of ethylene diamine (EDA) on the physical properties and adsorption of carbon dioxide were investigated.

Ludwinowicz and Jaroniec [19] performed a simple one-pot synthesis of carbon spheres and obtained very good CO_2 adsorption values. In this work, simple autoclave was replaced by microwave assisted solvothermal reactor. The use of such a reactor enabled a significantly shortened reaction time. Heating with microwaves avoids a variance in the temperature profile in the reactor volume, no local overheating, and the products obtained are of very good quality, with uniform shape and size of the produced particles.

Previous research obtained in this research community has been promising [33,34]. In the present work, we describe more in depth research on the influence of modificator and carbonization temperature on surface area, porosity, and carbon dioxide adsorption.

2. Experimental

2.1. Sample Preparation

The samples were synthesized as follows: First, an aqueous–alcohol solution consisting of 60 mL distilled water and 24 mL ethanol was prepared by mixing at an ambient temperature. Subsequently, 0.60 g of resorcinol and 0.30 mL of ammonium hydroxide (25 wt.%) was added to the mixture under continuous stirring for 10 min. After dissolving of the resorcinol, the proper amount of potassium oxalate was added and the mixture was stirred for 30 min. The weight ratio potassium: carbon was 7:1 and 9:1. For samples modified with EDA, 0.3 mL of EDA was added. Next, 0.9 mL of 37 wt.% formaldehyde was dropped into the solution and kept under magnetic stirring for 24 h. Afterwards, the solution was treated in a solvothermal microwave reactor ERTEC MAGNUM II (pressure 2 MPa, time—15 min). The resulting materials were dried at 80 °C for 48 h. The carbonization of the carbon nanospheres was performed in argon atmosphere at 350 °C for 2 h with a heating rate of 1 °C/min, then the temperature was raised to 600 °C, 700 °C, or 800 °C with the same heating rate and also for 2 h. The materials obtained were washed two times with 200 mL of distilled water and dried at 80 °C for 48 h.

2.2. Characterization

The morphology of the produced samples was determined using a Hitachi SU8020 Ultra-High Resolution Field Emission Scanning Electron Microscope (FE-SEM).

The density of the materials was determined using a helium pycnometer Micro-Ultrapyc 1200e.

The chemical composition of the samples' surface was studied by X-ray Photoelectron Spectroscopy (XPS). The measurements were conducted using Mg K_a (hv = 1253.6 eV) radiation in a Prevac (Poland) system equipped with a Scienta (Sweden) SES 2002 electron energy analyzer operating with constant transmission energy (E_p = 50 eV). The analysis chamber was evacuated to a pressure below $1 \cdot 10^{-9}$ mbar. A powdered sample of the material was placed on a stainless steel sample holder.

Thermal stability of the produced materials was investigated using Thermal Gravimetric Analysis (TGA). The thermogravimetric measurements were carried out with the use of STA 449 C thermobalance (Netzsch Company, Germany). Approximately 10 mg of the sample was heated at 10 °C/min to 950 °C under air atmosphere.

To determine textural properties of the carbon spheres, the low temperature physical adsorption of nitrogen was carried out at −196 °C using the Quadrasorb volumetric apparatus (Quantachrome Instruments). Carbon dioxide uptake was gathered at temperature 0 °C and 25 °C using the same apparatus.

3. Results and Discussion

3.1. Samples' Morphology

The morphology of the carbon spheres was studied using Scanning Electron Microscopy (SEM). The SEM images of the samples carbonized in 700 °C are shown in Figure 1. For the material without modification (Figure 1a), small, monodisperse spheres were obtained. The average diameter of the carbon spheres was determined by SEM to be about 400 nm. For the material prepared by the addition of the activator (potassium oxalate, Figure 1b), two classes of spheres were observed: smaller spheres, about 500 nm in diameter, and larger, about 2–3 μm. The large difference in the diameter of the spheres was the result of the addition of potassium oxalate. The resorcinol–formaldehyde spheres were influenced by the oxalate moieties, and thus larger spheres were formed. However, there was a fraction of the smaller spheres, where oxalate moieties were likely less present. The higher concentration of potassium oxalate (Figure 1c) resulted in higher saturation of the solution. The spheres containing more potassium oxalate were larger. Nonetheless, there was a large amount of small spheres, which did not contain oxalate moieties. Thus, a large amount of carbon material was not modified.

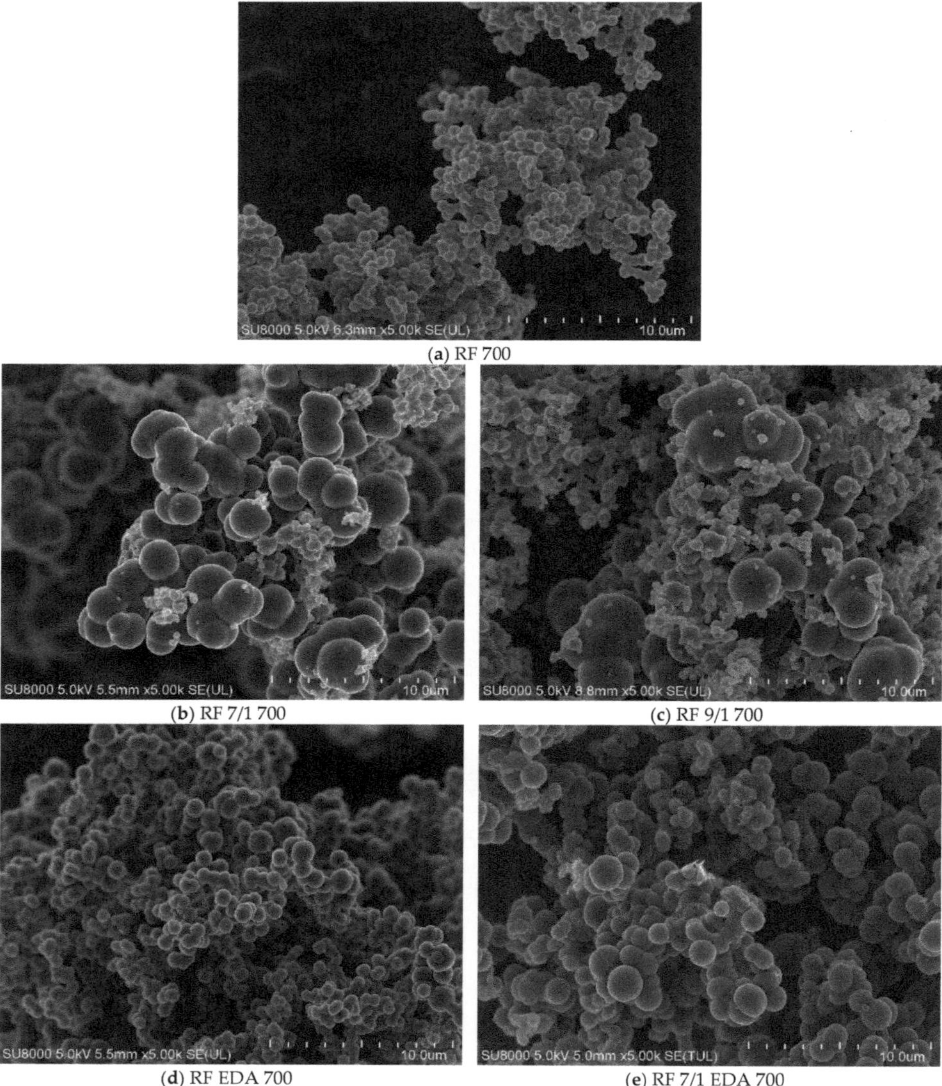

Figure 1. Scanning Electron Microscopy (SEM) images of the spheres: (**a**) without modification; (**b**) with activator concentration 7/1; (**c**) with activator concentration 9/1; (**d**) with ethylene diamine (EDA) modification; (**e**) with activator concentration 7/1 and EDA modification.

For the material modified with EDA only (Figure 1d), the monodispersity of the spheres was kept, however larger spheres (diameter ca. 800 nm) were formed. The larger diameter and the monodispersity of the spheres are evidence that EDA was well dispersed in the whole volume, and so, all carbon spheres contained EDA. In the case of the material modified with both EDA and potassium oxalate (Figure 1e), the influence of both modifiers can be noticed. The average diameter of the spheres was larger and the spheres were more uniform. EDA provided better dispersion of potassium oxalate in the reaction volume, thus potassium ions were present in a higher amount of the spheres. In the end, a much bigger specific surface area value was reached.

The size distribution of the produced particles was evaluated from the SEM images using the ImageJ software tool and is illustrated in Figure 2a–e. The quantity of spheres taken into account was 50 for every kind of the sample.

The results for the reference sample are given in Figure 2a. RF 700 exhibited the highest monodispersity among all the samples. The diameter of the spheres was about 600 nm.

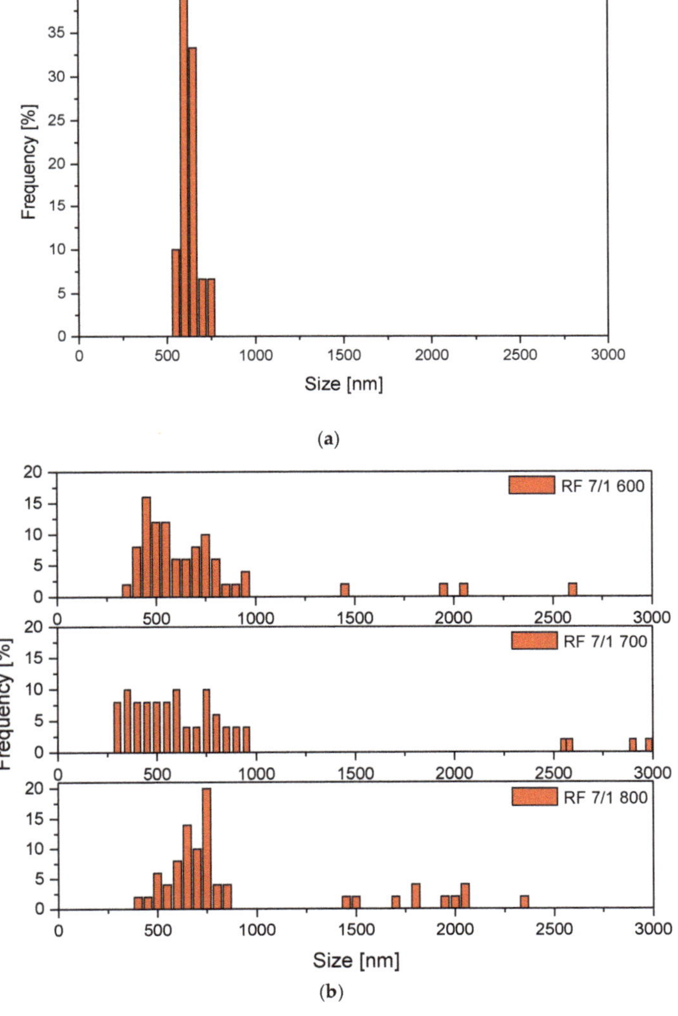

(a)

(b)

Figure 2. *Cont.*

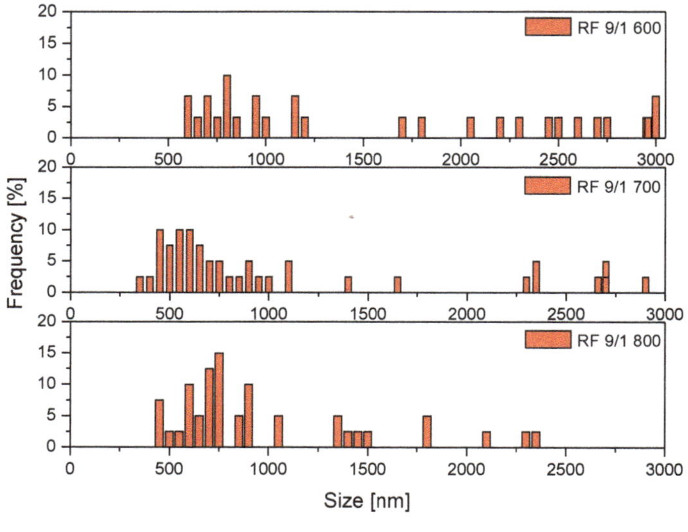

(**c**)

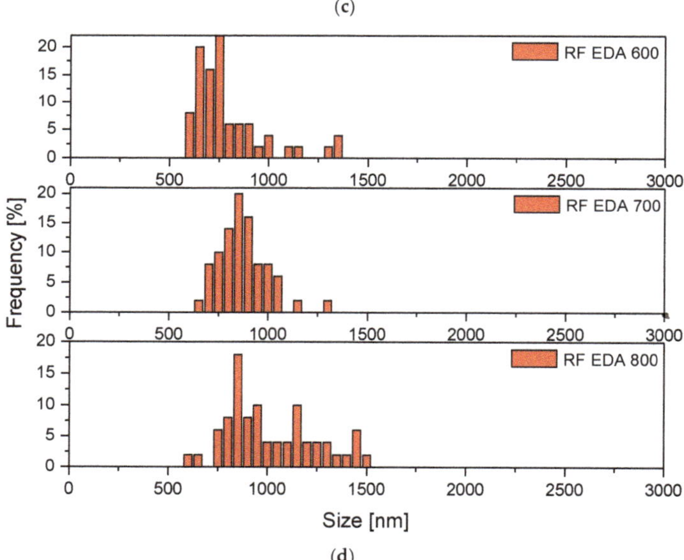

(**d**)

Figure 2. *Cont.*

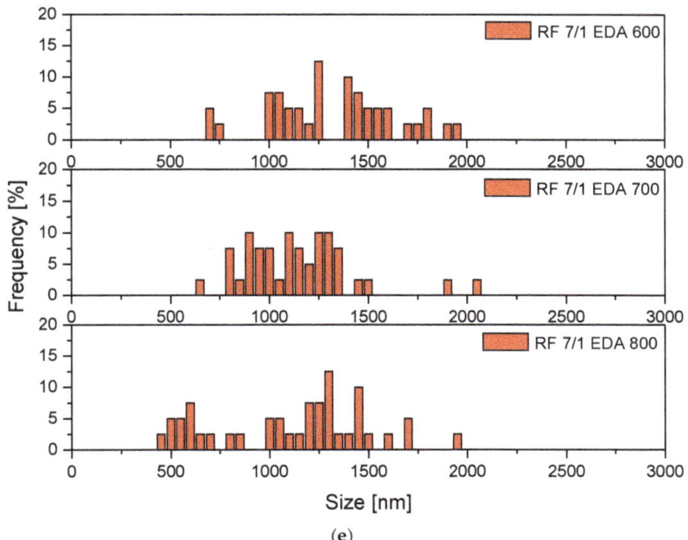

Figure 2. (**a**) Size distribution of the reference sample. (**b**) Size distributions of the samples with lower activator content (7/1). (**c**) Size distribution of the samples with higher content of the activator (9/1). (**d**) Size distribution of the samples modified with EDA only. (**e**) Size distribution of the samples modified with EDA and activator.

The modification of carbon materials with the lower content of potassium oxalate resulted in higher variation in the spheres' size distribution. A considerable amount of produced spheres had a diameter from 300 to 1000 nm, as shown in Figure 2b. With the increase of the carbonization temperature, the formation of larger spheres (about 2000 nm) was observed. For the material carbonized at 600 °C, the majority of spheres were about 500 nm, whereas on the other hand for the sample carbonized at 800 °C, this value shifted to about 700 nm.

Comparing the samples with different amounts of activator, the strong influence of the activator concentration on the spheres' size was noticed. Higher activator content in the samples resulted in the widest size distribution (Figure 2c). Moreover, the large spheres of diameter over 2000 nm were formed.

The size distribution of the samples modified with EDA only is presented in Figure 2d. Compared to the samples modified with potassium oxalate, the highest monodispersity of the spheres was gained. Nonetheless, increasing the carbonization temperature caused the distribution to be broader.

As can be seen in Figure 2e, the application of both the modificators limited the production of the large spheres (over 2000 nm). Unlike previous distributions, by increasing the carbonization temperature, the shift of distribution towards smaller spheres was noticed.

3.2. Surface Chemistry

The surface composition of materials was analyzed by X-ray Photoelectron Spectroscopy (XPS). The survey spectra acquired for all analyzed samples are shown in Figure 3. The evaluation of the elemental composition of the surface of all samples is presented in Table 2. In all samples, carbon and oxygen was present, and potassium was observed in samples prepared with potassium oxalate.

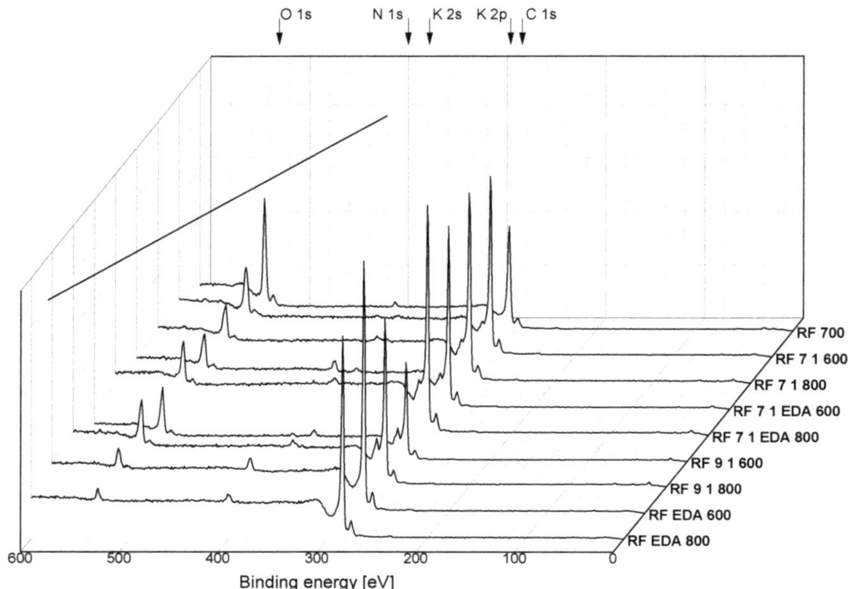

Figure 3. X-ray Photoelectron Spectroscopy (XPS) survey spectra.

Table 2. Elemental composition of the surface of the samples.

Sample	C	O	N	K
			at. %	
RF_700	**96**	**3**	~1	-
RF_7_1_600	88	11	<1	~1
RF_7_1_800	92	7	<1	~1
RF_7_1_EDA600	87	8	3	~2
RF_7_1_EDA800	88	9	<1	~3
RF_9_1_600	76	18	~2	~4
RF_9_1_800	85	11	<1	~3
RF_EDA600	92	4	~4	-
RF_EDA800	93	3	~4	-

The highest carbon content was observed for the pure carbon material (RF_700); oxygen constitutes only 3% of the surface atoms. The surface of the samples prepared with EDA only also contained a relatively small number of oxygen atoms (approximately 4%), however those surfaces also contained about 4% of nitrogen atoms. The presence of potassium in the internal structure of the material is associated with an increased concentration of oxygen atoms. The more potassium observed in the material, the higher the concentration of oxygen observed, as residual potassium atoms were bound with oxygen. In general, when the carbonization temperature was increased to 800 °C, this resulted in a lower oxygen concentration than that observed for samples carbonized at 600 °C. There is noticeably more potassium retained on the surface of the samples modified by both potassium oxalate and EDA in comparison to materials modified by potassium oxalate only. A possible explanation for this is that a reaction of potassium with amine groups occurred.

The analysis of high-resolution XPS data brings a more detailed view of the chemistry of the surface of the studied materials. In Figure 4, the spectral region of binding energy between 280 eV and 300 eV is displayed for two samples of carbon spheres obtained with the weight ratio potassium: carbon of 9:1 (RF_9_1_600 and RF_9_1_800). This region contains the spectrum components originating from C 1s and K 2p electrons. The peak maximum of K $2p_{3/2}$ is located at 293 eV and it is accompanied with a K $2p_{1/2}$ spin-orbit component at 295 eV. The peak maximum of XPS C 1s spectrum is located at 284.4 eV. This energy is characteristic for highly graphitized carbon materials. However, a distinctive shoulder at about 288 eV is present in the spectrum for both samples, though more prominent on the sample carbonized at 600 °C. This position is usually ascribed to the general group of carbon moieties containing O–C=O bindings. The intensity of the spectra is normalized in respect to the intensity of the main peak of carbon. It can be pointed out that the relative intensity of lines corresponding with potassium atoms as well as carbon atoms in O–C=O bindings decreases in comparison to C–C bonds, reflected by XPS C 1s peak at 284.4 eV. This shows that increased carbonization temperature results in a partial depletion of potassium atoms from the surface as well as a decomposition of a part of C–O bonds. This corresponds well with the quantitative analysis described above. Similar observations are also valid for samples with a lower potassium:carbon ratio.

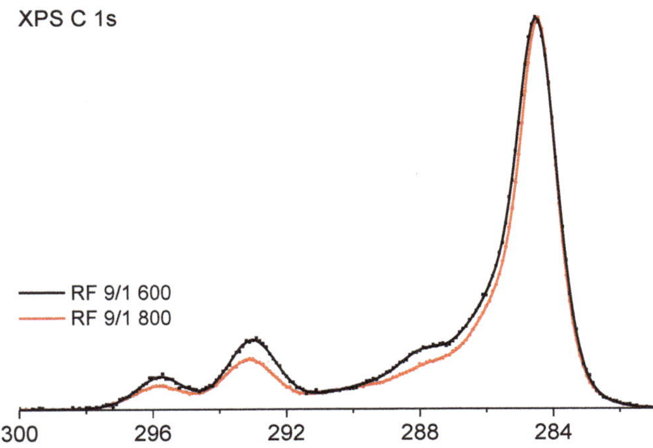

Figure 4. X-ray photoelectron spectrum of C 1s and K 2p regions for samples prepared with potassium oxalate with a potassium:carbon ratio of 9:1.

Slightly different behaviour of the surface species is observed for samples prepared with EDA. In Figure 5, the spectral region of the binding energy between 280 eV and 300 eV is displayed for two samples of carbon spheres obtained with the weight ratio potassium:carbon of 7:1 with the addition of EDA (RF_7_1_EDA600 and RF_7_1_EDA800). The position of the K 2p peaks is identical to samples without EDA admixture, indicating that the chemical state of potassium atoms is not changed by EDA presence during the preparation stage. However, the peak maximum of the C 1s line for sample RF_7_1_EDA600 is located at 284.6 eV, which is characteristic for C–C bonding in aliphatic sp^3 bonds or non-graphitic amorphous carbon. For the sample carbonized at 800 °C, the respective peak maximum of C 1s line is shifted to 284.4 eV. Similar to the samples prepared without EDA, this peak position is assigned to C–C bonds in graphitized carbon material. It is worth noting that the relative intensity of K 2p lines for 600 °C and 800 °C of carbonization is only slightly changed.

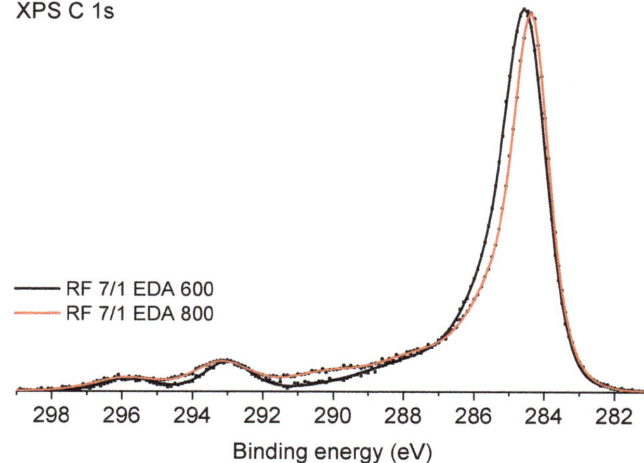

Figure 5. X-ray photoelectron spectrum of C 1s and K 2p regions for samples prepared with potassium oxalate with a potassium:carbon ratio of 7:1 with the addition of EDA.

3.3. Thermogravimetric Analysis

In order to investigate the thermal stability of the samples, thermogravimetric measurements were performed. The results of the TGA studies are shown in Figure 6 and Table 3. The reference sample began to lose mass at about 364 °C. This can be explained by the decomposition of the functional groups. Further, the decomposition of the carbon matrix occurred.

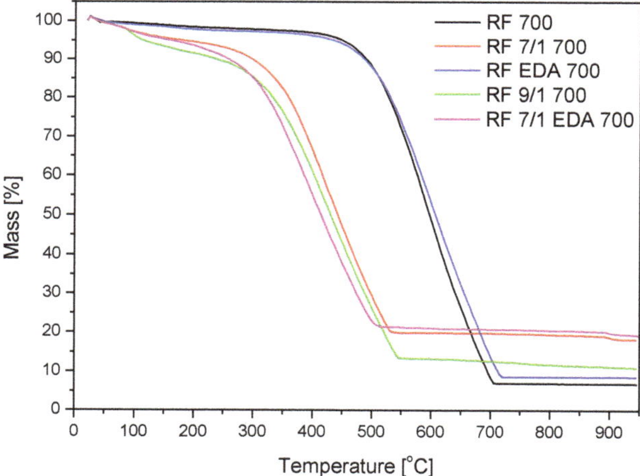

Figure 6. Results of thermogravimetric studies (heating in air).

Compared with the non-modified sample RF 700, the addition of the EDA alone did not affect the thermal stability of the sample. On the contrary, the addition of potassium oxalate led to a significant decrease of the thermal stability. All samples containing potassium oxalate were characterized by a lower decomposition temperature. Potassium ions are attracted to polar water molecules, thus the addition of potassium to the carbon matrix resulted in higher hydrophilicity of the material. The mass loss began at 100–150 °C because of the removal of water molecules. Due to mobile energized potassium ions, the depleted carbon matrix is less resistant to thermal decomposition (start of decomposition was detected at about 180 °C).

The thermal stability of the sample with the addition of both modificators was similar to that modified with potassium oxalate only.

Table 3. Results of the thermogravimetric studies.

Sample	Start of Mass Loss (°C)	Max of Mass Loss (°C)	End of Mass Loss (°C)	Residue (%)
RF 700	364	600	712	7.03
RF 7/1 600	181	420	547	16.71
RF 7/1 700	181	420	547	20
RF 7/1 800	181	432	573	14.53
RF + EDA 700	364	589	728	8.6
RF 7/1 + EDA 700	149	391	519	21

3.4. Adsorption Studies

According to the low-temperature nitrogen adsorption–desorption studies, for samples modified with oxalate only, the increase of carbonization temperature resulted in a higher volume of adsorbed nitrogen. The opposite effect was observed for samples modified with EDA only. The addition of both modificators gave a similar result to the use of oxalate only.

Some examples of low-temperature nitrogen adsorption isotherms are shown in Figure 7.

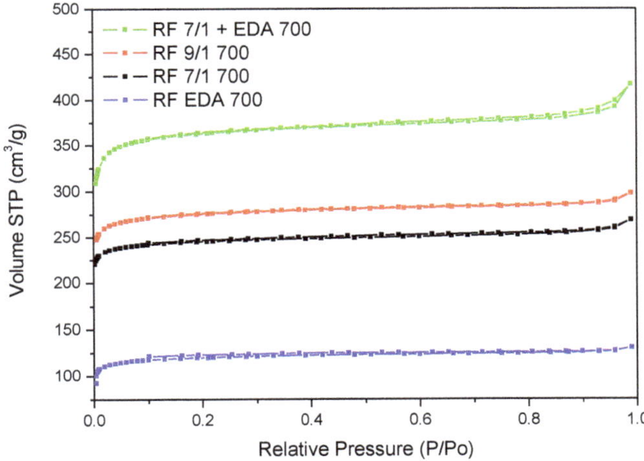

Figure 7. Low-temperature nitrogen adsorption isotherms of carbon spheres modified with potassium oxalate and/or with EDA.

The isotherms are of type Ia [35], characteristic for microporous materials, however a slight increase of adsorbed nitrogen volume at the highest P/P_0 can be attributed to the presence of macropores (type II). Spheres, modified with EDA only, adsorbed the lowest nitrogen volume. Modification with potassium oxalate resulted in higher nitrogen adsorption, slightly increasing with dopant concentration. The highest amount of nitrogen was adsorbed in the sample modified with both potassium oxalate and EDA.

Physico-chemical properties of the samples were measured, and the results are shown in Table 4. In almost all cases, except samples modified with EDA only, an increase in carbonization temperature resulted in an increase of the samples' density, specific surface area, total pore volume, and CO_2

adsorption. An unusual increase in density, simultaneously with an increase in surface area and porosity can be explained by the decomposition of modificators and removal of gaseous decomposition products. The same phenomenon was reported for activated carbon produced from palm shell and modified with potassium carbonate [36] or phosphoric acid [37].

An extremely high increase in specific surface area and CO_2 adsorption was observed for the samples modified with both potassium oxalate and EDA. In contrast, samples without potassium oxalate carbonized at higher temperatures did not exhibit larger surface area, and a lower amount of CO_2 was adsorbed (because of a lower microporosity). However, a higher concentration of the activator did not improve the specific surface area. Due to the higher saturation of the mixture, bigger spheres were formed, but oxalate moieties were not well dispersed within the volume of the sample (as shown before in SEM images).

Table 4. Physico-chemical properties of the obtained samples.

Sample	Density (g/cm^3)	S_{BET} (m^2/g)	Total Pore Volume (cm^3/g)	CO_2 Adsorption at 0 °C (mmol/g)	CO_2 Adsorption at 25 °C (mmol/g)
RF 700	1.79	444	0.25	3.25	2.43
RF 7/1 600	1.68	599	0.34	3.73	3.41
RF 7/1 700	1.97	645	0.43	5.15	3.67
RF 7/1 800	2.26	1331	0.74	5.52	3.96
RF 9/1 600	1.75	530	0.29	3.79	3.20
RF 9/1 700	1.80	843	0.46	4.43	3.74
RF 9/1 800	1.95	1252	0.68	5.56	3.87
RF EDA 600	1.59	396	0.24	2.66	1.94
RF EDA 700	1.67	369	0.20	2.57	2.01
RF EDA 800	1.72	341	0.19	3.01	2.28
RF 7/1 EDA 600	1.62	439	0.27	3.50	2.51
RF 7/1 EDA 700	1.74	1114	0.64	5.89	4.60
RF 7/1 EDA 800	1.99	1614	0.93	6.21	4.02

For adsorption of carbon dioxide, the presence of the micropores below 1 nm is considered to be most important, and the pore size distribution in this area was calculated from CO_2 adsorption at 0 °C and is shown in Figure 8.

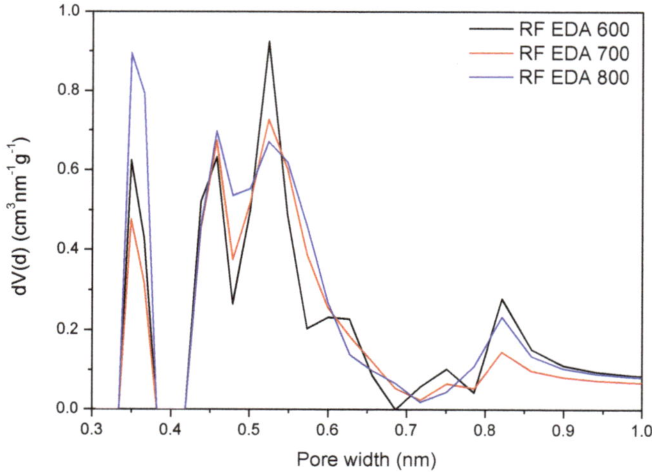

Figure 8. PSD (Pore Size Distribution) calculated from CO_2 adsorption at 0 °C for the samples modified with EDA only.

In the research paper [21], doping of the carbon spheres with EDA was described. Increasing the EDA ratio (from 0.2 mL to 0.8 mL for 0.4 g of resorcinol) led to an improvement of specific surface area and CO_2 uptake at 25 °C. The work of Sibera et al. [34] also reported a positive effect of a higher concentration of EDA as a modificator, improving the CO_2 uptake. In the present paper, more detailed studies on the influence of carbonization temperature on samples modified with EDA were performed.

The samples modified with EDA showed much lower surface areas and CO_2 adsorption than the reference sample RF 700. A higher carbonization temperature resulted in lower surface area (Table 4) and lower total pore volume, but an increase of the CO_2 uptake was observed. This observation can be explained by higher micropore volume, below 0.4 nm for the sample RF EDA 800 (Figure 8). At elevated temperatures, carbon spheres have a tendency to aggregate, thus the effective surface area decreased. Density measurements proved the increase of density (from 1.59 g/cm^3 for RF EDA 600 to 1.72 g/cm^3 for RF EDA 800). In contrast, an increase in the carbonization temperature increased the volume of the pores below 0.4 nm (Figure 8) and consequently the CO_2 adsorption capacity.

Significant differences were noticed for the samples modified with potassium oxalate. In the paper [19], Ludwinowicz and Jaroniec applied three potassium oxalate concentrations, with a K:C ratio of 3:1, 5:1, and 7:1. The growth in surface area (460 m^2/g for pure material and 2130 m^2/g for the highest concentration potassium oxalate) and CO_2 adsorption (2.8 mmol/g for pure material and 6.6 mmol/g for the highest concentration potassium oxalate) was observed. In order to investigate the influence of the activator concentration on the physico-chemical properties of the spheres, we employed a higher concentration of potassium oxalate monohydrate (weight ratio K:C = 9:1). The specific surface area values were similar to the values for samples with a lower activator concentration. A significant difference in CO_2 uptake at 0 °C and 25 °C was observed for sample RF 9/1 carbonized at 800 °C.

The microporosity of these samples carbonized in 700 °C is given in Figure 9. The sample RF 700, compared to the samples modified with potassium oxalate, had the lowest specific surface area. This was caused by a lack of energized potassium ions to interact with the carbon matrix and a lack of carbon dioxide released in the result of decomposition of potassium oxalate, creating porosity. For the sample RF 7/1 700, modified with the lower amount of activator, a significant increase of the microporosity in the range of width from 0.3 to 0.7 nm was observed. Application of the higher concentration of the activator in the sample RF 9/1 700 improved the specific surface area, but the lower amount of adsorbed CO_2 was noticed, which was in agreement with the lower volume of pores below 0.7 nm, as shown in Figure 9. The highest values of the specific surface area and CO_2 adsorption were obtained for the samples modified with potassium oxalate and EDA simultaneously. Energized potassium ions penetrated the nanocarbon material, but on the other hand, EDA improved the basicity of the material and distribution of the oxalate moieties throughout the nanocarbon sphere. For the sample RF 7/1 + EDA 700, the value of the specific surface area was twice as high as the sample modified with potassium oxalate only, but the CO_2 adsorption was only slightly higher. The microporosities of both samples with a diameter of 0.7 nm were comparable. The sample RF 7/1 + EDA 700 had a higher volume of pores from 0.7 to 0.9 nm, however this feature did not improve the CO_2 adsorption significantly. Despite the higher value of the specific surface area of the sample carbonized in 800 °C (500 m^2), the CO_2 uptake at 0 °C was only slightly better, however at 25 °C, a decrease of the adsorbed value for the sample RF 7/1 + EDA 700 was observed.

The adsorption capacity values of all samples are presented in Figure 10. The samples carbonized in 600 °C were more resistant to a decrease in CO_2 adsorption at the higher adsorption temperature. Mostly, the increase of carbonization temperature led to higher surface area and CO_2 adsorption, but a significant decrease of the adsorbed values of CO_2 at 25 °C compared to 0 °C was observed.

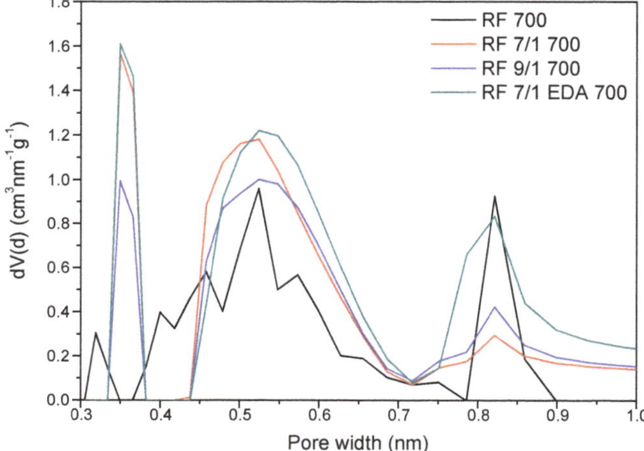

Figure 9. PSD calculated from CO_2 adsorption at 0 °C for the samples carbonized at 700 °C.

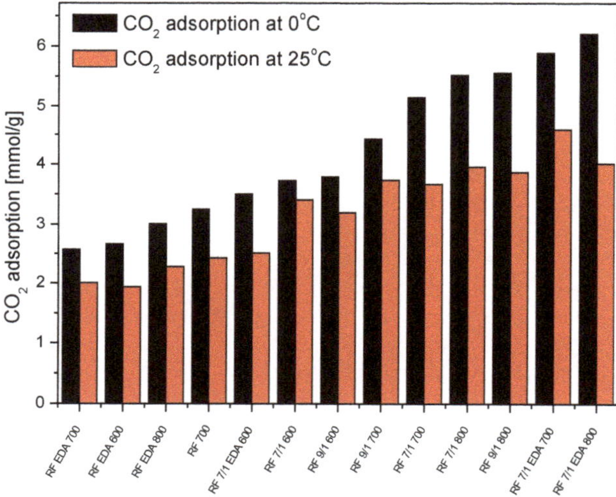

Figure 10. CO_2 adsorption values of the tested samples.

4. Discussion

The modificators played a double role in this reaction system: first creating more porosity and second, giving a basic chemical character to the produced material.

In our previous paper [33], chemical activation of carbon spheres using a similar amount of potassium oxalate monohydrate was achieved. In the case of the materials prepared with potassium oxalate monohydrate, two activation mechanisms can be distinguished. First, potassium ions penetrate the carbon material and a high porosity material was formed [38]. The effect of a higher carbonization temperature resulted in more energetic potassium ions migrating into the nanocarbon spheres. Second, decomposition of potassium oxalate monohydrate at about 570 °C resulted in the release of CO_2, which would help remove pyrolyzed volatile products from the carbon matrix and could also prevent an aggressive pore widening process, leading to better microporosity [39,40]. Potassium oxalate decomposes to release carbon dioxide and to form potassium carbonate. The latter decomposed above

700 °C, also with the release of carbon dioxide, and then, an increase of carbonization temperature from 700 to 800 °C resulted in an increase in specific surface area by over 100%, from 645 to 1331 m^2/g (Table 4). Nonetheless, the CO_2 adsorption values were only slightly enhanced (from 5.15 to 5.52 mmol/g at 0 °C).

EDA also decomposes at elevated temperatures, with the release of ammonia, carbon dioxide, carbon monoxide, nitrogen oxides, and/or volatile amines. However, the release of the gaseous decomposition products did not result in an increase of the porosity of the material. Thus, the use of EDA alone did not change the physical properties of the material.

The best results were obtained when both activators were applied. We posit that EDA reacted with potassium oxalate, forming the chelates, which improved the homogeneous distribution of potassium within the sample volume. This can be explained by the trapping of migrating potassium ions by amine groups. The ability of EDA to form chelates with metals ("amino acid metals") is well known. Half of EDA produced by the Dow Chemical Company [41] is used as a chelating agent, forming complexes with certain metal ions to prevent the ions from interfering with processing or to promote buffering, concentration, separation, or transport.

Carbonization at high temperatures caused decomposition of both modificators and of the formed chelates, nevertheless some potassium remained in the samples and had a positive effect on the adsorption properties towards carbon dioxide, increasing surface basicity.

Potassium can form carbides with carbon. According to the literature [42], there is a possibility of the formation of the following potassium carbides: KC_8, KC_{16}, KC_{24}, KC_{32}, KC_{48}, and KC_{60}. These carbides have graphite-like lattices in which the metal atoms are situated between the layers of carbon atoms. The metal atoms are located at the centers of the carbon hexagons.

Relevance of the presence of pores below 1 nm in the matter of CO_2 adsorption has been widely documented [43,44]. Presser and coworkers [45] claimed that under atmospheric pressure, the contribution of the pores below 0.8 nm to the CO_2 adsorption was the most significant. Pore size distributions of the modified samples showed that use of chemical activator was necessary, i.e., potassium oxalate monohydrate was responsible for the creation of micropores beneath 0.4 nm.

Activation with potassium led to the creation of irregularities on the surface of carbon materials [46].

Gadkaree and Jaroniec [47] investigated pore structure development in carbon materials produced from resins. They fabricated two types of carbon honeycomb structures: standard type A, which involved phenolic resin as the liquid precursor, and type B, which involved the same phenolic resin but containing cobalt acetate dissolved at 1 wt.%. Both kinds of samples were carbonized in nitrogen at a high temperature and then activated in CO_2. For samples of type A, only micropores (no mesopores at all) were formed and their volume increased as a result of the deepening of pores formed during carbonization. No pore broadening was observed for these samples. The introduction of a metallic catalyst (cobalt) in the precursor resin changed the pore structure dramatically (samples B). The pore structure on carbonization remained the same as that of the carbon without the catalyst (samples A). The difference between type A and B appeared upon activation. A bimodal pore size distribution was observed for samples B. Despite micropores of the same size range as in samples A, large meso and macropores were formed in samples B. In both kinds of samples, A and B, the micropore volume increase took place because of pore deepening, rather than pore broadening.

According to Casso et al. [48], the adsorption in pores depended on the applied pressure. With an increase of adsorption pressure, bigger and bigger pores govern the adsorption, nonetheless researchers state that at the atmospheric pressure, adsorption is contributed to by the narrow micropores (below 0.6 nm) with high adsorption potential. To use the whole adsorption potential of bigger micropores, higher fugacity of adsorbent is required. Hence, the higher pressure must be applied.

Investigation of Chen et al. [49] showed that the adsorption of CO_2 molecules in the 0.3 nm slit pores, due to the similar size of CO_2 molecules, was very poor. In contrast, the highest CO_2 adsorption was noticed in the 0.4 nm pores. Furthermore, strong decrease of the stabilization energy of CO_2 molecules in pores larger than 0.4 nm was noticed. With an increase of pore size, the interactions between CO_2 molecules and carbon pores were more and more weak.

To summarize, the presence of micropores below 0.7 nm is one of the essential traits for a good CO_2 adsorbent. Comparing the samples modified with potassium oxalate, the 7:1 weight ratio is an optimum one value, and there is no benefit to employ more.

According to the analysis of the obtained results, the adsorption properties of nanocarbon materials towards carbon dioxide increased with both increasing specific surface area and porosity (Figures 11 and 12).

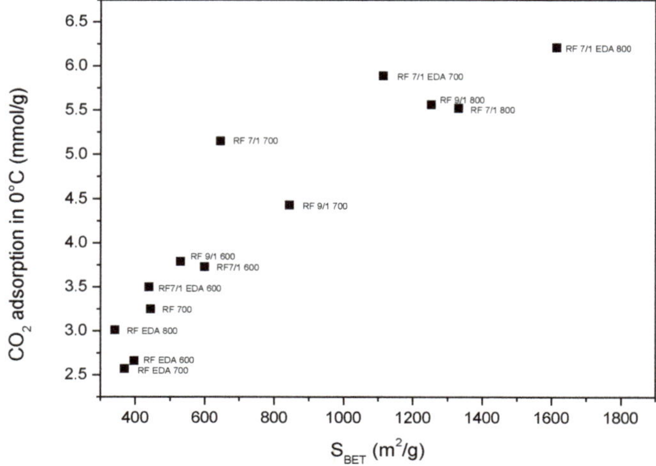

Figure 11. Relation between specific surface area and CO_2 uptake at 0 °C.

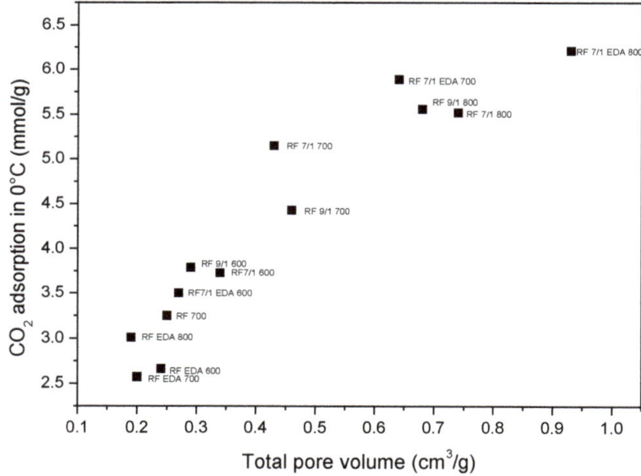

Figure 12. Relation between the total pore volume of the samples and CO_2 uptake at 0 °C.

The presence of potassium on the surface of the samples had a positive effect on the CO_2 adsorption (Figure 13), however the presence of surface oxygen had no apparent effect on adsorption. Surprisingly, the presence of surface nitrogen decreased the ability to adsorb carbon dioxide (Figure 14).

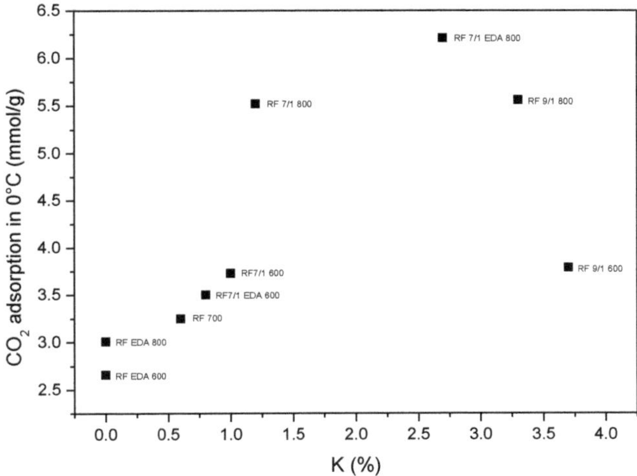

Figure 13. Relation between the surface concentration of potassium and CO_2 uptake at 0 °C.

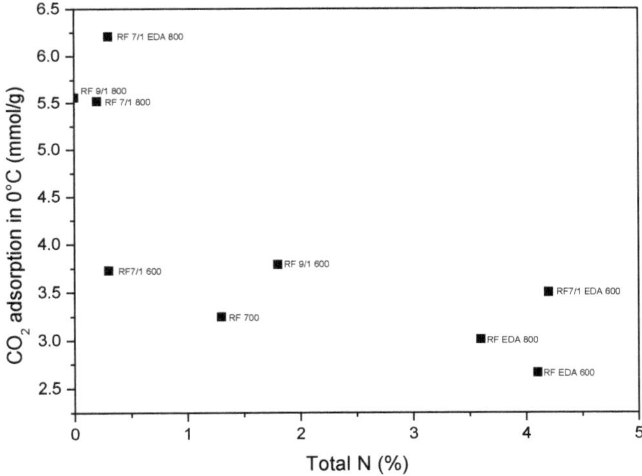

Figure 14. Relation between the surface concentration of nitrogen and CO_2 uptake at 0 °C.

5. Conclusions

Highly porous nanocarbon materials for CO_2 adsorption were produced through a novel synthesis method using a microwave assisted solvothermal reactor and varying the concentration of key reactants and modifactors. Replacement of an autoclave by a microwave assisted solvothermal reactor resulted in a significant shortening of reaction time (from several hours to minutes) and very good quality of the obtained product (uniform shape and narrow size distribution).

Using potassium oxalate monohydrate as an activator agent resulted in a high volume of micropores in the material, which are responsible for CO_2 adsorption at atmospheric pressure. EDA by itself did not improve the physicochemical properties of the carbon material, as shown in CO_2 uptake. However, use of both modificators led to the formation of a highly microporous material exhibiting both large specific surface areas and high CO_2 uptake. It is thus concluded that surface morphology (microporosity) and surface chemistry, especially an amine promotor, lead to the best CO_2

adsorption profile. Further, the differences between low temperature CO_2 adsorption (0 °C), where physisorption dominates, and higher temperature adsorption, where chemisorption is more dominant, highlights the importance of surface chemical engineering of nanocarbon materials and, additionally, the importance of surface analysis in process development and optimization using both carbonization and modificators.

Author Contributions: Investigation, P.S., D.S., D.M. and R.J.W.; Methodology, R.D.C.; Supervision, U.N.; Writing—original draft, P.S. and U.N.; Writing—review & editing, R.D.C.

Funding: This research was partially funded by the Polish-Norwegian Research Programme operated by the National Centre for Research and Development under the Norwegian Financial Mechanism 2009–2014 in the frame of Project Contract No. Pol-Nor/237761/98.

Conflicts of Interest: The authors declare no conflict of interest

References

1. Attahiru, Y.B.; Aziz, M.M.A.; Kassim, K.A.; Shahid, S.; Wan Abu Bakar, W.A.; NSashruddin, T.F.; Rahman, F.A.; Ahamed, M.I. A review on green economy and development of green roads and highways using carbon neutral materials. *Renew. Sus. Energy Rev.* **2019**, *101*, 600–613. [CrossRef]
2. Zhang, L.L.; Long, R.Y.; Chen, H.; Geng, J.C. A review of China's road traffic carbon emissions. *J. Clean. Prod.* **2019**, *207*, 569–581. [CrossRef]
3. Zhu, X.; Baran, S.; Cel, W.; Cao, Y. Sustainable approach to mitigation of CO_2 emission. *Ecol. Chem. Eng. S* **2015**, *21*, 617–622. [CrossRef]
4. D'Alessandro, D.M.; Smit, B.; Long, J.R. Carbon dioxide capture: Prospects for new materials. *Angew. Chem. Int. Ed.* **2010**, *49*, 6058–6082. [CrossRef] [PubMed]
5. OECD/IEA. *CO2 Emissions From Fuel Combustion Highlights (2017 Edition)*; OECD/IEA: Paris, France. Available online: www.iea.org (accessed on 15 April 2019).
6. Luis, P. Use of monoethanolamine (MEA) for CO_2 capture in a global scenario: Consequences and alternatives. *Desalination* **2016**, *380*, 93–99. [CrossRef]
7. Sema, T.; Naami, A.; Usubharatana, P.; Wang, X.; Gao, R.; Liang, Z.; Idem, R.; Tontiwachwuthikul, P. Mass transfer of CO_2 absorption in hybrid MEA-methanol solvents in packed column. *Energy Proc.* **2013**, *37*, 883–889. [CrossRef]
8. Samanta, A.; Zhao, A.; Shimizu, G.K.H.; Sarkar, P.; Gupta, R. Post-combustion CO_2 capture using solid sorbents: A review. *Ind. Eng. Chem. Res.* **2012**, *51*, 1438–1463. [CrossRef]
9. Hauchhum, L.; Mahanta, P. Carbon dioxide adsorption on zeolites and activated carbon by pressure swing adsorption in a fixed bed. *Int. J. Energy Environ. Eng.* **2014**, *5*, 349–356. [CrossRef]
10. Lu, C.; Bai, H.; Su, F.; Chen, W.; Hwang, J.F.; Lee, H.-H. Adsorption of Carbon Dioxide from Gas Streams via Mesoporous Spherical-Silica Particles. *J. Air Waste Manag. Assoc.* **2010**, *60*, 489–496. [CrossRef]
11. Lu, W.; Sculley, J.P.; Yuan, D.; Krishna, R.; Wei, Z.; Zhou, H.C. Polyamine-tethered porous polymer networks for carbon dioxide capture from flue gas. *Angew. Chem. Int. Ed.* **2012**, *51*, 7480–7484. [CrossRef]
12. Li, J.R.; Ma, Y.; McCarthy, M.C.; Sculley, J.; Yu, J.; Jeong, H.K.; Balbuena, P.B.; Zhou, H.C. Carbon dioxide capture-related gas adsorption and separation in metal-organic frameworks. *Coord. Chem. Rev.* **2011**, *255*, 1791–1823. [CrossRef]
13. Gui, M.M.; Yap, Y.X.; Chai, S.P.; Mohamed, A.R. Multi-walled carbon nanotubes modified with (3-aminopropyl)triethoxysilane for effective carbon dioxide adsorption. *Int. J. Green. Gas Control* **2013**, *14*, 65–73. [CrossRef]
14. Sevilla, M.; Fuertes, A.B. A general and facile synthesis strategy towards highly porous carbons: Carbonization of organic salts. *J. Mater. Chem. A* **2013**, *1*, 13738. [CrossRef]
15. Xu, B.; Zheng, D.; Jia, M.; Cao, G.; Yang, Y. Nitrogen-doped porous carbon simply prepared by pyrolyzing a nitrogen-containing organic salt for supercapacitors. *Electrochim. Acta* **2013**, *98*, 176–182. [CrossRef]
16. Zhang, X.; Li, W.; Lu, A. Designed porous carbon materials for efficient CO_2 adsorption and separation. *New Carbon Mater.* **2015**, *30*, 481–501. [CrossRef]
17. Wang, R.; Lang, J.; Yan, X. Effect of surface area and heteroatom of porous carbon materials on electrochemical capacitance in aqueous and organic electrolytes. *Sci. China Chem.* **2014**, *57*, 1570–1578. [CrossRef]

18. Choma, J.; Kloske, M.; Dziura, A.; Stachurska, K.; Jaroniec, M. Preparation and Studies of Adsorption Properties of Microporous Carbon Spheres. *Eng. Prot. Environ.* **2016**, *19*, 169–182. [CrossRef]

19. Ludwinowicz, J.; Jaroniec, M. Potassium salt-assisted synthesis of highly microporous carbon spheres for CO_2 adsorption. *Carbon* **2015**, *82*, 297–303. [CrossRef]

20. Wang, X.; Zhou, J.; Xing, W.; Liu, B.; Zhang, J.; Lin, H.; Cui, H.; Zhuo, S. Resorcinol–formaldehyde resin-based porous carbon spheres with high CO_2 capture capacities. *J. Energy Chem.* **2017**, *26*, 1007–1013. [CrossRef]

21. Wickramaratne, N.P.; Xu, J.; Wang, M.; Zhu, L.; Dai, L.; Jaroniec, M. Nitrogen enriched porous carbon spheres: Attractive materials for supercapacitor L. electrodes and CO_2 adsorption. *Chem. Mater.* **2014**, *26*, 2820–2828. [CrossRef]

22. Sevilla, M.; Fuertes, A.B. CO_2 adsorption by activated templated carbons. *J. Colloid Interface Sci.* **2012**, *366*, 147–154. [CrossRef] [PubMed]

23. Chiang, Y.-C.; Yeh, C.-Y.; Weng, C.-H. Carbon Dioxide Adsorption on Porous and Functionalized Activated Carbon Fibers. *Appl. Sci.* **2019**, *9*, 1977. [CrossRef]

24. Li, Y.; Xu, R.; Wang, B.; Wei, J.; Wang, L.; Shen, M.; Yang, J. Enhanced N-doped Porous Carbon Derived from KOH-Activated Waste Wool: A Promising Material for Selective Adsorption of CO_2/CH_4 and CH_4/N_2. *Nanomaterials* **2019**, *9*, 266. [CrossRef] [PubMed]

25. Sevilla, M.; Fuertes, A.B. Sustainable porous carbons with a superior performance for CO_2 capture. *Energy Environ. Sci.* **2011**, *4*, 1765–1771. [CrossRef]

26. Liu, J.; Qiao, S.Z.; Liu, H.; Chen, J.; Orpe, A.; Zhao, D.; Lu, G.Q. Extension of the Stöber method to the preparation of monodisperse resorcinol-formaldehyde resin polymer and carbon spheres. *Angew. Chem. Int. Ed.* **2011**, *50*, 5947–5951. [CrossRef] [PubMed]

27. Tian, H.; Liu, J.; O'Donnell, K.; Liu, T.; Liu, X.; Yan, Z.; Liu, S.; Jaroniec, M. Revisiting the Stöber method: Design of nitrogen-doped porous carbon spheres from molecular precursors of different chemical structures. *J. Colloid Interface Sci.* **2016**, *476*, 55–61. [CrossRef] [PubMed]

28. Zhao, J.; Niu, W.; Zhang, L.; Cai, H.; Han, M.; Yuan, Y.; Majeed, S.; Anjum, S.; Xu, G. A template-free and surfactant-free method for high-yield synthesis of highly monodisperse 3-aminophenol-formaldehyde resin and carbon nano/microspheres. *Macromolecules* **2013**, *46*, 140–145. [CrossRef]

29. Wickramaratne, N.P.; Jaroniec, M. Activated carbon spheres for CO_2 adsorption. *ACS Appl. Mater. Interfaces* **2013**, *5*, 1849–1855. [CrossRef] [PubMed]

30. Pari, G.; Darmawan, S.; Prihandoko, B. Porous Carbon Spheres from Hydrothermal Carbonization and KOH Activation on Cassava and Tapioca Flour Raw Material. *Procedia Environ. Sci.* **2014**, *20*, 342–351. [CrossRef]

31. Wickramaratne, N.P.; Jaroniec, M. Importance of small micropores in CO_2 capture by phenolic resin-based activated carbon spheres. *J. Mater. Chem. A* **2013**, *1*, 112–116. [CrossRef]

32. Okunev, A.G.; Sharonov, V.E.; Aristov, Y.I.; Parmon, V.N. Sorption of carbon dioxide from wet gases by K_2CO_3-in-porous matrix: Influence of the matrix nature. *React. Kinet. Catal. Lett.* **2000**, *71*, 355–362. [CrossRef]

33. Sibera, D.; Narkiewicz, U.; Kapica, J.; Serafin, J.; Michalkiewicz, B.; Wróbel, R.J.; Morawski, A.W. Preparation and characterisation of carbon spheres for carbon dioxide capture. *J. Por. Mater.* **2019**, *26*, 19–27. [CrossRef]

34. Sibera, D.; Sreńscek-Nazzal, J.; Morawski, W.A.; Michalkiewicz, B.; Serafin, J.; Wróbel, R.J.; Narkiewicz, U. Microporous carbon spheres modified with EDA used as carbon dioxide sorbents. *Adv. Mat. Lett.* **2018**, *9*, 432–435. [CrossRef]

35. Thommes, M.; Kaneko, K.; Neimark, A.V.; Olivier, J.P.; Rodriguez-Reinoso, F.; Rouquerol, J.; Sing, K.S.W. Physisorption of gases, with special reference to the evaluation of surface area and pore size distribution (IUPAC Technical Report). *Pure Appl. Chem.* **2015**, *87*, 1051–1069. [CrossRef]

36. Adinata, D.; Daud, W.M.A.W.; Aroua, M.K. Preparation and characterization of activated carbon from palm shell by chemical activation with K_2CO_3. *Bioresour. Technol.* **2007**, *98*, 145–149. [CrossRef]

37. Guo, J.; Lua, A.C. Textural and chemical properties of adsorbent prepared from palm shell by phosphoric acid activation. *Mater. Chem. Phys.* **2003**, *80*, 114–119. [CrossRef]

38. Musa, M.S.; Sanagi, M.M.; Nur, H.; Wan Ibrahim, W.A. Understanding pore formation and structural deformation in carbon spheres during KOH activation. *Sains Malays* **2015**, *44*, 613–618. [CrossRef]

39. Chen, Y.D.; Chen, W.Q.; Huang, B.; Huang, M.J. Process optimization of $K_2C_2O_4$-activated carbon from kenaf core using Box-Behnken design. *Chem. Eng. Res. Des.* **2013**, *91*, 1783–1789. [CrossRef]

40. Mohamed, M.A.; Galwey, A.K.; Halawy, S.A. The activities of some metal oxides in promoting the thermal decomposition of potassium oxalate. *Thermochim. Acta* **2002**, *387*, 63–74. [CrossRef]

41. The Dow Chemical Company. *Product Safety Assessment. DOWTM Ethylenediamine, Form No. 233-00318-MM-0515X*; The Dow Chemical Company: Midland, MI, USA, 2015.

42. Ya, T. *Kosolapova, Carbides, Properties, Production, and Applications*; Plenum Press: New York, NY, USA; London, UK, 1971.

43. Hu, X.; Radosz, M.; Cychosz, K.A.; Thommes, M. CO2-filling capacity and selectivity of carbon nanopores: Synthesis, texture, and pore-size distribution from quenched-solid density functional theory (QSDFT). *Environ. Sci. Technol.* **2011**, *45*, 7068–7074. [CrossRef]

44. Yin, G.; Liu, Z.; Liu, Q.; Wu, W. The role of different properties of activated carbon in CO_2 adsorption. *Chem. Eng. J.* **2013**, *230*, 133–140. [CrossRef]

45. Presser, V.; McDonough, J.; Yeon, S.H.; Gogotsi, Y. Effect of pore size on carbon dioxide sorption by carbide derived carbon. *Energy Environ. Sci.* **2011**, *4*, 3059–3066. [CrossRef]

46. Lee, S.Y.; Park, S.J. Determination of the optimal pore size for improved CO_2 adsorption in activated carbon fibers. *J. Coll. Interf. Sci.* **2013**, *389*, 230–235. [CrossRef] [PubMed]

47. Gadkaree, K.P.; Jaroniec, M. Pore structure development in activated carbon honeycombs. *Carbon* **2000**, *38*, 983–993. [CrossRef]

48. Casco, M.E.; Martínez-Escandell, M.J.; Silvestre-Albero, J.F.; Rodríguez-Reinoso, J.F. Effect of the porous structure in carbon materials for CO2 capture at atmospheric and high-pressure. *Carbon* **2014**, *67*, 230–235. [CrossRef]

49. Chen, L.; Watanabe, T.; Kanoh, H.; Hata, K.; Ohba, T. Cooperative CO_2 adsorption promotes high CO_2 adsorption density over wide optimal nanopore range. *Adsorpt. Sci. Technol.* **2018**, *36*, 625–639. [CrossRef]

 *applied*
sciences

Article

Carbon Footprint of the Agricultural Sector in Qinghai Province, China

Xiuhong Wang [1] and Yili Zhang [1,2,3,*]

[1] Key Laboratory of Land Surface Pattern and Simulation, Institute of Geographic Sciences and Natural Resources Research, CAS, Beijing 100101, China; wangxh@igsnrr.ac.cn
[2] CAS Center for Excellence in Tibetan Plateau Earth Sciences, Beijing 100101, China
[3] College of Resources and Environment, University of Chinese Academy of Sciences, Beijing 101408, China
* Correspondence: zhangyl@igsnrr.ac.cn; Tel.: +86-01-6485-6505

Received: 1 April 2019; Accepted: 16 May 2019; Published: 17 May 2019

Abstract: The agricultural sector has become an important emitter of greenhouse gases in China. The CO_2 emissions in the western undeveloped region have attracted less attention than those in the eastern developed region in China. In this paper, the change in carbon footprint (CF) caused by agrochemical and agricultural energy inputs, the contributions of various inputs to the total carbon footprint (TCF), and the different changing trends between carbon intensity in output value (CV) and carbon intensity in area (CA) in Qinghai province were studied based on the data for agrochemical and energy inputs over 1995–2016. The change in TCF had a roughly stable period over 1995–1999, a slowly decreasing period over 2000–2007, and a rapidly increasing period over 2008–2016, which generally synchronize with the periods of before the Grain for Green Policy (GFGP), during the GFGP, and after the GFGP, respectively. The chemical nitrogen fertilizer and energy inputs were the principal factors influencing the TCF. The N fertilizer was the highest contributor to the TCF and contributed more to the relatively lower TCF during the GFGP in the study area. The relative CF caused by plastic film and diesel input in the study area increased faster than that in the whole country. The CV declined, with a mean of 0.022 kg carbon equivalent (CE)/Chinese Yuan (CNY), which was 55.59% of the mean CV in China over 1995–2016; inversely, the CA obviously rose after 2007, with a mean of 5.11 kg CE/ha, which was only 1.94% of the mean CA in China from 1995 to 2016. Compared with the whole country, Qinghai province generally had a higher rate of increase of carbon efficiency accompanied by a higher rate of increase of CA. The improvements of local agricultural activities should aim to keep a balance between higher carbon efficiency and lower CA in the study area.

Keywords: carbon emissions; emission coefficient; agricultural land; agricultural inputs; agricultural policies; Qinghai province

---

1. Introduction

CO$_2$ emissions have kept increasing due to large-scale use of fossil fuels and have attracted global attention [1]; and recent human-induced emissions of greenhouse gases are reaching the highest levels in history [2]. Increasing human-induced CO_2 emissions may affect climate change and limit the sustainable development of human society and economy [3]. In order to study the trend and affecting factors of CO_2 emissions in different production sectors and to promote the green economic development, many researchers have used the evaluation methods of carbon footprint (CF) in different research fields [4–6].

China is a country with a long history of agricultural civilization. Because of the population increase, economic development, improvement of living standards, and limitation of agricultural land resources, the modern agricultural production mainly depends on agrochemicals and fossil fuels both in the economically developed and underdeveloped regions in China [7]. In order to combat serious ecological degradation in its economically underdeveloped regions, China has launched a few key ecological and agricultural policies or projects. Large-scale policies or projects have strongly affected the agrochemical and energy inputs. Previous studies have paid more attention to human-induced CO_2 emissions for agricultural production, their potential risks to the environment, and ways of reducing agricultural carbon emission by using CF evaluation in economically developed regions [8–11]; yet, relatively few studies have focused on the change in CF for agricultural production in its ecologically vulnerable and economically underdeveloped regions, which have also been strongly affected by the rapid development of the economy in China.

Human-induced CO_2 emissions for agricultural production and their potential risks to the environment are closely related to the area and location of agricultural land and the agrochemical and energy inputs. As a type of highly intensive agricultural land, cropland sometimes decreases due to the change of cropland into construction land for regional economic development and the change of low quality cropland to forest or grassland for ecological security, but sometimes it increases because of the change of grassland or forest to cropland for food security in the ecologically vulnerable regions in China [12–14]. The Chinese government always tries to keep a balance between decreasing the ecological degradation and increasing the agricultural efficiency. In order to reduce the adverse effects of ecological and agricultural policies on agricultural production and to support and benefit farmers, the Chinese government has launched some preferential agricultural policies after the Grain for Green Policy (GFGP) [15]. For example, focusing on "agriculture, rural areas, and farmers", 12 consecutive "Central First Documents" were released from 2004 to 2015 [16]. The implementation of the mentioned agricultural policies has effectively changed the agricultural land use structure and input structure in China, especially in the GFGP implementation area since the early 1990s. Thus, it is necessary to study the total carbon footprint (TCF) trend and its potential risks to the environment in its ecologically vulnerable and economically underdeveloped regions.

The above-mentioned ecologically vulnerable and economically underdeveloped regions are mainly located in western China, with physical geographical characteristics of the obvious change in altitude, more mountain and plateau landscapes, great daily or seasonal range of temperature, and frequently occurring droughts, floods, and strong storms [17]. The GFGP has mainly launched in western China, which mainly covers five topographic regions. Some studies have paid attention to the CO_2 emissions for agricultural production and their potential risks to the environment in northwest China [18], Loess Plateau [19], Sichuan Basin [20], and Yunnan–Guizhou Plateau [21]. In fact, the CO_2 emissions for agricultural production and their potential risks to the environment on the fragile Tibetan Plateau should be paid more attention, because the Tibetan Plateau is always regarded as a driver and amplifier of the global climate change and the birthplace of China's major rivers.

Taking Qinghai province on the Tibetan Plateau as a study area, and using carbon footprint analysis as the evaluation method, the authors have the following study goals: (1) To estimate the changing trend in TCF of the agricultural sector in Qinghai province associated with the GFGP implementation periods; (2) to recognize the main factors affecting CO_2 emissions associated with the structure of agricultural inputs; and (3) to assess the different changing trends between carbon intensity in output value and carbon intensity in area in order to hold a balance between higher carbon efficiency and lower carbon intensity in area (CA) in the study area.

2. Data Sources and Research Methodology

2.1. Study Area

Located in the northeast region of the Tibetan Plateau, Qinghai province has a geographical position of north latitude 31°39′–39°19′, east longitude 89°35′–103°04′. It covers an area of 7.18 × 10^5 km^2, ranking fourth in China (Figure 1). It has an annual average temperature between −5.6 and 8.6 °C and an annual precipitation ranging from 17.6 to 764.4 mm. The average elevation is over 3000 m, with a peak of 6860 m and a low point of 1650 m. As the birthplace of China's major rivers, it has a terrain higher in the southern and western region and lower in the eastern and middle region. Its sunshine is long and the solar radiation is strong. The annual sunshine reaches 2300–3600 h and the total radiation quantity in the province is 585.20–739.86 KJ cm^{-2}, second only to the Tibet autonomous region in China [22].

Figure 1. Division of western China and location of Qinghai province, China (The division of western China was based on Wang and Shen [17]).

Agricultural land use types in Qinghai province mainly included cropland, orchard, grassland, and forest land, with proportions of 0.76%, 0.01%, 56.25%, and 3.69%, respectively in 2006 [22]. Agricultural output value increased over 1995–2016, especially after 2007 (Figure 2). The output value from planting and stock raising contributed about 96% of the total agricultural output value. The output value from planting was basically equivalent to that from stock raising, although the latter was slightly higher than the former after the GFGP.

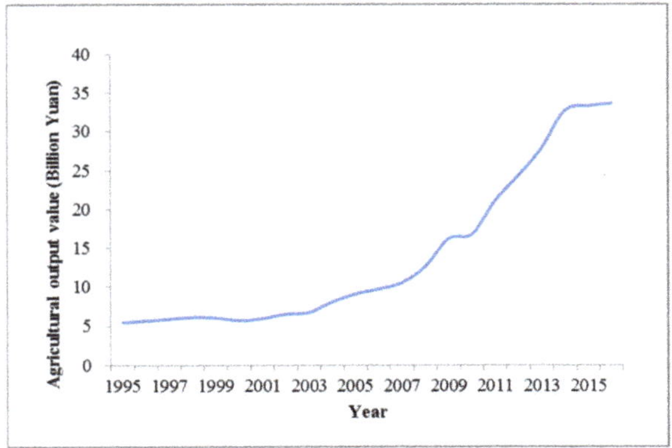

Figure 2. Agricultural output value in Qinghai province over 1995–2016.

2.2. Data Sources

The data for cropland, orchard, grassland, and forest land areas were based on the published data [22,23] and the Third Pole Environment Database [24] .The input data for chemical fertilizers, pesticides, and plastic film were obtained from the National Data for the period 1995–2016 [25]. The agricultural output value for the initial year of the study period and the annual increasing index of agricultural output value for Qinghai province over 1995–2016 were also obtained from the National Data for the period 1995–2016 [25]. The consumption data for raw coal, gasoline, diesel, and electricity were obtained from the energy consumption in the agricultural sector in Qinghai province from 1995 to 2016 [26].

2.3. Research Methodology

By considering each section of emissions of greenhouse gases in carbon equivalent (CE), the TCF in the agricultural sector was estimated over 1995–2016, which was caused by 10 main inputs of chemical fertilizers, pesticides, plastic film, raw coal, gasoline, diesel, and electricity. The TCF for agricultural production mainly consists of two parts, i.e., (1) indirect emissions of greenhouse gases from the manufacture of chemical fertilizers, pesticides, and plastic film and (2) direct emissions of greenhouse gases from the raw coal, gasoline, diesel, and electricity used by agricultural activities. Based on the published data [27], the annual emission coefficient for electricity use was estimated, which ranged from 0.23 to 0.18 kg C/kWh over 1995–2016. The coefficient for N_2O emissions from chemical fertilizer-N application in Qinghai province was based on the regional mean value of 0.0056 kg N_2O–N/kg N fertilizer in northwestern China [28], which was lower than the mean of the whole country [29]. The direct N_2O emissions from chemical fertilizer-N application in agricultural lands should be taken into account, because it was one of the major contributors to the TCF [8].

The CF of individual emission source input for agricultural production was estimated based on Equation (1):

$$CF = ES \times EC,\tag{1}$$

where *CF* is the carbon footprint in carbon equivalent (CE) for CO_2 emissions caused by individual emission source input; *ES* is the amount of individual emission source input; and *EC* is the individual emission coefficient. Table 1 lists the individual emission source and coefficient in the agricultural sector in Qinghai province.

Table 1. Main emission sources and emission coefficients in the agricultural sector in Qinghai province.

	Emission Source	Emission Coefficient (kg C/kg)	Reference
Agrochemicals	Nitrogen	1.74	[30]
	Phosphorus	0.20	[10]
	Potassium	0.15	[10]
	Pesticides	6.00	[10]
	Plastic film	2.58	[9]
Energy	Raw coal	0.52	[29,31]
	Gasoline	0.80	[29,31]
	Diesel	0.84	[29,31]

The CF for N_2O emissions caused by chemical N fertilizer use was estimated based on Equation (2):

$$CF = 127.21 \times 0.0056N,\tag{2}$$

where *CF* is the carbon footprint in carbon equivalent for N_2O emissions caused by N fertilizer use; *N* is the amount of N fertilizer used; 0.0056 is the emission coefficient for N_2O emissions caused by N fertilizer application in the study area; and 127.21 is the conversion coefficient [8,32].

The TCF caused by agrochemical and energy inputs was calculated based on Equation (3):

$$TCF = \sum_{i=1}^{10} ES_iEC_i,\tag{3}$$

where *TCF* is the total carbon footprint, ES_i is the input amount of the emission source *i*; EC_i is the emission coefficient of emission source *i*; and *i* means the main type of emission source, which ranges between 1 and 10.

Two types of carbon intensities were selected to describe the economic benefits and related environmental issues caused by CO_2 emissions for agricultural production [10]. Carbon intensity in output value (CV) means the total carbon footprint per unit of agricultural output value. Carbon efficiency can be used to directly reflect the benefit of agricultural production at the expense of agricultural CO_2 emissions, which is the reciprocal of CV. Carbon intensity in area (CA) means the total carbon footprint per unit of agricultural land area, which can be used to reflect the potential environmental problems.

In order to reflect the relative carbon efficiency and potential environmental problems in Qinghai province, comparative analysis of the changing trends of TCF, CV, and CA was used between the study area and the whole country.

3. Results and Analysis

3.1. Change in TCF for Agricultural Production

The TCF induced by 10 main types of agrochemical and energy inputs rose from 194.30 to 309.89 Kt CE (1 Kt = 10^3 t) over 1995–2016. It had a mean value of 222.22 Kt CE and a yearly rate of increase of 2.83% over the study period (Figure 3 and Table 2). Three time periods could be found for the TCF changing trend over 1995–2016, i.e., a five-year period over 1995–1999, an eight-year period over 2000–2007, and a nine-year period over 2008–2016 (Figure 3). The TCF was roughly stable over 1995–1999, with a lower yearly rate of increase of 1.57%; slowly declined over 2000–2007, with a yearly decreasing rate of 1.02%; and rapidly rose over 2008–2016, with a higher yearly increasing rate of 4.85%. The mean value of TCF decreased from the first period to the second period; yet, it increased from the second period to the third period. Distinct change in TCF occurred from 2007 to 2008. Indicated by Figure 3, the relative TCF was lower during the GFGP in the study area, which was obviously different from that in the whole country [3].

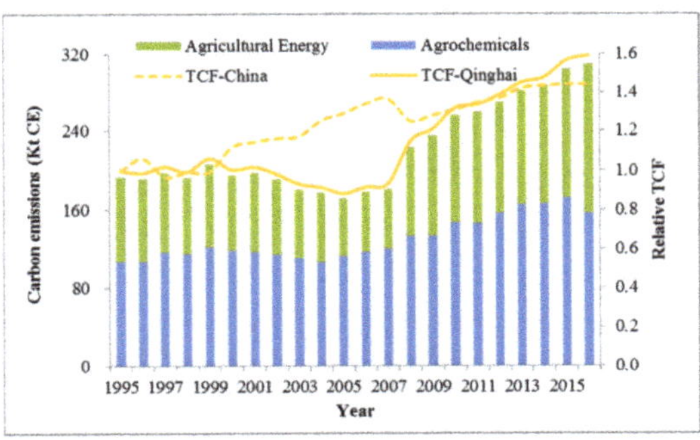

Figure 3. Changing trends of carbon footprints induce by agrochemical and agricultural energy inputs in Qinghai province over 1995–2016. TCF—total carbon footprint.

Table 2. Mean TCF and mean agricultural output values for four periods.

Year	Mean TCF (Kt CE)	TCF Growth Rate (%)	Mean AV (BCNY *)	AV Growth Rate (%)	Relative Contribution by N (%)	Relative Contribution by Energy (%)
1995–1999	196.92	1.57	5.89	2.61	49.57	42.36
2000–2007	184.19	−1.02	7.82	11.94	51.89	37.78
2008–2016	270.08	4.85	24.30	20.70	44.98	43.16
1995–2016	222.22	2.83	14.12	24.32	48.54	41.02

* Note: 1 BCNY = 10^9 CNY (Chinese Yuan); 1 Kt = 10^3 t.

3.2. Relative Contributions of Agricultural Inputs to the TCF

The relative contributions of carbon footprints induced by 10 main agricultural inputs to the TCF for four periods are displayed in Figure 4. Based on the values of relative contributions for various agricultural inputs from 1995 to 2016, the contributors could be divided into three groups, i.e., higher contributors, including nitrogen (48.54%), electricity (13.65%), and diesel (12.36%); moderate contributors, including raw coal (9.23%), gasoline (5.78%) and pesticides (4.99%); and lower contributors, including plastic film (2.56%), phosphorus (2.09%), and potassium (0.80%). The relative contributions indicated that 92.46% of the TCF was contributed by chemical fertilizers and energy consumption; while only 7.54% of the total CF was contributed by pesticides and plastic film use (Figure 4). The CF caused by nitrogen input induced N_2O was even greater than that caused by electricity use. The CF caused by energy consumption was 41.02% of the TCF. Particularly, the relative contribution of CF caused by electricity use roughly decreased from 1995 to 2011; however, it rapidly increased after 2011.

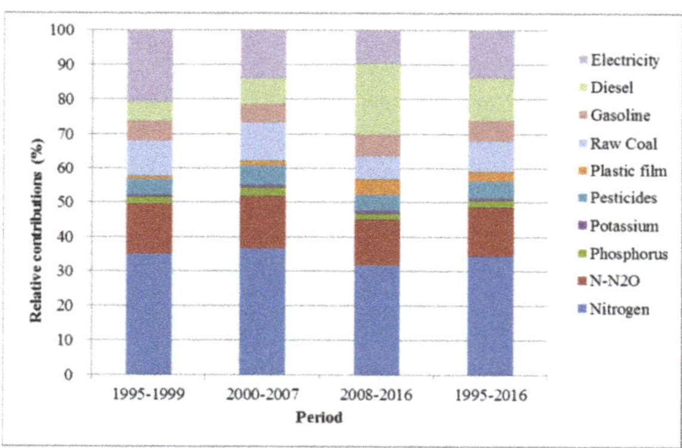

Figure 4. Relative carbon footprint (CF) contributions of 10 emission sources to the TCF.

The CF caused by plastic film was only 2.56% of the TCF; yet, it had a higher rate of increase than those caused by other agrochemical inputs over 1995–2016. The relative CF caused by plastic film in the study area increased even faster than that in China. The CF caused by diesel was 12.36% of the TCF, lower than that by electricity; yet, it had a higher rate of increase than those caused by other agricultural energy inputs over 1995–2016. The relative CF caused by diesel in the study area also increased faster than that in China (Figure 5).

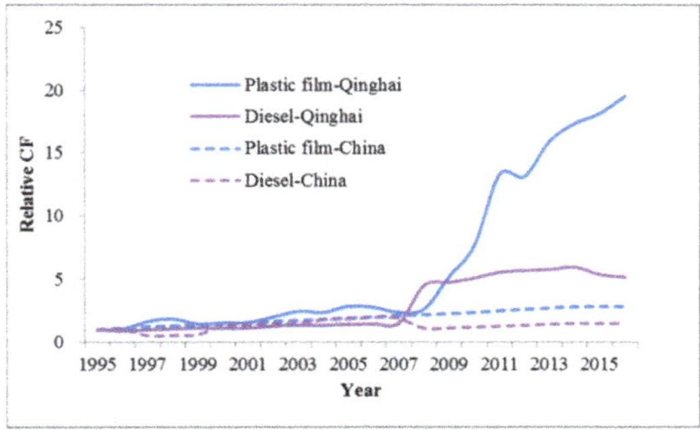

Figure 5. Changes in the relative CFs of plastic film and diesel in Qinghai province and China.

The relative CF contributions of 10 emission sources to the TCF for four time periods are shown in Table 2 and Figure 4. The changing trend of TCF was mainly controlled by that of carbon footprints induced by N fertilizer and energy inputs. The relative mean CF contribution of N fertilizer to the TCF increased from 49.57% during the period 1995–1999 to 51.89% during the period 2000–2007, then dropped to 44.98% during the period 2008–2016 (Table 2). The changing trend for relative mean CF contribution of agricultural energy to TCF was contrary to that of N fertilizer to the TCF. Therefore, higher proportion of N fertilizer input over 2000–2007 during GFGP and higher proportion of agricultural energy consumption over 2008–2016 after GFGP could be found in the study area. Compared with the whole country [3], the study area had a lower proportion of N fertilizer input and a higher proportion of agricultural energy consumption.

3.3. Changing Trends of Carbon Intensity in Output Value and Carbon Intensity in Area

The agricultural output value rapidly increased from 2007 to 2016, with an annual rate of increase of 24.32% over 1995–2016, which was about three times that for China. The CV declined from 0.035 to 0.009 kg CE/Chinese Yuan (CNY) over 1995–2016 (Figure 6). The changing trend of CV indicated that the carbon efficiency was improved, partly because both agricultural land use structure and agricultural input structure were gradually optimized according to the market demands. The average CV was 0.022 kg CE/CNY, which was only 55.59% of the average CV in China over 1995–2016 [3,33]. Based on the ratio of CV in the studied years to that in the initial year of the study period, the relative CV in Qinghai province was higher than that in China over 1995–2002; however, it was lower than that in China over 2003–2016 (Figure 7). The ratio of CV in Qinghai province to that in China was roughly decreased from 0.65 in 1995 to 0.33 in 2016 (Figure 8). Thus, the CV in Qinghai province was lower than that in China; however, the rate of increase of carbon efficiency in Qinghai province was higher than that in China.

A roughly increasing trend of CA could be found over 1995–2016; however, the CA was lower over 2000–2007 during the GFGP (Figure 6). The mean CA was 5.11 kg CE/ha, which was only 1.94% of the mean CA in China over 1995–2016. Based on the ratio of CA in the studied years to that in the initial year of the study period, the relative CA in Qinghai province was generally lower than that in China over 1995–2012; however, it was higher than that in China over 2014–2016 (Figure 7). The ratio of CA in Qinghai province to that in China was roughly decreased from 0.021 in 1995 to 0.014 in 2007; however, it was roughly increased from 0.014 in 2007 to 0.024 in 2016 (Figure 8). Thus, the CA in Qinghai province was much lower than that in China; however, the rate of increase of CA in Qinghai province accelerated after 2007, and became higher than that in China after 2013. It is clear that the increase in carbon efficiency was accompanied by an increase in CA, especially after 2007 in the study area.

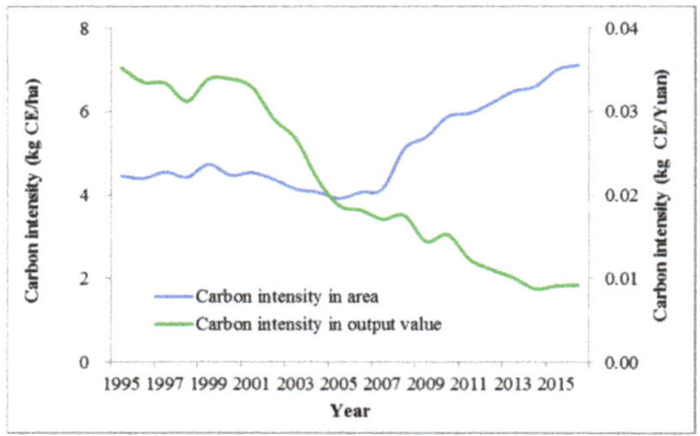

Figure 6. Change in carbon intensity in Qinghai province over 1995–2016.

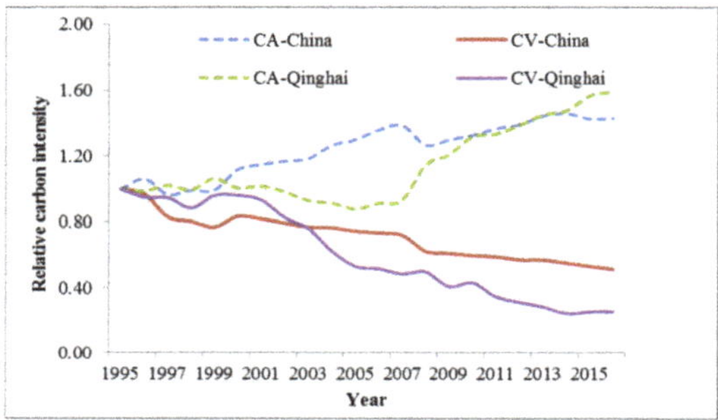

Figure 7. Change in relative carbon intensity in Qinghai province and China over 1995–2016.

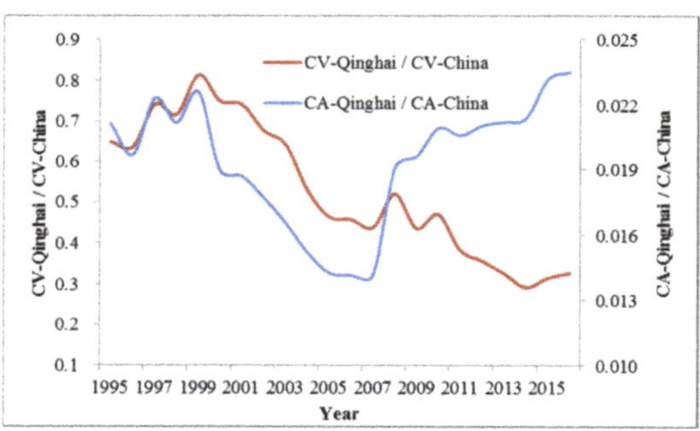

Figure 8. Change in ratio of carbon intensity in Qinghai province to that in China over 1995–2016.

4. Discussion

4.1. Related Agricultural Policies Affect the TCF

The changing trend of TCF was affected by the related national and local agricultural policies. The TCF had experienced roughly stable, slowly decreasing, and rapidly increasing periods, which were related to the periods 1995–1999 before the GFGP, 2000–2007 during the GFGP, and 2008–2016 after the GFGP. From 1995 to 1999, the TCF was roughly stable, because the food security should have been guaranteed by intensifying agricultural land use or expanding cropland area [25,26]. From 2000 to 2005, the TCF decreased, because the ecological security should have been guaranteed mainly by reducing cropland at high risk of degradation, and the agricultural inputs were relatively lower over 2000–2005 of the initial implementation period of GFGP in the study area, which was obviously different from other GFGP implementation areas in western China [34].

The TCF slightly increased between 2006 and 2007, probably because some contingency policies were implemented to keep the balance between the food security and ecological security in the late implementation period of the GFGP period [35]. In order to control the increasing TCF and to maintain the agricultural production, China adopted many specific low-carbon agricultural techniques [36]; however, the overall agricultural energy efficiency was low, especially in undeveloped western China [37]. In addition, the environmental pollution induced by higher agrochemical inputs has not attracted wide public attention because this type of concealed and dispersed pollution is difficult to detect [38]. The TCF in the study area rapidly increased between 2008 and 2016, and the CF proportion by energy input during that period was relatively higher. This was because that the cropland and orchard area had slightly increased and all agricultural inputs had generally increased after the implementation period of GFGP. The changes in agricultural policies and associated agricultural land use structure and input structure strongly affected the changing trend of TCF for agricultural production in the study area, which were also evident in China [3], Jordan [39], the Indo-Gangetic Plains [40], and the South American Chaco [41].

As a part of driver and amplifier of the global climate change, Qinghai province should control its relatively lower but increasing TCF. Based on the changing trends for CV and CA in Qinghai province over 1995–2016 (Figure 6), the input structure and intensity of agrochemical and energy during 2005–2007 may be comparatively reasonable to hold a balance between higher carbon efficiency and lower CA in the study area under certain natural and economic conditions. In order to limit the increasing TCF, it is necessary to control the application amount and use efficiency of N fertilizer, diesel, and plastic film.

4.2. Agricultural Carbon Efficiency and the Associated Environmental Problems

The decreasing trend of carbon intensity in output value and increasing trend of agricultural carbon efficiency in Qinghai province were mainly determined by the increasing proportion of the agricultural products with higher economic output value and the change in agricultural input structure over the study period. However, the population growth, increase in food and income demands, implementation of series of ecological and supporting agricultural policies, (especially the GFGP during 2000–2007), and the unique fragile plateau environment not only result in the increasing trend of carbon intensity in area but also cause other associated environmental problems.

The rural population in Qinghai province rose from 3.23 million to 3.96 million between 1995 and 2016, with an annual increment of 37.60 thousand. Meantime, the grain crop production increased from 1.42 million tons to 1.46 million tons over 1995–2016, with an annual increment of 3.52 thousand tons; while the cash crop production increased from 0.55 million tons to 2.00 million tons over the study period, with an annual increment of 77.98 thousand tons. Practically, the cash crop production exceeded the grain crop production in Qinghai province after 2009. Therefore, the increasing agricultural carbon efficiency mainly depended on the increasing cash crop production. Based on National Survey Data, compared with major grain production, vegetable production requires 2.1 times input of chemical fertilizers, 5.5 times input of pesticides, and 61.9 times input of plastic film in China, respectively [42]. Thus, more agrochemical inputs have to be used to increase the production of high-valued cash crops.

Therefore, the N fertilizer application had increased from 64.42 to 80.99 kg/ha between 1995 and 2016, with a mean of 73.54 kg/ha in Qinghai province, which was 33.91% of the mean in the whole country; while the P fertilizer application had increased from 30.51 to 45.02 kg/ha between 1995 and 2016, with a mean of 38.87 kg/ha in Qinghai province, which was 39.91% of the mean in the whole country. Based on the National Data [25], China has exceeded its safe N input limit of 170 kg/ha since 1995, and safe P_2O_5 input limit of 80 kg/ha since 2001 [43,44]. Although the chemical fertilizer applications were under the general safe environmental limits for N and P_2O_5 in Qinghai province, the indisputable fact is that all agrochemical inputs increased after 2007. Thus, the agrochemical input intensity during 2005–2007, which may be comparatively reasonable to hold a balance between higher carbon efficiency and lower CA, and the unique safe environmental limits of agrochemicals and energy on the plateau need further studying.

Currently, it is necessary to efficiently use high-quality agrochemicals and energy, and to develop more biological and eco-friendly agricultural additions on the fragile plateau in order to increase agricultural carbon efficiency and to control the associated environmental problems [45]. In addition, as the birthplace of China's major rivers, Qinghai province should impose heavy environment taxes on agrochemicals and energy and waive value-added taxes on biological and eco-friendly agricultural additions [46], which will be useful for the improvement of the local and downstream environmental conditions. New technologies and methods for carbon capture, storage and rational use (e.g., CO_2 injection enhanced oil recovery), and manufacture of high-quality and low-pollution agrochemicals and energy should be experimentally used to protect the fragile plateau [47–51].

4.3. Uncertainty of Evaluation of the TCF and the Related CV and CA

The TCF caused by 10 main types of agricultural inputs for agricultural production and the related CV and CA in Qinghai province on the Tibetan Plateau over 1995–2016 were evaluated. However, the authors should list some sources of uncertainty for evaluation of the TCF. Due to lack of relevant data, the construction, maintenance, and enhancement of agriculture infrastructure were not regarded as emission sources, which affected the evaluation of annual absolute TCF in the study area. Furthermore, the regional difference of emission coefficients for 10 main types of agrochemicals and agricultural energy also affected the TCF evaluation. Uncertainty of evaluation of CA could be unavoidable, because the 10 main types of agrochemicals and agricultural energy were mainly used in cropland and orchard. Thus, the TCF per unit of agricultural land area could not reflect the detailed CA in cropland, orchard, forest, and grassland. Uncertainty of evaluation of CV could also be unavoidable, because some new agricultural varieties were sometimes used, which affected the index for agricultural output value. However, the mentioned sources of uncertainty could not interfere with the general changing trend of CF caused by each main agricultural input in Qinghai province over 1995–2016. Referring to the study by Li et al. [52], more emission sources and detailed emission coefficients with regional characteristics should be taken into consideration to study the carbon footprint in future works. Located in the eastern agricultural area of Qinghai province, the Hehuang valley should be taken as a key study area to further study the carbon footprint of agricultural sector and its affecting factors. The valley has a land area accounting for about 26.3% of the total area of Qinghai province but a cropland area accounting for 60.8% of the province's total cropland area [53].

5. Conclusions

The socioeconomic development and implementation of a series of ecological and supporting agricultural policies since 1995 affected the changing trends of total carbon footprint, carbon intensity in output value, and carbon intensity in area for agricultural production over 1995–2016 in Qinghai province.

The changing trend of TCF for agricultural production could be divided into the three time periods of a roughly stable period 1995–1999, a slowly declining period 2000–2007, and a rapidly increasing period 2008–2016 in Qinghai province, which generally synchronized with the periods of before the GFGP, during the GFGP, and after the GFGP, respectively.

The chemical nitrogen fertilizer and energy inputs were the principal influencing factors in the change of the TCF. N fertilizer was the biggest contributor to the TCF and contributed more to the relatively lower TCF over 2000–2007 during the GFGP in the study area. In addition, the relative CF caused by plastic film and diesel in the study area increased faster than that in the whole country. Thus, the potential environmental risks caused by N fertilizer, diesel, and plastic film inputs should be monitored.

The decreasing CV generally synchronized with the increasing CA in Qinghai province. The CV declined, with a mean of 0.022 kg CE/CNY, which was 55.59% of the mean CV in China over 1995–2016; inversely, the CA obviously rose after 2007, with a mean of 5.11 kg CE/ha, which was only 1.94% of the mean CA in China from 1995 to 2016. The agrochemical input intensity during 2005–2007 may be comparatively reasonable to hold a balance between higher carbon efficiency and lower CA. Both CV and CA in the study area were lower than those in China; however, both the carbon efficiency and CA in the study area generally had a higher rate of increase than that in China, which might lead to increasing potential environmental risks in local or downstream regions.

Author Contributions: X.W. collected and analyzed the data and wrote the original draft. Y.Z. conducted the data analysis, reviewed, and edited the paper.

Funding: This study was supported by the Second Tibetan Plateau Scientific Expedition and Research (2019QZKK0600) and the Strategic Priority Research Program of the Chinese Academy of Sciences (Grant No. XDA20040200).

Conflicts of Interest: We declare no conflict of interest.

References

1. Machado, K.S.; Seleme, R.; Maceno, M.M.C.; Zattar, I.C. Carbon footprint in the ethanol feedstocks cultivation—Agricultural CO_2 emission assessment. *Agric. Syst.* **2017**, *157*, 140–145. [CrossRef]
2. IPCC. *Climate Change 2014: Synthesis Report*; Contribution of Working Groups I, II and III to the Fifth Assessment Report of the Intergovernmental Panel on Climate Change; Pachauri, R.K., Meyer, L.A., Eds.; IPCC: Geneva, Switzerland, 2014.
3. Wang, X.H. Changes in CO_2 Emissions Induced by Agricultural Inputs in China over 1991–2014. *Sustainability* **2016**, *8*, 414. [CrossRef]
4. Rowe, R.L.; Street, N.R.; Taylor, G. Identifying potential environmental impacts of large-scale deployment of dedicated bioenergy crops in the UK. *Renew. Sustain. Energy Rev.* **2009**, *13*, 271–290. [CrossRef]
5. Shen, L.; Gao, T.; Zhao, J.; Wang, L.; Wang, L.; Liu, L.; Chen, F.; Xue, J. Factory-level measurements on CO2 emission factors of cement production in China. *Renew. Sustain. Energy Rev.* **2014**, *34*, 337–349. [CrossRef]
6. Zhao, R.; Huang, X.; Zhong, T.; Peng, J. Carbon footprint of different industrial spaces based on energy consumption in China. *J. Geogr. Sci.* **2011**, *21*, 285–300. [CrossRef]
7. Wang, X. Spatio-temporal changes in agrochemical inputs and the risk assessment before and after the grain-for-green policy in China. *Environ. Monit. Assess.* **2013**, *185*, 1927–1937. [CrossRef] [PubMed]
8. Cheng, K.; Pan, G.; Smith, P.; Luo, T.; Li, L.; Zheng, J.; Zhang, X.; Han, X.; Yan, M. Carbon footprint of China's crop production-An estimation using agro-statistics data over 1993–2007. *Agric. Ecosyst. Environ.* **2011**, *142*, 231–237. [CrossRef]
9. Huang, Z.; Mi, S. Agricultural sector carbon footprint accounting: A case of Zhejiang, China. *Issues Agric. Econ.* **2011**, *11*, 40–47.
10. Dubey, A.; Lal, R. Carbon footprint and sustainability of agricultural production systems in Punjab, India, and Ohio, USA. *J. Crop Improv.* **2009**, *23*, 332–350. [CrossRef]
11. Wu, J.F.; Wang, X.H. Dynamic changes in the carbon intensity and sustainability of farmland use: A case study in Pingdu County, Shandong Province, China. *Acta Ecol. Sin.* **2017**, *37*, 2904–2912.
12. Feng, Z.; Yang, Y.; Zhang, Y.; Zhang, P.; Li, Y. Grain-for-green policy and its impacts on grain supply in West China. *Land Use Policy* **2005**, *22*, 301–312. [CrossRef]
13. Lin, Y.; Yao, S. Impact of the Sloping Land Conversion Program on rural house hold income: An integrated estimation. *Land Use Policy* **2014**, *40*, 56–63. [CrossRef]
14. Wang, X.; Lu, C.; Fang, J.; Shen, Y. Implications for development of grain-for-green policy based on cropland suitability evaluation in desertification-affected north China. *Land Use Policy* **2007**, *24*, 417–424. [CrossRef]
15. Liu, H. Evolution of Chinese agricultural supporting policies and rural development. *China Agric. Inf.* **2012**, *15*, 24–25.
16. Sun, T. Study of Chinese agricultural policy changes. *J. Anhui Agri. Sci.* **2015**, *43*, 346–348.
17. Wang, X.; Shen, Y. Ecological restoration in West China: Problems and Proposals. *AMBIO* **2009**, *38*, 177–180. [CrossRef] [PubMed]

18. Xiong, C.; Yang, D.; Huo, J. Spatial-temporal characteristics and LMDI-Based impact factor decomposition of agricultural carbon emissions in Hotan prefecture, China. *Sustainability* **2016**, *8*, 262. [CrossRef]

19. Chen, R.; Xu, C.; Deng, Y.; Jiang, Z. Research on low carbon agricultural production modes in Loess Plateau. *J. Northwest A F Univ. (Soc. Sci. Ed.)* **2017**, *17*, 55–65.

20. Hu, A.; Wang, Y. Research on low carbon agriculture development in Sichuan under the background of low carbon economy. *Sichuan Environ.* **2015**, *34*, 110–114.

21. Zhang, J.; Liu, L. Decoupling analysis between agricultural economic development and agricultural production energy consumption in Yunnan. *Environ. Sci. Surv.* **2018**, *37*, 29–34.

22. Si, H.; Yuan, C.; Zhou, W. Effect of land-use on ecosystem service values in Qinghai province. *Agric. Res. Arid Areas* **2016**, *34*, 254–260.

23. Liu, Y. *Investigation Datasets of China Land Resources*; Office of National Land Resources Survey: Beijing, China, 2000.

24. Third Pole Environment Database. Available online: http://www.tpedatabase.cn (accessed on 15 March 2019).

25. National Data, 1995–2016. Available online: http://data.stats.gov.cn (accessed on 15 March 2019).

26. National Bureau of Statistics of China (NBSC). *China Energy Statistical Yearbook, 1996–2017*; China Statistics Press: Beijing, China, 1996–2017.

27. Huo, M.; Han, X.; Shan, B. Empirical Study on Key Factors of Carbon Emission Intensity of Power Industry. *Electr. Power* **2013**, *46*, 122–126.

28. National Development and Reform Commission of China (NDRCC). Guidelines for Provincial Greenhouse Gas Inventories (Trial). Available online: http://www.doc88.com/p-9819327912648.html (accessed on 15 March 2019).

29. Lu, Y.; Huang, Y.; Zou, J.; Zheng, X. An inventory of N_2O emissions from agriculture in China using precipitation-rectified emission factor and background emission. *Chemosphere* **2006**, *65*, 1915–1924. [CrossRef]

30. Lu, F.; Wang, X.; Han, B. Assessment on the availability of nitrogen fertilization in improving carbon sequestration potential of China's cropland soil. *Chin. J. App. Ecol.* **2008**, *19*, 2239–2250.

31. General Administration of Quality Supervision, Inspection and Quarantine of the People's Republic of China (AQSIQ); Standardization Administration of the People's Republic of China (SAC). *General Principles for Calculation of the Comprehensive Energy Consumption (GB/T 2589-2008)*; China Standard Press: Beijing, China, 2008.

32. IPCC. *2006 IPCC Guidelines for National Greenhouse Gas Inventories, Prepared by the National Greenhouse Gas Inventories Programme*; Eggleston, H.S., Buendia, L., Miwa, K., Ngara, T., Tanabe, K., Eds.; IGES: Tokyo, Japan, 2006.

33. FAOSTAT. 2016. Available online: http://faostat3.fao.org/download/R/RL/E (accessed on 18 March 2019).

34. Wang, X.; Shen, J.; Zhang, W. Emergy evaluation of agricultural sustainability of Northwest China before and after the grain-for-green policy. *Energy Policy* **2014**, *67*, 508–516. [CrossRef]

35. He, J.; Feng, J. The stage features and policy choices in China's agriculture development: A comparative analysis under the perspective of "Four Stages Theory" in international agricultural development process. *Chin. Agric. Sci. Bull.* **2010**, *26*, 439–444.

36. Luan, Y.; Ren, J. Agricultural total factor energy efficiency of China and its convergence. *Chin. J. Agric. Resour. Reg. Plan.* **2014**, *35*, 20–24.

37. Zhang, X. Evaluation of China agriculture energy efficiency based on energy method. *Spec. Zone Econ.* **2014**, *10*, 143–147.

38. Song, J.; Li, Y.; Song, Y.; Yan, J.; Zhou, L. Research and prospect on non-point pollution from agriculture. *Chin. Agric. Sci. Bull.* **2010**, *26*, 362–365.

39. Ismael, M.; Srouji, F.; Mohamed, B.A. Agricultural technologies and carbon emissions: Evidence from Jordanian economy. *Environ. Sci. Pollut. Res.* **2018**, *25*, 10867–10877. [CrossRef]

40. Benbi, D.K. Carbon footprint and agricultural sustainability nexus in an intensively cultivated region of Indo-Gangetic Plains. *Sci. Total Environ.* **2018**, *644*, 611–623. [CrossRef]

41. Baumann, M.; Gasparri, I.; Piquer-Rodríguez, M.; Gavier, P.G.; Griffiths, P.; Hostert, P.; Kuemmerle, T. Carbon emissions from agricultural expansion and intensification in the Chaco. *Glob. Chang. Biol.* **2017**, *23*, 1902–1916. [CrossRef]

42. Wang, X. Agricultural material inputs and the potential risk assessment for vegetable production in China. *J. Resour. Ecol.* **2016**, *7*, 269–274.

43. Yang, S.; Han, R.; Liu, C. Study on the given amount per unit field and load capacity of livestock and poultry manure at provincial scale. *J. China Agric. Univ.* **2016**, *21*, 142–151.

44. Zhu, Z. Loss of fertilizer N from plants-soil system and the strategies and techniques for its reduction. *Soil Environ. Sci.* **2000**, *9*, 1–6.

45. Wang, X. Sustainable development in Tibet requires control of agricultural nonpoint pollution. *Environ. Sci. Technol.* **2014**, *48*, 8944–8945. [CrossRef]

46. Luo, Y.; Long, X.; Wu, C.; Zhang, J. Decoupling CO_2 emissions from economic growth in agricultural sector across 30 Chinese provinces from 1997 to 2014. *J. Clean. Prod.* **2017**, *159*, 220–228. [CrossRef]

47. Jia, B.; Tsau, J.; Barati, R. Role of molecular diffusion in heterogeneous, naturally fractured shale reservoirs during CO_2 huff-n-puff. *J. Pet. Sci. Eng.* **2018**, *164*, 31–42. [CrossRef]

48. Jia, B.; Tsau, J.; Barati, R. A review of the current progress of CO_2 injection EOR and carbon storage in shale oil reservoirs. *Fuel* **2019**, *236*, 404–427. [CrossRef]

49. Plasynski, S.I.; Litynski, J.T.; McIlvried, H.G.; Srivastava, R.D. Progress and New Developments in Carbon Capture and Storage. *Crit. Rev. Plant Sci.* **2009**, *28*, 123–138. [CrossRef]

50. Mi, J.; Ren, J.; Wang, J.; Bao, W.; Xie, K. Ultrasonic and Microwave Desulfurization of Coal in Tetrachloroethylene. *Energy Sources Part A Recovery Util. Environ. Eff.* **2007**, *29*, 1261–1268. [CrossRef]

51. Bui, M.; Adjiman, C.S.; Bardow, A.; Anthony, E.J.; Boston, A.; Brown, S.; Fennell, P.S.; Fuss, S.; Galindo, A.; Hackett, L.A.; et al. Carbon capture and storage (CCS): The way forward. *Energy Environ. Sci.* **2018**, *11*, 1062–1176. [CrossRef]

52. Li, S.; Zhang, Y.; Wang, Z.; Li, L. Mapping human influence intensity in the Tibetan Plateau for conservation of ecological service functions. *Ecosyst. Serv.* **2018**, *30*, 276–286. [CrossRef]

53. Wang, X.; Zhang, H. On the development and utilization of land resources in Hehuang valley in Qinghai province. *New Heights* **2013**, *32*, 113–115.

Article

Air Purification Performance of Photocatalytic Concrete Paving Blocks after Seven Years of Service

Hubert Witkowski [1,*], Wioletta Jackiewicz-Rek [2], Karol Chilmon [2], Janusz Jarosławski [3], Anna Tryfon-Bojarska [4,5] and Arkadiusz Gąsiński [6]

[1] Department of Building Physics and Building Materials, Lodz University of Technology, 90-924 Lodz, Poland
[2] Faculty of Civil Engineering, Warsaw University of Technology, 00-637 Warsaw, Poland;
w.jackiewicz-rek@il.pw.edu.pl (W.J.-R.); k.chilmon@il.pw.edu.pl (K.C.)
[3] Institute of Geophysics, Polish Academy of Science, 01-452 Warsaw, Poland; januszj@igf.edu.pl
[4] Skanska CDE, CEE Market, 00-877 Warsaw, Poland;anna.tryfon@skanska.pl
[5] SGH Warsaw School of Economics, 02-554 Warsaw, Poland
[6] Institute of Ceramics and Building Materials, 02-676 Warsaw, Poland; akgasin@gmail.com
* Correspondence: eng.hubert.witkowski@gmail.com; Tel.: +48-502-746-797

Received: 28 March 2019; Accepted: 15 April 2019; Published: 26 April 2019

Abstract: This paper presents the results of laboratory tests on photocatalytic pavement blocks from a bicycle lane in Poland after seven years of service. Air purification performance was tested on dusty and clean samples using different light sources and setups, with non-laminar gas circulation. Secondary Electrons Secondary Ions (SESI) and InLens detectors combined with SEM–EDS and X-ray analyses were applied to confirm the presence of TiO_2 in the studied blocks. The obtained results show that TiO_2 was present in the form of agglomerates with a diameter of 0.25–5 μm and was bonded to the cement matrix components. The tested samples still maintained nitric oxide (NO) removal capability with a NO reduction rate of 4–45%, depending on light source and surface cleanliness.

Keywords: photocatalytic concrete pavement; NO reduction; SEM analysis

1. Introduction

The problem of deteriorating air quality in urban areas has become one of the major challenges of recent times. With the rapid growth of metropolises, the problem of airborne pollution increases [1]. Considerable attention is therefore given to solutions that may reduce the concentration of harmful compounds, such as nitrogen oxides (NO_x), in the air. Short-term exposure to nitrogen dioxide (NO_2) leads to irritation of the upper respiratory tract, and long-term exposure to NO_2 leads to chronic diseases. Nitric oxide (NO) is significantly less harmful to human health, however, in contact with air it oxidizes to form NO_2.

The use of photocatalytic concrete in urban areas can contribute to a reduction in the concentration of NO_x in the air. Concrete is the most common construction material in use, hence its adoption to reduce the concentration of NO_x is a very promising solution.

The implementation of photocatalytic cement-based materials has been the subject of a number of research projects and applications across Europe and North America [2–6]. The photocatalytic properties of these materials are provided by the use of TiO_2 nanoparticles in cement, or in surface suspension. The mechanism of the photocatalytic reaction of TiO_2 has been described by Fujishima and Honda [7]. When TiO_2 (semiconductor) is illuminated with high-energy photons, whose energy is equal to or greater than semiconductor band–gap energy, electrons transfer from the valence band to the conduction band. This starts a series of oxidation–reduction reactions with substances adsorbed on the semiconductor surface, which leads to the creation of hydroxyl radicals (OH) (Figure 1).

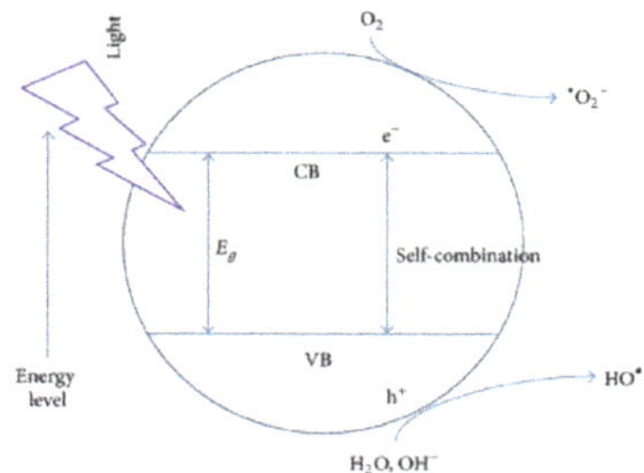

Figure 1. Photocatalyticprocess scheme [8].

The following process of NOx reduction can be illustrated with the following equations:

$$NO + OH \xrightarrow{hV,\ TiO_2} NO_2 + H^+ \tag{1}$$

$$NO_2 + OH \xrightarrow{hV,\ TiO_2} NO_3^- + H^+ \tag{2}$$

On a concrete surface, NO_3^- is created, which reacts with the cementitious compounds of concrete. Products of this reaction (nitric soils) are removed by rainwater from the concrete surface. As a result of this reaction, the concentration of nitric oxides in the vicinity of the surface is reduced, and the effect of air purification is observed.

The photocatalytic properties of concrete with the addition of TiO_2 have been proved in a number of laboratory studies [9–14]. The results of a research described by Beeldens [15] indicated that the major factors affecting the reaction are UV light intensity, surface exposure to UV light, pollutant concentration, ambient temperature, and air flow rate. As the research showed, the best results were obtained at high UV intensity, with a temperature above 25°C, with low relative humidity and low air flow.

Another important aspect is the durability of the air purifying capacity over time, especially in the case of pavement materials exposed to abrasion and soil. A case study of a street in Bergamo [16] indicated that paving blocks may still show a satisfactory air purification performance after two years of service. The same study also indicated that the effectiveness of the blocks was strongly dependent on surface cleanliness. Research showed that the reduction of NOx was significantly higher on days when the paving blocks were cleaned.

Unfortunately, the number of studies on air purification performance over time is very limited, particularly in relation to the effectiveness of the blocks after long-term usage in a moderate climate, where pavement materials are exposed not only to abrasion and soil, but also to cyclic freezing and thawing through the presence of deicing salt.

This paper presents the results of laboratory tests on pavement blocks collected from a bicycle lane in Zielona Gora (Poland). The aim of the study was to verify the air purification performance of photocatalytic concrete paving blocks after long-term service (seven years) in a moderate climate.

2. Experimental Procedure

Laboratory tests were conducted on two 330 × 150 × 80 mm exposed aggregate pavement blocks (Figure 2) collected after seven years of service from a bicycle lane that runs along one of the main roads in the city. The pavements blocks were collected from the $2m^2$ area of the bicycle lane. The top layer (5 mm thick) of each sample was made of concrete containing CEM II/A–S 42.5 R (EN 197-1) with nano-TiO_2. The pavement blocks were produced in accordance with the EN 1339:2005 standard. The characteristic bending strength declared by the producer was 5.0 MPa (each single bending test resulted in no less than 4.0 MPa according to EN 1339:2005), and the declared water absorbability was less than 6%.

Figure 2. Photocatalytic concrete pavement block.

Samples were tested to determine photocatalytic activity with a special test setup. First, the reduction of NO was measured in controlled conditions. The concrete blocks were then examined with a Scanning Electron Microscope (SEM) using Energy Dispersive X-ray Spectroscopy (EDS) and SEM elemental mapping. The main objective of the research was to verify the photocatalytic properties after seven years of service.

2.1. Test for Air Purification Performance

To determine the efficiency of photocatalytic materials in air purification, a number of different test procedures have been developed. The most often applied methods are ISO 22197-1:2016 [17], UNI–11247:2010 [18], and JIS TR Z 0018 [19]. Test methods assume laminar flow of the gas, as the distance between the sample and the glass window is very narrow (about 5 mm). Although such a set–up provides ideal conditions for measuring the reduction of NO_x concentration at a constant gas flow, in real conditions the reaction of NO_x reduction does not have the laminar character of a gas flow. Therefore, a novel setup (Figure 3) was developed to measure the air purification performance of the collected sample.

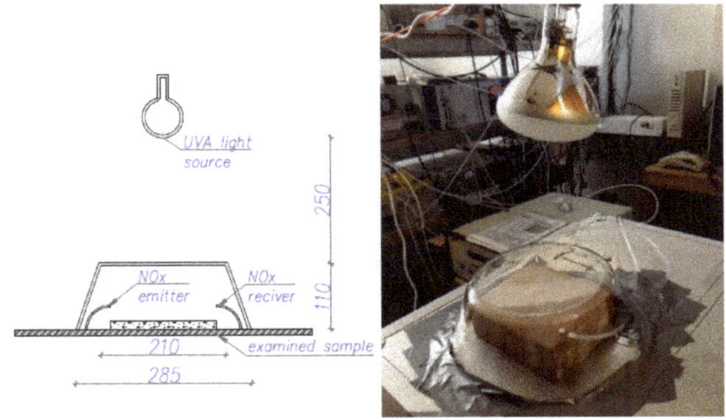

Figure 3. Test for air purification performance of photocatalytic concrete pavement blocks.

The developed setup provided gas circulation analogous to the conditions in the real project. The applied gas was a mixture of synthetic air (20% oxygen, 80% nitrogen) and NO in concentration of 100 ppb. The evaluation of the measurement was analogous to the procedure described by Husken et al. [10]. The NO abatement was defined as the ratio of the average NO concentration after turning the UV light on to the NO concentration before turning the UV light on. The flow rate in the research was 120 L/hour. The concentration of NO was measured with an API Model 200A NO_x Monitor with an accuracy of ±5%. To provide UV light, two types of light source were applied: 70 W and

300 W. The reduction efficiency depends on the intensity of UV light, with a wave length in the range of 300–400 nm. The spectra obtained from each light source were therefore measured, including a measure of light spectra after passing the glass (Figure 4).

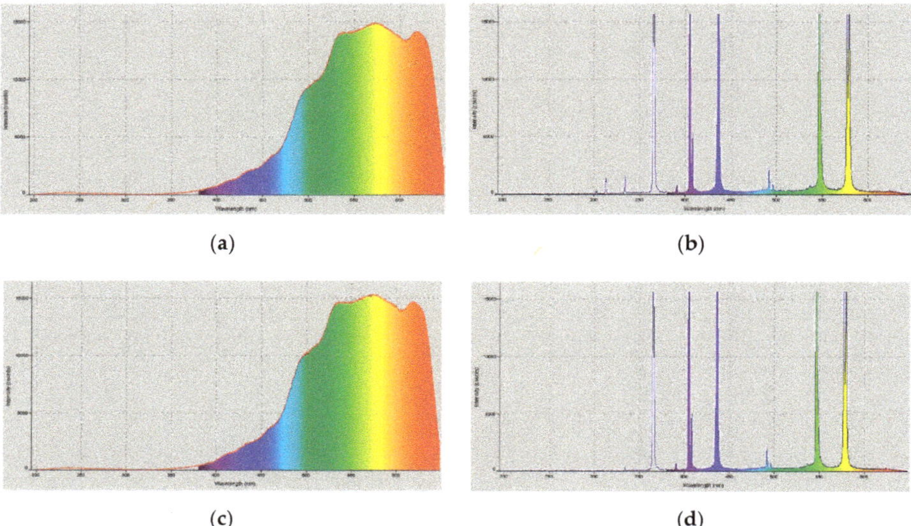

(a)

(b)

(c)

(d)

Figure 4. Spectraobtained by applying two UV light sources. (**a**) Original spectra of 70 W bulb, (**b**) original spectra of 300 W bulb, (**c**) spectra after passing the desiccator glass, 70 W bulb, and (**d**) spectra after passing the desiccator glass, 300 W bulb.

The tested concrete samples had to be cut to fit to the desiccators and had a diameter of 180×150 mm (270 cm^2). The test procedure assumed achieving a maximum constant gas concentration of approximately 95 ppb in the desiccator, then the UV light was turned on for at least 20 min, and the reduction of NO was measured until a constant gas concentration was achieved. The average temperature of the measurement was 25°C with an average relative humidity of 50%.

2.2. SEM Analysis

For the SEM analysis, the Sigma VP (Zeiss), equipped with two EDS XFlash 6/10 (Bruker) detectors, was applied. As a standard, a voltage-accelerating electron beam of 15 kV was used. In the study of chemical composition with the EDS method, a 120 μm aperture was used, while for the high-resolution imaging, a 30 μm aperture was applied. The studied samples were collected from the concrete block surface. In order to ensure the discharge of electric charges on the surface of the test sample, the samples were sprayed with carbon and secured with a special strip of electrical charge. SEM tests were carried out in high-vacuum conditions (pressure $< 1 \times 10^{-5}$ Pa).

3. Results

The samples were tested according to the procedure and setup described. Therefore, the results cannot be compared to the results of standard tests.

3.1. NO Reduction

At the beginning, the samples were tested with a 70 W light source without any treatment (the samples were dusty). Then, the samples were cleaned with pure water, dried, placed again into the glass desiccator, and tested with two different light sources. The results are presented in Figures 5–7 and Table 1.

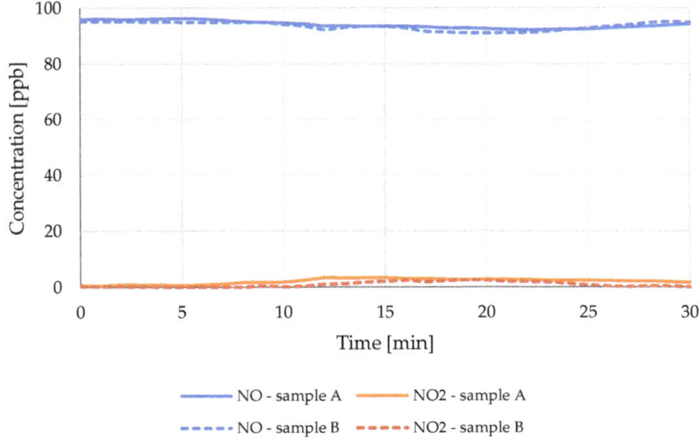

Figure 5. Concentration of nitrogen oxides (NO_x) over time (dusty samples, 70 W light source).

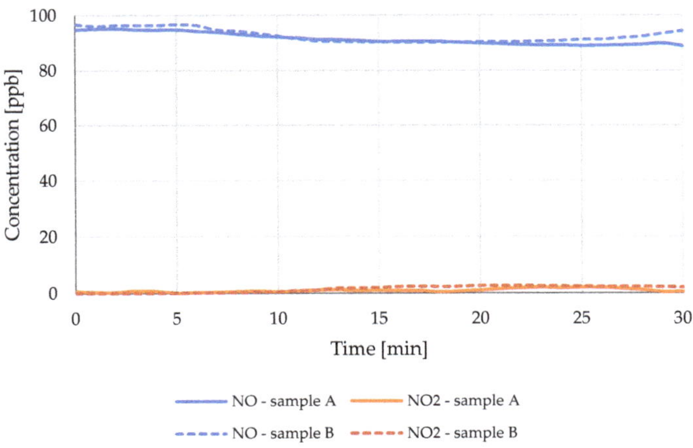

Figure 6. Concentration of NO_x over time (cleaned samples, 70 W light source).

Table 1. Air purification performance test results.

Light Source	70 W	70 W	300 W
Surface Cleanliness	**Dusty**	**Cleaned**	**Cleaned**
ΔNO *	−4 ppb ± 1.3	−5 ppb ± 0.3	−43 ppb ± 1.0
ΔNO_2 *	+3 ppb ± 0.6	+2 ppb ± 0.5	+9 ppb ± 3.1
ΔNOx * ($\Delta NO + \Delta NO_2$)	−1 ppb	−4 ppb	−34 ppb
NO reduction * %	4%	5%	45%

* mean results (±standard deviation) calculated from the maximum observed values during experiments for samples A and B.

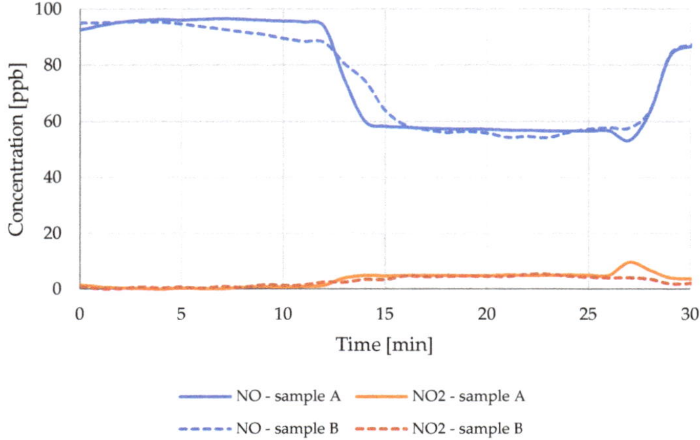

Figure 7. Concentration of NO_x over time (cleaned sample, 300 W light source).

The obtained results indicate that after seven years of service the tested paving blocks were still able to reduce NO content in the air (Table 1). The effectiveness of this phenomenon strongly depended on UV light intensity. The percentage of NO abatement was significantly higher when a 300 W light source was used. The surface cleanliness had a minor influence on NO abatement when the 70 W light source was used.

During the experiments, an increase of NO_2 content was observed. This phenomenon occurred, because of the complex characteristics of the photocatalytic process. Before NO finally oxidizes to NO_3^- or HNO_4, the formation of HNO_3 and NO_2 occurs. NO_2 is about 5 to 25 times more toxic than NO [20]. This indicates that in some cases, the air quality may be degraded rather than improved. Therefore, when studying the air purification performance of photocatalytic materials, both NO removal and NO_2 formation must be considered [21]. The maximum average increase of NO_2 content was 8 ppb compared with a 43 ppb average reduction in NO content. Only in the case of weak light and a dusty sample was the $|\Delta NO_2|$ equal to $|\Delta NO|$.

3.2. SEM Analysis

SEM analysis allows for a precise examination of a sample surface. However, in the case of concrete with TiO_2 nanoparticles, such analysis is difficult. In the study, a Secondary Electrons Secondary Ions Detector (SESI) and an InLens detector were applied. An InLens detector was used to map the surface of the sample, and the SESI detector was used for images with fine detail. Using these detectors, it was possible to make morphological studies of particles and agglomerates, although the chemistry of these particle had to be confirmed by EDS analysis. EDS mapping was therefore performed first to indicate areas with a higher and lower content of titanium. Morphological studies of both types of area were then carried out using secondary electrons techniques. This approach enabled an investigation of the content of titanium dioxide and the morphology of its particles. Results are presented in Figures 8–10. The obtained images confirmed the presence of TiO_2 particles in the studied sample. Titanium dioxide was present in the form of agglomerates with a diameter of 0.25–5 μm. Titanium dioxide agglomerates were bonded to the compounds of the cement matrix.

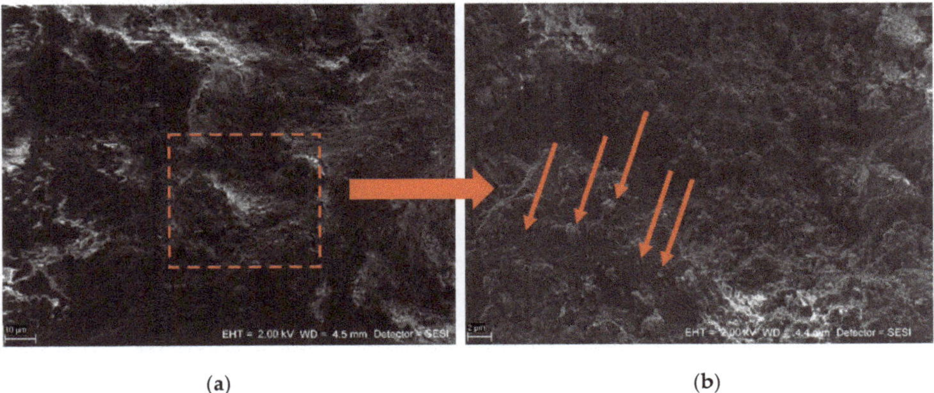

(a)   (b)

Figure 8. SEMimage of the sample surface (SESI detector at ETH = 2.00 kV). (**a**) General view, (**b**) magnification of the selected area. Red arrows indicate TiO_2 agglomerates on the observed surface.

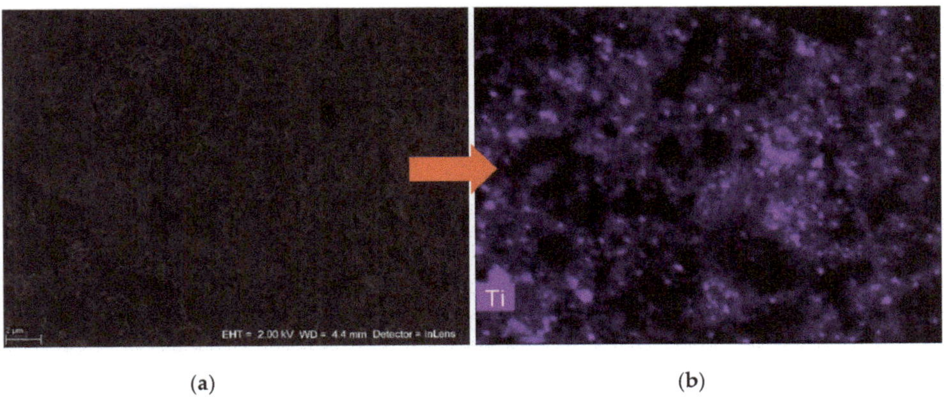

(a)   (b)

Figure 9. SEM image of the sample surface (InLens detector at ETH = 2.00 kV). (**a**) Morphology, (**b**) EDS maps of titanium in the presented region.

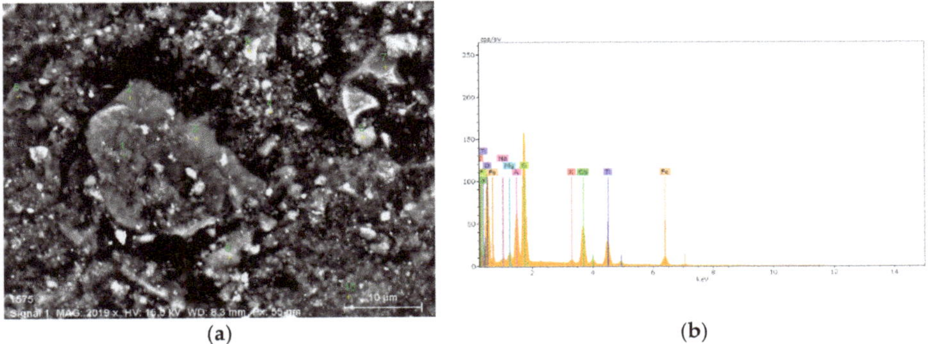

(a)   (b)

Figure 10. (**a**) SEM image with indicated region of the X-ray analysis, (**b**) X-ray analysis of the indicated region.

For further analysis, scanning electron microscopy coupled with energy-dispersive spectrometry (SEM–EDS) was used to conduct quantitative X-ray analyses of the sample. To determine the places of X-ray analysis, an accurate sample EDS mapping of titanium was performed. The results of this analysis are presented in Figure 10. The obtained images and results indicate a steady distribution of TiO_2 agglomerates.

4. Conclusions

Laboratory tests confirmed the air purification performance of pavement blocks after seven years of service in a moderate climate. In accordance with the findings of Boonen and Beeldens [3], the ability to reduce the concentration of NO_x in the air strongly depended on the UV light source. The results of the tests on samples with different levels of surface cleanliness did not differ significantly. This was due to a UV light source which was not efficient (70 W). After cleaning the sample and applying a more intense UV light source (300 W), with better UV spectra, pronounced abatement of NO was observed.

During the tests, an increase of NO_2 content was observed. Only in the case of the 70 W light source and a dusty sample was $|\Delta NO_2|$(3 ppb) similar to $|\Delta NO|$(4 ppb), which indicated that weak lighting with a combination of dirt on the photocatalytic material surface and low nitrate selectivity may contribute to the degradation of air quality.

Images and mapping of SEM analysis confirmed the presence of TiO_2 in the form of agglomerates with a steady distribution in the tested sample. A chemical characterization of the sample with X-ray analysis also confirmed the presence of titanium dioxide.

Photocatalytic cementitious materials are a promising solution for the problem of deteriorating air quality, but taking into account the long service life of this type of elements (often much longer than the seven years of the tested samples), the efficiency of air purification should also be tested after longer intervals so as to confirm the usability of the technology throughout the period of use. The efficiency of this solution should also be verified on larger scale projects and in-situ studies, especially in the case of countries with a low UV index during most of the year.

Author Contributions: Conceptualization, H.W. and A.T.-B.; methodology and investigation, H.W., J.J., W.J.-R., K.C., A.G.; Writing—original draft preparation, H.W., W.J.-R., K.C.; Writing—review and editing, J.J., A.G.

Funding: This research was funded by Skanska Poland and Górażdże Cement, as a part of research and implementation project on photocatalytic concrete.

Conflicts of Interest: The authors declare no conflict of interest.

References

1. Muilwijk, C.; Schrijvers, P.J.C.; Werz, S.; Kenjeres, S. Simulations of photochemical smog formation in complex urban areas. *Atmos. Environ.* **2016**, *147*, 470–484. [CrossRef]
2. Guerrini, G.L.; Beldens, A.; Crispino, M.; D'Ambarosio, G.; Vismaro, S. Environmental benefits of innovative photocatalytic cementitous road material. In Proceedings of the 10th International Conference on Concrete Pavement, Quebec City, QC, Canada, 8–12 July 2012. [CrossRef]
3. Boonen, E.; Beeldens, A. Recent Photocatalytic Applications for Air Purification in Belgium. *Coatings* **2014**, *4*, 553–573. [CrossRef]
4. Guerrini, G.L. Photocatalytic performance in a city tunnel in Rome: NOx monitoring results. *Constr. Build. Mater.* **2012**, *27*, 165–175. [CrossRef]
5. Boonen, E.; Akylas, V.; Barmpas, F.; Boreave, A.; Bottalico, L.; Cazaunau, M.; Chen, H.; Daele, V.; De Marco, T.; Doussin, J.F.; et al. Construction of a photocatalytic de–polluting field site in the Leopold II tunnel in Brussels. *J. Environ. Manag.* **2015**, *155*, 136–144. [CrossRef] [PubMed]
6. George, C.; Beeldens, A.; Barmpas, F.; Doussin, J.F.; Manganelli, G.; Herrmann, H.; Kleffmann, J.; Mellouki, A. Impact of photocatalytic remediation of pollutnats on urban air quality. *Front. Environ. Sci. Eng.* **2016**, *10*, 1–11. [CrossRef]
7. Fujishima, A.; Honda, K. Electrochemical Photolysis of Water at a Semiconductor Electrode. *Nature* **1972**, *238*, 37–38. [CrossRef] [PubMed]

8. Wang, C.; Liu, H.; Qu, Y. TiO$_2$—Based Photocatalytic Process for Purification of Polluted Water: Bridging Fundamentals to Applications. *J. Nanomater.* **2013**, *2013*, 1–14. [CrossRef]
9. Shen, W.; Zang, C.; Li, Q.; Zhang, W.; Cao, L.; Ye, P. Preparation of titanium dioxide nano particle modified photocatalytic self–cleaning concrete. *J. Clean. Prod.* **2015**, *87*, 762–765. [CrossRef]
10. Husken, G.; Hunger, M.; Bruwers, H.J.H. Experimental study of photocatalytic concrete products for air purification. *Build. Environ.* **2009**, *44*, 2463–2474. [CrossRef]
11. Macphee, D.E.; Folli, A. Photocatalytic concretes–The interface between photocatalysis and cement chemistry. *Cem. Concr. Res.* **2016**, *85*, 48–54. [CrossRef]
12. Poon, C.S.; Cheung, E. NO removal efficiency of photocatalytic paving blocks prepared with recycled materials. *Constr. Build. Mater.* **2007**, *21*, 1746–1753. [CrossRef]
13. Cassar, L.; Beeldens, A.; Pimpinelli, N.; Guerrini, G.L. Photocatalysis of cementitous materials. In Proceedings of the International RILEM Symposium on Photocatalysis, Environment and Construction Materials, Florence, Italy, 8–9 October 2012.
14. Chen, J.; Poon, C. Photocatalytic construction and building materials: From fundamentals to applications. *Build. Environ.* **2009**, *44*, 1899–1906. [CrossRef]
15. Beeldens, A. An environmental friendly solution for air purification and self-cleaning effect: The application of TiO$_2$ as photocatalyst in concrete. In Proceedings of the 8th International Conference on Concrete Blocks Paving, San Francisco, CA, USA, 6–8 November 2006.
16. Guerrini, G.L. Some observations regarding in–service performance Photocatalytic paving block surfaces. *Betonwerk Fertigteil-Technik BFT* **2009**, *5*, 16–25.
17. *ISO 22197-1:2016, Fine Ceramics (Advanced Ceramics, Advanced Technical Ceramics)—Test Method for Air-Purification Performance of Semiconducting Photocatalytic Materials—Part 1: Removal of Nitric Oxide*; ISO: Geneva, Switzerland, 2016.
18. *UNI-11247:2010, Determination of the Degradation of Nitrogen Oxides in the Air by Inorganic Photocatalytic Materials: Continuous Flow Test Method*; Ente Nazionale Italiano di Unificazione: Milano, Italy, 2010. (Italian Standard).
19. *JIS TR Z 0018, Photocatalytic Materials—Air Purification Test Procedure*; Japanese Standards Association: Tokyo, Japan, 2002. (Japan Standard).
20. Folli, A.; Macphee, D.E. Future challenges for photocatalytic concrete technology. In Proceedings of the 34th Cement and Concrete Science Conference, University of Sheffield, Sheffield, UK, 14–17 September 2014.
21. Bloh, J.; Folli, A.; Macphee, D. Photocatalytic NO$_x$ abatement: Why the selectivity matters. *RSC Adv.* **2014**, *4*, 45726–45734. [CrossRef]

 *applied sciences*

Article

Soil-Gas Concentrations and Flux Monitoring at the Lacq-Rousse CO$_2$-Geological Storage Pilot Site (French Pyrenean Foreland): From Pre-Injection to Post-Injection

Frédérick Gal [1,*], Zbigniew Pokryszka [2], Nadège Labat [3], Karine Michel [1], Stéphane Lafortune [2] and André Marblé [3]

[1]   Bureau de Recherches Géologiques et Minières (BRGM), 3 Avenue Claude Guillemin, 45060 Orléans, France; k.michel@brgm.fr

[2]   Institut national de l'environnement industriel et des risques (INERIS), Parc Technologique Alata, 60550 Verneuil-en-Halatte, France; zbigniew.POKRYSZKA@ineris.fr (Z.P.); Stephane.LAFORTUNE@ineris.fr (S.L.)

[3]   TOTAL, Centre Scientifique et Technique Jean Féger (CSTJF), Avenue Larribau, 64000 Pau, France; nadege.labat@total.com (N.L.); andre.marble@total.com (A.M.)

*   Correspondence: f.gal@brgm.fr; Tel.: +33-(0)2-38-64-38-86

Received: 23 December 2018; Accepted: 12 February 2019; Published: 14 February 2019

Abstract: Soil-gas concentrations and flux were measured during 20 separate measurement campaigns at the TOTAL Lacq-Rousse carbon capture and storage (CCS) pilot site, southern France, where 51,000 tons of CO$_2$ were injected in a depleted natural gas field. Baseline data (September 2008 to December 2009) are compared to monitoring data from the injection (March 2010 to March 2013) and post-injection (February 2014 to December 2015) periods. CO$_2$ soil-gas concentrations varied from atmospheric concentrations to more than 16% vol. with 1.4% as median value. Summer data showed high CO$_2$ concentrations in the soil that remained quite high during winter. Median CO$_2$ flux at the soil/atmosphere interface was close to 4.4 cm^3·min^{-1}·m^{-2}. Carbon-isotope ratios measured on CO$_2$ in soil gas had a mean value of $-23.5 \pm 3.1‰$, some deviation being due to atmospheric CO$_2$. Comparison between different gas species and the influence of temperature, pressure and soil-water content suggest that gases in near-surface environments are produced locally and naturally, and are unrelated to CO$_2$ ascending from the storage reservoir. Monitoring of CO$_2$ injection and the use of threshold levels is discussed as part of a practical approach considering specific regulations for the Lacq-Rousse CCS pilot experiment and constraints for the site operator.

Keywords: CO$_2$ storage; depleted gas field; soil-gas monitoring; baseline; injection; post-injection

1. Introduction

Mitigation of CO$_2$ emissions in the atmosphere has been a great challenge since the end of the 20th century. One of the options is to store CO$_2$ below ground at depths where confinement can be reached with a sufficient level of confidence, a process commonly known as CO$_2$ Capture and Storage (CCS). In Europe, CCS has received additional attention since adoption of the European Directive 2009/31/EC on the geological storage of carbon dioxide that established a legal framework for CCS.

Much research was done before this pivotal date [1,2] in order to gain confidence in CCS activities and related monitoring, mitigation and verification (MMV) approaches for an environmentally safe storage of CO$_2$ [3,4]. Pioneering work was done to evaluate the storage potential in various geological settings, including deep saline aquifers [5–8] depleted oil/gas reservoirs [8–10] and unminable coal seams [11,12], although storage in oceans as an ultimate sink was envisaged as well [13,14]. The first commercial injection of CO_2 in deep formations took place in 1996 in the Sleipner oil and gas field, where CO_2—a by-product of hydrocarbon exploitation—is captured, compressed and injected into the Utsira sandstone strata 1000 m below the seabed [15]. Since then, tens of projects have been launched all over the world (an exhaustive overview is found in the IEAGHG database [16]), either dealing with capture or storage aspects, or with both [17]. In the meantime, regulation and monitoring work has progressed as well [18].

Many storage, pilot or research projects use CO_2 injection in deep saline aquifers, such as at Compostilla [19], Gorgon [20], Ketzin [21], Sleipner [6] and Snøhvit [22], or in depleted reservoirs such as In Salah [22], K12-B [21], Lacq-Rousse [23] and Otway [24], without a direct aim of enhancing hydrocarbon production. Other projects deal with the direct injection of CO_2 in quiescent hydrocarbon reservoirs in order to enhance oil/gas production, as in Weyburn [25] and many US projects [26]. Finally, many smaller-scale projects focus on shallow monitoring [18,27].

Among operations that inject CO_2 into depleted hydrocarbon fields, our study focuses on the Lacq-Rousse CCS project. For this pilot project, the CO_2 is produced from gas burned at the Lacq plant through an oxyfuel combustion process, then compressed, transported by existing pipeline and injected into the depleted Rousse field, thus having complete control over the entire CO_2 capture, transport and storage process [28]. The project was thought of and designed in 2006. At that time, the EU Directive 2009/31/EC did not yet exist and the regulatory aspects were implemented under the French Mining Code for the subsurface aspects [29] and the Environmental Protection Code regarding the surface installation aspects [30,31].

From mid-2008 to the end of 2009, baseline data acquisitions concerned geophysical and geochemical aspects [32], from the surface down to 4,200 m, the depth of the depleted Rousse reservoir. The permit for injecting CO_2 was obtained in May 2009, allowing the injection of 120,000 tons of CO_2 over two years [31]. This work started in January 2010. An extension of the injection period was obtained in late 2011, together with a reduction to 90,000 tons of the CO_2 quantity that could be stored [31]. At the stop of injection in mid-March 2013, a total of 51,000 tons of CO_2 had been injected into the reservoir, leading to an over-pressurization of the reservoir from 30 to 80 bars, the initial pressure before natural-gas exploitation being 480 bars [33,34]. According to regulatory requirements, post-injection monitoring was then carried out from March 2013 to March 2016 as the site was shifted to a post-injection surveillance phase. The site is now closed.

The present study reviews some aspects of the geochemical monitoring on the surface. Pre-injection characterization was described in a previous paper [35], and here we focus on soil-gas concentrations and soil-gas flux measured at the soil/atmosphere interface during nearly 8 years of monitoring. One objective is to report the variation ranges of soil-gas concentrations and fluxes to highlight the site dependency [36,37]. A second objective is to discuss the way the data were included in the operating program, which is designed under a regulatory framework. Emphasis is put on the use of threshold levels, a framework rarely used for CCS pilots [18]. Another objective is to consider the monitoring work under cost/benefit considerations, which is mandatory for reaching the economic feasibility of a project.

2. Monitoring Protocols and Methods

2.1. Defining the Monitoring Area

The Lacq-Rousse pilot was part of a MMV program, established by the site operator to satisfy regulatory requirements developed for this experimental project in accordance with French government requirements. Monitoring was designed according to a preliminary risk analysis and the specific configuration of the storage site (deep reservoir; thick caprock; small amount of CO_2 injected; final pressure far below initial reservoir pressure; only one well intersecting the reservoir). The monitoring should provide information on site integrity, well injectivity and storage performance, as required by the French administration, and was designed to cover ten "parameters": (1) continuous monitoring of the injected CO_2 stream, (2) pressure and (3) temperature monitoring along the injection well, (4) microseismic monitoring in six dedicated geophysical monitoring boreholes (Figure 1—points 1 to 6), (5) periodical monitoring of surface water and (6) groundwater quality, as well as (7) of biodiversity of the ecosystems [31]. Monitoring of (8) soil-gas concentrations and (9) soil flux is part of site-integrity verification, aimed at addressing leakage detection, adverse effects on the environment, and assessment of the safety and integrity of the storage complex in the short term. The surveillance network set up and used during the 2008–2015 period was intended to monitor the natural geochemical background and to act as a warner in case the thresholds of natural CO_2-flux concentration levels and their variability were exceeded. Last, (10) particular attention was paid to social aspects, as the area is urbanized and sensitive to environmental aspects since the start of natural gas exploitation from the Lacq reservoir in the 1960 s [28,38]. The importance of considerations on social acceptance is now well established as being one of the key parts of CCS projects [39].

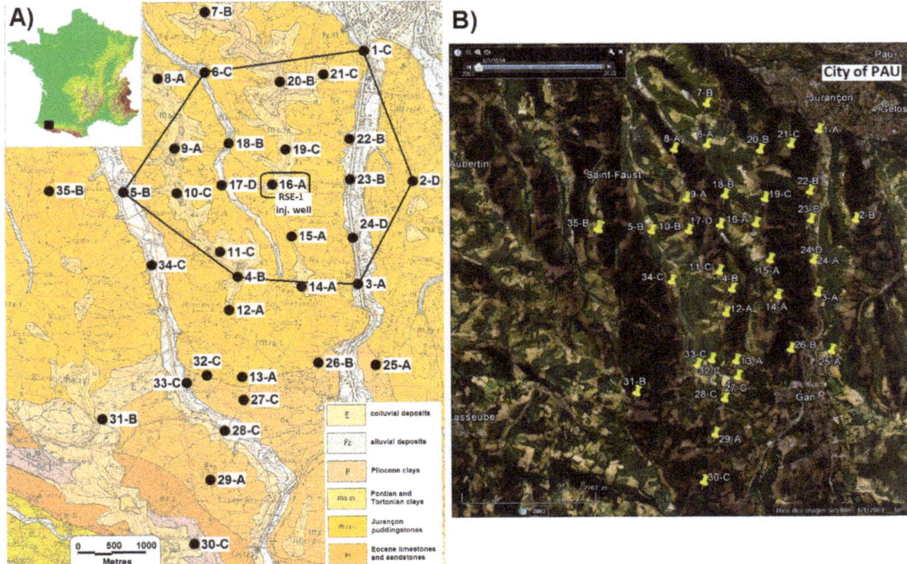

Figure 1. (**A**) Locations of the 36 monitoring points (35 different locations) plotted onto a geological map. The hexagon shape links the six geophysical observation boreholes for passive seismic monitoring down to 2 km-depth. Point 16-A is close to the location of the injection well (RSE-1). (**B**) Monitoring points plotted on Google Earth satellite view using shaded relief (afternoon view).

When the project was designed, monitoring with remote technology of a large geographical area—several square kilometres—was not really feasible with sufficient confidence and at a reasonable economic cost. It was thus decided to focus the geochemical surface characterization on selected locations that could be regularly monitored. The surface projection of the Rousse reservoir (close to 4 km^2) is included in the hexagon figure around the geophysical monitoring boreholes (1 to 6, Figure 1). The CO_2 was injected at the RSE-1 borehole near monitoring point 16A (Figure 1). A much larger area was used for surface measurements, including 35 monitoring locations (36 measurement points; Figure 1) selected by the site operator in accordance with the authorities. These 35 locations lie on a 7 × 5-km grid southwest of Pau City and mostly in the Jurançon hills, which have a steep topography so that the points range in height from 180 to 350 m a.s.l. (mean elevation of 270 m).

The locations remained the same throughout the entire project period, and were chosen to be representative of the local geological context as known from previous geological and geophysical work, and to largely exceed the surface projection of the storage reservoir. The monitoring points were also selected in locations that could remain the same throughout the monitoring period, as the region contains many vineyards and other agricultural plots (corn fields and pastures). Additional information and discussion on the representativeness of such a geographical coverage is given in [35].

The number of monitoring points results from a compromise between the operator's point of view, oriented towards a best fit between costs and benefits to ensure storage integrity, and the scientific point of view that might have required a larger number of monitoring locations to establish a more statistically reliable database. A third aspect to be considered was public acceptance of the Lacq-Rousse pilot at the time the project was launched, as some opposition existed to the project [31]. The site operator insisted that surveillance should occur at well-defined locations, the access to which is managed through specific agreements with each landowner. No other monitoring points could be created. Consequently, the 35 sites pre-selected in spring 2008 were first surveyed in September 2008 to define the exact location of measurement points, and then surveyed during subsequent data acquisitions.

The monitoring points cover six different geological units. Most measurements were made in the Jurançon hills 'puddingstone' (7 points; m2a-l on Figure 1), Pontian to Tortonian clays (9 points; m3-2b) and alluvial deposits (10 points; Fz). Other measurements were located over Eocene limestone and sandstone (5 points; e4), Pliocene clays (3 points; p) and colluvial deposits (2 points; E). These 35 locations were monitored as part of an evolutionary scheme:

- Six baseline measurement campaigns, from September 2008 to December 2009, were done quarterly as required by the MMV plan, including the September 2008 campaign that was initially designed to select the monitoring locations and thus is only partially included in the database used for defining baseline values.
- Measurements during the CO_2 injection period (January 2010 to March 2013) were set by the site operator at two sessions per year, one at the end of fall and the other at the end of winter. Seven campaigns, referred to as "low season" campaigns, were carried out during this phase. In order to complete the dataset, other data acquisitions—not required by French law—maintained the quarterly frequency of data acquisitions. Funding for these two additional campaigns was obtained from a related research project, leading to four additional "high season" campaigns in spring and summer.
- Finally, post-injection monitoring (March 2013 to December 2015) was restricted to "low season" data acquisitions performed three times as required by French law.

Site measurements thus took place from September 2008 to December 2015, resulting in a dataset of 20 different campaigns available for the duration of the entire project.

2.2. Choices of the Methods: What has Been Done and What has Not

The monitoring methods considered in this paper are based on surface geochemical characterization of gas species. The way the monitoring was designed must be viewed from the perspective of what was technically and economically realistic in the mid-2000 s (overall budget of the pilot site was 60 M€ plus 3 M€ for additional research; [31]), rather than from today's viewpoint based on then non-existing monitoring techniques and protocols [18]. Most of the surface geochemical monitoring then relied on robust methods based on spot measurements of soil-gas concentrations at 1m-depth and of soil-gas flux at the soil/atmosphere interface at regular intervals [35]. Gas flux measurements concerned CO_2 and CH_4 emissions on the soil surface. Soil-gas concentration measurements covered CO_2, O_2 and CH_4. Regulatory requirements only concerned CO_2 monitoring (soil concentration and flux). Other gas species which may have been surveyed (e.g., ^4He; [35]) are not considered. Biases that may be inherent in these monitoring methods will affect the entire dataset, whose interpretation thus is unaffected by internal inconsistencies linked to changes of the monitoring equipment. Given the limited number of monitoring points, it is clear that no spatial integration of the data is possible. Data interpolation for locating leakage points with a high confidence level requires many more monitoring locations [40] and less than 100–200 m interspace between points for statistical soundness. With this spatial density of sampling, only small areas can be surveyed at limited scale; to date, examples rely on induced CO_2 leakage in near-surface environments [27,41,42], or on the spots of natural CO_2 seeps [40,41].

This monitoring scheme can be criticized especially if it is considered through the prism of knowledge today, e.g., considering the use of long-term monitoring methods or the use of methods of larger geographical coverage. Some of these methods were nevertheless used at the Lacq-Rousse pilot, being part of research applications because the feedbacks on their uses were often too limited to allow uses under regulated applications. These uses concerned:

- The monitoring of temperature, pressure and radon activity time changes (and not CO_2) at one location (Point 24, Figure 1). Nowadays, under a cost/benefit approach, the use of such continuous monitoring systems is more developed [42] but these systems were only in use by the research groups who developed them in 2007/2008 [43].
- The monitoring of the atmosphere was done at one location (point 16A; Figure 1) by the eddy-covariance technique or by passive-infrared remote sensing [31,44,45]. Although the eddy-covariance technique was robust in the 2000s, the radius of the monitored area (1 to 2 km with a 10 m height pole) would have implied deploying numerous devices in order to monitor the entire 35 km^2. A single eddy-covariance system cannot detect leakage in an area it does not cover, or where the wind direction is not oriented toward the detection system. The use of such systems to locate leaks within a pre-defined surveillance area is of more recent application, relying on data from risk assessment of the plausible location of leaks [46]. The passive-infrared remote sensing system showed that the CO_2 gas concentration in the atmosphere close to the RSE-1 pad was influenced mostly by photosynthetic processes and by particular wind dynamics.

2.3. Soil-Gas Concentrations

Soil-gas concentrations were measured at 1m-depth after drilling a small hole in the soil with a battery-powered drill (1m-depth, 1 cm diameter) and inserting a sampling probe (copper tube, 1 m long and 1 cm diameter). Due to land-use constraints, it was not possible to keep the sampling probe in place between each monitoring campaign, which would have been the best option. The holes were surface-capped with locally present soil. For each new measurement, probes were re-inserted after a new hole was drilled. Soil-gas concentrations were directly monitored using with an LFG20 Infra-Red Gas Analyser (IRGA—ADC Gas Analysis Ltd., UK) using low-rate pumping (200 mL·min^{-1}) to avoid disturbing the soil atmosphere. After purging of the tubes and pipes, stable readings were reached within tens of seconds. The stability of the readings, during measurement and gas-sample collection, was regularly checked and, if necessary, the site was resampled. The analytical precision for CO_2 and CH_4 was ±0.5% (of the reading) at low concentrations (0.01–10% vol.) and ±3% (of the reading) at higher ranges (10–50% vol.). Oxygen precision for was ±0.4% (of the reading) in the range 0–25%. About 27% of the field measurements were cross-checked by gas-chromatography analyses at laboratory allowed determining that the overall error (sampling error induced by re-drilling + instrumental error + error on replicates) is less than 10% for CO_2 concentrations greater than 8% and less than 5% for CO_2 concentrations lesser than 6%.

2.4. Soil-Gas Flux

Soil-gas flux was measured using a direct method: the accumulation-chamber method with external recirculation [47], a protocol that is intermediate between the static and the dynamic methods. A 0.25 m^2 chamber hermetically encloses the soil surface after which gases emitted by the soil surface accumulate in the chamber. During the measurement period (typically 3 minutes), gas accumulation is characterized by a roughly linear increase of the gas concentration inside the chamber [48]. Afterwards, the flux is calculated using the slope of the concentration versus the time regression line. The gas fluxes are expressed in $cm^3 \cdot min^{-1} \cdot m^{-2}$ STP (standard conditions for temperature and pressure). Conversion factors are for CO_2: $1\ cm^3 \cdot min^{-1} \cdot m^{-2}$ STP = 0.033 $mg \cdot s^{-1} \cdot m^{-2}$; for CH_4: $1\ cm^3 \cdot min^{-1} \cdot m^{-2}$ STP = 0.012 $mg \cdot s^{-1} \cdot m^{-2}$.

CO_2 is detected with a Licor Li800 infrared device (detection limit of 1 ppm), which can be used for the monitoring of very low fluxes around 0.1 $cm^3 \cdot min^{-1} \cdot m^{-2}$ to very high gas emissions around 4000 $cm^3 \cdot min^{-1} \cdot m^{-2}$ with a detection limit of 0.03 to 0.5 $cm^3 \cdot min^{-1} \cdot m^{-2}$ [40].

CH_4 flux was measured with the same accumulation chamber, associated with a Gazomat Inspectra infrared laser detector (detection limit of 0.5 ppm). This system permits detecting an extremely low CH_4 emission level of around 0.01 $cm^3 \cdot min^{-1} \cdot m^{-2}$.

2.5. Other Field Measurements

The field measurements were completed by measuring local temperature and pressure in the atmosphere using a Kestrel handheld monitor, by measuring soil temperature at 15 cm-depth with a Pt100 thermocouple (Ecoscan, EUTECH) and by evaluating the soil-water content at the same depth. This water content is expressed as the ratio of the mass of wet soil to that of dry soil. Drying was done at 65 °C, below the normalized temperature (105 °C) required by the gravimetric method, but the water loss was regularly checked over a much longer period (minimum 3 days vs. 24 hours for the gravimetric method) and the soil was sieved so that confidence was good for this evaluation of the water content.

Atmospheric parameters allow normalization of the measured gas flux. Comparison of the soil and atmospheric temperatures gives information on the thermal influence prevailing at the time of sampling. Coupled with knowledge on water content in the upper soil horizons, it provides insights into the influences of external parameters on gas emission from soil [49] and thus on gas flux from soil to atmosphere.

Finally, a Barasol probe (Algade, France) was used to get temperature and pressure data on an hourly basis (precision: ± 0.1 °C and ± 1 mbar respectively). These data are used to provide local control on temperature and pressure changes and as a basis of comparison with meteorological data obtained from Pau-Uzein airport (about 10 km away). The probe was set at 60 cm-depth in the soil at monitoring point 24. The probe was installed after drilling a hole with a hand auger, and was laterally protected by a PVC pipe; it was installed so that its end was in contact with bare soil at the hole bottom. On surface, the PVC pipe was protected by a GoreTex®membrane and then covered by a thin layer of soil.

2.6. Laboratory Analyses

Carbon-isotope ratios of the CO_2 soil-gas were determined on the same glass bulbs. After purifying the CO_2, the gas is analysed through a thermostated chromatographic column connected to the inlet of a mass spectrometer (Gas Bench system associated with a Delta + XP mass spectrometer). The $\delta^{13}C_{CO2}$ isotopic ratio is expressed with reference to the "Vienna Pee Dee Belemnite" standard (‰ VPDB). Analytical precision is better than 0.2‰.

2.7. Regulatory Aspects

The Lacq-Rousse CCS pilot was implemented under a well-constrained monitoring scheme referring to threshold values, authorized by the French administration [23,31]. Surface monitoring of soil-gas concentrations and flux is only one aspect of the 10 parameters described in the monitoring programme agreed by the French administration (Decree FR 2011-1411 and transcription of EC Directive 2009/31/CE). Like other aspects, it serves to establish that the storage at depth confines the injected CO_2 and that no gas migrates upward.

As mentioned above, there is a clear distinction between the baseline data acquisitions operated over four seasons (2008 and 2009) for establishing the environmental background, the monitoring during the injection process (January 2010 to March 2013), and the post-injection monitoring (March 2013 to December 2015). The regulatory obligations of the three project phases are described in more detail hereafter.

2.7.1. Regulations Governing Baseline Monitoring

A set of gas species (CO_2, O_2, CH_4, ^4He and ^{222}Rn for soil concentrations, CO_2 and CH_4 for soil flux) was regularly monitored at the 36 monitoring points. These data were used as input for defining seasonal changes in the soils overlying the detrital rock formations that originated from the dismantling of the Pyrenees forefront.

2.7.2. Regulations Governing Injection Monitoring

The baseline data were then used—in accordance with French Administration guidelines—to define threshold levels for a subset of gas species (CO_2 soil-gas concentration and CO_2 soil flux). The baseline data were compared with literature data in soils under a temperate climate [50–56]. This analysis—by the site operator—allowed for defining concentration and flux levels that can be considered as normal or anomalous ones (threshold levels); French regulations mention that the operator shall determine how information is processed. In particular, the operator determines the threshold values that enable detecting an anomaly in the environment that may be caused by a leak from the reservoir. The operator then decides if further investigations are necessary, or even if the injection process must be modified.

The thresholds were defined based on simple statistical considerations as there were little data available for defining universal threshold values that could be applied univocally to the Lacq-Rousse site. For example, in 2009, the potential leakage reported at the Kerr Farm, Weyburn oilfield, was not known [57,58], so that the criticality of threshold definition was not as clear. Nonetheless, it was agreed with the French authorities that these thresholds could be debated later and that they probably needed adaptation during the pilot experiment. The selected approach thus defined a "vigilance" mode as the mean value (M) plus twice the standard deviation (σ) of the baseline dataset ("vigilance" mode = M + 2σ), and an "anomaly" mode as the mean value plus three times the standard deviation of the dataset ("anomaly" mode = M + 3σ). The first attempts at defining the thresholds used the September 2008 to September 2009 dataset, as the December 2009 data were then thought to postdate the start of injection. The result was a set of high "vigilance" and "anomaly" levels (respectively 8.3% vol. and 11.2% vol. for CO_2 concentration and 16.6 $cm^3 \cdot min^{-1} \cdot m^{-2}$ and 21.5 $cm^3 \cdot min^{-1} \cdot m^{-2}$ for CO_2 flux). The operator then refined these levels by only considering the data acquired during "low season", but this meant that the redefined threshold levels were only valid for the "low season" monitoring. This was approved by the administration. The question of what would happen if leakage was reported during the "high season" was not specifically addressed; further work would have been necessary, such as required when a threshold is exceeded during "low season". Fortunately, soil-gas investigations and the use of a process-based method (defined as such by Romanak et al. [59], but used before) would certainly have provided sufficient evidence of the origin of the gas to reject any allegations of leakage. From analysis of the "low season" baseline data, the following cases were defined and used for data acquired during the injection monitoring period:

- "Vigilance" mode (M + 2σ): This mode is activated if 5 different locations of the 35 show CO_2-gas concentrations >5.4% vol. and CO_2 fluxes >7.4 $cm^3 \cdot min^{-1} \cdot m^{-2}$. When this happens, the measurements are repeated to check if the threshold levels are permanently exceeded or not. The regulations did not provide for repeating the measurements, and repeat measurements were scheduled to be done during a different part of the day than the first one. If a vigilance situation occurred in the morning, the measurement was repeated during the afternoon or the day after if necessary, and if it happened during the afternoon, the repeat took place the day after. As concentration and flux are likely to vary throughout the day, measuring at different times may show whether the CO_2 contribution is variable over time—probably indicating a near-surface origin—or not, and thus potentially indicating another contributing endmember.

- "Anomaly" mode (M + 3σ): This mode is activated if 5 different locations of the 35 show CO_2 gas concentrations >7.3% vol. and CO_2 fluxes >9.5 $cm^3 \cdot min^{-1} \cdot m^{-2}$. Another configuration that may indicate massive leakage, is if 1 location shows a CO_2 concentration >50% vol. or if the CO_2 flux is >100 $cm^3 \cdot min^{-1} \cdot m^{-2}$. Regardless of how the "anomaly" mode is reached, the measurements are repeated, to check the durability of the gas signals. A soil-gas sample is also taken for laboratory measurement of the $\delta^{13}C$ isotope ratio of the CO_2. A target value of $-33.6‰$ was calculated as a reference, corresponding to the mean value between the mean $\delta^{13}C$ isotope ratio measured in the soil in 2008 and 2009, and the $\delta^{13}C$ isotope ratio of the CH_4 produced by the Lacq natural gas field. This carbon-13 isotope ratio has no regulatory significance and as the $\delta^{13}C$ isotope ratio is measured in the laboratory, there may be a delay in knowing that the threshold was exceeded. If the values are confirmed, more measurements are made at neighbouring monitoring points to determine if the CO_2-level increase is confined to a small area (where the anomaly is defined), or if the increase occurs at a wider geographic scale suggesting a leak from the reservoir.
- "Normal" mode: None of the above defined threshold levels is reached.

2.7.3. Regulations Governing Post-Injection Monitoring

The threshold levels were redefined in relation with third-party expert assessment of the injection-period data. This third-party expertise concerned the whole project and not only the soil-gas measurements, and was requested by the French authorities when injection stopped. As mentioned before, such adaptations of the monitoring protocol were foreseen from the beginning of the project, at the request of the French authorities or at the request of the site operator. The third-party expert used the soil gas concentration and flux data to calculate new thresholds. These new threshold levels were only valid for "low season" monitoring the frequency of which was reduced to one campaign per year, ideally in December. These new levels applied only to the post-injection monitoring data.

The authorities approved a new scheme. Monitoring is restricted to 35 monitoring points: the third-party expert recommended one point to be discarded as it is too frequently water-saturated and thus no monitoring of the soil-gas concentrations is possible (point 23-B, Figure 1). There, of the 10 "low season" surveys between December 2008 and March 2013, no soil-gas concentrations could be measured 5 times because of flooding, and a further 3 campaigns gave CO_2 concentrations at atmospheric level. Only 2 campaigns allowed measuring a concentration well below the "vigilance" threshold (max. concentration of 1.9% vol.). Removal of this point thus had very little influence on the overall dataset.

The set of monitored gas species was reduced to the sole monitoring of CO_2 and CH_4 (both soil-gas concentration and flux). In addition, the acquisition of O_2 soil-gas concentrations continued as the IRGA provides the opportunity to obtain this information at no extra cost.

The "vigilance" mode was redefined and is now activated if 3 different locations of the 35 show CO_2 gas concentrations >3.6% vol. and CO_2 fluxes >5.9 $cm^3 \cdot min^{-1} \cdot m^{-2}$. The third-party expert assessment—referring to the methodology initially used—helped calculating the new threshold level, which now corresponds to the mean value plus twice the standard deviation of the CO_2 concentration and the CO_2 flux measured in December 2008, 2009, 2010 and 2011 (M + 2σ). If repeated measurements show exceeding of the "vigilance" mode, a sample is taken for measuring the $\delta^{13}C$ isotope ratio of CO_2 gas in the soil.

The "anomaly" mode was also redefined and is now activated if, at a single location, CO_2 gas concentration is >4.7% vol. and CO_2 flux is >7.3 $cm^3 \cdot min^{-1} \cdot m^{-2}$. The thresholds were calculated from the mean value plus three times the standard deviation of the December data (M + 3σ).

An "isotopic" mode was introduced. It implies determination of the $\delta^{13}C$ isotope ratio of the CO_2 gas phase in soil for 7 monitoring points. One of these points (Point 28-C, Figure 1) was surveyed during 11 campaigns out of 13 and a good estimate of its isotope signature is available together with its variability (-23.7 ± 1.3‰). Some of the other points showed—since the start of measurements—more depleted isotope ratios together with gas concentrations over 2% vol. Therefore, for each CO_2 soil-gas concentration >2% vol., a sample must be taken for laboratory measurement of this isotope ratio. The third expertise evaluated indicative levels based on the $\delta^{13}C$ isotope ratio in the injected CO_2 (-39.6‰). These suggested levels are -26‰ ("vigilance") and -28.4‰ ("anomaly"), respectively. The site operator set its own levels at -29‰ and -31‰, respectively, but none of these levels is part of the regulatory scheme so that the values are only given as indicative.

The changes induced by this modification of the threshold levels will be discussed hereafter. As the measurements were made above a depleted natural-gas field, CH_4 concentrations and flux levels were specifically checked in the soil gas and on surface. This was done to ensure that CO_2 injection would not induce migration of residual CH_4 from the depleted natural-gas reservoir.

3. Results

3.1. Influences of Meteorological Parameters (Temperature, Atmospheric Pressure, Rainfall)

For such long-term monitoring periods, getting a representative view of the evolution of external parameters is important. In a temperate climate, soil properties—specifically water content and temperature—can have an influence on soil-gas dynamics, especially considering the geology of surface formations (mainly detrital rocks with high clay content). Estimating the relative influence of one parameter to the other on the soil gas concentration and/or the flux and their evolutions through time, is not straightforward using the existing database. Some characteristics can nevertheless be highlighted using the dataset.

Figure 2 presents the temperature, pressure and soil water content evolutions during the 8 years of monitoring as monitored at each point during field acquisitions. The right insert details the atmospheric temperature data from Pau-Uzein airport, the data from the Barasol probe and the temperatures measured during sampling in the soil. As the thermal state of the atmosphere influences soil temperature, especially in the present case where soil temperatures were measured at shallow depth, the soil temperatures should be lower during winter than during summer, as shown by the whisker data. "Low season" soil temperatures mainly fall in a narrow range from 8.3 to 9.4 °C mean value. However, colder climatic conditions during the field surveys can cause strong temperature decreases in soil, as in March 2010 (mean value 4.3 °C); "Indian summer" conditions, on the contrary, can lead to soil temperatures over 10 °C, as in December 2012 (mean value 10.7 °C) and in December 2014 (mean value 10.3 °C). Annual climatic variations linked to the local conditions (temperate climate with oceanic influence), are thus reflected in the database, with statistically lower temperatures during winter, but these long-term trends can be overlapped by variations of higher frequency. These variations are partly linked to the geographical location of each monitoring point (e.g., hill slope oriented north and covered by trees vs. southwest oriented glade) and the time the measurements are performed.

During the 2008 to 2015 period, the mean atmospheric pressure was very close to 1015 hPa (calculated value: 1017 ± 8 hPa). The variability is higher during winter as a result of more frequent low-pressure events. Information from the pressure sensor of the Barasol probe is similar to that measured in the atmosphere (mean value close to 1021 ± 8 hPa).

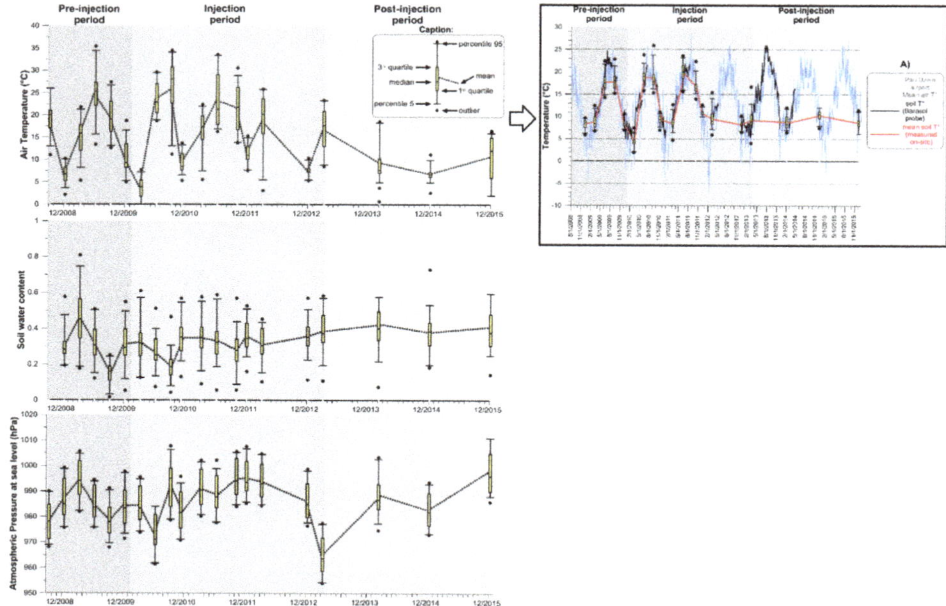

Figure 2. Air temperature, soil water content and atmospheric pressure measured during each field session at each monitoring location. Data are shown as box plots presenting median value, quartiles and outliers. The inset on the right details the evolution of soil temperature measured at each point and measured at Point 24 using a Barasol probe (in black), and the mean air temperature measured at Pau-Uzein airport (in blue; http://french.wunderground.com/).

Rainfall amounts and their derivate the soil-water contents were variable one year to the other (Figure 2). Generally, strong events occur in November (2009,2010,2013) with monthly amounts greater than 200 mm, but thunderstorms may lead to even larger amounts at other times, such as >250 mm in May 2013. The rainfall regime is variable because of the location of the study area, near the Pyrenees Mountains to the south and not far from the sea in the west. The result is that the general rule of winter being wetter than summer is not always true here. If "low season" monitoring pointed to water contents varying between 30 and 40%, similar values were measured in June 2009 or June 2011. However, soil-water content was generally low after summer and below 20% in September.

3.2. Soil Gas Concentrations—Spot Data

The distribution of data (CO_2, O_2 in soil gas) follows neither a normal nor a lognormal distribution. The use of Kolmogorov–Smirnov or Khi2 statistical tests always gives p-values <0.0001. The only exception to this rule is for CO_2 fluxes that are likely to have a lognormal distribution (p-value of 0.271 with the Kolmogorov–Smirnov test and a p-value of 0.157 with the Khi2 test). Therefore, the use of arithmetic mean or median values, used here although not ideal, is preferred as soil-gas studies usually refer to the use of such averaging methods.

CO_2 concentration changes with time are shown on Figure 3. O_2-concentration variations will be discussed later in the Discussion section, together with CO_2-concentration variations. A total of 710 single measurements is available for the entire monitoring period. CO2 concentrations varied from near atmospheric levels (0.03 to 0.04%) up to 16.7% (Point 24-A, September 2008). Mean value is $2.1 \pm 2.2\%$ for the whole dataset and the median value is close to 1.4%. Oxygen soil-gas concentrations ranged from 1 to 21% with a mean value of $17.8 \pm 3.5\%$ and a median value of 19%.

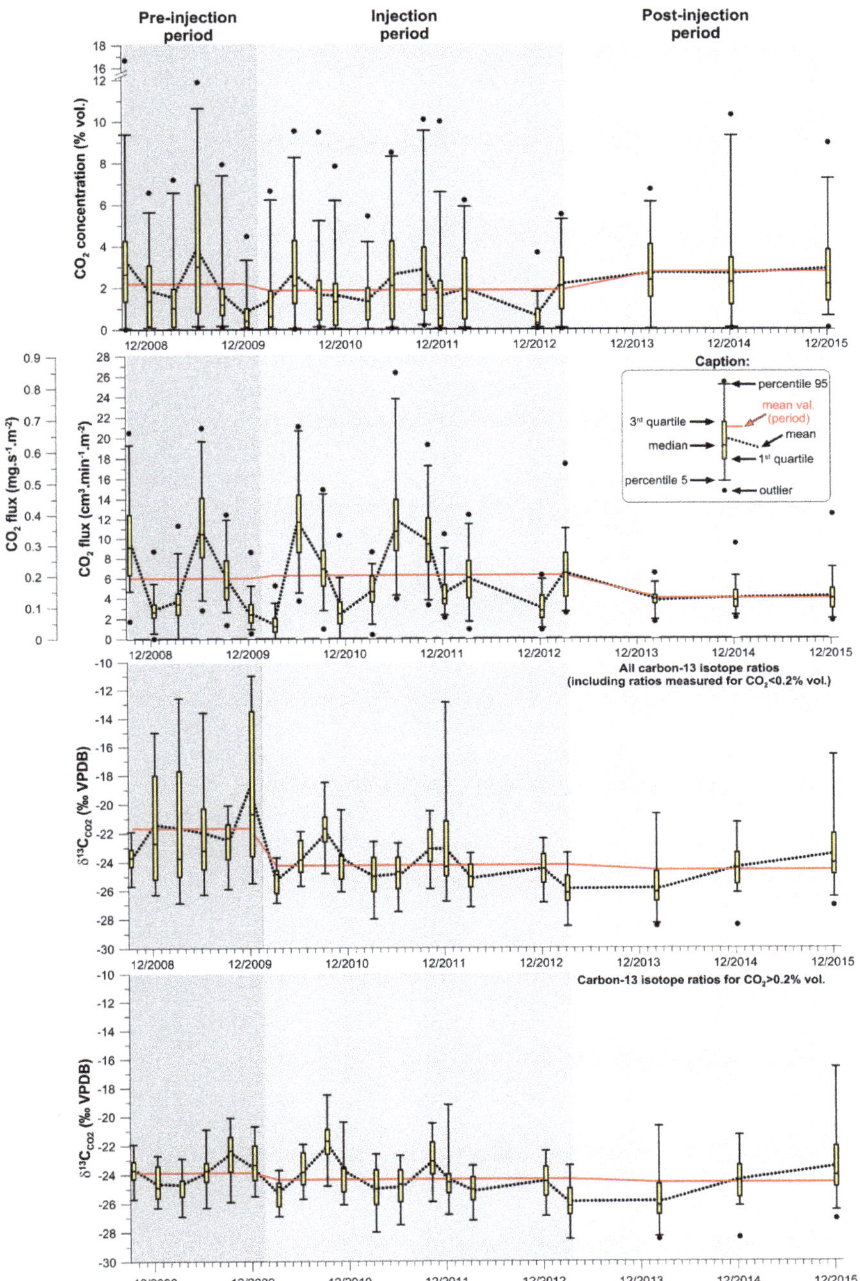

Figure 3. From top to bottom: soil CO_2 concentration, soil-gas flux at the soil/atmosphere interface, $\delta^{13}C$ isotope ratios of the CO_2 in soil gas (all data), and $\delta^{13}C$ isotope ratios of CO_2 in soil gas (excluding CO_2 <0.2%). Data are shown as box plots presenting median value, quartiles and outliers. Mean values indicated by the red line correspond to the mean values calculated for each pilot phase (pre-injection, injection and post-injection).

There are some differences between "low season" and "high season" measurements. During summer, CO_2 concentrations tended to be higher (mean value 2.7 ± 2.7%; median value 1.7%; 252 data) as a result of biological and microbial activity in soils. This statement has to be balanced by reference to the temperatures and especially the soil-water contents (Figure 2). In September 2009 and 2010, soil-water contents were at their lowest levels (<30%). Soil dryness has adverse effects on soil organisms and reduces their activity. This may have led to a decrease of CO_2 production in soil so that soil CO_2 concentrations had low mean and median values, mimicking those measured during winter. The highest mean and median CO_2 concentrations were measured when soils were warm and humid, which was mainly encountered during June acquisitions. On the contrary, winter-time acquisitions do not necessarily imply a strong decrease in CO_2 soil-gas concentrations. The mean value calculated for the whole dataset is lower (1.8 ± 1.9%; 468 data), but the median value (1.3%) is relatively close to that calculated for summer months (only a 0.4% decrease in the median value). This suggests that soil biological activity decreases during winter, but that there is still enough activity to produce significant CO_2 gas concentrations in soil. The influence of soil humidity on soil CO_2 production was less obvious as no water stress was found during winter months. Soil temperature influenced the measurements; when they dropped below 10 °C (e.g., December 2009, March 2010 and December 2012), the soil CO_2 concentrations were at very low median levels, well below 1% in volume. However, when monitoring during more clement climatic conditions, such as in December 2014 and 2015, the CO_2 concentrations mimicked those of "high season" conditions (median values >2%).

Another potential variability is much more short-term: the distinction between morning and afternoon sampling is far from obvious. During one monitoring session, the routes between points were randomly chosen so there was no additional bias linked to preferential occupation of locations during the day. There certainly is a daily variation of the soil CO_2 concentrations but these changes cannot be assessed with the present dataset.

The statement suggesting that the best period for site monitoring is when the biological activity is low in soils cannot be verified in the present case. Although winter is generally referred to as the best period for highlighting the presence of gas with a deep origin [55], in our study CO_2 concentrations were not always at their lowest levels during winter. For example, another survey in southeast France showed lower CO_2 mean concentrations during spring or summer [60]. It is thus of crucial importance to have other markers (monitoring of other gas species and/or isotope data) in order to correctly attribute gas concentrations to surface processes or to other origins.

Monitoring of methane using IRGA was not informative (no deviation of the reading) but the cross-checking measurements done at the lab by gas-chromatography allowed us to detect very low amounts on some samples (66 of 195 samples). Minimum and maximum concentrations were 2 ppm (instrumental detection limit) and 13 ppm, respectively, compared to the 1.8 ppm concentration of the atmosphere. Mean and median values were 3 ± 5 ppm and 2 ppm, respectively. CH_4 detection in soil was rather infrequent for most points. Of the 36 monitoring points, 10 points never showed detectable CH_4 concentrations, including point 16-A close to the injection borehole RSE-1, and for eight further points CH4 was only monitored once, including point 17-D close to the other borehole reaching the former CH_4 reservoir. However, CH_4 was measured 10 times at points 1-C and 28-C, and 8 times at point 24-A; for these three points, there was no apparent relation with time, i.e., no increase of CH_4 concentration that could to be related to the CO_2-injection process. The hypothetical origin of this CH_4 will be discussed in Section 4.

3.3. CH_4 and CO_2 Gas Flux

CH_4 flux was searched for in parallel with all CO_2-flux measurements, but none exceeded the instrumental detection limit (0.01 $cm^3 \cdot min^{-1} \cdot m^{-2}$). This agrees with similar measurements in normally drained European soils [51]. CH_4 flux was thus considered as insignificant.

CO_2 flux changes over time are shown on Figure 3. A total of 717 single measurements are available for the whole monitoring period. CO_2 flux varied from 0 up to 26.4 $cm^3 \cdot min^{-1} \cdot m^{-2}$ (Point 4-B,

June 2011). Mean and median values for the whole dataset were 5.8 ± 4.2 cm^3·min^{-1}·m^{-2} and 4.4 cm^3.min^{-1}.m^{-2}, respectively. These values are in the upper range of bibliographic values for biological flux in the European temperate climate zone, outside any known geochemical anomaly and on a normally drained soil (flux less than 4 cm^3.min^{-1}.m^{-2} [51]; flux up to 14 cm^3·min^{-1}·m^{-2} [61]).

However, the recent analysis of the biogenic CO_2 and CH_4 emissions data in France (24 various natural sites, more than 2600 measurements) shows that the most probable CO_2 flux ranges are respectively from 0 to 6 cm^3·min^{-1}·m^{-2} in winter and 3 to 20 cm^3·min^{-1}·m^{-2} in summer. The average values are respectively 2.7 cm^3·min^{-1}·m^{-2} in winter, and 9.3 cm^3·min^{-1}·m^{-2} in summer and 5.8 cm^3·min^{-1}·m^{-2} for all seasons' data [62].

CO_2 gas flux measured between 2008 and 2015 on the surface of the Rousse site are mostly in the high range of this all reference flux, because they were measured on very fertile land in the oceanic climate of SW France that is characterized by hot and wet summers and mild winters.

Regardless of the measurement period, the ratio between highest and lowest measured values is always highly significant. Here it was 4 to 60, without considering zero-flux measurements, but the same observation is made on reference sites and is due to the heterogeneity of biological activity and to the permeability of soil to gas.

Quarterly measurements from September 2008 until March 2012 highlighted a clear seasonal variation of the gas flux between summer and winter conditions (Figure 3). Mean CO_2 flux values in June or September (6 to 11.8 cm^3·min^{-1}·m^{-2}) were always higher than those in December or March (1.4 to 5.9 cm^3·min^{-1}·m^{-2}). The same trend was observed for maximum and minimum values, except in one case: zero-flux and values close to zero occurred only in winter, when soil biological activity is low, or rain can saturate soil porosity with meteoric water, causing reduced permeability. The observed seasonal variation of the gas flux is well-known when measuring CO_2 flux from biological origin; it is due to natural changes throughout the year of bacterial and root activity in the soil.

No significant difference of gas flux was observed between the baseline study from September 2008 to December 2010, and the injection period from January 2010 to March 2013. During these two periods, CO_2 flux followed the expected seasonal pattern, without any significant deviation. The mean values measured during the baseline-study and the injection periods were almost the same, about 6 cm^3·min^{-1}·m^{-2}. During the post-injection period, three field campaigns took place in winter; the mean value was about 4 cm^3·min^{-1}·m^{-2}, which was very similar to the flux measured in winter during the baseline and the injection periods.

3.4. Carbon-Isotope Ratios of Soil-Gas CO_2 ($\delta^{13}C_{CO2}$)

A set of 260 carbon-13 isotope ratios was determined from September 2008 to December 2015 (Figure 3). One location (point 28) was monitored each time in order to evaluate the temporal changes that may affect isotope ratios. $\delta^{13}C_{CO2}$ isotope ratios ranged between -11‰ and -28.5‰ VPDB, with a mean value of -23.5 ± 3.1‰ and a median value of -24.2‰. These mean and median values fall well within the field of carbon-isotope ratios originating from soil biological activity in a temperate climate. These ratios are more depleted than those reported for the sole baseline acquisitions. An average ratio of -20.6‰ was at that time calculated for five campaigns between December 2008 and December 2009, thus excluding the first campaign in September 2008 [35]. As the sampling strategy during injection and post-injection monitoring was oriented to the sampling of points with a significant CO_2 concentration in the gas phase, biases linked to the dilution of soil air with low CO_2 concentration by atmospheric CO_2 were therefore reduced, as will be discussed later.

4. Discussion

4.1. Relations between Parameters

The results presented in the previous sections indicate that the variability of soil-gas concentrations or CO_2 soil flux is mainly related to annual seasonal changes. The whole dataset can thus be considered in terms of time-related changes, suggesting that such changes are unrelated to the CO_2-injection process. This agrees with the baseline data [35]. Figure 4 shows some of the relationships between gas phases and external parameters.

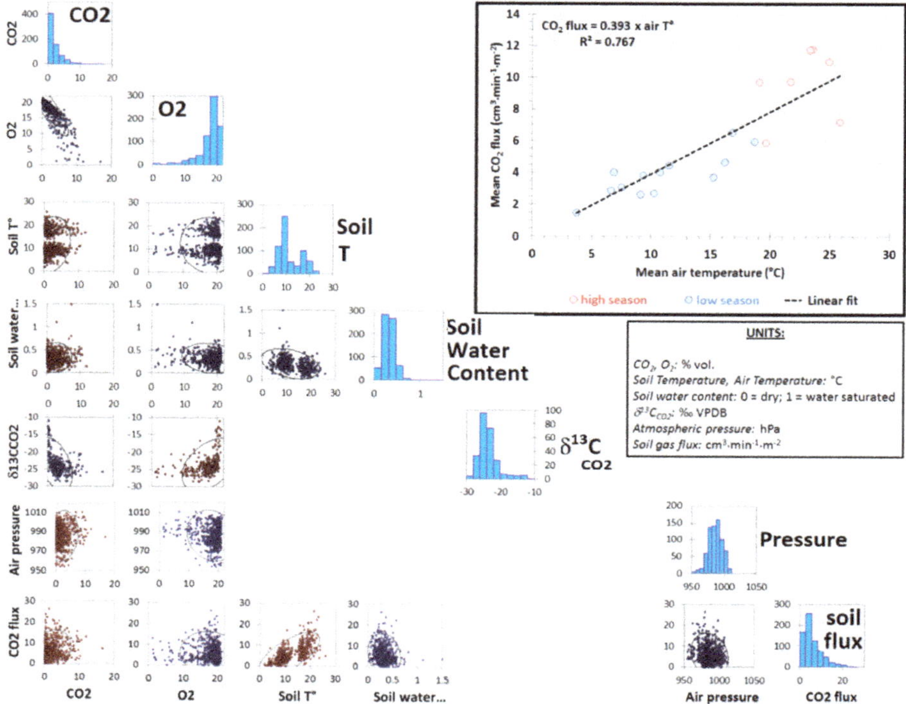

Figure 4. Correlation matrix; inset shows the relation between air temperature and CO_2 flux taken as mean values during each campaign.

4.1.1. Influences of Temperature, Pressure and Soil Water Content

Plots representing soil temperature as a function of CO_2 and O_2 have the same shape. Two groups exist, one corresponding to "low season" data acquisitions, with soil temperature around 10 °C and less, and the other corresponding to "high season" data with a higher temperature (over 15 °C). When the data are considered as a whole, the relation between soil-gas concentrations and the soil temperature is masked: e.g., CO_2 concentrations around 10% were monitored during both winter and summer. This is not contradictory with previous statements (see Section 3.2) that have established the mean CO_2 concentrations were lower during the "low season". This is only the result of considering the entire dataset in which local effects (topography, soil temperature, soil water content) are hidden. When the data are considered more in detail, the influence of temperature can be highlighted, e.g., for CO_2-flux measurements that show a tendency to be higher when the temperature is higher (inset in Figure 4). Flux equal to zero is often measured during winter, when the soil temperature is only a few degrees Celsius and the air temperature hovers around 0 °C. Depending on the soil biological activity, gas flux is generally higher during "high season" but high flux, over 10 $cm^3 \cdot min^{-1} \cdot m^{-2}$, may be measured locally during the "low season". This illustrates the effect of thermal draw on gas exchanges between soil and surface.

At the opposite, the relationships with pressure or with soil water content (Figure 4) are non-informative in the present case study.

4.1.2. Relationships between Gas Species

As part of a "process-based approach" [60], the relationship linking CO_2 and O_2 soil-gas concentrations (Figure 5) is the most frequently used. There is a strongly inverse correlation between CO_2 and O_2 (O_2 = 20.7 − 1.32 x CO_2; R^2 = 0.724), a rise in CO_2 being balanced by a depletion in O_2, though the slope of the regression line is not equal to −1. The ratio of −1 would suggest that each appearance of 1 CO_2 mole induces the disappearance of 1 O_2 mole (pure respiration process: $CH_2O + O_2 = CO_2 + H_2O$; in green on Figure 5A). The regression line calculated for summer acquisitions ("high season") points to a slope close to −1 and thus points to a dominant influence of the respiration process in soil to account for the CO_2/O_2 relationship. The CO_2/O_2 relationship during winter ("low season") has a different slope of −1.6 (Figure 5A), closer to that of the CH_4 oxidation line ($CH_4 + 2O_2 = CO_2 + 2H_2O$; in black, Figure 5A).

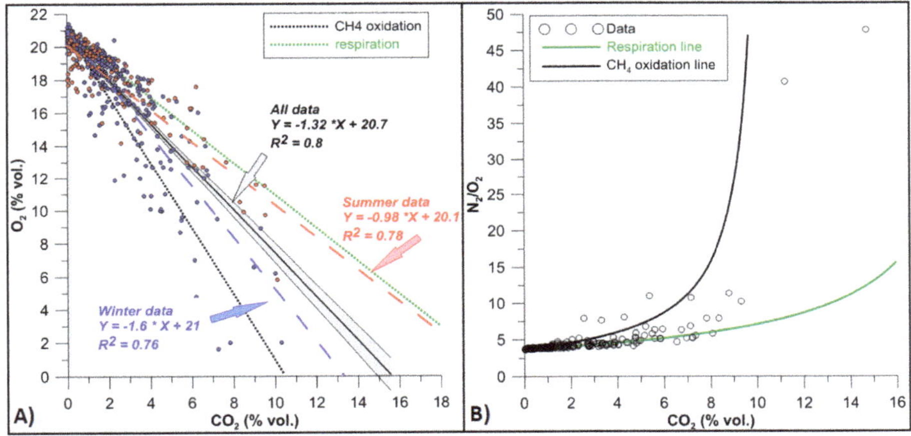

Figure 5. (**A**) Relationship between CO_2 and O_2 in soil gas and comparison to source attribution protocol defined by [59]; (**B**) Relationship between CO_2 and N_2/O_2 in soil gas. Data fits are compared with control lines corresponding to CH_4 oxidation and respiration processes.

Some "high season" data have CO_2 concentrations >2% vol. and plot above the respiration line (i.e., slope less than 1), in an area where the CO_2 concentration in gas rises with only minor depletion in O_2 (CO_2/O_2 ratio <1). Such behaviour may correspond to the influence of exogenous CO_2. The presence of CO_2 from mantle/crustal origin is unlikely; such seeps from depth are not reported locally – mineral waters circulating at depth are N_2-rich and depleted in CO_2. The presence of naturally occurring CO_2 from the Lacq reservoir, or of CO_2 resulting from the injection process, is not compatible with carbon-13 isotope data. The origin of data falling above the respiration line thus remains unclear, but may be due to specific processes occurring in soil in late summer, when soils are under hydric stress.

Some data, especially during the "low season", may show some link with the CH_4 oxidation line, or with processes such as pyrite oxidation [63] that deplete the O_2 with a CO_2/O_2 concentration changes ratio >1. Nevertheless, pyrite oxidation was discarded during the baseline acquisitions [35], as this mineral is uncommon within the detrital sedimentary formations. Information from the N_2/O_2 ratio vs. CO_2 concentrations plot [59] given on Figure 5B also suggests CH_4 oxidation may have some contribution for some measurements, albeit the main driver of CO_2/O_2 concentration changes in soil remains respiration processes.

As mentioned before, no CH_4 flux was measured but some soil gas measurements did point to the presence of a few ppm of CH_4. The presence of CH_4 at concentrations over the atmospheric concentration is not common, as most soils in a temperate climate show CH_4 oxidation in the soil column [64,65]. Natural leakage from the former CH_4 reservoir may thus be the culprit, even though no isotope ratios were measured ($\delta^{13}C_{CH4}$ and δD_{CH4}) to clearly attribute a CH_4 origin. Methane concentrations of a few tens of ppm are likely to be found in faulted areas (vertical migration of CH_4; [66]) but deep-rooted faults are not known to affect the thick sedimentary pile of molasse formations [31,67]. Methane emission to the atmosphere can also occur as a purely diffuse mechanism [68] to form the so-called CH_4 micro-seepage, defined as the slow and continual loss and upward flux of CH_4 and light alkanes from depths of 2–5 km in sedimentary basins [69]. Methanotrophic oxidation processes in soil consume CH_4 and lead to non-existent or negative CH_4 flux at the soil/atmosphere interface – this is congruent with our observations (no CH_4 flux).

The occurrence of such seepage is yet unproven and seems hypothetical in view of other information. Reactional processes can consume methane, but this is unlikely for helium. The Lacq-Rousse reservoir is ^{4}He-rich (65 ppm; [67]), and if micro-seepage occurs then ^{4}He will also diffuse and, as a chemically very stable gas, should be detected in the soil. No ^{4}He enrichment is reported in soil gas (unpublished data) neither in a local aquifer sampled near injection well RSE-1 [31]. This aquifer has also very low concentrations in dissolved CH_4 (10^{-8} mol·L^{-1} when measurable; unpublished data) and the supply of CH_4 by the aquifer is highly unlikely. Natural micro-seepage from the reservoir is thus not certain and it must be assumed that the presence of CH_4 is rather linked to surface biological processes in non-anthropogenic soils. The fact is that soil CO_2 seems to be dependent from production through CH_4 oxidation in winter during periods of lower biological activity in soil. This effect vanishes—or is non-existent—during period of higher CO_2 production from biological processes in soil during summer.

The relation between CO_2 as soil gas and CO_2 as soil flux is presented in Figure 4. Baseline data did not highlight a relation between concentration and flux, only a slight positive tendency for "high season" acquisitions [35]. The situation did not change during injection and post-injection monitoring. "Low season" measurements are strictly uncorrelated (R^2 = 0.006; 458 data) and "high season" ones are barely correlated (R^2 = 0.018; 252 data), but this may be only related to the number of data that is smaller for summer acquisitions. There is no correlation between CO_2 flux and CO_2 soil-gas concentrations, suggesting a near-surface origin of the CO_2 gas (no CO_2 flux from the storage reservoir and only CO_2 production in the soil). In addition, this highlights the poor vertical connectivity of the soil, probably due to high clay-mineral content.

4.1.3. Carbon Isotope Ratios ($\delta^{13}C_{CO2}$)

Another way to infer the origin of the CO_2 gas phase is to refer to carbon-13 isotope-ratio determinations (Figure 6). As shown during the baseline site characterization, carbon-isotope ratios mainly fall within a range of −19 to −28‰ VPDB. A larger range was established by Garcia [70] (−15 to −25‰). The −19 to −28‰ range is the result of CO_2 isotope equilibration with that of biological organisms in the soil under steady-state conditions [71] and more specifically the equilibration with vegetation whose photosynthetic cycle produces molecules with three carbon atoms (Calvin cycle). Molecules produced under the Calvin cycle have carbon-isotope ratios ranging from −21 to −33‰ [72] with a mean value of −27‰ VPDB. The CO_2 produced from these molecules is enriched by a factor close to +4.4‰ [73], leading to $\delta^{13}C_{CO2}$ ranging from −17 to −29‰, the most common values in a temperate climate being near −22 to −25‰ [74] with possible extension to more depleted ratios (down to −28‰; [75]) or to more enriched ratios (up to −13‰; [76]). Applied to our dataset, this suggests that the presence of CO_2 in soil over the Lacq-Rousse storage site is essentially the result of natural biological processes occurring in soil.

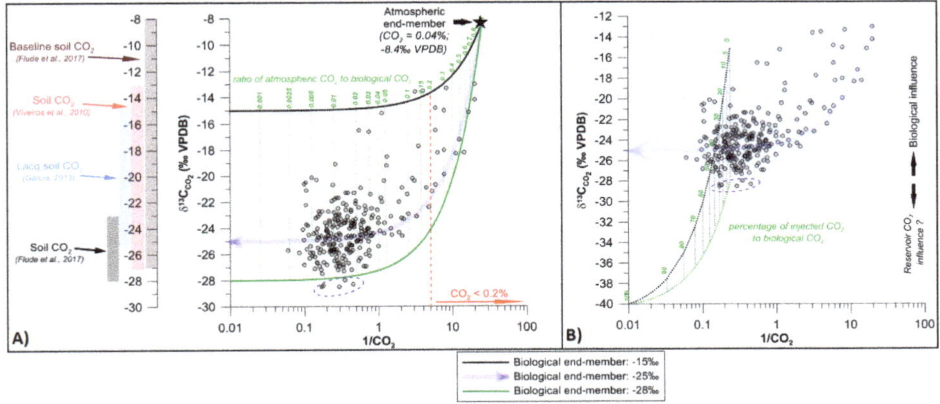

Figure 6. Carbon-isotope ratios ($\delta^{13}C_{CO2}$) as a function of the inverse of CO_2 concentration in soil gas (semi-logarithmic scale). (**A**) three atmospheric-biological mixing lines are calculated, using 0.04% CO_2 and −8.4‰ Vienna Pee Dee Belemnite (VPDB) standard for the atmospheric end-member, and 10% CO_2 and −15 to −28‰ VPDB for the biological end-member. Variation ranges of literature data are reported on the Y-axis. (**B**) mixing curves between two biological end-members (−15 and −28‰) and hypothetical deep CO_2 end-member (taken at −40‰).

If the CO_2 were produced from the remaining CH_4 in the storage reservoir, it should have a lower $\delta^{13}C_{CO2}$, as equilibration with CH_4 having a lower $\delta^{13}C_{CH4}$ ($\delta^{13}C_{CH4}$ of the Lacq-Rousse field: −41 to −43‰ VPDB; [67]) will produce depleted $\delta^{13}C_{CO2}$ [75]. Similarly, the injected CO_2, produced by CH_4 oxy-combustion, has a $\delta^{13}C_{CO2}$ of −40‰ [67]. Although some re-equilibration and/or dissolution processes in water may occur in a 4500-m-thick sedimentary pile, it is very unlikely that the monitored carbon-isotope ratios of the CO_2 originate from the storage reservoir. A difference larger than 10‰ between injected CO_2 and baseline CO_2 is recognized as representing favorable conditions for the use of carbon isotope ratios as a tracer [77]. The hypothesis of CO_2 originated from the reservoir is tested in Figure 6B for the five samples having ratios lower than −28‰. The isotope composition of the CO_2 that may originate from the storage is taken at −40‰ and mixing with soil gas is calculated (using the −15‰ and −28‰ endmembers). Three samples are outside binary mixing relations – to plot under such relations, the soil gas component should have a ratio more depleted than −28‰—founding such a depleted ratio in soil is very unlikely. Only two samples may plot on a binary relation between CO_2 from the storage and CO_2 from the soil (at −23‰) but this is not in agreement with other data. Analyses of noble gases in soil gas highlight the presence of atmospheric endmembers (neither ^{4}He nor ^{84}Kr enrichment), whereas the gas phase from the reservoir has a very different noble-gas signature [70].

Nevertheless, some $\delta^{13}C_{CO2}$ ranging from −18‰ to −11‰ VPDB are not explained—or are partly explained—by the previous statements; they are less depleted in the heavy isotope (^{13}C) than the majority of the data. Such deviations may be attributed to contamination by atmospheric CO_2 ($\delta^{13}C_{CO2}$ of −8.4‰; [78]) at least for data that plot close to the dilution lines reported in Figure 6A and that have low CO_2 concentrations (<0.2% vol. i.e., $1/CO_2$ >5). Under 0.2% of CO_2, the atmospheric endmember has a contribution that may become significant as may be deduced from the relative contribution of soil CO_2 and atmospheric CO_2 reported on Figure 6A. Such contamination was mainly stated during winter acquisitions (2/3 of the data) and may result from sampling bias linked to defective sealing of the sampling probe or the presence of water in soil that hampers the correct pumping of soil gas (half of the anomalies are from the December 2009 campaign). Hypothetical mantle or the lower crust CO_2 degassing ($\delta^{13}C_{CO2}$ ranging from −4 to −8‰ VPDB; [79]) is not supported by ^{4}He measured values. Finally, CO_2 produced by the dissolution of carbonate rocks or soil carbonates will have a $\delta^{13}C_{CO2}$ around 0‰ VPDB and a contribution of carbonate dissolution may also explain part of the low depleted data but this hypothesis cannot be tested using our data.

Scattering of the $\delta^{13}C_{CO2}$ data is strongly reduced when low CO_2 data are filtered from the measurements (Figure 3). There was only little variability throughout the monitoring period: the mean baseline value is -23.9‰, the mean injection value is -24.3‰ and the mean post-injection value is -24.6‰. The apparently progressive decrease of ratios that is observed when using the whole $\delta^{13}C_{CO2}$ dataset is thus the result of more frequent atmospheric contamination of the samples during the first sampling sessions, as no selection based on CO_2 concentration was made. The selectivity used afterwards, especially during post-injection acquisitions ($CO_2 > 2\%$), strongly limits this risk.

In conclusion, the data provided by carbon-isotope ratio measurements point to a near-surface biological origin for CO_2 gas in the soil, and thus to CO_2 flux at the soil/atmosphere interface.

4.2. General Discussion

4.2.1. Monitoring Strategy—The Aims

Geochemistry, including surface and near-surface investigations, is one of the many tools available for CCS site monitoring [37,80] and was extensively used at the Lacq-Rousse CCS pilot [23]. Among these investigations, soil-gas monitoring–concentration and soil flux measurements—was implemented under a scheme approved by the French administration. Soil-gas monitoring is a major geochemical tool for detecting gas micro-seepage from geologic sequestration [81] and should be included in MMV protocols [18]. A key parameter about the impact of leakage from deep storage formations, commonly discussed in recent years, was highlighted by Jones et al. [82]: "The impacts from many lower level fault—or well—related leakage scenarios are likely to be limited spatially and temporarily and recovery may be rapid". However, Beaubien et al. [83] stated that "leakage detection can be complicated by the natural background of near-surface gas geochemistry and by its variability as a consequence of temporal, seasonal, geological, biological ... changes".

Consequently, each near-surface strategy for gas monitoring is site specific and must be thoroughly evaluated in order to provide reliable information during the injection and post-injection periods. The definition of baseline monitoring is important [37]. Once injection takes place, no reverse process exists for determining background levels of soil gas. Even if no leakage occurs, detractors may argue that the storage formation could have experienced some changes after which it has become difficult to prove site integrity. The controversy about the Weyburn field and the allegations of leakage after gas injection illustrate the importance of correct baseline measurements, even though in this case seepage allegations were dropped in the light of data and methods that were not specifically considered during baseline, if any, data acquisitions [57,84].

The soil-gas monitoring of the Lacq-Rousse field relies on these facts. The initial ideas in 2006 for this work focused on studying the feasibility of the pilot itself, but the importance of a correct design of the baseline monitoring and its strategy became rapidly clear and was adopted in early 2007 when the European Directive 2009/31/EC was not yet operative. This matched French regulatory requirements as well as requirements of the site operator. Compared to, for instance, the Weyburn area [84], the Lacq-Rousse site has a hilly relief (slopes of 15% are common), even though hilltops culminate at only $+360$ m above sea level. This had a direct effect on monitoring: valleys are more accessible and have a thinner sedimentary succession than hills. A third of the monitoring points were thus located in valleys as they may be earlier affected by CO_2 leakage.

The monitoring locations covered the whole surface projection of the storage reservoir and were equally distributed around the CO_2 injection well. Their number is higher near the surface projection of the reservoir and includes locations on topographic highs. Six wells (locations 1 to 6, Figure 1) were drilled for seismic monitoring. As they might represent potential leakage pathways [82], some monitoring points were located close to these geophysical boreholes. Similarly, the injection well (point 16) and a former CH_4 exploitation borehole (point 17) were surveyed as well. In order to consider the geological structure of the Pyrenees forefront, some points covered different geological units, such as to the south of the reservoir. Although their number is low, these points are near outcrops of deep-rooted geological units whose internal discontinuities may affect horizons lying 2000 to 3000 m above the storage reservoir at a depth of 4500 m.

Finally, the baseline characterization was done at high frequency albeit not continuous [82]. It is established that leakage is easier to detect during winter, when biological activity is low [55]. In summer, potential CO_2 leaks may be hidden by high CO_2 production in soil by biological processes. Nevertheless, summer baseline data allow an evaluation of the natural variability of soil-gas signals, which can be large [37,82,84].

4.2.2. Monitoring Strategy—Ways of Improvement

The surface monitoring of the Lacq-Rousse pilot site does not pretend to be exhaustive. It is part of a larger-scale project that was entirely funded by the operator, which is not so common [17]. It should be mentioned again that the French Administration approved all operations at the time the actions were planned, when these operations were considered to provide sufficient information to ensure storage safety and conformance. However, since those days progress has been made in several fields of knowledge that merit further discussion.

The first field concerns the use of monitoring mostly restricted to characterization of the CO_2 phase in the near-surface environment, using a spot-sampling approach. Gas-source attribution techniques [85,86] rely on complementary gas phases to account for the specificities of the CO_2-gas phase (non-uniqueness of the source, ability to interact with water, etc.). Similarly, the use of methods with a wider geographical coverage and/or allowing continuous monitoring is commonly included in compliance monitoring [18].

The second field concerns the restricted number of monitoring locations. Site selection was based on several technical factors, but also on the cost of the monitoring actions, using the ratio 'number of locations' to 'information collected' as an economic parameter. The will to deploy several monitoring points per monitoring site, which enhances scientific relevance, did not match this economic constraint. Today, conformance monitoring would certainly refer more to the use of continuous monitoring and the use of systems surveying the atmosphere [37].

Other fields of potential improvement concern the lack of monitoring deep aquifers, if present close to the injection site, or the absence of monitoring of mineral-water sites at a larger geographical scale. Some local mineral waters that are of sulphide type, such as Eaux Chaudes or Eaux Bonnes [87]. CO_2 intrusion in water has a direct effect on pH and delayed effects on water mineralization, especially in the case of mineral waters that naturally have low CO_2 content.

A final field in which progress has been made concerns the planning of monitoring sessions, particularly the focus on "low season" (winter) monitoring, recommended by Klusman [55] and others, leakage certainly being independent from climatic conditions. The way baseline data were used for computing initial threshold levels and the way data from the injection period were used for defining new threshold levels are also questionable. This last point is discussed in the next section.

4.2.3. Are Baseline Acquisitions Important?

Soil-gas and soil-flux monitoring during summer and winter times showed that major changes can occur from one year to the next at the study site [35]. Changes of mean CO_2 concentration and mean CO_2 flux between winter and summer were from 200% to 280% (Figure 7). This situation continued during the injection period with changes ranging from 160% to 240%. At the opposite, the mean values during the same season (winter/summer) were less variable: e.g., winter CO_2 concentrations and fluxes only varied between 15% and 30% from baseline to injection periods. The fluxes remained at similar mean values during the post-injection monitoring albeit the soil CO_2 concentrations were higher. An evaluation of the dispersal of soil-gas concentrations and fluxes is then important for obtaining sound threshold values, but the validity of the latter over long periods should be questioned.

Biological activity varies throughout the year and from year to year, because of variable rainfall, variable climatic conditions, or changes in land use. This is well illustrated for soil CO_2 flux and soil temperature (Figure 7B). In the present study, data acquisitions before the injection period extended over 18 months, which is in line with hydrogeological methods that consider a similar period for defining the natural variability of water resources.

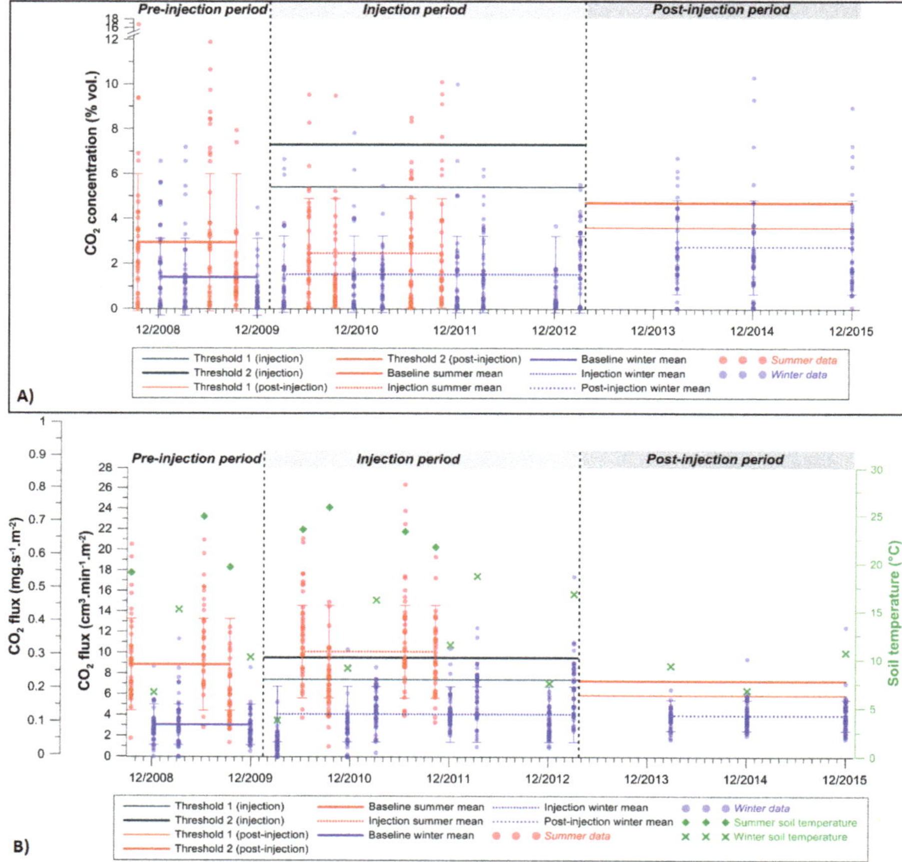

Figure 7. (**A**) CO_2 concentrations as a function of time: winter and summer data are treated separately as a function of the period of site activity (baseline, injection, post-injection). Mean values for each period are indicated (blue and red lines respectively) with error bar corresponding to the standard deviation. Threshold levels in use during injection and post-injection periods are reported for comparison. (**B**) similar plot for CO_2-fluxes; green points correspond to mean soil temperature at the time of the sampling session.

The resulting dataset showed strong variability over this initial monitoring period, but the data were nonetheless used by the site operator for defining the geochemical thresholds as approved by the French authorities. This dataset provided a good description of the soil-gas behaviour over several seasons, as it showed that soil-gas and soil-flux variability may be high from one point to the next and between campaigns. Such evaluations of the natural variability are important, if only for demonstrating that each CCS site has its own characteristics and its own degree of variability. Comparison with data from the Weyburn site [83] shows that the amplitude of CO_2 concentrations, or flux, is only similar to that of Lacq-Rousse during summer, and that monitoring in September or October gives much lower values as a whole.

It is thus very important to evaluate the amplitude of natural changes that can occur in a future storage site, and this not only for future checking of site conformance, but also for public acceptance as people may be opposed to CCS [88,89]. The presence of a surface monitoring network that is used at regular intervals or, even better, for continuous monitoring, is intrinsically reassuring for people living in the area of interest [37]. Its geographical coverage may not be sufficient to detect leaks, but since it exists it gives precious information on gas emissions at the surface. It will also provide control data for future use in case of deviations noted by third parties, by simply showing how the ecosystem constantly evolves.

The ultimate questions are the obligation of getting baseline data and the pertinence of surface geochemical monitoring. The Weyburn example shows that leakage allegations can be disproved even if baseline data are lacking or insufficient [58], but feedback from the US shale-gas industry shows that consensus and public confidence on results from monitoring studies can be hard to obtain if baseline data are missing. Their comparison with operating-phase data remains the easiest way for people to understand expert conclusions.

Surface geochemical monitoring at Lacq-Rousse is thus certainly a valid option, but the way it has been done is perfectible, with e.g., more emphasis on specific locations such as near boreholes or faults. The monitoring scheme as devised for Lacq-Rousse helped, proving that, compared to baseline data, no major changes occurred during injection and post-injection. A baseline dataset, whose internal scattering is mostly linked to densely vegetated surface formations and correlative strong biological activity in soil, thus adds value to an MMV program.

4.2.4. Are Threshold Levels Required?

At reservoir scale, the Lacq-Rousse CCS pilot has proven its integrity throughout the injection period [90]. Data from surface monitoring also suggest site integrity is preserved: any thresholds that were exceeded were attributed to surface processes referring to gas source attribution techniques [59,85,86]. At least, no leak event is reported as having occurred near monitoring locations.

The threshold values were defined after completing the baseline acquisitions, and were the only references that were used throughout storage activity. It must be stressed that the definition of the thresholds was not under the responsibility of the scientific team performing the site acquisitions. Therefore, other ways to compute thresholds, such as calculating one threshold for each geological formation, were beyond the scope of the monitoring actions. The thresholds were then redefined for the closure and post-closure phases. The option chosen by the site operator was to define warning levels using the whole baseline database and to interpret it using statistical mean value and standard deviation. This procedure was approved by the French administration, and "vigilance" and "anomaly" thresholds were defined and referred to throughout the "low season" injection periods. Feedback from local authorities and inhabitants suggests that the literal sense of the word 'threshold' may induce unnecessary fears. A better way would have been to refer only to the sole mathematical definition. From that perspective, the use of "statistical difference from baseline values" and "significant or notable difference from baseline values" for levels 1 and 2, respectively, would have been more suitable.

The question of the sole use of wintertime monitoring data is also open to debate. From March 2010 to March 2013, "low season" data acquisitions used the first definition of threshold levels, and CO_2 concentrations and CO_2 flux were considered separately (Figure 7). This led to few occurrences of level exceeded for concentrations. The situation was different for soil flux, with high soil degassing in March 2012 and March 2013, the hottest "low season" months of the entire dataset (Figure 2). In addition, the confrontation between relatively high air temperatures and biological processes in soils that have just begun to reactivate after winter causes normal CO_2 concentrations and high CO_2 flux. From December 2013 to December 2015, "low season" data were acquired using the redefined threshold levels. The change in CO_2-flux threshold levels had almost no impact on threshold exceeding, but the situation was greatly different for CO_2 concentrations in soil. Numerous cases of threshold exceeding were noted during the field work, covering both "vigilance" and "anomaly" levels, which would not have occurred when using only baseline levels. Fortunately, the coupling of soil-gas and flux data did not lead to the exceeding of warning levels, but this could have occurred as well. A consequence of these observations is that the redefinition of threshold levels during the activity on a site may lead to an exceeding of warning levels, thus requiring additional measurements to ensure that no adverse event affects the storage. However, such a change of threshold values may also be seen as a proactive approach to gain more benefit from earlier acquired data. As a CCS site generally is scheduled to operate over decades, any opportunity to adapt the monitoring scheme during the life of the site will be meaningful for considering the potential effects of climate change [85].

The downside is that such a modification of threshold values can have a financial impact on the project budget, because additional measurements may be needed. Such a change in monitoring activity is likely to be questioned so as to avoid triggering warning levels whose origin is only related to surface- or near-surface processes.

Further comments can be made on the above-defined thresholds whose use is restricted to winter times. Figure 7 well illustrates that CO_2-flux threshold levels cannot be used during the 'high season' as most of the data would be anomalous or at least exceeding the "vigilance" level. In that case, if thresholds were to be used during summer times, then they should have been adapted to this time period (to a value at least two times greater). This is less true for CO_2 concentrations because these data had, during baseline acquisitions, a larger range of variation thus inducing a larger standard deviation (typically of the order of the mean value whereas flux data had only a SD of half of the mean value). This points out another drawback based on the univocal definition of thresholds considering the whole dataset: when the data are less related to phenomenological parameters such as temperature then the threshold definition may induce the calculation of levels that are too high.

Such bias has been taken into account when the levels were re-evaluated at the end of the injection period (Figure 7). The thresholds for the flux had minor decrease (minus 25–30%) whereas the thresholds for the gas concentrations were more drastically reduced (minus 50–55%). As the CO_2 fluxes remained at low levels during winter acquisitions, the change of thresholds had limited influence onto level exceeding (1 or 2 exceeds with new "vigilance" threshold vs. 0 to 1 exceed with the former threshold). At the opposite, the CO_2 concentrations were high and thresholds were often exceeded (7 to 11 exceeds with new "vigilance" threshold vs. 4 exceeds with the former threshold). Without the help of other data (O_2 or C-isotopes) the attribution of so many exceeding thresholds would have been more complicated.

In conclusion, it appears important to define threshold levels in connection with seasonal, meteorological and biological conditions, ensuring that the statistical definition of threshold levels covers as many parameters as possible. With our dataset, the influence of local geological conditions appears to be of secondary importance especially when considering CO_2 concentrations. Concentrations greater than 5% were indeed measured on half of the locations (18 points over 36) and these points were located over all the geological formations reported in the area and were indifferently set in valleys or top hill. Such a definition linked to other parameters will avoid a future occurrence of false positives that only result from normal climate-driven changes in near-surface formations. Nevertheless, the question of the validity of threshold levels for the entire period of activity of the storage remains open, as climate change and its potential long-term impact on soil-gas concentrations and soil flux cannot be controlled. Two options can thus be envisaged, either separately or complementarily.

The first is to complement the data on soil-gas concentration and/or flux with the monitoring of other tracers. However, though such process-based methods are of great use, a conclusive detection of seepage may remain problematic [91]. The use of additional monitoring, such as of carbon-13 isotope ratios, is important as such ratios are not likely to change rapidly because of climate change, except if the consequence of climate change is a quick shift from photosynthetic C3 cycle to C4 cycle.

Nevertheless Ehleringer et al. [92] reported that millions of years were necessary for CO_2 levels to decline to concentrations that were sufficiently low (from a few thousand to a few hundred ppm) to give an advantage to C4 photosynthesis over C3 photosynthesis. The reverse process (slight increase of the CO_2 concentrations) may thus have limited impact over short time scales. As a consequence, a change of carbon-13 isotope ratios will be rather linked to changes of the CO_2 sources (surface processes vs. deeper origin) which may have specific carbon isotope signature [75,78]. Other co-injected tracers or intrinsic tracers [91] may also be monitored in order to strengthen the conclusions drawn from monitoring of the $\delta^{13}C_{CO2}$. Noble gases originating from the CO_2-production process and injected along with CO_2 at the Lacq-Rousse pilot represent one of these types of tracers that are important to monitor [67].

The second option is to adopt another strategy for defining threshold levels, if necessary. Such levels are not systematically required for ensuring the viability of storage containment [59,86], but are often desired by the authorities. The interpretation of soil-gas and soil-flux data obtained in geothermal and volcanic areas is commonly based on the use of graphical statistical analysis (GSA) and of log probability plots [93], or on the use of more complex conditional sequential Gaussian simulations (sGs; [94]). These methods can be adapted in case the dataset is skewed and does not fall under normal or lognormal distribution, which is the case for our soil-gas data.

Figure 8 presents some insights using the GSA approach. The calculation refers to winter and summer data acquired during the baseline period and to winter data acquired during the injection period. This corresponds to the time periods used as references to define the different threshold levels during the pilot activity. As a consequence of high soil CO_2 concentrations in the Lacq-Rousse area, the thresholds that maybe defined using the GSA approach have little scatter (4.5 to 5.2%). The situation is different with CO_2-fluxes because of great variations between summer and winter times. The GSA-defined CO_2 thresholds are also in good agreement with the "vigilance" threshold used at Lacq-Rousse suggesting that even basic statistics may have some soundness. The CO_2-flux threshold that may be postulated from injection data—or from the mean of the summer and winter baseline data—is less in agreement with the "vigilance" threshold that was used because it includes larger variability related to meteorological forcing. Another threshold can be defined for isotope data; it ranges in the lower part of the biological pool (−25 to −27‰; see Figure 6). As it is mainly defined by referring to carbon isotope data of CO_2 that existed prior to any CO_2 injection in the reservoir, it highlights that the first threshold that was suggested to be used during the post-injection phase (−26‰) does not represent an anomalous value by itself and only a biological end-member.

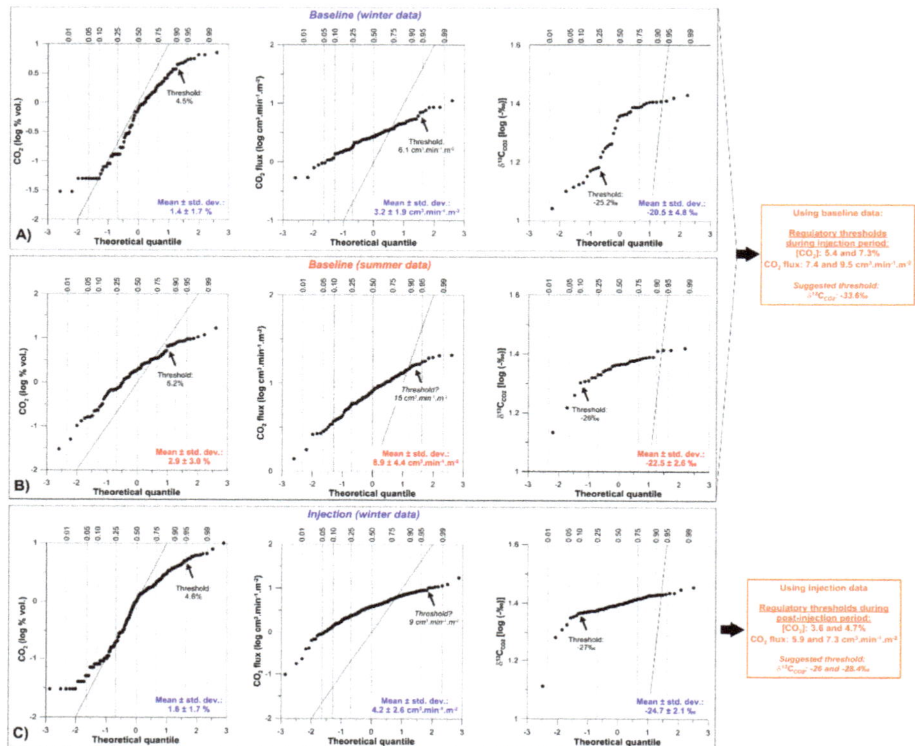

Figure 8. Log-probability plots for CO_2 concentrations, CO_2-fluxes and $\delta^{13}C_{CO2}$ isotope ratios. (**A**) winter baseline data; (**B**) summer baseline data; (**C**) winter injection data. Mean value and standard deviation of the dataset are indicated in each plot. The threshold definition that was used during pilot activity is indicated for comparison.

5. Conclusions

Measurements of soil-gas concentration and soil flux are reported for the TOTAL Lacq-Rousse CO_2 CCS pilot site. Baseline data, collected from September 2008 to December 2009, are compared with the data obtained during the injection (March 2010 to March 2013) and post-injection (February 2014 to December 2015) periods. Twenty separate monitoring sessions are available for the 8-year duration of the pilot project. Although not continuous, these methods did allow a correct evaluation of the integrity of the site while adequately attributing the origin of the gas signals. The French administration approved the planned strategy of monitoring (spot measurements strategy, number of sampling locations, frequency of monitoring and changes of this frequency).

The main source of variations over time in soil-gas concentrations or CO_2 fluxes was found to be related to yearly seasonal changes. CO_2 soil-gas concentrations are roughly the same on an annual basis, but tend to be higher during summer and lower during winter. CO_2 flux is high compared to other sites, because of strong biological activity in humid and fertile soils under a temperate climate. Fluxes tend to be higher when the temperature is higher, well illustrating the thermal draw effect on local gas exchanges between shallow soil and the atmosphere.

CO_2 and O_2 soil-gas concentrations are strongly reversely correlated and are greatly influenced by respiration processes in soil, especially during summer and for CO_2 concentrations >2%. The occurrence of few ppm of CH_4 in the soil by diffuse degassing from the former methane reservoir is yet unproven. The comparison between CO_2 concentration in soil and CO_2 flux shows no interdependence, advocating for a near-surface origin of the CO_2 gas and highlighting the poor vertical connectivity of the soils. Carbon-isotope ratios in CO_2 soil gas point also to a biological origin for this gas.

Geochemical threshold values were used for defining "vigilance" and "anomaly" modes that were only valid for data acquisition during winter, as approved by the French authorities. This reference to winter times may question the validity of leakage detection during summer, and the ways of confirming or discarding the occurrence of possible leakage. Changes of the threshold levels when the site shifted to post-injection monitoring did not answer this point, but more information comes from the inclusion of carbon-13 isotope-ratio measurements. Threshold redefinition is a way to account for background variability over longer time scales but gas source attribution techniques may be better suited. The comparison of baseline and new thresholds reveals that the occurrence of threshold exceeding is more common for the new ones, even though no significant change occurred in the environmental parameters. From an operating viewpoint, this may indicate the need for additional measurements, to ensure that site integrity is preserved.

This points out the difficulty of constructing monitoring scenarios and how to define the limits within which parameters can vary. It also points out the need for having complementary data, in this case carbon-isotope ratios of CO_2. From the viewpoint of a CCS site operator, the monitoring of gas phases that are not directly linked to CO_2, or to products injected with the CO_2, may seem to be of little interest because of cost/benefit considerations. Moreover, the baseline data acquisitions essential for this type of work imply uncompressible costs that may further reduce its attractiveness. In that case, other methods, such as isotope measurements, may be better suited.

Author Contributions: Conceptualization, F.G., Z.P., N.L., K.M., S.L., A.M.; data curation, F.G., Z.P.; formal analysis, F.G., Z.P., S.L.; resources, Z.P., N.L., K.M., A.M.; investigation, F.G., Z.P., K.M., S.L.; methodology, F.G., Z.P., K.M., S.L.; project administration, F.G., Z.P., N.L., K.M.; validation, F.G., Z.P., visualization, F.G., Z.P.; writing: F.G., Z.P., S.L., A.M.

Funding: "Low season" data acquisitions were funded by TOTAL. "High season" sampling was done through funding of the "Sentinelle" research project supported by the French Research Agency (ANR-07-PCO2-007).

Conflicts of Interest: The authors declare no conflict of interest.

References

1. Benson, S.M.; Dorchak, T.; Jacobs, G.; Ekmann, J.; Bishop, J.; Grahame, T. Carbon Dioxide Reuse and Sequestration: The State of the Art Today. 2000. Available online: http://escholarship.org/uc/item/30f3r8sf (accessed on 14 April 2017).

2. Gentzis, T. Subsurface sequestration of carbon dioxide—An overview from an Alberta Canada perspective. *Int. J. Coal Geol.* **2000**, *43*, 287–305. [CrossRef]

3. Bachu, S.; Gunter, W.D. Storage capacity of CO_2 in geological media in sedimentary basins with application to the Alberta basin. In *Greenhouse Gas Control Technologies*; Riemer, P., Eliasson, B., Wokaun, A., Eds.; Elsevier: Amsterdam, The Netherlands, 1999; pp. 195–200.

4. Chen, Z.; Huang, G.H.; Chakma, A.; Tontiwachwuthikul, P. Environmental risk assessment for aquifer disposal of carbon dioxide. In *Greenhouse Gas Control Technologies*; Riemer, P., Eliasson, B., Wokaun, A., Eds.; Elsevier: Amsterdam, The Netherlands, 1999; pp. 223–228.

5. Kaarstad, O. Emission-free fossil energy from Norway. *Energy Convers. Manag.* **1992**, *33*, 781–786. [CrossRef]

6. Korbol, R.; Kaddour, A. Sleipner Vest CO_2 disposal—Injection of removed CO_2 into the Utsira Formation. *Energy Convers. Manag.* **1995**, *36*, 509–512. [CrossRef]

7. Gupta, N.; Sass, B.; Sminchak, J.; Naymik, T.; Bergman, P. Hydrodynamics of CO_2 disposal in a deep saline formation in the midwestern United States. In *Greenhouse Gas Control Technologies*; Riemer, P., Eliasson, B., Wokaun, A., Eds.; Elsevier: Amsterdam, The Netherlands, 1999; pp. 157–162.

8. Bachu, S.; Brulotte, M.; Grobe, M.; Stewart, S. *Suitability of the Alberta Subsurface for Carbon-Dioxide Sequestration in Geological Media, Alberta Energy and Utilities Board and Alberta Geological Survey*; Earth Sciences Report 00-11; Alberta Geological Survey: Edmonton, AB, Canada, 2000.

9. Barnhart, W.; Coulthard, C. Weyburn CO_2 miscible flood conceptual design and risk assessment. *Pet. Soc. Can.* **1999**. [CrossRef]

10. Hattenbach, R.P.; Wilson, M.; Brown, K.R. Capture of carbon dioxide from coal combustion and its utilization for enhanced oil recovery. In *Greenhouse Gas Control Technologies*; Riemer, P., Eliasson, B., Wokaun, A., Eds.; Elsevier: Amsterdam, The Netherlands, 1999; pp. 217–222.

11. Byrer, C.W.; Guthrie, H.D. Coal deposits: Potential geological sink for sequestering carbon dioxide emissions from power plants. In *Greenhouse Gas Control Technologies*; Riemer, P., Eliasson, B., Wokaun, A., Eds.; Elsevier: Amsterdam, The Netherlands, 1999; pp. 181–187.

12. Stevens, S.H.; Kuuskraa, V.A.; Spector, D.; Riemer, P. CO_2 sequestration in deep coal seems: Pilot results and worldwide potential. In *Greenhouse Gas Control Technologies*; Riemer, P., Eliasson, B., Wokaun, A., Eds.; Elsevier: Amsterdam, The Netherlands, 1999; pp. 175–180.

13. Riemer, P. Greenhouse gas mitigation technologies, an overview of the CO_2 capture, storage and future—Activities of the IEA Greenhouse Gas R&D Programme. *Energy Convers. Manag.* **1996**, *37*, 665–670.

14. Herzog, H.J. Ocean sequestration of CO_2—An overview. In *Greenhouse Gas Control Technologies*; Riemer, P., Eliasson, B., Wokaun, A., Eds.; Elsevier: Amsterdam, The Netherlands, 1999; pp. 237–242.

15. Herzog, H.J. What future for Carbon Capture and Sequestration? *Environ. Sci. Technol.* **2001**, *35*, 148–153. [CrossRef]

16. IEAGHG Database. Available online: http://www.ieaghg.org/ccs-resources/rd-database (accessed on 14 April 2017).

17. Herzog, H.J. Lessons Learned from CCS Demonstration and Large Pilot Projects. An MIT Energy Initiative Working Paper 2016. Available online: https://sequestration.mit.edu/bibliography/CCS%20Demos.pdf (accessed on 16 October 2017).

18. Jenkins, C.; Chadwick, A.; Hovorka, S. The state of the art in monitoring and verification—Ten years on. *Int. J. Greenh. Gas Control* **2015**, *40*, 312–349. [CrossRef]

19. Lupion, M.; Diego, R.; Loubeau, L.; Navarrete, B. CIUDEN CCS Project: Status of the CO_2 capture technology development plant in power generation, GHGT-10. *Energy Procedia* **2011**, *4*, 5639–5646. [CrossRef]

20. Flett, M.; Brantjes, J.; Gurton, R.; McKenna, J.; Tankersley, T.; Trupp, M. Subsurface development of CO_2 disposal for the Gorgon Project, GHGT-9. *Energy Procedia* **2009**, *1*, 3031–3038. [CrossRef]

21. Wildenborg, T.; Bentham, M.; Chadwick, A.; David, P.; Deflandre, J.P.; Dillen, M.; Groenenberg, H.; Kirk, K.; Le Gallo, Y. Large-scale CO_2 injection demos for the development of monitoring and verification technology and guidelines (CO2ReMoVe), GHGT-9. *Energy Procedia* **2009**, *1*, 2367–2374. [CrossRef]

22. Eiken, O.; Ringrose, P.; Hermanrud, C.; Nazarian, B.; Torp, T.A.; Høier, L. Lessons learned from 14 years of CCS operations: Sleipner, In Salah and Snøhvit, GHGT-10. *Energy Procedia* **2011**, *4*, 5541–5548. [CrossRef]

23. Lescanne, M.; Hy-Billiot, J.; Aimard, N.; Prinet, C. The site monitoring of the Lacq industrial CCS reference project, GHGT-10. *Energy Procedia* **2011**, *4*, 3518–3525. [CrossRef]

24. Sharma, S.; Cook, P.; Berly, T.; Lees, M. The CO2CRC Otway Project: Overcoming challenges from planning to execution of Australia's first CCS project, GHGT-9. *Energy Procedia* **2009**, *1*, 1965–1972. [CrossRef]

25. Preston, C.; Monea, M.; Jazrawi, W.; Brown, K.; Whittaker, S.; White, D.; Law, D.; Chalaturnyk, R.; Rostron, B. IEA GHG Weyburn CO_2 monitoring and storage project. *Fuel Process. Technol.* **2005**, *86*, 1547–1568.

26. Carbon Capture and Sequestration Technologies Program at MIT. Available online: https://sequestration.mit.edu/index.html (accessed on 21 October 2018).

27. Feitz, A.J.; Leamon, G.; Jenkins, C.; Jones, D.G.; Moreira, A.; Bressan, L.; Melo, C.; Dobeck, L.M.; Repasky, K.; Spangler, L.H. Looking for leakage or monitoring for public assurance? *Energy Procedia* **2014**, *63*, 3881–3890. [CrossRef]

28. Monne, J.; Prinet, C. Rousse-Lacq Industrial CCS reference project: Description and operational feedback after two and half years of operation, GHGT-11. *Energy Procedia* **2013**, *37*, 6444–6457. [CrossRef]

29. French Mining Code. Available online: www.legifrance.gouv.fr/affichCode.do?cidTexte=LEGITEXT000023501962&dateTexte=20110511 (accessed on 16 October 2017).

30. Environmental Protection Code. Available online: www.legifrance.gouv.fr/affichCode.do?cidTexte=LEGITEXT000006074220 (accessed on 16 October 2017).

31. TOTAL. Carbon Capture and Storage: The Lacq Pilot. Project and Injection Period 2006–2013. Available online: https://www.globalccsinstitute.com/publications/carbon-capture-and-storage-lacq-pilot-project-and-injection-period-2006-2013 (accessed on 16 October 2017).

32. Prinet, C.; Thibeau, S.; Lescanne, M.; Monne, J. Lacq-Rousse CO_2 Capture and Storage demonstration pilot: Lessons learnt from two and a half years monitoring, GHGT-11. *Energy Procedia* **2013**, *37*, 3610–3620. [CrossRef]

33. Gapillou, C.; Thibeau, S.; Mouronval, G.; Lescanne, M. Building a geocellular model of the sedimentary column at Rousse CO_2 geological storage site (Aquitaine, France) as a tool to evaluate a theoretical maximum injection pressure, GHGT9. *Energy Procedia* **2009**, *1*, 2937–2944. [CrossRef]

34. Thibeau, S.; Chiquet, P.; Prinet, C.; Lescanne, M. Rousse-Lacq CO_2 Capture and Storage demonstration pilot: Lessons learnt from reservoir modelling studies, GHGT-11. *Energy Procedia* **2013**, *37*, 6306–6316. [CrossRef]

35. Gal, F.; Michel, K.; Pokryszka, Z.; Lafortune, S.; Garcia, B.; Rouchon, V.; de Donato, P.; Pironon, J.; Barrès, O.; Taquet, N.; et al. Study of the environmental variability of gaseous emanations over a CO_2 injection pilot—Application to the French Pyrenean foreland. *Int. J. Greenh. Gas Control* **2014**, *21*, 177–190. [CrossRef]

36. Feitz, A.; Jenkins, C.; Schacht, U.; McGrath, A.; Berko, H.; Schroder, I.; Noble, R.; Kuske, T.; George, S.; Heath, C.; et al. An assessment of near surface CO_2 leakage detection techniques under Australian conditions. *Energy Procedia* **2014**, *63*, 3891–3906. [CrossRef]

37. Schloemer, S.; Furche, M.; Dumke, I.; Poggenburg, J.; Bahr, A.; Seeger, C.; Vidal, A.; Faber, E. A review of continuous soil gas monitoring related to CCS—Technical advances and lessons learned. *Appl. Geochem.* **2013**, *30*, 148–160. [CrossRef]

38. Ha-Duong, M.; Gaultier, M.; De Guillebon, B. Social aspects of Total's Lacq CO_2 capture, transport and storage pilot project. In Proceedings of the 10th International Conference on Greenhouse Gas Control Technologies GHGT-10, Amsterdam, The Netherlands, 19–23 September 2010.

39. Boyd, A.D.; Hmielowski, J.D.; Prabu, D. Public perceptions of carbon capture and storage in Canada: Results of a national survey. *Int. J. Greenh. Gas Control* **2017**, *67*, 1–9. [CrossRef]

40. Pokryszka, Z.; Charmoille, A.; Bentivegna, G. Development of methods for gaseous phase geochemical monitoring on the surface and in the intermediate overburden strata of geological CO_2 storage sites. *Oil Gas Sci. Technol.* **2010**, *65*, 653–666. [CrossRef]

41. Spangler, L.H.; Dobeck, L.M.; Repasky, K.S.; Nehrir, A.R.; Humphries, S.D.; Barr, J.L.; Keith, C.J.; Shaw, J.A.; Rouse, J.H.; Cunningham, A.B.; et al. A shallow subsurface controlled release facility in Bozeman, Montana, USA, for testing near surface CO_2 detection techniques and transport models. *Environ. Earth Sci.* **2010**, *60*, 227–239. [CrossRef]

42. Jones, D.; Barlow, T.; Barkwith, A.; Hannis, S.; Lister, T.; Strutt, M.; Beaubien, S.; Bellomo, T.; Annunziatellis, A.; Graziani, S.; et al. Baseline variability in onshore near surface gases and implications for monitoring at CO_2 storage sites, GHGT12. *Energy Procedia* **2014**, *63*, 4155–4162. [CrossRef]

43. Carapezza, M.L.; Ricci, T.; Ranaldi, M.; Tarchini, L. Active degassing structures of Stromboli and variations in diffuse CO_2 output related to the volcanic activity. *J. Volcanol. Geotherm. Res.* **2009**, *182*, 231–245. [CrossRef]

44. De Donato, P.; Pironon, J.; Mouronval, G.; Hy-Billiot, J.; Lescanne, M.; Garcia, B.; Lucas, H.; Pokryszka, Z.; Lafortune, S.; Flamant, P.H.; et al. SENTINELLE: Research project on development of combined geochemical monitoring from geosphere to atmosphere. In Proceedings of the Experimentation on the site of Rousse (TOTAL CCS Pilot), GHGT-10, Amsterdam, The Netherlands, 19–23 September 2010.

45. Taquet, N. Monitoring géochimique de la géosphère et de l'atmosphère: Application au stockage géologique du CO_2. Ph.D. Thesis, University of Lorraine, Lorraine, France, 2012.

46. Jenkins, C.; Kuske, T.; Zegelin, S. Simple and effective atmospheric monitoring for CO_2 leakage. *Int. J. Greenh. Gas Control* **2016**, *46*, 158–174. [CrossRef]

47. Tauzière, C.; Pokryszka, Z.; Jodart, A. Measurement of the Gas Flux from a Surface. European Patent EP 0 807 822 B1, 5 September 2001.

48. Pokryszka, Z.; Tauzière, C. Evaluation of gas emission from closed mines surface to atmosphere. In *Environmental Issues and Management Waste in Energy and Mineral Production*; Balkema: Rotterdam, The Netherlands, 2000; pp. 327–329. ISBN 9789058090850.

49. Toutain, J.P.; Baubron, J.C. Gas geochemistry and seismotectonics: A review. *Tectonophysics* **1999**, *304*, 1–27. [CrossRef]

50. Kiefer, R.H.; Amey, R.G. Concentrations and controls of soil carbon dioxide in sandy soil in the North Carolina coastal plain. *Catena* **1992**, *19*, 539–559. [CrossRef]

51. Raich, J.W.; Schlesinger, W.H. The global carbon dioxide flux in soil respiration and its relationship with vegetation and climate. *Tellus* **1992**, *44B*, 81–99. [CrossRef]

52. Hinkle, M.E. Environmental conditions affecting concentrations of He, CO_2, O_2 and N_2 in soil gases. *Appl. Geochem.* **1994**, *9*, 53–63. [CrossRef]

53. Koerner, B.; Klopatek, J. Anthropogenic and natural CO_2 emission sources in an arid urban environment. *Environ. Pollut.* **2002**, *116*, S45–S51. [CrossRef]

54. Jassal, R.; Black, A.; Novak, M.; Morgenstern, K.; Nesic, Z.; Gaumont-Guay, D. Relationship between soil CO_2 concentrations and forest-floor CO_2 effluxes. *Agric. For. Meteorol.* **2005**, *130*, 176–192. [CrossRef]

55. Klusman, R.W. Baseline studies of surface gas exchange and soil-gas composition in preparation for CO_2 sequestration research: Teapot Dome, Wyoming. *AAPG Bull.* **2005**, *89*, 981–1003. [CrossRef]

56. von Arnold, K.; Weslien, P.; Nilsson, M.; Svensson, B.H.; Klemedtsson, L. Fluxes of CO_2, CH_4 and N_2O from drained coniferous forests on organic soils. *For. Ecol. Manag.* **2005**, *210*, 239–254. [CrossRef]

57. Romanak, K.D.; Sherk, G.W.; Hovorka, S.D.; Yang, C. Assessment of alleged CO_2 leakage at the Kerr farm using a simple process-based soil gas technique: Implications for carbon capture, utilization, and storage (CCUS) monitoring, GHGT-11. *Energy Procedia* **2013**, *37*, 4242–4248. [CrossRef]

58. Romanak, K.D.; Wolaver, B.; Yang, C.; Sherk, G.W.; Dale, J.; Dobeck, L.M.; Spangler, L.H. Process-based soil gas leakage assessment at the Kerr Farm: Comparison of results to leakage proxies at ZERT and Mt. Etna. *Int. J. Greenh. Gas Control* **2014**, *30*, 42–57. [CrossRef]

59. Romanak, K.D.; Bennett, P.C.; Yang, C.; Hovorka, S.D. Process-based approach to CO_2 leakage detection by vadose zone gas monitoring at geologic CO_2 storage sites. *Geophys. Res. Lett.* **2012**, *39*, L15405. [CrossRef]

60. Gal, F.; Joublin, F.; Haas, H.; Jean-Prost, V.; Ruffier, V. Soil gas (^{222}Rn, CO_2, ^{4}He) behaviour over a natural CO_2 accumulation, Montmiral area (Drôme, France): Geographical, geological and temporal relationships. *J. Environ. Radiat.* **2011**, *102*, 107–118. [CrossRef]

61. Gal, F.; Le Pierrès, K.; Brach, M.; Braibant, G.; Bény, C.; Battani, A.; Tocqué, E.; Benoît, Y.; Jeandel, E.; Pokryszka, Z.; et al. Surface gas geochemistry above the natural CO_2 reservoir of Montmiral (Drôme-France), source tracking and gas exchange between soil, biosphere and atmosphere. *Oil Gas Sci. Technol.* **2010**, *65*, 635–652. [CrossRef]

62. Pokryszka, Z. Reference Values of Naturally Occurring Biogenic CO_2 and CH_4 Fluxes from Soils, Ineris Study Report DRS-17-164646-05731A. 2017. Available online: https://www.ineris.fr (accessed on 13 March 2018).

63. Gregory, R.G.; Durrance, E.M. Helium, carbon dioxide and oxygen soil gases: Small-scale variations over fractured ground. *J. Geochem. Explor.* **1985**, *24*, 29–49. [CrossRef]

64. Born, M.; Dörr, H.; Levin, I. Methane consumption in aerated soils of the temperate zone. *Tellus* **1990**, *42B*, 2–8. [CrossRef]

65. Le Mer, J.; Roger, P. Production, oxidation, emission and consumption of methane by soils: A review. *Eur. J. Soil Biol.* **2001**, *37*, 25–50. [CrossRef]

66. Etiope, G. Subsoil CO_2 and CH_4 and their advective transfer from faulted grassland to the atmosphere. *J. Geophys. Res.* **1999**, *104*, 16889–16894. [CrossRef]

67. Garcia, B.; Hy-Billiot, J.; Rouchon, V.; Mouronval, G.; Lescanne, M.; Lachet, V.; Aimard, N. A geochemical approach for monitoring a CO_2 Pilot Site: Rousse, France. A major gases, CO_2-carbon isotopes and noble gases combined approach. *Oil Gas Sci. Technol.* **2012**, *67*, 341–353. [CrossRef]

68. Etiope, G.; Martinelli, G. Migration of carrier and trace gases in the geosphere: An overview. *Phys. Earth Planet. Inter.* **2002**, *129*, 185–204. [CrossRef]

69. Etiope, G.; Klusman, R.W. Geologic emissions of methane to the atmosphere. *Chemosphere* **2002**, *49*, 777–789. [CrossRef]

70. Garcia, B. A geochemical approach for monitoring a CO_2 pilot site—Rousse, France. In Proceedings of the EAGE Meeting "Sustainable Earth Sciences 2013", Pau, France, 30 September–4 October 2013.

71. Amundson, R.; Stern, L.; Baisden, T.; Wang, Y. The isotopic composition of soil and soil respired CO_2. *Geoderma* **1998**, *82*, 83–114. [CrossRef]

72. Deines, P.; Langmuir, D.; Harmon, R.S. Stable carbon isotope ratios and the existence of a gas phase in the evolution of carbonate ground waters. *Geochim. Cosmochim. Acta* **1974**, *38*, 1147–1164. [CrossRef]

73. Cerling, T.E. The stable isotopic composition of modern soil carbonate and its relationship to climate. *Earth Planet. Sci. Lett.* **1984**, *71*, 229–240. [CrossRef]

74. Bakalowicz, M. Contribution de la géochimie des eaux à la connaissance de l'aquifère karstique et de la karstification. Ph.D. Thesis, Université Paris VI, Paris, France, 1979.

75. Flude, S.; Györe, D.; Stuart, F.M.; Zurakowska, M.; Boyce, A.J.; Haszeldine, R.S.; Chalaturnyk, R.; Gilfillan, S.M.V. The inherent tracer fingerprint of captured CO_2. *Int. J. Greenh. Gas Control* **2017**, *65*, 40–54. [CrossRef]

76. Viveiros, F.; Cardellini, C.; Ferreira, T.; Caliro, S.; Chiodini, G.; Silva, C. Soil CO_2 emissions at Furnas volcano, São Miguel Island, Azores archipelago: Volcano monitoring perspectives, geomorphologic studies, and land use planning application. *J. Geophys. Res.* **2010**, *115*, B12208. [CrossRef]

77. Isotopic Data Gallery. Available online: http://scrippsco2.ucsd.edu/graphics_gallery/isotopic_data (accessed on 3 December 2018).

78. Mayer, B.; Humez, P.; Becker, V.; Dalkhaa, C.; Rock, L.; Myrttinen, A.; Barth, J.A.C. Assessing the usefulness of the isotopic composition of CO_2 for leakage monitoring at CO_2 storage sites: A review. *Int. J. Greenh. Gas Control* **2015**, *37*, 46–60. [CrossRef]

79. Gerlach, T.M.; Taylor, B.E. Carbon isotope constraints on degassing of carbon dioxide from Kilauea Volcano. *Geochim. Cosmochim. Acta* **1990**, *54*, 2051–2058. [CrossRef]

80. May, F.; Waldmann, S. Tasks and challenges of geochemical monitoring. *Greenh. Gas Sci. Technol.* **2014**, *4*, 176–190. [CrossRef]

81. Klusman, R.W. Comparison of surface and near-surface geochemical methods for detection of gas microseepage from carbon dioxide sequestration—Review. *Int. J. Greenh. Gas Control* **2011**, *5*, 1369–1392. [CrossRef]

82. Jones, D.G.; Beaubien, S.E.; Blackford, J.C.; Foekema, E.M.; Lions, J.; De Vittor, C.; West, J.M.; Widdicombe, S.; Hauton, C.; Queirós, A.M. Developments since 2005 in understanding potential environmental impacts of CO_2 leakage from geological storage. *Int. J. Greenh. Gas Control* **2015**, *40*, 350–377. [CrossRef]

83. Beaubien, S.E.; Ruggiero, L.; Annunziatellis, A.; Bigi, S.; Ciotoli, G.; Deiana, P.; Graziani, S.; Lombardi, S.; Tartarello, M.C. The importance of baseline surveys of near-surface gas geochemistry for CCS monitoring, as shown from onshore case studies in northern and southern Europe. *Oil Gas Sci. Technol.* **2015**, *70*, 615–633. [CrossRef]

84. Beaubien, S.E.; Jones, D.G.; Gal, F.; Barkwith, A.K.A.P.; Braibant, G.; Baubron, J.C.; Cardellini, C.; Ciotoli, G.; Graziani, S.; Lister, T.R.; et al. Monitoring of near-surface gas geochemistry at the Weyburn, Canada, CO_2-EOR site, 2001-2011. *Int. J. Greenh. Gas Control* **2013**, *16S*, S236–S262. [CrossRef]

85. Dixon, T.; Romanak, K.D. Improving monitoring protocols for CO_2 geological storage with technical advances in CO_2 attribution monitoring. *Int. J. Greenh. Gas Control* **2015**, *41*, 29–40. [CrossRef]

86. Anderson, J.S.; Romanak, K.D.; Yang, C.; Lu, J.; Hovorka, S.D.; Young, M.H. Gas source attribution techniques for assessing leakage at geologic CO_2 storage sites: Evaluating a CO_2 and CH_4 soil gas anomaly at the Cranfield CO_2-EOR site. *Chem. Geol.* **2017**, *454*, 93–104. [CrossRef]

87. Llauro, G.; Toutain, J.P.; Desfours, J.P.; Dupuy, C.; Baubron, J.C. A new device for monitoring of spring gases: Laboratory tests and in-situ validation at Pyrenean Eaux Chaudes spring. *Sep. Purif. Technol.* **2001**, *21*, 279–283. [CrossRef]

88. Jenkins, C.R.; Cook, P.J.; Ennis-King, J.; Undershultz, J.; Boreham, C.; Dance, T.; de Caritat, P.; Etheridge, D.M.; Freifeld, B.M.; Hortle, A.; et al. Safe storage and effective monitoring of CO_2 in depleted gas fields. *PNAS* **2012**, *109*, E35–E41. [CrossRef] [PubMed]

89. Braun, C. Not in My Backyard: CCS sites and public perception of CCS. *Risk Anal.* **2017**, *37*, 2264–2275. [CrossRef] [PubMed]

90. Pourtoy, D.; Onaisi, A.; Lescanne, M.; Thibeau, S.; Viaud, C. Seal integrity of the Rousse depleted gas field impacted by CO_2 injection (Lacq industrial CCS reference project France). *Energy Procedia* **2013**, *37*, 5480–5493. [CrossRef]

91. Flude, S.; Johnson, G.; Gilfillan, S.M.V.; Haszeldine, R.S. Inherent tracers for carbon capture and storage in sedimentary formations: Composition and applications. *Environ. Sci. Technol.* **2016**, *50*, 7939–7955. [CrossRef]

92. Ehleringer, J.R.; Sage, R.F.; Flanagan, L.B.; Pearcy, R.W. Climate change and the evolution of C4 photosynthesis. *Trends Ecol. Evol.* **1991**, *6*, 95–99. [CrossRef]

93. Chiodini, G.; Cioni, R.; Guidi, M.; Raco, B.; Marini, L. Soil CO_2 flux measurements in volcanic and geothermal areas. *Appl. Geochem.* **1998**, *13*, 543–552. [CrossRef]
94. Cardellini, C.; Chiodini, G.; Frondini, F. Application of stochastic simulation to CO_2 flux from soil: Mapping and quantification of gas release. *J. Geophys. Res.* **2003**, *108*, 2425. [CrossRef]

MDPI

St. Alban-Anlage 66

4052 Basel

Switzerland

Tel. +41 61 683 77 34

Fax +41 61 302 89 18

www.mdpi.com

Applied Sciences Editorial Office

E-mail: applsci@mdpi.com

www.mdpi.com/journal/applsci

www.ingramcontent.com/pod-product-compliance
Lightning Source LLC
La Vergne TN
LVHW071525170726
843515LV00008B/2064